Lecture Notes in Computer Scienc

T0237960

Commenced Publication in 1973
Founding and Former Series Editors:
Gerhard Goos, Juris Hartmanis, and Jan van Leeuwen

Editorial Board

David Hutchison
  Lancaster University, UK
Takeo Kanade
  Carnegie Mellon University, Pittsburgh, PA, USA
Josef Kittler
  University of Surrey, Guildford, UK
Jon M. Kleinberg
  Cornell University, Ithaca, NY, USA
Alfred Kobsa
  University of California, Irvine, CA, USA
Friedemann Mattern
  ETH Zurich, Switzerland
John C. Mitchell
  Stanford University, CA, USA
Moni Naor
  Weizmann Institute of Science, Rehovot, Israel
Oscar Nierstrasz
  University of Bern, Switzerland
C. Pandu Rangan
  Indian Institute of Technology, Madras, India
Bernhard Steffen
  University of Dortmund, Germany
Madhu Sudan
  Microsoft Research, Cambridge, MA, USA
Demetri Terzopoulos
  University of California, Los Angeles, CA, USA
Doug Tygar
  University of California, Berkeley, CA, USA
Gerhard Weikum
  Max-Planck Institute of Computer Science, Saarbruecken, Germany

Andrea Corradini   Ugo Montanari (Eds.)

Recent Trends in Algebraic Development Techniques

19th International Workshop, WADT 2008
Pisa, Italy, June 13-16, 2008
Revised Selected Papers

 Springer

Volume Editors

Andrea Corradini
Ugo Montanari
Università di Pisa
Dipartimento di Informatica
Largo Bruno Pontecorvo 3, 56127 Pisa, Italy
E-mail: {andrea,ugo}@di.unipi.it

Library of Congress Control Number: Applied for

CR Subject Classification (1998): F.3, D.2.4, D.3.1, F.4, I.1

LNCS Sublibrary: SL 1 – Theoretical Computer Science and General Issues

ISSN   0302-9743
ISBN-10   3-642-03428-4 Springer Berlin Heidelberg New York
ISBN-13   978-3-642-03428-2 Springer Berlin Heidelberg New York

Typesetting: Camera-ready by author, data conversion by Scientific Publishing Services, Chennai, India
Printed on acid-free paper   SPIN: 12728596   06/3180   5 4 3 2 1 0

Preface

This volume contains selected papers from WADT 2008, the 19th International Workshop on Algebraic Development Techniques. After having joined forces with CMCS for CALCO 2007 in Bergen, WADT took place in 2008 as an individual workshop and in its traditional format.

Like its predecessors, WADT 2008 focussed on the algebraic approach to the specification and development of systems, which encompasses many aspects of formal design. Originally born around formal methods for reasoning about abstract data types, WADT now covers new specification frameworks and programming paradigms (such as object-oriented, aspect-oriented, agent-oriented, logic and higher-order functional programming) as well as a wide range of application areas (including information systems, concurrent, distributed and mobile systems). The main topics are: foundations of algebraic specification and other approaches to formal specification, including process calculi and models of concurrent, distributed and mobile computing; specification languages, methods and environments; semantics of conceptual modelling methods and techniques; model-driven development; graph transformations, term rewriting and proof systems; integration of formal specification techniques; formal testing and quality assurance; and validation and verification.

The Steering Committee of WADT consists of Michel Bidoit, José Fiadeiro, Hans-Jörg Kreowski, Till Mossakowski, Peter Mosses, Fernando Orejas, Francesco Parisi-Presicce, and Andrzej Tarlecki.

WADT 2008 took place during June 13–16, 2008, at Hotel Santa Croce in Fossabanda, a former monastery in the center of Pisa, and was organized by a committee chaired by Andrea Corradini and including Filippo Bonchi, Roberto Bruni, Vincenzo Ciancia and Fabio Gadducci. The scientific program consisted of 33 presentations selected on the basis of submitted abstracts, as well as invited talks by Egon Börger, Luca Cardelli and Stephen Gilmore.

The workshop took place under the auspices of IFIP WG 1.3 (Foundations of System Specification), and it was organized by the Dipartimento di Informatica of the University of Pisa. It was sponsored by IFIP TC1 and by the University of Pisa.

All the authors were invited to submit a full paper for possible inclusion in this volume. An Evaluation Committee was formed which consisted of the Steering Committee of WADT with the additional members Andrea Corradini (Co-chair), Fabio Gadducci, Reiko Heckel, Narciso Martí-Oliet, Ugo Montanari (Co-chair), Markus Roggenbach, Grigore Roşu, Don Sannella, Pierre Yves Schoebbens and Martin Wirsing.

All submissions underwent a careful refereeing process. We are also grateful to the following additional referees for their help in reviewing the submissions: Cyril Allauzen, Dénes Bisztray, Paolo Baldan, Filippo Bonchi, Artur Boronat,

Vincenzo Ciancia, Razvan Diaconescu, Renate Klempien-Hinrichs, Alexander Kurz, Sabine Kuske, Alberto Lluch Lafuente, Carlos Gustavo Lopez Pombo, Christoph Lüth, Hernan Melgratti, Giacoma Valentina Monreale, Miguel Palomino, Marius Petria, Laure Petrucci, Andrei Popescu, Florian Rabe, Pierre-Yves Schobbens, Lutz Schröder, Traian Şerbănuţă, Gheorghe Stefanescu and Paolo Torrini. This volume contains the final versions of the contributions that were accepted.

March 2009   Andrea Corradini
                                                        Ugo Montanari

Table of Contents

Concurrent Abstract State Machines and ^{+}CAL Programs

Michael Altenhofen[1] and Egon Börger[2]

[1] SAP Research, Karlsruhe, Germany
Michael.Altenhofen@sap.com
[2] Università di Pisa, Dipartimento di Informatica, I-56125 Pisa, Italy
boerger@di.unipi.it

Abstract. We apply the ASM semantics framework to define the **await** construct in the context of concurrent ASMs. We link ^{+}CAL programs to concurrent control state ASMs with turbo ASM submachines.

1 Introduction

In recent work we made use of the Abstract State Machines (ASM) method [8] to analyze a given cluster protocol implementation. We extracted from the code a high-level model that could be used for the analysis. We also refined the abstract model to an executable CoreASM [10,11] model so that we could run scenarios in the model. The imperative to keep the abstract models succinct and graspable for the human eye led us to work with the **await** construct for multiple agent asynchronous ASMs. In this paper we define this construct for ASMs by a conservative extension of basic ASMs.

1.1 Problem of Blocking ASM Rules

As is well known, **await** *Cond* can be programmed in a non-parallel programming context as **while not** *Cond* **do skip**, where *Cond* describes the wait condition; see the flowchart definition of the control state ASM in Fig. 1, where as usual the circles represent control states (called internal states for Finite State Machines, FSMs) and the rhombs a test.

One has to be careful when using this construct in an asynchronous multi-agent (in the sequel shortly called concurrent) ASM, given that the semantics of each involved single-agent ASM is characterized by the synchronous parallelism of a basic machine step, instead of the usual sequential programming paradigm or interleaving-based action system approaches like the B method [1], where for each step one fireable rule out of possibly multiple applicable rules is chosen for execution. The problem is to appropriately define the scope of the blocking effect of **await**, determining which part of a parallel execution is blocked where **await** occurs as submachine. One can achieve this using control states, which play the role of the internal states of FSMs; see for example Fig. 1 or the following control state ASM, which in case $ctl\_state = wait$ **and not** *Cond* holds produces the empty update set and remains in $ctl\_state = wait$, thus 'blocking'

A. Corradini and U. Montanari (Eds.): WADT 2008, LNCS 5486, pp. 1–17, 2009.
© Springer-Verlag Berlin Heidelberg 2009

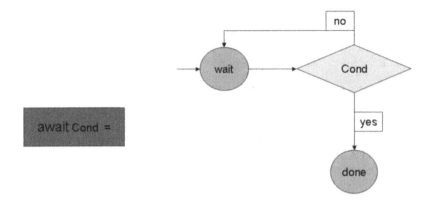

visualized also as follows, skipping
mentioning the control states:

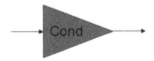

Fig. 1. Control State ASM for **await** *Cond*

the execution (under the assumption that in the given ASM no other rule is guarded by $ctl\_state = wait$)[1]:

> **await** *Cond* =
>   **if** $ctl\_state = wait$ **then**
>   **if** *Cond* **then** $ctl\_state := done$

However, when the underlying computational framework is not the execution of one rule, but the synchronous parallel execution of multiple transition rules, the explicit use of control states leads quickly to hard to grasp complex combinations of conditions resulting from the guards of different rules where an **await** *Cond* has to be executed. The complexity of the corresponding flowchart diagrams can be reduced up to a certain point using the triangle visualization in Fig. 1. It represents the right half of the traditional rhomb representation for check points in an FSM flowchart—where the right half represents the exit for *yes* (when the checked condition evaluates to true) and the left half the exit for *no* (when the checked condition evaluates to false). Is there a simple definition for the semantics of the **await** construct within the context of synchronous parallel basic ASMs?

---

[1] 'Blocking' here means that as long as the empty update set is produced, the state of the machine—in particular its control state—does not change.

Otherwise stated, can a definition of the semantics of the **await** *Cond* **do** M machine, namely to wait until *Cond* becomes true and then to proceed with executing M, be smoothly incorporated into the usual semantical definition of ASMs based on constructing update sets?

Such a definition would avoid the need for control states in the high-level definition of **await** (without hindering its implementation by control states). Above all it would preserve a main advantage of the update set construction in defining what is an ASM step, namely to elegantly capture what is intended to be considered as a basic machine step. This is important where one has to work with different degrees of granularity of what constitutes a basic machine step, usually called 'atomic' step to differentiate it from the typical sequence of actions in a standard program with ";" (sequential execution). For example basic ASMs, which consist only of rules **if** *Cond* **then** *Updates*, have been equipped in [7] with a notation for operators to **seq**uentialize or call submachines. It is defined within the ASM semantics framework in such a way that the computation performed by M **seq** N appears to the outside world as one atomic step, producing the overall effect of first executing an atomic M-step and in the thus produced state an atomic N-step; analogously for submachine execution. Machines with these constructs are called turbo ASMs because they offer two levels of analysis, the macro step level and the view of a macro step as a sequence of micro steps (which may contain again some macro steps, etc.).

We provide in this paper a similar application of the ASM semantics framework to define the meaning of **await** for concurrent ASMs, in the context of the synchronous parallelism of single-agent ASMs.

1.2   Atomicity in Control State ASMs and ^+CAL

When discussing this issue with Margus Veanes from Microsoft Research at the ABZ2008 conference in London, our attention was drawn to the recently defined language ^+CAL for describing concurrent algorithms. It is proposed in [21] as "an algorithm language that is designed to replace pseudo-code" (op.cit., abstract), the idea being that describing algorithms in the ^+CAL language provides two advantages over traditional pseudo-code whose "obvious problems ... are that it is imprecise and ... cannot be executed" (op.cit. p.2): a) "a user can understand the precise meaning of an algorithm by reading its TLA^+ translation", and b) "an algorithm written in ^+CAL ... can be executed—either exhaustively by model checking or with non-deterministic choices made randomly" (ibid.), using the TLC model checker.

These two features advocated for ^+CAL are not new. ASMs have been used successfully since the beginning of the 1990'ies as an accurate model for pseudo-code, explicitly proposed in this function in [2,3]. A user can *understand the precise meaning of ASMs directly*, namely as a natural extension of Finite State Machines (FSMs). Defining rigorously the operational semantics of ASMs uses only standard algorithmic concepts and no translation to a logic language. Furthermore, comprehensive classes of ASMs have been made *executable, using various interpreters*, the first two defined in 1990 at Quintus and at the university of

Dortmund (Germany) for experiments with models of Prolog,[2] the more recent ones built at Microsoft Research (AsmL [12]) and as open source project (Core-ASM [10]). ASMs have also been *linked to various standard and probabilistic model checkers* [9,14,15,16,17,22,23,25]. Last but not least from ASMs reliable executable code can be compiled, for a published industrial example see [6].

What we find interesting and helpful in ^+CAL is (besides the link it provides to model checking) the succinct programming notation it offers for denoting groups of sequentially executed instructions as atomic steps in a concurrent algorithm, interpreted as basic steps of the underlying algorithm. As mentioned above, such a malleable atomicity concept allowing sequential subcomputations and procedure calls has already been provided through the **seq**, **iterate** and submachine concepts for turbo ASMs [7], which turn a sequence or iteration of submachine steps into one atomic step for the main machine. However the label-notation of ^+CAL, which is a variation of the control state notation for FSMs, will be more familiar to algorithm designers and programmers than the **seq** notation (in AsmL [12] the name **step** is used instead of **seq**) and is more concise. We will illustrate that one can exploit the ^+CAL notation in particular as a convenient textual pendant to the FSM flowchart diagram description technique for control state ASMs which contain turbo ASM submachines. As side effect of linking corresponding features in ^+CAL and in concurrent ASMs one obtains an ASM interpretation for ^+CAL programs, which supports directly the intuitive understanding of ^+CAL constructs and is independent of the translation of ^+CAL to the logic language TLA^+.[3]

In Section 2 we extend the standard semantics of ASMs to concurrent ASMs with the **await** construct. For the standard semantics of ASMs we refer the reader to [8]. In Section 2 we link the corresponding constructs in ^+CAL and in the class of concurrent constrol state ASMs with **await**.

2   Concurrent ASMs with the await Construct

A basic ASM consists of a signature, a set of initial states, a set of rule declarations and a main rule. A rule is essentially a parallel composition of so-called transition rules **if** *Cond* **then** *Updates*, where *Updates* is a finite set of assignment statements $f(e_1, \ldots, e_n) := e$ with expressions e_i, e. In each step of the machine all its transition rules that are applicable in the given state are executed simultaneously (synchronous parallelism); a rule is applicable in state S if its guard *Cond* evaluates in S to true.

In more detail, the result of an M-step in state S can be defined in two parts. First one collects the set U of all updates which will be performed by any of the rules **if** *Cond* **then** *Updates* that are applicable in the given state S; by update we understand a pair (l, v) of a location l and its to be assigned value v, read:

---

[2] see the historical account in [4].

[3] For the sake of completeness we remark that there is a simple scheme for translating basic ASMs to TLA^+ formulae which describe the intended machine semantics. Via such a translation one can model check ASMs using the TLC model checker of TLA^+.

the value the expression *exp* of an assignment $f(e_1, \ldots, e_n) := e$ evaluates to in S. The reader may think of a location as an array variable (f, val), where *val* consists of a sequence of parameter values to which the expressions e_i in the left side $f(e_1, \ldots, e_n)$ of the assignment evaluate in S. Then, if U is consistent, the next (so-called internal) state $S + U$ resulting from the M-step in state S is defined as the state that satisfies the following two properties:

- for every location (f, val) that is not element of the update set U, its value, written in standard mathematical notation as $f(val)$, coincides with its value in S (no change outside U),
- each location (f, val) with update $((f, val), v) \in U$ gets as value $f(val) = v$ (which may be its previous value in S, but usually will be a new value).

In case U is inconsistent no next state is defined for S, so that the M-computation terminates abruptly in an error state because M can make no step in S. For use below we also mention that in case the next internal state $S + U$ is defined, the next step of M takes place in the state $S + U + E$ resulting from $S + U$ by the environmental updates of (some of) the monitored or shared locations as described by an update set E.

The reader who is interested in technical details can find a precise (also a formal) definition of this concept in the AsmBook [8, Sect.2.4]. In particular, there is a recursive definition which assigns to each of the basic ASM constructs[4] P its update set U such that $yield(P, S, I, U)$ holds (read: executing transition rule P in state S with the interpretation I of the free variables of P yields the update set U), for each state S and each variable interpretation I. We extend this definition here by defining $yield(\textbf{await } Cond, S, I, U)$. The guiding principle of the definition we are going to explain is the following:

- (the agent which executes) an ASM M becomes blocked when at least one of its rules, which is called for execution and will be applied simultaneously with all other applicable rules in the given state S, is an **await** *Cond* whose *Cond* evaluates in S to false,
- (the agent which executes) M is unblocked when (by actions of other executing agents which constitute the concurrent environment where M is running) a state S' is reached where for each **await** *Cond* rule of M that is called for execution in S', *Cond* evaluates to true.

In the sequential programming or interleaving-based action systems context there is at each moment at most one applicable rule to consider and thus at most one **await** *Cond* called for execution. In the ASM computation model, where at each moment each agent fires simultaneously all the rules that are applicable, it seems natural to block the computation at the level of agents, so that possibly multiple **await** *Cond* machines have to be considered simultaneously. Variations of the definition below are possible. We leave it to further experimentation to evaluate which definition fits best practical needs, if we do not want to abandon the advantage of the synchronous parallelism of basic ASMs (whose benefit is to force the designer to avoid sequentiality wherever possible).

---

[4] **skip, par, if then else, let, choose, forall, seq** and machine call.

The technical idea is to add to the machine signature a location *phase* with values *running* or *wait*,[5] which can be used to prevent the application of the internal state change function $S + U$ in case *Cond* of an **await** *Cond* called for execution does not evaluate to true in S. In this case we define **await** *Cond* to yield the *phase* update (*phase*, *wait*) and use the presence of this update in U to block the state change function $S + U$ from being applied. This leads to the following definition:

$yield(\textbf{await } Cond, S, I, U) =$
$\quad \emptyset \qquad\qquad\qquad \textbf{if } Cond$ is true in S // proceed
$\quad \{(phase, wait)\} \textbf{ else } $ //change *phase* to *wait*

We now adapt the definition of the next internal state function $S + U$ to the case that U may contain a *phase* update. The intuitive understanding of an **await** *Cond* statement is that the executing agent starts to wait, continuously testing *Cond* without performing any state change, until *Cond* becomes true (through actions of some other agents in the environment). In other words, a machine M continues to compute its update set, but upon encountering an **await** *Cond* with false *Cond*ition it does not trigger a state change. We therefore define as follows (assuming $yields(M, S, I, U)$):

- If *phase* = *running* in S and U contains no update (*phase*, *wait*), it means that no **await** *Cond* statement is called to be executed in state S. In this case the definition of $S + U$ is taken unchanged from basic ASMs as described above and *phase* = *running* remains unchanged.
- If *phase* = *running* in S and the update (*phase*, *wait*) is an element of U, then some **await** *Cond* statement has been called to be executed in state S and its wait *Cond*ition is false. In this case we set $S+U = S+\{(phase, wait)\}$. This means that the execution of any **await** *Cond* statement in P whose *Cond* is false in the given state S blocks the (agent who is executing the) machine P as part of which such a statement is executed. Whatever other updates—except for the *phase* location—the machine P may compute in U to change the current state, they will not be realized (yet) and the internal state remains unchanged (except for the *phase* update).
- If in state S *phase* = *wait* holds and the update (*phase*, *wait*) is element of U, we set the next internal state $S + U$ as undefined (blocking effect without internal state change). This definition reflects that all the **await** *Cond* statements that are called for execution in a state S have to succeed simultaneously, i.e. to find their *Cond* to be true, to let the execution of P proceed (see the next case). In the special case of a sequential program without parallelism, in each moment at most one **await** *Cond* statement is called for execution so that in this case our definition for ASMs corresponds to the usual programming interpretation of **await** *Cond*.

---

[5] *phase* could be used to take also other values of interest in the concurrency context, besides *running* and *wait* for example *ready*, *suspended*, *resumed*, etc., but here we restrict our attention to the two values *running* and *wait*.

- If $phase = wait$ holds in S and U contains no $phase$ update $(phase, wait)$, it means that each **await** $Cond$ statement that may be called to be executed in state S has its $cond$ evaluated to true. In this case the next internal state is defined as $S + U + \{(phase, running)\}$, i.e. the internal state $S + U$ as defined for basic ASMs with additionally $phase$ updated to $running$. Otherwise stated when all the waiting conditions are satisfied, the machine continues to run updating its state via the computed set U of updates.

The first and fourth case of this definition imply the conservativity of the resulting semantics with respect to the semantics of basic ASMs: if in a state with $phase = running$ no **await** $Cond$ statement is called, then the machine behaves as a basic ASM; if in a state with $phase = wait$ only **await** $Cond$ statements with true waiting $Cond$ition are called, then the machine switches to $phase = running$ and behaves as a basic ASM.

One can now define **await** $Cond$ M as parallel composition of **await** $Cond$ and M. Only when $Cond$ evaluates to true will **await** $Cond$ yield no $phase$ update $(phase, wait)$ so that the updates produced by M are taken into account for the state change obtained by one M-step.

> **await** $Cond$ $M =$
> **await** $Cond$
> M

Remark. The reason why we let a machine M in a state S with $phase = wait$ recompute its update set is twofold. Assume that **await** $Cond$ is one of the rules in the **then** branch M_1 of $M =$ **if** $guard$ **then** M_1 **else** M_2, but not in the **else** branch. Assume in state S $guard$ is true, $phase = running$ and $Cond$ is false, so that executing **await** $Cond$ as part of executing M_1 triggers the blocking effect. Now assume that, due to updates made by the environment of M, in the next state $guard$ changes to false. Then **await** $Cond$ is not called for execution any more, so that the blocking effect has vanished. Symmetrically an **await** $Cond$ that has not been called for execution in S may be called for execution in the next state S', due to a change of a $guard$ governing **await** $Cond$; if in S' $Cond$ition is false, a new reason for blocking M appears that was not present in state S. Clearly such effects cannot happen in a sequential execution model because there, a program counter which points to an **await** $Cond$ statement will point there until the statement proceeds because $Cond$ became true.

The above definition represents one possibility to incorporate waiting into the ASM framework. We are aware of the fact that its combination with the definition of turbo ASMs may produce undesired effects, due to the different scoping disciplines of the two constructs.[6] Other definitions of **await** $Cond$ M with different scoping effect and combined with non-atomic sequentialization concepts should be tried out.

---

[6] Consider for example M **seq await** $Cond$ or **await** $Cond$ **seq** M, where M contains no **await**, applied in a state where $Cond$ evaluates to false. The update $(phase, wait)$ of **await** $Cond$ will be overwritten by executing M even if $Cond$ remains false.

3 Linking Concurrent ASMs and ^+CAL Programs

The reader has seen in the previous section that in the ASM framework a basic step of a machine M is simply a step of M, which computes in the given state with given variable interpretation the set U of updates such that $yield(M, S, I, U)$ and applies it to produce the next state $S + U$. Once an update set U has been used to build $S + U$, there remains no trace in $S + U$ about which subsets of $Updates$, via which M-rule if $Cond$ then $Updates$ that is executable in the given state, have contributed to form this state. Thus each step of an ASM is considered as atomic and the grain of atomicity is determined by the choice made for the level of abstraction (the guards and the abstract updates) at which the given algorithm is described by M. Typically precise links between corresponding ASMs at different levels of abstraction are established by the ASM refinement concept defined in [5].

The ASM literature is full of examples which exploit the atomicity of the single steps of an ASM and their hierarchical refinements. A simple example mentioned in the introduction is the class of turbo ASMs, which is defined from basic ASMs by allowing also the sequential composition $M = M_1$ **seq** M_2 or iteration $M =$ **iterate** M_1 of machines and the (possibly recursive) call $M(a_1, \ldots, a_n)$ of submachines for given argument values a_i. In these cases the result of executing in state S a sequential or iterative step or a submachine call is defined by computing the comprehensive update set, produced by the execution of all applicable rules of M, and applying it to define the next state. The definition provides a big-step semantics of turbo ASMs, which has been characterized in [13] by a tree-like relation between a turbo ASM macro step and the micro steps it hides (see [8, 4.1]).

Another way to describe the partitioning of an ASM-computation into atomic steps has been introduced in [3] by generalizing Finite State Machines (FSMs) to control state ASMs. In a control state ASM M every rule has the following form:

$$\text{Fsm}(i, cond, rule, j) =$$
$$\quad \textbf{if } ctl\_state = i \textbf{ then}$$
$$\quad\quad \textbf{if } cond \textbf{ then}$$
$$\quad\quad\quad rule$$
$$\quad\quad\quad ctl\_state := j$$

Such rules are visualized in Fig. 2, which uses the classical graphical FSM notation and provides for it a well-defined textual pendant for control state ASMs with a rigorously defined semantics. We denote by $dgm(i, cond, rule, j)$ the diagram representing $\text{Fsm}(i, cond, rule, j)$. We skip $cond$ when it is identical to $true$. In control state ASMs each single step is controlled by the unique current control state value and a guard. Every M-step leads from the uniquely determined current ctl_state value, say i, to its next value j_k (out of $\{j_1, \ldots, j_n\}$, depending on which one of the guards $cond_k$ is true in the given state[7])—the way FSMs change their internal states depending on the input they read. Otherwise stated the control state pair (i, j_k) specifies the desired grain of atomicity, namely any $rule_k$ constituting a single machine step. This is essentially what is used in ^+CAL to indicate atomicity, except for the notational difference that the control states are written as labels and the fact that ^+CAL is based upon the sequential programming paradigm (see the details below).

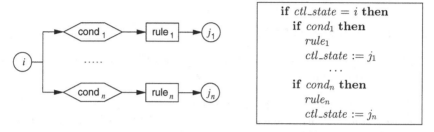

Fig. 2. Flowchart for control state ASMs

However, each $rule_k$ may be a complex ASM, for example another control-state ASM whose execution may consists of multiple, possibly sequential or iterated substeps.[8] If their execution has to be considered as one step of the main machine M where $rule_k$ appears, namely the step determined by going from control state i to control state j_k, a notation is needed to distinguish the control states within $rule_k$ from those which define the boundary of the computation segment that is considered as atomic.

There are various ways to make such a distinction. The M **seq** N operator can be interpreted as composing control state ASMs with unique *start* and *end* control state, namely by identifying $end_M = start_N$. Suppressing the visualization of this intermediate control state, as indicated in Fig. 3, provides a way to render also graphically that the entire machine execution leading from control state *start* $= start_M$ to control state *end* $= end_N$ is considered as atomic. This corresponds to the distinction made in the ^+CAL language: it is based upon sequential control (denoted by the semicolon) the way we are used from programming languages, but specific sequences of such sequential steps can be aggregated using labels, namely by defining as (atomic) *step* each "control path that starts at a label, ends at a label, and passes through no other labels" (op.cit., p.19). ^+CAL programs appear to be equivalent to control state ASMs with turbo submachines; the labels play the role of the control states and the turbo submachines the role of sequentially executed nonatomic ^+CAL code between two

---

[7] If two guards $cond_k, cond_l$ have a non empty intersection, in case $ctl\_state_k \neq ctl\_state_l$ an inconsistent update set U is produced so that by definition $S + U$ is undefined. If instead a non-deterministic interpretation of FSM rules is intended, this non-determinism can be expressed using the ASM **choose** construct.

[8] Replacing FSM$(i, rule, j)$ by FSM(i, M, j) with a new control state ASM M is a frequent ASM refinement step, called procedural in [5].

Fig. 3. Sequential composition of control state ASMs

labels. This is explained by further details in Sect. 3.1 and illustrated in Sect. 3.2 by writing the major example from [21] as a turbo control state ASM.

3.1 Control State ASM Interpretation of ^+CAL Programs

^+CAL is proposed as an algorithm language to describe multiprocess algorithms. The chosen concurrency model is interleaving:

> A multiprocess algorithm is executed by repeatedly choosing an arbitrary process and executing one step of that process, if that step's execution is possible. [20, p.26]

This can be formalized verbatim by an ASM-scheme MULTIPROCESS for multi-agent ASMs, parameterized by a given set *Proc* of constituting processes. The execution behavior of each single process is defined by the semantics of basic (sometimes misleadingly also called sequential) ASMs. The ASM **choose** construct expresses choosing, for executing one step, an arbitrary process out of the set $CanExec(Proc)$ of those $P \in Proc$ which can execute their next step:[9]

> MULTIPROCESS(*Proc*) =
>   **choose** $P \in CanExec(Proc)$
> P

Therefore for the rest of this section we focus on describing the behavior of single ^+CAL programs by ASMs, so that MULTIPROCESS becomes an interpreter scheme for ^+CAL programs. The program behavior is determined by the execution of the statements that form the program body (called algorithm body in ^+CAL), so that we can concentrate our attention on the operational description of ^+CAL statements and disregard here the declarations as belonging to the signature definition.

We apply the FSM flowchart notation to associate to each ^+CAL program body P a diagram $dgm(P)$ representing a control state ASM $asm(P)$ which defines the behavior of P (so that no translation of P to TLA^+ is needed to define the semantics of P). Each label in P is interpreted as what we call a *concurrent*

---

[9] There are various ways to deal with the constraint "if that step's execution is possible". Using the ASM **choose** operator has the effect that in case $CanExec(P)$ is empty, nothing happens (more precisely: an empty update set is produced whose application does not change the current state). If one wants this case to be interpreted as explicitly blocking the scheduler, it suffices to add the **await** $CanExec(P) \neq \emptyset$ machine. The predicate $CanExec(P)$ is defined inductively.

control state. The other control states in $dgm(P)$ are called *sequential control states* because they serve to describe the sequentiality of micro-steps (denoted in P by the semicolon), the constituents of sequences which are considered as an atomic step. They are the control states that are hidden by applying the **seq**, **while** and submachine call operators. Since the construction of $dgm(P)$ uses only standard techniques we limit ourselves here to show the graphical representation for each type of ^+CAL statements. Out of these components and adding rules for the evaluation of expressions one can build a ^+CAL interpreter ASM, using the technique developed in [24] to construct an ASM interpreter for Java and JVM programs. We leave out the print statement and the assert statement; the latter is of interest only when model checking a ^+CAL program.

As basic statements a ^+CAL program can contain assignment statements or the empty statement **skip**. Program composition is done via the structured programming constructs sequencing (denoted by the semicolon), **if then else**, **while** together with **await** (named **when**) statements (a concurrent pendant of **if** statements), two forms of choice statements (nondeterminism), statements to call or return from subprograms. Since statements can be labeled, also *Goto l* statements are included in the language, which clearly correspond to simple updates of *ctl_state* resp. arrows in the graphical representation.

For the structured programming constructs the associated diagram $dgm(stm)$ defining the normal control flow consists of the traditional flowchart representation of FSMs, as illustrated in Fig. 3 and Fig. 4. One could drop writing "yes" and "no" on the two exits if the layout convention is adopted that the "yes" exit is on the upper or hight half of the rhomb and the "no" exit on the lower or left half, as is usually the case. In Fig. 4 we explicitly indicate for each diagram its control state for begin (called *start*) and end (called *done*), where each $dgm(stm)$ has its own begin and end control state, so that *start* and *done* are considered as implicitly indexed per *stm* to guarantee a unique name. Most of these control states will be sequential control states in the diagram of the entire program that is composed from the subdiagrams of the single statements. These sequential control states can therefore be replaced by the turbo ASM operator **seq** as done in Fig. 3, which makes the atomicity of the sequence explicit.

await statements are written **when** *Cond*; in ^+CAL. The semantical definition of $asm(\textbf{await } Cond)$ has been given within the ASM framework in Sect. 2, extending the sequential programming definition in Fig. 1. For the visualization we define $dgm(\textbf{when } Cond;)$ by the triangle shown in that figure.

Basic statements *stm* are represented by the traditional FSM graph of Fig. 2, in Sect. 3 denoted by $dgm(start, asm(stm), done)$. The statement **skip**; "does nothing" and thus has the behavior of the homonymous ASM $asm(\textbf{skip };) = \textbf{skip}$, which in every state yields an empty update set. An assignment statement in ^+CAL is a finite sequence $stm = lhs_1 := exp_1 \;||\; \dots \;||\; lhs_n := exp_n$ of assignments, executed by "first evaluating the right-hand sides of all its assignments, and then performing those assignments from left to right". This behavior is that of the following ASM, where the expression evaluation is performed in

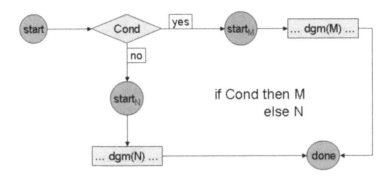

Identify: done = done$_M$ = done$_N$

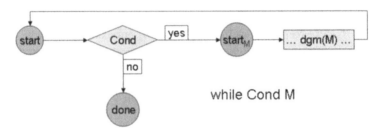

Fig. 4. Control State ASM for Structured Programming Concepts

parallel for all expressions in the current state, whereafter the assignment of the computed values is done in sequence.[10]

$$asm(lhs_1 := exp_1 \; || \; \dots \; || \; lhs_n := exp_n \; ;) =$$
$$\textbf{forall } 1 \le i \le n \textbf{ let } x_i = exp_i$$
$$lhs_1 := x_1 \textbf{ seq} \dots \textbf{seq } lhs_n := x_n$$

The behavior of statements $Goto \; l$; is to "end the execution of the current step and causes control to go to the statement labeled l" [20, p.25], wherefor such statements are required to be followed (in the place where they occur in the program) by a labeled statement. Thus the behavior is defined by $asm(Goto \; l \; ;) = (ctl\_state := l)$ and $dgm(Goto \; l \; ;)$ as an arrow leading to control state l from the position of the $Goto \; l$;, which is a control state in case the statement is labeled.

The two constructs expressing a non determinstic choice can be defined as shown in Fig. 5.

---

[10] It seems that this treatment of assignment statements is related to the semantics of nested EXCEPTs in TLA^+ and thus permits a simple compilation.

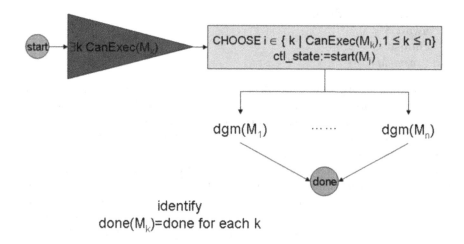

Fig. 5. Control State ASMs for Choice Statements

The first type of non deterministic statement **either** M_1 **or** M_2 **or** ... **or** M_n; chooses an executable statement among finitely many statements M_i. It is defined to be executable if and only if one of M_i is executable. This means that the execution of the statement has a blocking character, namely to wait as long as none of the substatements is executable. Similarly the second type of non deterministic statement, written **with** $id \in S$ **do** M;, is meant to choose an element in a set S if there is one and to execute M for it. The statement is considered as not executable (blocking) if the set to choose from is empty.

Since the ASM **choose** construct is defined as non blocking, but yields an empty update set in case no choice is possible, we make use of the **await** construct to let **choose** have an effect only when there is something to choose from. We write *CanExec* for the executability predicate.

$asm(\textbf{either } M_1 \textbf{ or } M_2 \textbf{ or } \ldots \textbf{ or } M_n;) =$
  $\textbf{await forsome } 1 \leq j \leq n \ CanExec(M_j)$
  $\textbf{choose } i \in \{j \mid CanExec(M_j) \textbf{ and } 1 \leq j \leq n\}$
    $asm(M_i)$

$$asm(\textbf{with } id \in S \textbf{ do } M;) =$$
$$\textbf{await } S \neq \emptyset$$
$$\textbf{choose } id \in S$$
$$asm(M(id))$$

The remaining ^+CAL statements deal with procedure call and return in a standard stack machine like manner. Define *frame* as quadruple consisting of a control state, the values of the procedure's arguments respectively of its local variables and the procedure name. We denote the frame stack by a location *stack* and the current frame by a quadruple of four locations *ctl_state*, *args* (which is a sequence, of any finite length, of variables standing for the parameters), *locals* (which is a sequence, of any finite length, of local variables) and *proc* (which denotes the currently executed procedure). For a call statement $P(expr_1, \ldots, expr_n)$;, "executing this call assigns the current values of the expressions $expr_i$ to the corresponding parameters $param_i$, initializes the procedure's local variables, and puts control at the beginning of the procedure body", which "must begin with a labeled statement" [20, p.27]. As preparation for the return statement one has also to record the current frame on the frame *stack*. We denote the sequence of local variables of P by $locVars(P)$ and their initial values by $initVal$. For the sake of brevity, for sequences *locs* of locations and *vals* of values we write $locs := vals$ for the simultaneous componentwise assignment of the values to the corresponding locations, to be precise for the machine $asm(locs_1 := vals_1 \parallel \ldots \parallel locs_n := vals_n ;)$ defined above where $locs = (locs_1, \ldots, locs_n)$ and $vals = (vals_1, \ldots, vals_n)$. Let $start_P$ denote the label of the first statement of P.

$$asm(P(exp_1, \ldots, exp_n)) =$$
$$\text{PUSHFRAME}(P, (exp_1, \ldots, exp_n))$$
$$\textbf{where}$$
$$\text{PUSHFRAME}(P, exps) =$$
$$stack := stack.[ctl\_state, args, locals, proc] \text{ // push current frame}$$
$$proc := P$$
$$args := exps \text{ // pass the call parameters}$$
$$locals := initVal(locVars(P)) \text{ // initialize the local variables}$$
$$ctl\_state := start_P \text{ // start execution of the procedure body}$$

A return statement consists in the inverse machine PUSHFRAME. If *ctl* is the point of a call statement in the given program, let $next(ctl)$ denote the point immediately following the call statement.

$$asm(return) =$$
$$\textbf{let } stack = stack'.[ctl, prevArgs, prevLocs, callingProc] \textbf{ in}$$
$$ctl\_state := next(ctl) \text{ // go to next stm after the call stm}$$
$$args := prevArgs \text{ // reassign previous values to args}$$
$$locals := prevLocs \text{ // reassign previous values to locals}$$
$$proc := callingProc$$
$$stack := stack' \text{ // pop frame stack}$$

3.2   Fast Mutex Example

Fig. 6 illustrates the diagram notation explained in Sect. 3.1 for ^+CAL programs.

The ^+CAL program for the fast mutual exclusion algorithm from [19] is given in [21]. Fig. 6 does not show the declaration part, which is given in the signature definition. We write Ncs and Cs for the submachines defining the non critical resp. critical section, which in ^+CAL are denoted by an atomic **skip** instruction describing—via the underlying stuttering mechanism of TLA—a nonatomic program. Given the structural simplicity of this program, which says nothing about the combinatorial complexity of the runs the program produces, there is only one sequential subprogram. It corresponds to two simultaneous updates, so that the sequentialization can be avoided and really no turbo ASM is needed because there are no control states which do not correspond to ^+CAL labels. This case is a frequent one when modeling systems at an abstract level, as the experience with ASMs shows. In general, the synchronous parallelism of ASMs drives the model designer to avoid sequentialization as much as possible and to think instead about orthogonal components which constitute atomic steps.

Comparing the two representations the reader will notice that even the layouts can be made to be in strict correspondence, so that each of the labeled lines in the

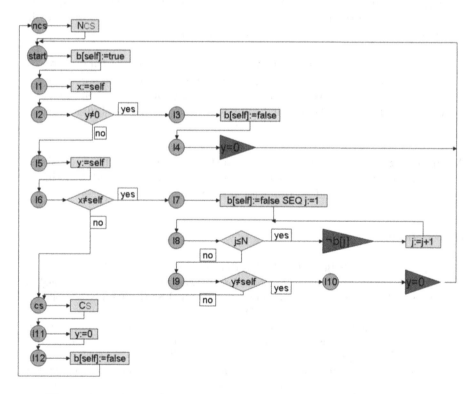

Fig. 6. Control state ASM for the Fast Mutual Exclusion ^+CAL Program

textual description corresponds to a line starting a new control state subdiagram. This is in line with the following well-known fact we quote from [18, p.72]:

> The visual structure of **go to** statements is like that of flowcharts, except reduced to *one* dimension in our source languages.

We can confirm from our own work the experience reported in [21] that the notation works well for programs one can write on a couple of pages, making judicious use of procedures where possible to cut down the size of each single program one has to analyze. Such programs seem to be the main target for ^+CAL code and model checkable TLA^+ translations for concurrent algorithms with combinatorially involved behaviour. For larger programs flowcharts present some small advantage over the representation of programs as strings, however, as Knuth continues op.cit.:

> ... we rapidly loose our ability to understand larger and larger flowcharts; some intermediate levels of abstraction are necessary.

The needed abstractions can be provided in control state ASMs by using separately defined complex submachines, which in the flowcharts appear as simple rectangles to be executed when passing from one to the next control state. This follows an advice formulated by Knuth op.cit. as one of the conclusions of his discussion of structured programming with **go to** statements:

> ... we should give meaningful names for the larger constructs in our program that correspond to meaningul levels of abstraction, and we should define those levels of abstraction in one place, and merely use their names (instead of including the detailed code) when they are used to build larger concepts.

Acknowledgement. We thank L. Lamport and S.Merz for the discussion of a preliminary version of this paper. We also thank M. Butler and A. Prinz for some helpful comment and S. Tahar for pointing out a relevant reference.

References

1. Abrial, J.-R.: The B-Book. Cambridge University Press, Cambridge (1996)
2. Börger, E.: Why use Evolving Algebras for hardware and software engineering? In: Bartosek, M., Staudek, J., Wiedermann, J. (eds.) SOFSEM 1995. LNCS, vol. 1012, pp. 236–271. Springer, Heidelberg (1995)
3. Börger, E.: High-level system design and analysis using Abstract State Machines. In: Hutter, D., Stephan, W., Traverso, P., Ullmann, M. (eds.) FM-Trends 1998. LNCS, vol. 1641, pp. 1–43. Springer, Heidelberg (1999)
4. Börger, E.: The origins and the development of the ASM method for high-level system design and analysis. J. Universal Computer Science 8(1), 2–74 (2002)
5. Börger, E.: The ASM refinement method. Formal Aspects of Computing 15, 237–257 (2003)

6. Börger, E., Päppinghaus, P., Schmid, J.: Report on a practical application of ASMs in software design. In: Gurevich, Y., Kutter, P.W., Odersky, M., Thiele, L. (eds.) ASM 2000. LNCS, vol. 1912, pp. 361–366. Springer, Heidelberg (2000)
7. Börger, E., Schmid, J.: Composition and submachine concepts for sequential ASMs. In: Clote, P.G., Schwichtenberg, H. (eds.) CSL 2000. LNCS, vol. 1862, pp. 41–60. Springer, Heidelberg (2000)
8. Börger, E., Stärk, R.F.: Abstract State Machines. In: A Method for High-Level System Design and Analysis. Springer, Heidelberg (2003)
9. Del Castillo, G., Winter, K.: Model checking support for the ASM high-level language. In: Schwartzbach, M.I., Graf, S. (eds.) TACAS 2000. LNCS, vol. 1785, pp. 331–346. Springer, Heidelberg (2000)
10. Farahbod, R., et al.: The CoreASM Project, http://www.coreasm.org
11. Farahbod, R., Gervasi, V., Glässer, U.: CoreASM: An Extensible ASM Execution Engine. Fundamenta Informaticae XXI (2006)
12. Foundations of Software Engineering Group, Microsoft Research. AsmL. Web pages (2001), http://research.microsoft.com/foundations/AsmL/
13. Fruja, N.G., Stärk, R.F.: The hidden computation steps of turbo Abstract State Machines. In: Börger, E., Gargantini, A., Riccobene, E. (eds.) ASM 2003. LNCS, vol. 2589, pp. 244–262. Springer, Heidelberg (2003)
14. Gargantini, A., Riccobene, E., Rinzivillo, S.: Using Spin to generate tests from ASM specifications. In: Börger, E., Gargantini, A., Riccobene, E. (eds.) ASM 2003. LNCS, vol. 2589, pp. 263–277. Springer, Heidelberg (2003)
15. Gawanmeh, A., Tahar, S., Winter, K.: Interfacing ASMs with the MDG tool. In: Börger, E., Gargantini, A., Riccobene, E. (eds.) ASM 2003. LNCS, vol. 2589, pp. 278–292. Springer, Heidelberg (2003)
16. Gawanmeh, A., Tahar, S., Winter, K.: Formal verification of asms using mdgs. Journal of Systems Architecture 54(1-2), 15–34 (2008)
17. Glässer, U., Rastkar, S., Vajihollahi, M.: Computational Modeling and Experimental Validation of Aviation Security Procedures. In: Mehrotra, S., Zeng, D.D., Chen, H., Thuraisingham, B., Wang, F.-Y. (eds.) ISI 2006. LNCS, vol. 3975, pp. 420–431. Springer, Heidelberg (2006)
18. Knuth, D.: Structured programming with goto statements. Computing Surveys 6 (December 1974)
19. Lamport, L.: A fast mutual exclusion algorithm. ACM Transactions of Computer Systems 5(1), 1–11 (1987)
20. Lamport, L.: A +CAL user's manual.P-syntax version. (June 29, 2007), http://research.microsoft.com/users/lamport/tla
21. Lamport, L.: The +CAL algorithm language. (February 14, 2008), http://research.microsoft.com/users/lamport/tla/pluscal.html
22. Plonka, C.N.: Model checking for the design with Abstract State Machines. Diplom thesis, CS Department of University of Ulm, Germany (January 2000)
23. Slissenko, A., Vasilyev, P.: Simulation of timed Abstract State Machines with predicate logic model-checking. J. Universal Computer Science 14(12), 1984–2007 (2008)
24. Stärk, R.F., Schmid, J., Börger, E.: Java and the Java Virtual Machine: Definition, Verification, Validation. Springer, Heidelberg (2001)
25. Winter, K.: Model checking for Abstract State Machines. J. Universal Computer Science 3(5), 689–701 (1997)

Molecules as Automata

Luca Cardelli

Microsoft Research, Cambridge, U.K.
luca@microsoft.com

Extended Abstract

Molecular biology investigates the structure and function of biochemical systems starting from their basic building blocks: macromolecules. A macromolecule is a large, complex molecule (a protein or a nucleic acid) that usually has inner mutable state and external activity. Informal explanations of biochemical events trace individual macromolecules through their state changes and their interaction histories: a macromolecule is endowed with an identity that is retained through its transformations, even through changes in molecular energy and mass. A macromolecule, therefore, is qualitatively different from the small molecules of inorganic chemistry. Such molecules are stateless: in the standard notation for chemical reactions they are seemingly created and destroyed, and their atomic structure is used mainly for the bookkeeping required by the conservation of mass.

Attributing identity and state transitions to molecules provides more than just a different way of looking at a chemical event: it solves a fundamental difficulty with chemical-style descriptions. Each macromolecule can have a huge number of internal states, exponentially with respect to its size, and can join with other macromolecules to from even larger state configurations, corresponding to the product of their states. If each molecular state is to be represented as a stateless chemical species, transformed by chemical reactions, then we have a huge explosion in the number of species and reactions with respect to the number of different macromolecules that actually, physically, exist. Moreover, macromolecules can join to each other indefinitely, resulting in situations corresponding to infinite sets of chemical reactions among infinite sets of different chemical species. In contrast, the description of a biochemical system at the level of macromolecular states and transitions remains finite: the unbounded complexity of the system is implicit in the potential molecular interactions, but does not have to be written down explicitly. Molecular biology textbooks widely adopt this finite description style, at least for the purpose of illustration.

Many proposal now exist that aim to formalize the combinatorial complexity of biological systems without a corresponding explosion in the notation. Macromolecules, in particular, are seen as stateful concurrent agents that interact with each other through a dynamic interface. While this style of descriptions is (like many others) not quite accurate at the atomic level, it forms the basis of a formalized and growing body of biological knowledge.

The complex chemical structure of a macromolecule is thus commonly abstracted into just internal states and potential interactions with the environment.

A. Corradini and U. Montanari (Eds.): WADT 2008, LNCS 5486, pp. 18–20, 2009.
© Springer-Verlag Berlin Heidelberg 2009

Each macromolecule forms, symmetrically, part of the environment for the other macromolecules, and can be described without having to describe the whole environment. Such an open system descriptive style allows modelers to extend systems by composition, and is fundamental to avoid enumerating the whole combinatorial state of the system (as one ends up doing in closed systems of chemical reactions). The programs-as-models approach is growing in popularity with the growing modeling ambitions in systems biology, and is, incidentally, the same approach taken in the organization of software systems. The basic problem and the basic solution are similar: programs are finite and compact models of potentially unbounded state spaces.

At the core, we can therefore regard a macromolecule as some kind of automaton, characterized by a set of internal states and a set of discrete transitions between states driven by external interactions. We can thus try to handle molecular automata by some branch of automata theory and its outgrowths: cellular automata, Petri nets, and process algebra. The peculiarities of biochemistry, however, are such that until recently one could not easily pick a suitable piece of automata theory off the shelf. Many sophisticated approaches have now been developed, and we are particularly fond of stochastic process algebra. In this talk, however, we do our outmost to remain within the bounds of a much simpler theory. We go back, in a sense, to a time before cellular automata, Petri nets and process algebra, which all arose from the basic intuition that automata should interact with each other. Our main criterion is that, as in finite-state automata, we should be able to easily and separately draw the individual automata, both as a visual aid to design and analysis, and to emulate the illustration-based approach found in molecular biology textbooks.

With those aims, we investigate stochastic automata collectives. Technically, we place ourselves within a small fragment of a well-know process algebra (stochastic pi-calculus), but the novelty of the application domain, namely the mass action behavior of large numbers of well-mixed automata, demands a broader outlook. By a collective we mean a large set of interacting, finite state automata. This is not quite the situation we have in classical automata theory, because we are interested automata interactions. It is also not quite the situation with cellular automata, because our automata are interacting, but not necessarily on a regular grid. And it is not quite the situation in process algebra, because we are interested in the behavior of collectives, not of individuals. And in contrast to Petri nets, we model separate parts of a system separately. By stochastic we mean that automata interactions have rates. These rates induce a quantitative semantics for the behavior of collectives, and allow them to mimic chemical kinetics. Chemical systems are, physically, formed by the stochastic interactions of discrete particles. For large number of particles it is usually possible to consider them as formed by continuous quantities that evolve according to deterministic laws, and to analyze them by ordinary differential equations. However, one should keep in mind that continuity is an abstraction, and that sometimes it is not even a correct limit approximation.

In biochemistry, the stochastic discrete approach is particularly appropriate because cells often contain very low numbers of molecules of critical species: that is a situation where continuous models may be misleading. Stochastic automata collectives are hence directly inspired by biochemical systems, which are sets of interacting macromolecules, whose stochastic behavior ultimately derives from molecular dynamics. Some examples of the mismatch between discrete and continuous models are discussed.

Service-Level Agreements for Service-Oriented Computing

Allan Clark, Stephen Gilmore, and Mirco Tribastone

Laboratory for Foundations of Computer Science
The University of Edinburgh, Scotland

Abstract. Service-oriented computing is dynamic. There may be many possible service instances available for binding, leading to uncertainty about where service requests will execute. We present a novel Markovian process calculus which allows the formal expression of uncertainty about binding as found in service-oriented computing. We show how to compute meaningful quantitative information about the quality of service provided in such a setting. These numerical results can be used to allow the expression of accurate service-level agreements about service-oriented computing.

1 Introduction

Dynamic configuration is the essence of service-oriented computing. Service providers publish their services in a public registry. Service consumers discover services at run-time and bind to them dynamically, choosing from the available service instances according to the criteria which are of most importance to them. This architecture provides robust service in difficult operational conditions. If one instance of a service is temporarily unavailable then another one is there to take its place. It is likely though that this replacement is not fully functionally identical. It might have some missing functionality, or it might even offer additional functionality not found in the temporarily unavailable service instance.

However, even in the case of a functionally-identical replacement matters are still not straightforward when non-functional criteria such as availability and performance are brought into the picture. It is frequently the case that the functionally-equivalent replacement for the temporarily unavailable service will exhibit different performance characteristics simply because it hosts a copy of the service on another hardware platform. This impacts on essentially all performance measures which one would think to evaluate over the system configuration.

The world of distributed systems in which service-oriented computing resides is resource-sharing in nature. In such systems we have the additional complication that services may only be partially available in the sense that they are operational, but heavily loaded. In principle, all of their functionality is available, but only at a fraction of the usual level of performance. This becomes a pressing concern when service providers wish to advertise service-level agreements which

A. Corradini and U. Montanari (Eds.): WADT 2008, LNCS 5486, pp. 21–36, 2009.

provide service consumers with formal statements about the quality of service offered. For example, a service provider might believe that 90% of requests receive a response within 3 seconds, but how can they check this?

Analytical or numerical performance evaluation provides valuable insights into the timed behaviour of systems over the short or long run. Prominent methods used in the field include the numerical evaluation of continuous-time Markov chains (CTMCs). These bring a controlled degree of randomness to the system description by using exponentially-distributed random variables governed by rate constants to characterise activities of varying duration. Often generated from a high-level description language such as a Petri net or a process algebra, CTMCs are applied to study fixed, static system configurations with known subcomponents with known rate parameters. This is far from the operating conditions of service-oriented computing where for critical service components a set of replacements with perhaps vastly different performance qualities stand ready to substitute for components which are either unavailable, or the consumer just simply chooses not to bind to them. How can we bridge this gap and apply Markovian performance evaluation to the assessment of service-level agreements about service-oriented computing?

In the present paper we propose a new Markovian process calculus which includes language constructs for the formal expression of uncertainty about binding and parameters (in addition to the other dimension of uncertainty about durations modelled in the Markovian setting through the use of exponentially-distributed random variables). We put forward a method of numerical evaluation for this calculus which scales well with increasing problem size to allow precise comparisons to be made across all of the possible service bindings and levels of availability considered. Numerical evaluation is supported inside a modelling environment for the calculus. We demonstrate the approach by considering an example of a (fictional) virtual university formed by bringing together the resources of several (real) universities. Our calculus is supported by a freely-available software tool.

Structure of this paper: In Section 2 we introduce our new Markovian calculus. In Section 3 we present an example service-oriented computing system, a "virtual university". In Section 4 we describe the analysis which can be performed on our process calculus models. In Section 5 we explain the software tools which we use. We discuss related work in Section 6 and present conclusions in Section 7.

2   SRMC: Sensoria Reference Markovian Calculus

SRMC is a Markovian process calculus in the tradition of PEPA [1], Stochastic KLAIM [2], and Stochastic FSP [3]. On top of a classical process calculus, SRMC adds *namespaces* to allow the structured description of models of large size, and *dynamic binding* to represent uncertainty about component specification or the values of parameters. As a first step in machine processing, namespaces and dynamic binding can be resolved in order to map into a Markovian calculus without these features such as PEPA (for performance analysis [4,5]). Going further, rate information can also be erased in order to map into an untimed process calculus such as FSP (for analysis of safety and liveness properties [6]).

Namespaces in SRMC may be nested. Dynamic binding is notated by writing in the form of a set all of the possible values which may be taken. The binding records that the value is one of the values in the set (but we are not sure which one). The following example uses the name UEDIN for a location, the name Server for the server located there, the constant processors for the number of processors which the Edinburgh server has, and the constant availability for the availability of the server (which is between 50% and 100%).

```
UEDIN::{
   Server::{
      processors = 2;
      availability = { 0.5, 0.6, 0.7, 0.8, 0.9, 1.0 };
   }
   ...
}
```

Outside the namespace scope one refers to the first constant using the fully qualified name UEDIN::Server::processors and to the second using the name UEDIN::Server::availability.

In addition to being able to give names to numerical constants and values it is also possible to give names to processes (in order to describe recursive behaviour). Process terms are built up using prefix (.) and choice (+). The following process definition describes a lossy buffer which loses, on average, one datum in every ten. As the example shows, activity rates can be conditioned by probabilities (0.1 and 0.9 here).

```
LossyBuffer::{
   Empty = (put, 0.1 * r).Empty + (put, 0.9 * r).Full;
   Full = (get, s).Empty;
}
```

Processes of the SRMC language give rise to labelled transition systems which are converted to Continuous-Time Markov Chain (CTMC) representations in the way which is familiar from PEPA [1].

Process expressions can be defined conditionally in SRMC depending on the values obtained in the resolution of dynamic binding. For example, a server might allow additional sessions to be opened if availability is above 70% and forbid the creation of new sessions otherwise.

```
if availability > 0.7 then (openSession, r).ServeClient
```

An equivalent effect can be obtained using *functional rates* [7] which can allow the use of space-efficient state-space representation using Kronecker methods. The equivalent process expression using functional rates is below.

```
(openSession, if availability > 0.7 then r else 0.0).ServeClient
```

In stochastic Petri nets functional rates are termed "marking dependent rates".

Dynamic service binding is described by associating a name with a set of processes. The example below records that the server is either the Edinburgh server (UEDIN) or the Bologna server (UNIBO).

```
Server = { UEDIN::Server, UNIBO::Server };
```

2.1 Discussion

It might seem that it is not necessary to have the ability to describe sets of processes, binding to one of these later because it would be possible to implement the idea of dynamic binding instead using well-known process calculus primitives. For example, one could use a silent, internal τ transition at the start of the lifetime of one of the components to choose to behave as one of the binding sites, thereafter ignoring all of the possible behaviour described by the other components from the other sites. While this is possible, we do not favour this approach because it leads to the consideration of the full state space for every evaluation of parameters of the system. In contrast, the method of first projecting down to a particular binding and then evaluating this leads to the smallest possible state-space for each evaluation run, with attendant benefits for run-times and stability of the results. Further, the binding projection method allows the problem to be decomposed in a larger number of smaller problems, each of which can be solved independently and the results combined. We wish to perform scalable analysis of scalable systems and so this approach suits us well.

2.2 Numerical Evaluation

We have been keen to decompose the analysis problem so that we can ensure that the analysis can be performed as a large number of numerical evaluations of small size. Our preference for problems of this form stems from the fact that they are easy to distribute across a network of workstations. Thus, we use a distributed computing platform (Condor [8]) to accelerate the numerical evaluation work by distributing the computation across a cluster of workstations (a Condor "pool"). In this way we can greatly increase the speed of generation of results. In practice we have found that our Condor pool of 70 machines gives a speedup over sequential evaluation close to 70-fold. Because we are aware that others may wish to use our software but may not have a local Condor pool we also provide a purely sequential evaluation framework which does not depend on Condor.

We know that numerical linear algebra is not to everyone's taste so we will just give an outline of what we do here and refer the curious to [9]. Investigation of SLAs requires the transient analysis of a CTMC, represented as an $n \times n$ state transition matrix Q (the "generator matrix"). We are concerned with finding the transient state probability row vector $\pi(t) = [\pi_1(t), \ldots, \pi_n(t)]$ where $\pi_i(t)$ denotes the probability that the CTMC is in state i at time t. Transient and passage-time analysis of CTMCs proceeds by a procedure called *uniformisation* [10,11]. The generator matrix, Q, is "uniformized" with:

$$P = Q/q + I$$

where $q > \max_i |Q_{ii}|$ and I is the identity matrix. This process transforms a CTMC into one in which all states have the same mean holding time $1/q$.

Passage-time computation is concerned with knowing the probability of reaching a designated target state from a designated source state. It rests on two key sub-computations. First, the time to complete n hops ($n = 1, 2, 3, \ldots$), which is an Erlang distribution with parameters n and q. Second, the probability that the transition between source and target states occurs in exactly n hops.

3   Example: Distributed e-Learning Case Study

Our general concern is with evaluating quality of service in the presence of uncertainty such as that caused by dynamic binding but as a lighthearted example to illustrate the approach we consider a (fictional) Web Service-based distributed e-Learning and course management system run by the Sensoria Virtual University (SVU).

The SVU is a virtual organisation formed by bringing together the resources of the universities at Edinburgh (UEDIN), Munich (LMU), Bologna (UNIBO), Pisa (UNIPI) and others not listed in this example. The SVU federates the teaching and assessment capabilities of the universities allowing students to enrol in courses irrespective of where they are delivered geographically. Students download *learning objects* from the content download portals of the universities involved and upload archives of their project work for assessment. By agreement within the SVU, students may download from (or upload to) the portals at any of the SVU sites, not just the one which is geographically closest.

Learning objects may contain digital audio or video presentation of lecture courses and students may be required to upload archives of full-year project work. Both of these may be large files so the *scalability* of such a system to support large numbers of students is a matter of concern. We have addressed this issue previously [12,13].

3.1   The Servers

We start by describing the servers which are available for use. Dedicated upload and download portals are available at each site. At Edinburgh the portals sometimes fail and need to be repaired before they are available to serve content again. They are usually relatively lightly loaded and availability is between 70% and 100%. The portals at Edinburgh are described in SRMC thus.

```
UEDIN::{
    lambda = 1.65;   mu = 0.0275;   gamma = 0.125;   delta = 3.215;
    avail = { 0.7, 0.8, 0.9, 1.0 };
```

```
UploadPortal::{
   Idle = (upload, avail * lambda).Idle + (fail, mu).Down;
   Down = (repair, gamma).Idle;
}
DownloadPortal::{
   Idle = (download, avail * delta).Idle + (fail, mu).Down;
   Down = (repair, gamma).Idle;
}
}
```

The portals at Munich are so reliable that it is not worth modelling the very unlikely event of their failure. However, they are slower than the equivalent portals at Edinburgh and availability is more variable and usually lower, because the portals are serving a larger pool of local students.

```
LMU::{
   lambda = 0.965;   delta = 2.576;
   avail = { 0.5, 0.6, 0.7, 0.8, 0.9 };
   UploadPortal::{
      Idle = (upload, avail * lambda).Idle;
   }
   DownloadPortal::{
      Idle = (download, avail * delta).Idle;
   }
}
```

Because it is running a more recent release of the portal software the Bologna site offers secure upload and download also. Availability is usually very good. To maintain good availability the more expensive operations of secure upload and secure download are not offered if the system seems to be becoming heavily loaded.

```
UNIBO::{
   lambda = 1.65;   mu = 0.0275;   gamma = 0.125;   delta = 3.215;
   slambda = 1.25; sdelta = 2.255;   avail = { 0.8, 0.9, 1.0 };
   UploadPortal::{
      Idle = (upload, avail * lambda).Idle + (fail, mu).Down
            + if avail > 0.8 then (supload, avail * slambda).Idle;
      Down = (repair, gamma).Idle;
   }
   DownloadPortal::{
      Idle = (download, avail * delta).Idle  + (fail, mu).Down
            + if avail > 0.8 then (sdownload, avail * sdelta).Idle;
      Down = (repair, gamma).Idle;
   }
}
```

The Pisa site is just like the Bologna site, but uses a higher grade of encryption, meaning that secure upload and download are slower (slambda = 0.975, sdelta

= 1.765). We can list the possible bindings for upload and download portals in the following way.

```
UploadPortal =
   { UEDIN::UploadPortal::Idle, LMU::UploadPortal::Idle,
     UNIBO::UploadPortal::Idle, UNIPI::UploadPortal::Idle };

DownloadPortal =
   { UEDIN::DownloadPortal::Idle, LMU::DownloadPortal::Idle,
     UNIBO::DownloadPortal::Idle, UNIPI::DownloadPortal::Idle };
```

3.2 The Clients

We now describe two typical clients of the system, Harry and Sally. Both Harry and Sally wish to accomplish the same task, which is to download three sets of learning materials and to upload two coursework submissions. They perform this behaviour cyclically. Harry is unconcerned about security and never uses secure upload or download even if it is available. Sally uses secure upload and secure download sometimes when it is available, and uses non-secure upload and download when it is not. We are interested in the passage of time from start to finish for both Harry and Sally. Clients do not determine the rates of activities: others do (we write "_" for the rate here).

```
Harry::{
   Idle = (start, 1.0).Download;
   Download = (download, _).(download, _).(download, _).Upload;
   Upload = (upload, _).(upload, _).Disconnect;
   Disconnect = (finish, 1.0).Idle;
}

Sally::{
   Idle = (start, 1.0).Download;
   Download = (download, _).(download, _).(download, _).Upload
            + (sdownload, _).(sdownload, _).(sdownload, _).Upload;
   Upload = (upload, _).(upload, _).Disconnect
          + (supload, _).(supload, _).Disconnect;
   Disconnect = (finish, 1.0).Idle;
}
```

The client is either Harry or Sally, both initially idle.

```
Client = { Harry::Idle, Sally::Idle };
```

Finally, the complete system is formed by composing the client with the two portals, cooperating over upload and download. The upload and download portals do not communicate with each other (<>).

```
System = Client <upload, download, supload, sdownload>
               (UploadPortal <> DownloadPortal);
```

4   Analysis

The analysis applied to SRMC models is a staged computation:

Resolving service bindings: Each possible service binding is chosen in turn. This involves selecting one element of each set of possibilities for service providers.

Model minimisation: The model is reduced to remove unused definitions of processes and rate constants. This is a necessary economy applied to make the next stage more productive.

Parameter sweep: Parameter sweep is performed over the remaining rate values, executing processes in a distributed fashion on a Condor pool, or sequentially on a single machine.

Analysis and visualisation: The results are collected and summarised using statistical procedures. We visualise the results to aid in model interpretation and analysis.

4.1   Qualitative Analysis

On the way towards the quantitative results which we seek our state-space analysis delivers qualitative insights about the function of the system being modelled. We list three of the things which we learn here:

1. The system is *deadlock-free* for all configurations. No binding of service instances to service parameters gave rise to a model with a deadlock.
2. The system is *livelock-free* for all configurations. No binding of service instances to service parameters gave rise to a model where states could be visited only a finite number of times (a *transient state*, in Markov chain terminology).
3. All activities in the model are *weakly live*. That is, for each activity (such as `supload`) there is some configuration which allows that activity to occur, although it may be blocked in other configurations. Put more plainly, the SRMC model has no "dead code" (activities which can never occur).

4.2   Sensitivity Analysis

We are here concerned generally with lack of certainty about parameters such as rates but even in the case where rate information can be known with high confidence the framework which we have available for performing a parameter sweep across the rate constants can be used to perform *sensitivity analysis*. One way in which the results obtained by sensitivity analysis can be used is to determine which activities of the system are bottlenecks. That is, to discover which rate or rates should we alter to ensure that the user sees the greatest improvement in performance. We have an evaluation function which assigns a score to each solution of the underlying Markov chain. In this case, the less is the response time then the higher is the score.

It might seem that the results obtained from sensitivity analysis are likely to be pretty unsurprising and that it will turn out to be the case that increasing the

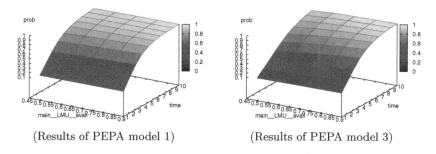

(Results of PEPA model 1)   (Results of PEPA model 3)

Fig. 1. Graphs showing sensitivity analysis over the rates in the produced models. The basic plot is a cumulative distribution function showing how the probability of completion of the uploads and downloads increases as a function of time. The surface plot is obtained from this because we vary one of the parameters. Here in both cases we vary the availability of the Munich server from 50% availability to 90% availability. Expressed as a scaling factor this becomes 0.5 to 0.9.

rate of any activity brings about a proportional decrease in response time. To see that this is not the case, we will compare two sets of results. Recall that SRMC generates many PEPA models; we number these. The first set of results shown in Fig. 1 comes from PEPA model 1, where Edinburgh is the upload portal, Munich the download portal, and Harry is the client. In model 3 they swap around so that Munich is the upload portal, Edinburgh the download, and Harry is again the client. In the latter case low availability of the Munich server makes a noticeable impact on response time (the curve takes longer to get up to 1) but in the former case the low availability of the Munich server has negligible impact. This is made clear in the results but it is unlikely that a modeller would see this trend just by inspecting the model; we needed to see the results to get this insight. We have generated many results so we have been able to get many such insights.

4.3   Computing Response-Time Percentiles

The results shown in Fig. 1 show ten of the nearly 250 cumulative distribution functions which we computed for the possible configurations of the example. We wanted to produce a simple statistical summary which brought together all of the results obtained. We computed *percentiles* of the results which declare that in (say) 90% of the possible configurations of the system the response-time will be in this region. This tells us about the experience which most users will have (where here, "most" means "90% of"). Some will see better response times, and some with see worse, but it is usually interesting to consider the common case response times.

To illustrate how percentiles can be used to summarise the results we show in Fig. 2(a) forty sets of results in the form of the cumulative distribution functions which we computed. These give a sense of the "envelope" in which the results are contained. Most configurations of the system produced by resolving the service

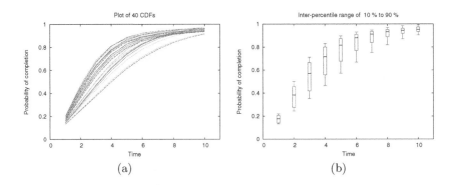

Fig. 2. Sub-figure (a) shows 40 of the response-time distributions computed for the Sensoria Virtual University example. Sub-figure (b) shows the 10% to 90% percentile of the results over all of the runs. The median value is also marked as a horizontal line cutting across the thick bar in the candlestick. From sub-figure (b) we can report results of the form "All uploads and downloads will have completed by time $t = 10$ with probability between 0.90 and 0.97, in 90% of configurations".

instance bindings are very likely to have completed the work to be done by $t = 10$. The majority of configurations give response-time distributions which put them towards the top of the "envelope" but there are a few configurations which perform quite a bit worse (and our analysis has identified which configurations these are).

The graph in Fig. 2(b) is known as a "candlestick" graph and is a summary of all of the solutions produced. It shows that 90% of the time the response time distribution will lie within the area described by the thick bar of the candlestick, but it has been seen to be as high as the top of the candlestick, and it has been seen to be as low as the bottom of the candlestick.

4.4   Comparisons across All Runs

Even for traditional computer systems without dynamic binding, service-level agreements are already quite complex because they relate a path through the system behaviour, a time bound, and a probability bound. (A typical example of an SLA is "We guarantee that 97.5% of requests will receive a response within three seconds". Here "from request to response" is the path through the system, three seconds is the time bound, and 97.5% gives the probability bound.) In the service-oriented computing setting we have yet another dimension of complication because we must add a qualifier speaking about the quantile of system configurations being considered ("... in 90% of the possible configurations"). Complicated service-level agreements of this form are unattractive.

We have found that an alternative presentation of the results can be easier to interpret in some cases and so the SRMC software supports a presentation mode where we show the probability of completion by a particular point in time, across all possible configurations. Having all of the results to hand, we are

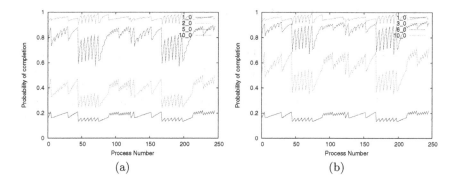

Fig. 3. Probability of completion of all uploads and downloads against time across all (nearly 250) possible configurations of the example. In sub-figure (a) the times considered are $t = 1.0, 2.0, 5.0$, and 10.0. In sub-figure (b) $t = 1.0$ and 10.0 are repeated for reference and $t = 3.0$ and 6.0 are also presented. By time $t = 10.0$ we are able to make meaningful comments about all configurations. For example, we can say that there is at least a 90% chance of having completed the uploads and downloads by time $t = 10.0$, irrespective of the system configuration. The greatest variability is seen at times around $t = 3.0$. Here for the best configurations the system has a 70% chance of having completed the work for the worst configurations there is less than a 40% chance of having completed.

able to reduce the dimension of the problem and make statements about the probability of completion of the work at a particular point in time, irrespective of the configuration of the system.

In reference to Fig. 3 we can see not a statistical summary (as we saw in Fig. 2(b) before) but the actual results of all runs at a particular point in time. This makes clear the difference between the best-performing configurations at time t and the worst-performing configurations at time t. For low values of t such as 1.0 there is little chance that any user has completed all uploads and downloads. For high values of t such as 10.0 there is little chance that they have not.

5   Software Tool Support

SRMC is supported by a tool chain whose main design goal has been to provide a friendly and rich graphical user interface as well as a set of efficient model solvers. The software comprises a graphical front-end written in Java for the Eclipse framework and a back-end implemented in Haskell and C++. The latter exposes its functionality via a command-line interface, and thus can be used as a stand-alone application in headless environments such as Condor or to reduce the tool's overall memory footprint. This section provides an overview of both modules; further information is available at the SRMC Web site [14], which also provides a download link to the tool.

5.1   Analysis Tools in the Back-end

The analysis back-end is implemented as a series of three applications: the Sensoria Markovian Compiler (`smc`), the Imperial PEPA Compiler (`ipc`) and the Hypergraph-based Distributed Response-Time Analyser (`hydra`). `smc` accepts SRMC models as input and generates the intermediate PEPA descriptions that represent all the possible configurations of the system. The main tasks performed by `smc` are resolving binding instantiations, name-resolution and flattening of

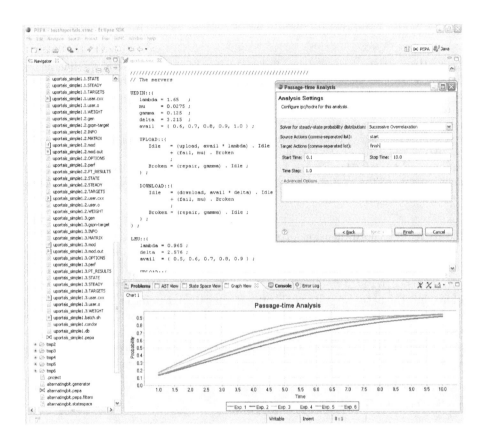

Fig. 4. Screenshot showing the SRMC Eclipse plug-in processing the SVU example. Displayed in the screenshot are (i) the workspace navigator showing compiled representations of the SRMC model as PEPA models, Hydra models and compiled Hydra C++ files; (ii) the SRMC model editor; (iii) the user-interface dialogue used for setting parameters on the analyser and running the transient analysis repeatedly; and (iv) a graphical display showing the results of all passage-time analysis runs expressed in the form of the cumulative distribution functions computed numerically by the Markov chain solver. In addition to providing user-interface widgets, the plug-in exposes SRMC tools to the framework through an application programming interface for third-party Eclipse plug-ins.

the SRMC model's namespaces, and generation of PEPA models for analysis. A database file produced by srmc maintains associations between the original SRMC model and the underlying PEPA models.

Such models are the basic units on which analysis is to be carried out. As PEPA benefits from extensive software support, a number of analysis tools are readily available for re-use in this context. Here, each PEPA model is run through ipc [15]. It translates the description into a format suitable for hydra [16], which performs passage-time analysis and stores the results to disk. Such results can be related back to the SRMC description via the database file from smc.

5.2   Presentation Layer at the Front-end

The graphical user interface is implemented as a contribution (*plug-in*) to Eclipse, a popular extensible cross-platform development framework. The plug-in provides an editor and a standard Eclipse contribution to the *Outline* view to concisely display information about the model. The plug-in also adds a top-level menu item through which SRMC features are accessible. In particular, a wizard dialogue guides the user through the set-up of passage-time analysis. Upon completion, the wizard schedules an array of background processes that run the back-end tool chain as described above. All the intermediate resources such as the PEPA model instances and the hydra description files are available in the user's workspace for further inspection via the Eclipse *Navigator* view. When the analysis is complete, the results are collected and presented to the user as a plot in the *Graph* view. Figure 4 shows a screenshot of an Eclipse session running the SRMC plug-in.

6   Related Work

The SRMC language builds on the PEPA language and tools. PEPA has been applied to a wide range of modelling problems across computer science including software [17,18,19], hardware [20,21,22], and services [23,24]. We see our work on modelling with SRMC as being similar in style but with an increased emphasis on experimentation.

In our numerical evaluation of the many possible system configurations which are described by an SRMC model we have essentially used the "brute force" solution of solving for all possible bindings. This has the advantage that it ensures that all of the bindings are considered, and is trivially parallelisable, but still costs a lot of computation time. It is possible that we could do fewer numerical evaluations and still explore the space of all possibilities well by applying methods which are well-known in the field of design of experiments. Similar strategic exploration of the solution space is found in state-of-the-art modelling platforms such as Möbius [25].

7 Conclusions

For software engineering to improve as a well-managed discipline we believe that it is critical to have access to a modelling process which can make sound quantitative predictions about the performance of complex systems. We have addressed the problem of how virtual organisations can defend any quantitative statements about their quality of service as expressed in service-level agreements given that their operation is founded on service-oriented computing. The essential function of dynamic binding brings uncertainty to the model concerning both functional and non-functional aspects. We have been able to control this uncertainty by considering all possible bindings, undertaking separate numerical evaluations of these, and combining the results to correctly quantify the uncertainty induced by dynamic binding and degree of availability.

We decomposed the computations needed into a large number of independent numerical evaluations each of which has modest memory requirements. We distributed the independent runs across a network of workstations. The distributed computing platform which we chose, Condor, makes use of the idle cycles on networked workstations meaning that we could perform all of the computations which were needed on typical desktop PCs when they were unused in our student computing laboratories. Widely-used in computational science, this approach uses stock hardware and scales well to apply to more complex problem cases with a greater range of possible configurations and parameter values. More computing power can be deployed on larger problems simply by adding more machines to the Condor pool. We hope that this is a "real-world" approach to a "real-world" problem.

Acknowledgements. The authors are supported by the EU FET-IST Global Computing 2 project SENSORIA ("Software Engineering for Service-Oriented Overlay Computers" (IST-3-016004-IP-09)). The Imperial PEPA Compiler was developed by Jeremy Bradley of Imperial College, London. The Hydra response-time analyser was developed by Will Knottenbelt and Nick Dingle of Imperial College, London. We extended both of these software tools for the present work.

References

1. Hillston, J.: A Compositional Approach to Performance Modelling. Cambridge University Press, Cambridge (1996)
2. De Nicola, R., Katoen, J.P., Latella, D., Massink, M.: STOKLAIM: A stochastic extension of KLAIM. Technical Report ISTI-2006-TR-01, Consiglio Nazionale delle Ricerche (2006)
3. Ayles, T.P., Field, A.J., Magee, J., Bennett, A.: Adding Performance Evaluation to the LTSA Tool. In: Kemper, P., Sanders, W.H. (eds.) TOOLS 2003. LNCS, vol. 2794. Springer, Heidelberg (2003)
4. Clark, A.: The ipclib PEPA Library. In: Harchol-Balter, M., Kwiatkowska, M., Telek, M. (eds.) Proceedings of the 4th International Conference on the Quantitative Evaluation of SysTems (QEST), September 2007, pp. 55–56. IEEE, Los Alamitos (2007)

5. Tribastone, M.: The PEPA Plug-in Project. In: Harchol-Balter, M., Kwiatkowska, M., Telek, M. (eds.) Proceedings of the 4th International Conference on the Quantitative Evaluation of SysTems (QEST), September 2007, pp. 53–54. IEEE, Los Alamitos (2007)
6. Magee, J., Kramer, J.: Concurrency: State Models and Java Programming, 2nd edn. Wiley, Chichester (2006)
7. Hillston, J., Kloul, L.: An efficient Kronecker representation for PEPA models. In: de Alfaro, L., Gilmore, S. (eds.) PROBMIV 2001, PAPM-PROBMIV 2001, and PAPM 2001. LNCS, vol. 2165, pp. 120–135. Springer, Heidelberg (2001)
8. Thain, D., Tannenbaum, T., Livny, M.: Distributed computing in practice: the Condor experience. Concurrency – Practice and Experience 17(2–4), 323–356 (2005)
9. Knottenbelt, W.: Performance Analysis of Large Markov Models. PhD. thesis, Imperial College of Science, Technology and Medicine, London, UK (February 2000)
10. Grassmann, W.: Transient solutions in Markovian queueing systems. Computers and Operations Research 4, 47–53 (1977)
11. Gross, D., Miller, D.: The randomization technique as a modelling tool and solution procedure for transient Markov processes. Operations Research 32, 343–361 (1984)
12. Gilmore, S., Tribastone, M.: Evaluating the scalability of a web service-based distributed e-learning and course management system. In: Bravetti, M., Núñez, M., Zavattaro, G. (eds.) WS-FM 2006. LNCS, vol. 4184, pp. 214–226. Springer, Heidelberg (2006)
13. Bravetti, M., Gilmore, S., Guidi, C., Tribastone, M.: Replicating web services for scalability. In: Barthe, G., Fournet, C. (eds.) TGC 2007 and FODO 2008. LNCS, vol. 4912, pp. 204–221. Springer, Heidelberg (2008)
14. SRMC Team: Sensoria Reference Markovian Calculus Web Site and Software (October 2008), http://groups.inf.ed.ac.uk/srmc
15. Bradley, J., Dingle, N., Gilmore, S., Knottenbelt, W.: Derivation of passage-time densities in PEPA models using IPC: The Imperial PEPA Compiler. In: Kotsis, G. (ed.) Proceedings of the 11th IEEE/ACM International Symposium on Modeling, Analysis and Simulation of Computer and Telecommunications Systems, University of Central Florida, pp. 344–351. IEEE Computer Society Press, Los Alamitos (2003)
16. Dingle, N., Harrison, P., Knottenbelt, W.: HYDRA: HYpergraph-based Distributed Response-time Analyser. In: Proc. International Conference on Parallel and Distributed Processing Techniques and Applications (PDPTA 2003), Las Vegas, Nevada, USA, June 2003, pp. 215–219 (2003)
17. Hillston, J., Kloul, L.: Performance investigation of an on-line auction system. Concurrency and Computation: Practice and Experience 13, 23–41 (2001)
18. Hillston, J., Kloul, L., Mokhtari, A.: Active nodes performance analysis using PEPA. In: Jarvis, S. (ed.) Proceedings of the Nineteenth annual UK Performance Engineering Workshop, July 2003, pp. 244–256. University of Warwick (2003)
19. Buchholtz, M., Gilmore, S., Hillston, J., Nielson, F.: Securing statically-verified communications protocols against timing attacks. Electr. Notes Theor. Comput. Sci. 128(4), 123–143 (2005)
20. Holton, D.: A PEPA specification of an industrial production cell. In: Gilmore, S., Hillston, J. (eds.) Proceedings of the Third International Workshop on Process Algebras and Performance Modelling, Special Issue of The Computer Journal 38(7), 542–551 (1995)
21. Gilmore, S., Hillston, J., Holton, D., Rettelbach, M.: Specifications in stochastic process algebra for a robot control problem. International Journal of Production Research 34(4), 1065–1080 (1996)

22. Console, L., Picardi, C., Ribaudo, M.: Diagnosis and diagnosability analysis using PEPA. In: Proc. of 14th European Conference on Artificial Intelligence, Berlin (August 2000); A longer version appeared in the Proc. of 11th Int. Workshop on Principles of Diagnosis (DX 2000), Morelia, Mexico (June 2000)

23. Clark, A., Gilmore, S.: Evaluating quality of service for service level agreements. In: Brim, L., Leucker, M. (eds.) Proceedings of the 11th International Workshop on Formal Methods for Industrial Critical Systems, Bonn, Germany, August 2006, pp. 172–185 (2006)

24. Argent-Katwala, A., Clark, A., Foster, H., Gilmore, S., Mayer, P., Tribastone, M.: Safety and response-time analysis of an automotive accident assistance service. In: Proceedings of the 3rd International Symposium on Leveraging Applications of Formal Methods, Verification and Validation (ISoLA 2008), Porto Sani, October 2008. Communications in Computer and Information Science (CCIS), vol. 17. Springer, Heidelberg (2008)

25. Courtney, T., Gaonkar, S., McQuinn, M., Rozier, E., Sanders, W., Webster, P.: Design of Experiments within the Möbius Modeling Environment. In: Harchol-Balter, M., Kwiatkowska, M., Telek, M. (eds.) Proceedings of the 4th International Conference on the Quantitative Evaluation of SysTems (QEST), pp. 161–162. IEEE, Los Alamitos (2007)

Tiles for Reo[*]

Farhad Arbab[1], Roberto Bruni[2], Dave Clarke[3], Ivan Lanese[4], and Ugo Montanari[2]

[1] CWI, Amsterdam, The Netherlands
farhad@cwi.nl
[2] Dipartimento di Informatica, Università di Pisa, Italy
{bruni,ugo}@di.unipi.it
[3] Department of Computer Science, Katholieke Universiteit Leuven, Belgium
Dave.Clarke@cs.kuleuven.be
[4] Dipartimento di Scienze dell'Informazione, Università di Bologna, Italy
lanese@cs.unibo.it

Abstract. Reo is an exogenous coordination model for software components. The informal semantics of Reo has been matched by several proposals of formalization, exploiting co-algebraic techniques, constraint-automata, and coloring tables. We aim to show that the Tile Model offers a flexible and adequate semantic setting for Reo, such that: (i) it is able to capture context-aware behavior; (ii) it is equipped with a natural notion of behavioral equivalence which is compositional; (iii) it offers a uniform setting for representing not only the ordinary execution of Reo systems but also dynamic reconfiguration strategies.

1 Introduction

Reo [1,7,8] is an exogenous coordination model for software components. It is based on channel-like connectors that mediate the flow of data and signals among components. Notably, a small set of point-to-point primitive connectors is sufficient to express a large variety of interesting constraints over the behavior of connected components, including various forms of mutual exclusion, synchronization, alternation, and context-dependency. In fact, components and primitive connectors can be composed in a circuit fashion via suitable attach points, called Reo nodes. Typical primitive connectors are the synchronous / asynchronous / lossy channels and the asynchronous one-place buffer. The informal semantics of Reo has been formalized in several ways, exploiting co-algebraic techniques [2], constraint-automata [3], and coloring tables [5]. However all the formalizations in the literature that we are aware of are unsatisfactory from some points of view. In fact, both [2] and [3] provide detailed characterizations of the behavior of connectors, allowing to exploit coinductive techniques, but they do not support context-awareness, and, in particular, they are not able to faithfully model the LossySync connector. Up to now, the only approach that takes context-awareness into account is the 3-color semantics presented in [5]. This semantics, however, describes only a single computational step, thus it does not describe the evolution of the state

---

[*] Research supported by the project FET-GC II IST-2005-16004 SENSORIA, by the Italian FIRB project TOCAI, by the Dutch NWO project n. 612.000.316 C-Quattro, and by the bilateral German-Dutch DFG-NWO project n. 600.643.000.05N12 SYANCO.

A. Corradini and U. Montanari (Eds.): WADT 2008, LNCS 5486, pp. 37–55, 2009.

of a connector. Also, none of these semantics allows reconfiguration, which then, for instance as in [12], has to be added on top of them. The interplay between the dataflow semantics of Reo circuits and their reconfiguration has been considered in [11] and [10] using graph transformations triggered by the 3-color semantics.

We aim to show that the Tile Model [9] offers a flexible and adequate semantic setting for Reo. The name 'tile' is due to the graphical representation of such rules (see Fig. 5 in Section 3). The tile α states that the *initial configuration s* can be triggered by the event a to reach the *final configuration t*, producing the *effect b*. Tiles resemble Gordon Plotkin's SOS inference rules [17], but they can be composed in three different ways to generate larger proof steps: (i) horizontally (synchronization), when the effect of one tile matches the trigger for another tile; (ii) vertically (composition in time), when the final configuration of one tile matches the initial configuration of another tile; and (iii) in parallel (concurrency). Tiles take inspiration from Andrea Corradini and Ugo Montanari's Structured Transition Systems [6] and generalise Kim Larsen and Liu Xinxin's context systems [13], by allowing for more general rule formats. The Tile Model also extends José Meseguer's rewriting logic [15] (in the non-conditional case) by taking into account rewrite with side effects and rewrite synchronization. As rewriting logic, the Tile Model admits a purely logical formulation, where tiles are seen as sequents subject to certain inference rules.

Roughly, in our tile encoding, Reo nodes and primitive connectors are represented as hyper-edges (with typed incoming and outgoing tentacles) that can be composed by connecting their tentacles. The one-step semantics of each primitive connector C is defined by suitable basic tiles whose initial configuration is the hyper-edge C (we use the same notation for primitive connectors and corresponding hyper-edges) and whose triggers and effects define how the data can flow through C.

A mapping of a fragment of Reo into the Tile Model has been already presented in [4]. There the emphasis was on exploiting for Reo connectors the normalization and axiomatization techniques developed therein for the used algebra of tile connectors. For this reason the mapping concentrated only on the synchronization connectors, i.e., data values were abstracted away, and data-sensitive connectors such as filters or stateful connectors such as buffers were not considered. The reason was that axiomatization for those more complex connectors was not available. The induced semantics corresponded to the data-insensitive 2-color semantics of Reo [5].

In this paper we extend the mapping in [4] to deal with all Reo connectors, and we concentrate on the 3-color semantics [5], the only one which captures context-awareness. The 3-color semantics for Reo that we propose in Section 6 recovers the good properties of the semantics in the literature, and provides also some additional benefits:

- it allows to model context dependency, and models faithfully the 3-color semantics of [5] as far as a single computational step is concerned;
- it is data-sensitive, describing the actual data that flow inside the connector;
- it can model whole computations, keeping into account the evolution of the state;
- it has a natural notion of behavioral equivalence, tile bisimilarity, that allows to exploit coinductive techniques similar to the ones in [2,3];
- the provided notion of bisimilarity is a congruence, i.e. the behavioral semantics is compositional;

- the congruence property can be easily proved by exploiting standard meta-theoretical results;
- it can be smoothly extended to deal with some form of reconfiguration (Section 7), and the extension also specifies in a formal way the interplay between computation and reconfiguration.

To clarify the approach we first model the simpler 2-color semantics and then show how to handle the 3-color case. In both cases we consider a data-sensitive semantics. Interestingly, the two semantics can be expressed in the same setting (and in a very similar way). Also, they give rise in a natural way to a notion of behavioral equivalence called tile bisimilarity, which is compositional. Finally, we hint at how the same setting can be exploited to model Reo reconfigurations, an aspect that is not considered by the standard Reo semantics. A more detailed treatment of this complex task is left for future work.

Structure of the paper. In Sections 2 and 3 we give some minimal background on Reo and Tile Logic. In Section 4 we define the representation of Reo graphs of connectors in terms of tile configurations. Sections 5 and 6 are dedicated respectively to the modeling of the 2-color and the 3-color semantics. Section 7 outlines the modeling of Reo reconfiguration. Concluding remarks are given in Section 8, together with some hints on future work we have in mind.

2   Reo Connectors

Reo [1,7,8] allows compositional construction of complex connectors with arbitrary behavior out of simpler ones. The simplest (atomic) connectors in Reo consist of a user defined set of *channels*. A channel is a binary connector: a medium of communication with exactly two directed ends. There are two types of channel ends: source and sink. A source channel end accepts data into its channel. A sink channel end dispenses data out of its channel. Every channel (type) specifies its own particular behavior as *constraints* on the flow of data through its ends. These constraints relate, for example, the content, the conditions for loss and/or creation of data that pass through the ends of a channel, as well as the atomicity, exclusion, order, and/or timing of their passage.

Although all channels used in Reo are user-defined and users can indeed define channels with any complex behavior (expressible in the semantic model) that they wish, a very small set of channels, each with very simple behavior, suffices to construct useful Reo connectors with significantly complex behavior [8]. Figure 1 shows a common set of primitive channels often used to build Reo connectors.

The Sync channel takes a data item from its source end and synchronously makes it available at its sink end. This transfer can succeed only if both ends are ready to communicate. The LossySync has the same behavior, except that it does not block its

Fig. 1. A typical set of Reo channels

Source node   Sink node   Mixed node

Fig. 2. Reo nodes

writer if its reader end cannot accept data. In this and only this case, the channel accepts the written data item and loses it. The FIFO1 is an asynchronous channel that has a buffer of size one. Unlike the prior channels, FIFO1 is a stateful channel: its behavior depends on whether its buffer is empty or full. The SyncDrain channel has two source ends (and no sink end) through which it can only consume data. It behaves as follows: if and only if there are data items available at both ends, it consumes (and loses) both of them atomically. The AsyncDrain is the asynchronous counterpart of the SyncDrain: it consumes and loses data items from either of its two ends only one at a time, but never from both ends together at the same time. Filter(P) is a synchronous channel with a data-sensitive behavior: it accepts through its source end and loses any data items that do *not* match its filter pattern *P*; it accepts a data item that matches P only if it can synchronously dispose of it through its sink end (exactly as if it were a Sync channel).

A channel end can be composed with other channel ends into Reo *nodes* to build more complex connectors. Reo nodes are logical places where channel ends coincide and coordinate their dataflows as prescribed by node types. Figure 2 shows the three possible node types in Reo. A node with only source channel ends is a *source node*; a node with only sink channel ends is a *sink node*; and a node with both source and sink channel ends is a *mixed node*. The term *boundary nodes* is also sometimes used to collectively refer to source and sink nodes. Boundary nodes define the interface of a connector. Components connect to the boundary nodes of a connector and interact anonymously with each other through this interface by performing I/O operations on the boundary nodes of the connector: *take* operations on sink nodes, and *write* operations on source nodes.

Reo fixes the semantics of (i.e., the constraints on the dataflow through) Reo nodes. Data flow through a source node only if a write operation *offers* a data item on this node and every one of its source channel ends can *accept* a copy of this data item. A source node, thus, behaves as a synchronized replicator. Data flow through a sink node only if at least one of its sink channel ends *offers* a data item and an input operation pending on this node can *accept* this data item. If more than one sink channel end offers data, the node picks one non-deterministically and excludes the offers of all the rest. A sink node, thus, behaves as a non-deterministic merger. The behavior of a mixed node is a combination of that of the other two: data flow through a mixed node only if at least one of its sink channel ends *offers* a data item and every one of its source channel ends can *accept* a copy of this data item. If more than one sink channel end offers data, the node picks one non-deterministically and excludes the offers of all the rest. Because a node has no buffer, data cannot be stored in a node. Hence, nodes instigate the propagation of synchrony and exclusion constraints on dataflow throughout a connector.

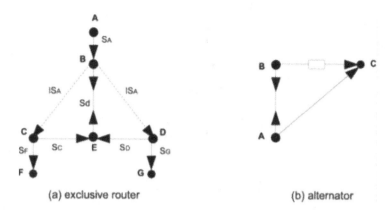

Fig. 3. Reo circuit for (a) exclusive router (from *A* to either *F* or *G*) and (b) Alternator

The simplest formalization of this behavior is the 2-color semantics presented in [5]. The two colors ■/□ model the flow/absence-of-flow of data at each node respectively (this is the so-called data-insensitive semantics; instead if different colors are used to distinguish the kind of data one obtains a data-sensitive semantics). This coloring must satisfy the constraint conditions imposed by connectors. Each connector determines the possible color combinations on its ends. For instance, both ends of Sync must have the same color (i.e. either the datum flows through the whole connector or no data flow at all), while AsyncDrain allows any coloring but (■, ■), which would represent data flowing synchronously at both of its ends. All channel ends connected to a □ node must be colored by □, while for ■ nodes, exactly one of the incoming channel ends, and all the outgoing channel ends, must have the ■ color. Deriving the semantics of a Reo connector amounts to resolving the composition of the constraints of its constituent channels and nodes. Given a connector C a coloring c for C is a function associating a color to each node in C. The 2-color semantics of C is given by its coloring table T_C, which contains all of its allowed colorings. For instance the coloring table of a connector with two nodes A and B connected by a Sync connector is $T = \{[A \mapsto \blacksquare, B \mapsto \blacksquare], [A \mapsto \square, B \mapsto \square]\}$.

In Fig. 3 we present two examples of Reo connectors that illustrate how non-trivial dataflow behavior emerges from composing simple channels using Reo nodes. The local constraints of individual channels propagate through (the synchronous regions of) a connector to its boundary nodes. This propagation also induces a certain context-awareness in connectors. See [5] for a detailed discussion of this.

The connector shown in Fig. 3(a) is an *exclusive router*: it routes data from A to either F or G (but not both). This connector can accept data only if there is a write operation at the source node A, and there is at least one taker at the sink nodes F and G. If both F and G can dispense data, the choice of routing to F or G follows from the non-deterministic decision by the mixed node E: E can accept data only from one of its sink ends, excluding the flow of data through the other, which forces the latter's respective LossySync to lose the data it obtains from A, while the other LossySync passes its

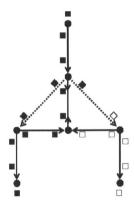

Fig. 4. A 2-coloring example for the exclusive router

data as if it were a Sync. A valid coloring of the exclusive router is shown in Fig. 4. The case shown in Fig. 4 corresponds to the forwarding of the data available on node A to the node F but not to G. There are two other possible 2-colorings for the exclusive router: one representing the case where the flow goes from A to G and not to F (i.e. the mirrored diagram w.r.t. Fig. 4) and one representing no dataflow (all the boxes are empty).

The connector shown in Fig. 3(b) is an *alternator* that imposes an ordering on the flow of the data from its input nodes A and B to its output node C. The SyncDrain enforces that data flow through A and B only synchronously. The empty buffer together with the SyncDrain guarantee that the data item obtained from A is delivered to C while the data item obtained from B is stored in the FIFO1 buffer. After this, the buffer of the FIFO1 is full and data cannot flow in through either A or B, but C can dispense the data stored in the FIFO1 buffer, which makes it empty again.

3   Tile Logic

Reo connectors are naturally represented as graphs. The advantage of using (freely generated) symmetric monoidal categories for representing configuration graphs is two-fold. First, it introduces a suitable notion of (observable) interfaces for configurations. Second, the natural isomorphism defined by symmetries allows to take graphs up to interface-preserving graph isomorphisms.

We recall that a *(strict) monoidal category* [14] $(C, \otimes, e)$ is a category C together with a functor $\otimes \colon C \times C \to C$ called the *tensor product* and an object e called the *unit*, such that for any arrows $\alpha_1, \alpha_2, \alpha_3 \in C$ we have $(\alpha_1 \otimes \alpha_2) \otimes \alpha_3 = \alpha_1 \otimes (\alpha_2 \otimes \alpha_3)$ and $\alpha_1 \otimes id_e = \alpha_1 = id_e \otimes \alpha_1$. The tensor product has higher precedence than the categorical composition ;. Note that we focus only on "strict" monoidal categories, where the monoidal axioms hold as equalities and not just up to natural isomorphisms. By functoriality of $\otimes$ we have, e.g., $\alpha_1 \otimes \alpha_2 = \alpha_1 \otimes id_{a_2}; id_{b_1} \otimes \alpha_2 = id_{a_1} \otimes \alpha_2; \alpha_1 \otimes id_{b_2}$ for any $\alpha_i \colon a_i \to b_i, i \in \{1, 2\}$.

Definition 1 (symmetric monoidal categories). *A symmetric (strict) monoidal category* $(C,\otimes,e,\gamma)$ *is a (strict) monoidal category* $(C,\otimes,e)$ *together with a family of arrows* $\{\gamma_{a,b}: a \otimes b \to b \otimes a\}_{a,b}$, *called* symmetries, *indexed by pairs of objects in C such that for any two arrows* $\alpha_1,\alpha_2 \in C$ *with* $\alpha_i: a_i \to b_i$, *we have* $\alpha_1 \otimes \alpha_2; \gamma_{b_1,b_2} = \gamma_{a_1,a_2}; \alpha_2 \otimes \alpha_1$ *(that is,* γ *is a natural isomorphism) that satisfies the coherence equalities (for any objects* a,b,c*):*

$$\gamma_{a,b};\gamma_{b,a} = id_{a\otimes b} \qquad\qquad \gamma_{a\otimes b,c} = id_a \otimes \gamma_{b,c}; \gamma_{a,c} \otimes id_b.$$

The categories we are interested in are those freely generated from a sorted (hyper)signature Σ, i.e., from a sorted family of operators $f: \tau_i \to \tau_f$. The objects are words on some alphabet S expressing the sorts of interfaces (we use ε to denote the empty word). Consider, e.g., $S = \{\bullet, \circ\}$. Then $f: \bullet\circ \to \bullet\bullet$ means that f has two "attach points" on both the interfaces, with types $\bullet\circ$ for the initial one and $\bullet\bullet$ for the final one. The operators $\sigma \in \Sigma$ are seen as basic arrows with source and target defined according to the sort of σ. Symmetries can always be expressed in terms of the basic sorted symmetries $\gamma_{x,y}: x \otimes y \to y \otimes x$. Intuitively, symmetries can be used to rearrange the input-output interfaces of graph-like configurations.

In this paper, we choose the Tile Model [9] for defining the operational and observational semantics of Reo connectors. In fact, tile configurations are particularly suitable to represent the above concept of connector, which includes input and output interfaces where actions can be observed and that can be used to compose configurations and also to coordinate their local behaviors.

A tile $\alpha : s \xrightarrow[b]{a} t$ is a rewrite rule stating that the *initial configuration s* can evolve to the *final configuration t* via α, producing the *effect b*; but the step is allowed only if the 'arguments' of s can contribute by producing a, which acts as the *trigger* of α (see Fig. 5(i)). Triggers and effects are called *observations* and tile vertices are called *interfaces*.

Tiles can be composed horizontally, in parallel, or vertically to generate larger steps (see Fig. 5). Horizontal composition $\alpha;\beta$ coordinates the evolution of the initial configuration of α with that of β, yielding the 'synchronization' of the two rewrites. Horizontal composition is possible only if the initial configurations of α and β interact cooperatively: the effect of α must provide the trigger for β. Vertical composition $\alpha * \beta$ is sequential composition of computations. The parallel composition $\alpha \otimes \beta$ builds concurrent steps.

The operational semantics of concurrent systems can be expressed via tiles if system configurations form a monoidal category $\mathcal{H}$, and observations form a monoidal category

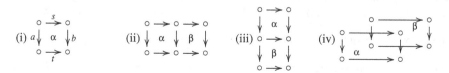

Fig. 5. Examples of tiles and their composition

$\mathcal{V}$ with the same underlying set of objects as $\mathcal{H}$. Abusing the notation, we denote by $\_\otimes\_$ both monoidal functors of $\mathcal{H}$ and $\mathcal{V}$ and by $\_;\_$ both sequential compositions in $\mathcal{H}$ and $\mathcal{V}$.

Definition 2 (tile system). *A* tile system *is a tuple* $\mathcal{R} = (\mathcal{H}, \mathcal{V}, N, R)$ *where* $\mathcal{H}$ *and* $\mathcal{V}$ *are monoidal categories with the same set of objects* $O_{\mathcal{H}} = O_{\mathcal{V}}$, N *is the set of rule names and* $R: N \to \mathcal{H} \times \mathcal{V} \times \mathcal{V} \times \mathcal{H}$ *is a function such that for all* $A \in N$, *if* $R(A) = \langle s, a, b, t \rangle$, *then the arrows* s, a, b, t *can form a tile like in Fig. 5(i).*

Like rewrite rules in rewriting logic, tiles can be seen as sequents of *tile logic*: the sequent $s \xrightarrow[b]{a} t$ is *entailed* by the tile logic associated with $\mathcal{R}$, written $\mathcal{R} \vdash s \xrightarrow[b]{a} t$, if it can be obtained by horizontal, parallel, and/or vertical composition of some basic tiles in R, plus possibly some auxiliary tiles such as identities $id \xrightarrow[a]{a} id$ which propagate observations, and horizontal symmetries $\gamma \xrightarrow[b\otimes a]{a\otimes b} \gamma$ which swap the order in which concurrent observations are attached to the left and right interfaces. The "borders" of composed sequents are defined in Fig. 6.

The main feature of tiles is their double labeling with triggers and effects, allowing to observe the input-output behavior of configurations. By taking $\langle$trigger, effect$\rangle$ pairs as labels one can see tiles as a labeled transition system. In this context, the usual notion of bisimilarity is called *tile bisimilarity*.

Definition 3 (tile bisimilarity). *Let* $\mathcal{R} = (\mathcal{H}, \mathcal{V}, N, R)$ *be a tile system. A symmetric relation* $\sim_t$ *on configurations is called a* tile bisimulation *if whenever* $s \sim_t t$ *and* $\mathcal{R} \vdash s \xrightarrow[b]{a} s'$, *then* t' *exists such that* $\mathcal{R} \vdash t \xrightarrow[b]{a} t'$ *and* $s' \sim_t t'$.
   The maximal tile bisimulation is called tile bisimilarity *and it is denoted by* $\simeq_t$.

$$\frac{s \xrightarrow[b]{a} t \quad h \xrightarrow[c]{b} f}{s;h \xrightarrow[c]{a} t;f} \text{ (hor)} \qquad \frac{s \xrightarrow[b]{a} t \quad h \xrightarrow[d]{c} f}{s \otimes h \xrightarrow[b\otimes d]{a\otimes c} t \otimes f} \text{ (par)} \qquad \frac{s \xrightarrow[b]{a} t \quad t \xrightarrow[d]{c} h}{s \xrightarrow[b;d]{a;c} h} \text{ (ver)}$$

Fig. 6. Inference rules for tile logic

Note that $s \simeq_t t$ only if s and t have the same input-output interfaces.

The basic source property is a syntactic criterion ensuring that tile bisimilarity is a congruence.

Definition 4 (basic source property). *A tile system* $\mathcal{R} = (\mathcal{H}, \mathcal{V}, N, R)$ *enjoys the basic source property if for each* $A \in N$ *if* $R(A) = \langle s, a, b, t \rangle$, *then* s *is an operator in* Σ.

The following result from [9] can be used to ensure that tile bisimilarity is a congruence.

Lemma 1. *If a tile system* $\mathcal{R}$ *enjoys the basic source property, then tile bisimilarity is a congruence for* $\mathcal{R}$.

4   From Reo Connectors to Tile Configurations

In order to give semantics to Reo connectors using tile logic, we first need to map them into tile configurations. The basic entities of Reo connectors are nodes and channels, which are then composed by plugging channels into nodes. Here we consider Reo nodes as composed out of replicators, mergers, and basic nodes, as in [5], since this will simplify our mapping. A replicator is a ternary atomic connector with one source and two sink ends. A merger is a ternary atomic connector with two source and one sink ends. A basic node is one that has at most one source and at most one sink ends. Essentially, a node N with $n > 1$ incoming and $m > 1$ outgoing channel ends will be represented by a basic node with one incoming tree of $n - 1$ mergers and one outgoing tree of $m - 1$ replicators. Incoming channel ends of N will be connected to the leaves of the tree of mergers, and outgoing channel ends of N will be connected to the leaves of the tree of replicators.

The horizontal signature of the tile system for modeling Reo connectors, thus, includes operators for basic nodes, mergers, replicators, and channels. As usual, when modeling graphs with tiles (see, e.g., [16]), nodes are on the left, with their interfaces heading toward right, and channels are on the right with their interfaces toward left. The two interfaces are joined using symmetries, mergers and replicators. Notice that this technique for representing graphs is fully general, i.e. any graph can be represented in this style. Since we do not model components explicitly, boundary nodes are nodes with a non-connected element of their right interface (the sink for source nodes, the source for sink nodes). Interfaces are typed according to the direction of flow of data: • for data going from left to right (from nodes to channels) and ∘ for data going from right to left (from channels to nodes). Thus, e.g., the Merger operator has sort Merger : ∘ → ∘∘. This denotes the fact that data flow in the Merger operator from right to left. Similarly the Sync channel has sort Sync : •∘ → ε, with an empty right interface as for all channels. Note that the order of elements in the interface matters. However, symmetries can be used to reorder the elements in an interface as necessary. A basic node, with one sink end and one source end has sort Node : ε → ∘•. The full horizontal signature of our tile system (for a sample set of basic connectors) is presented in Fig. 7: on the left-hand side in textual notation, and on the right-hand side in graphical notation, where for simplicity we abbreviate the names of operators using their initials.

The tile model for a general node, with n sink and m source ends is obtained by composing the tiles of a basic node, $n - 1$ mergers, and $m - 1$ replicators, as explained above. For instance, a node with 2 sinks and 3 sources is: Node; Merger $\otimes$ Replicator; $id_{\circ\circ\bullet} \otimes$ Replicator : ε → ∘ ∘ • • • (see Fig. 8).

We can now define the mapping $[\![\cdot]\!]$ from Reo connectors to tile configurations. If a connector C has n boundary nodes, then $[\![C]\!] : ε → \omega$ where $\omega \in \{\bullet, \circ\}^n$ is a word of length n. The mapping is parametric with respect to an interface function I_n associating to each boundary node in C an element in ω, i.e. I is a bijection between nodes of C and $\{1, \ldots, n\}$.

Definition 5 (from Reo to tile configurations). *Given a Reo connector C with n boundary nodes and an interface function I_n, the tile configuration $[\![C]\!]_{I_n}$ is defined as follows:*

Replicator : • → ••   Merger : ∘ → ∘∘
Node : ε → ∘•   Sync : •∘ → ε
SyncDrain : •• → ε   LossySync : •∘ → ε
SyncSpout : ∘∘ → ε   AsyncDrain : •• → ε
AsyncSpout : ∘∘ → ε   Filter(p) : •∘ → ε
FIFO1 : •∘ → ε   FIFO1(x) : •∘ → ε

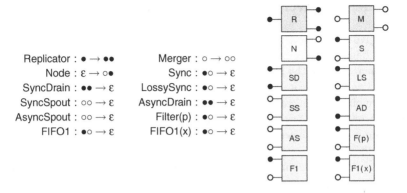

Fig. 7. Signature for Reo configurations

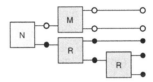

Fig. 8. A tile configuration representing a mixed node

- *on the left, it has a parallel composition of **Node** operators, one for each node in C, with the two ends connected to trees composed by $n-1$ mergers and $m-1$ replicators respectively, if the Reo node has n sources and m sinks (the trees may be empty); for boundary nodes one of the two attach points has no connected tree, and will be connected to the outside interface;*
- *on the right, it has a parallel composition of channel operators, one for each channel in C;*
- *the two parts are connected via identities and symmetries, so that each incoming channel is connected to the **Merger** tree of the corresponding node, and similarly for outgoing channels and **Replicator** trees;*
- *for each boundary node A, its free attach point is connected to the interface element $I_n(A)$ via identities and symmetries.*

The tile configurations corresponding to the Reo connectors that define the exclusive router and the alternator are presented in Fig. 9. The corresponding textual notation for the alternator is below (where *Perm* is a composition of identities and symmetries):

Node ⊗ Node ⊗ Node; $id_\circ$ ⊗ Replicator ⊗ $id_\circ$ ⊗ Replicator ⊗ Merger ⊗ $id_\bullet$; *Perm*;

$$id_{\circ\otimes\circ} \otimes \text{SyncDrain} \otimes \text{Sync} \otimes \text{FIFO1} \otimes id_\bullet : \varepsilon \to \circ\circ\bullet$$

Now we can give semantics to Reo connectors via tiles.

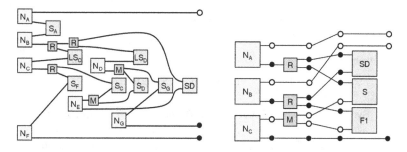

Fig. 9. Exclusive router and alternator as tile configurations

5 Modeling the 2-Color Semantics of Reo

The one-step tile semantics of a connector C is the set of all tiles that have C as their starting configuration. In order to give semantics to Reo connectors we need to provide the basic tiles defining the semantics of each operator, and then tile composition operations allow the derivation of the semantics of general connectors. We begin by presenting the 2-color data-sensitive semantics, which cannot express context-dependent behavior, but which is simpler than the corresponding 3-color semantics that we will introduce in the next section.

We choose as basic observations the data communicated at the interfaces of connectors, to model data-sensitive semantics, and we consider also a special observation untick to denote no data communication. For instance, the tile $\text{Merger} \xrightarrow[a \otimes \text{untick}]{a} \text{Merger}$ allows a Merger connector to get an action a from the first element in its right interface and propagate it to its left interface, provided that there is no piece of data on the other element of the right interface.

The basic tiles are described in Fig. 10, assuming an alphabet **Act** for basic actions. We also assume that x and y range over $\textbf{Act} \cup \{\text{untick}\}$ and a and b range over **Act**. A graphical representation of the tile that models the filling of a FIFO1 buffer is in Fig. 11. Note that observations on the interface are drawn along the vertical dimension.

These tiles define an LTS semantics for Reo, where states are tile configurations and observations are $\langle \text{trigger}, \text{effect} \rangle$ pairs. This semantics recovers all the information in the 2-color tile semantics for Reo described in [5]. Furthermore it adds to it: (i) the possibility of observing the actual data flowing in the connector, allowing to model data-sensitive primitive connectors such as filters, (ii) the possibility to consider full computations instead of single steps, keeping track also of how the state evolves (particularly, whether buffers get full or become empty). The theorem below shows how the information provided by the 2-color semantics can be recovered from the tile semantics. We call a connector data-insensitive if its behavior (i.e., whether or not it allows data to flow) does not depend on data values. Specifically, every connector built using any of the primitive connectors described above, excluding filters, is a data-insensitive connector. To formalize the correspondence between our tile model and the 2-color semantics, we must restrict tiles to the one-step semantics of the connectors, and therefore

$$\gamma \xrightarrow[y\otimes x]{x\otimes y} \gamma \qquad \text{Replicator} \xrightarrow[x\otimes x]{x} \text{Replicator} \qquad \text{Node} \xrightarrow[x\otimes x]{id_\varepsilon} \text{Node}$$

$$\text{Merger} \xrightarrow[a\otimes untick]{a} \text{Merger} \qquad \text{Merger} \xrightarrow[untick\otimes a]{a} \text{Merger} \qquad \text{Merger} \xrightarrow[untick\otimes untick]{untick} \text{Merger}$$

$$\text{Sync} \xrightarrow[id_\varepsilon]{x\otimes x} \text{Sync} \qquad \text{SyncDrain} \xrightarrow[id_\varepsilon]{a\otimes b} \text{SyncDrain} \qquad \text{SyncDrain} \xrightarrow[id_\varepsilon]{untick\otimes untick} \text{SyncDrain}$$

$$\text{SyncSpout} \xrightarrow[id_\varepsilon]{a\otimes b} \text{SyncSpout} \qquad \text{SyncSpout} \xrightarrow[id_\varepsilon]{untick\otimes untick} \text{SyncSpout}$$

$$\text{AsyncDrain} \xrightarrow[id_\varepsilon]{x\otimes untick} \text{AsyncDrain} \qquad \text{AsyncDrain} \xrightarrow[id_\varepsilon]{untick\otimes x} \text{AsyncDrain}$$

$$\text{AsyncSpout} \xrightarrow[id_\varepsilon]{x\otimes untick} \text{AsyncSpout} \qquad \text{AsyncSpout} \xrightarrow[id_\varepsilon]{untick\otimes x} \text{AsyncSpout}$$

$$\text{LossySync} \xrightarrow[id_\varepsilon]{x\otimes x} \text{LossySync} \qquad \text{LossySync} \xrightarrow[id_\varepsilon]{a\otimes untick} \text{LossySync}$$

$$\text{FIFO1} \xrightarrow[id_\varepsilon]{a\otimes untick} \text{FIFO1}(a) \qquad \text{FIFO1} \xrightarrow[id_\varepsilon]{untick\otimes untick} \text{FIFO1}$$

$$\text{FIFO1}(a) \xrightarrow[id_\varepsilon]{untick\otimes a} \text{FIFO1} \qquad \text{FIFO1}(a) \xrightarrow[id_\varepsilon]{untick\otimes untick} \text{FIFO1}(a)$$

$$\text{Filter(P)} \xrightarrow[id_\varepsilon]{a\otimes a} \text{Filter(P) if } P(a) \qquad \text{Filter(P)} \xrightarrow[id_\varepsilon]{a\otimes untick} \text{Filter(P) if } \neg P(a)$$

Fig. 10. Tiles for data-sensitive, 2-color semantics

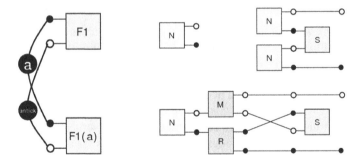

Fig. 11. The tile for filling a FIFO1 buffer (left) and three bisimilar configurations (right)

we do not need vertical composition of tiles. However, including vertical composition does not add any one-step transition either.

Theorem 1 (correspondence between 2-color coloring tables and tiles). *Let T_C be the 2-color coloring table of a data-insensitive Reo connector C with n boundary nodes. T_C contains a coloring c iff for each interface function I_n there exists a tile obtained without using vertical composition having as initial configuration $[\![C]\!]_{I_n}$ such that, for each node A, $c(A) = \square$ iff the observation at the interface $I_n(A)$ in the tile is* untick.

Proof (Sketch). First notice that there is a bijection between colorings for channels, mergers and replicators and the basic tiles that have the corresponding operators as their starting configurations, i.e. a basic connector allows a coloring c iff there is a basic tile with that operator as its starting configuration and observation untick on an interface iff the corresponding node has color $\square$ in the coloring.

One has to prove that the correspondence is preserved while composing colorings on one side and tiles on the other side. We consider the left-to-right implication, the other being simpler. Colorings can be composed iff they agree on the color of their common nodes (see Definition 3 in [5]). In order to compose the corresponding tiles to derive a tile with the desired starting configuration, observations on matching interfaces have to coincide. Let us consider the case of just one possible data value. Then the possibility of composing the tiles follows from the hypothesis if connectors are connected directly (e.g., channels to mergers and replicators), and from the properties of the auxiliary tiles for identities and symmetries and the basic tiles for nodes if connectors are connected via them.

Let us now consider the general case of an arbitrary set of data values. Note that for data-insensitive connectors, if a tile for a certain data flow exists, then a tile with the same data flow, but where all the data are equal can be built (this can be easily proved by induction on the number of operators in the starting configuration of the tile), thus the case of an arbitrary set of data values can be reduced to the one data value case. Notice that the above property does not hold for data-sensitive connectors. $\square$

As we have seen, all information provided by the coloring tables can be deduced from the tile semantics. Furthermore, the final configuration of a tile represents the state of the connector after the data flow has been performed. This can be used also to recover information provided by the constraint-automata or coalgebraic semantics of Reo. However a detailed comparison with those semantics is left for future work.

The theorem below ensures that the tile semantics is compositional w.r.t. the operators of parallel and sequential composition provided by tiles.

Theorem 2 (2-coloring congruence). *Tile bisimilarity is a congruence for the 2-color semantics of Reo connectors.*

Proof. Straightforward by inspection, using Lemma 1. $\square$

Note that the compositionality is proved w.r.t. the operators of tile composition, however this can be extended also to Reo composition operators. Composition in Reo is obtained by merging boundary nodes. In the tile model this can be obtained by connecting them via Sync channels (this corresponds to compose them in parallel and then sequentially with the Sync channel and some identities). The example below shows that the additional channel does not influence the behavior of the composition.

Example 1. Consider the simple Reo connector C_1 composed out of a mixed node with one source end and one sink end, Node : $\varepsilon \rightarrow \circ\bullet$ (see Fig. 11, top-center). We can show that this is bisimilar to a Reo connector C_2 composed out of two such nodes connected by a Sync channel: Node $\otimes$ Node; $id_\circ \otimes$ Sync $\otimes id_\bullet$: $\varepsilon \rightarrow \circ\bullet$ (see Fig. 11, top-right). First, note that the two connectors have the same interface. Then, observe that for both connectors the only possible tiles are vertical compositions of tiles $C_i \xrightarrow[x]{x} C_i$ (with $i = 1$ or $i = 2$). Thus, from the definition of bisimilarity $C_1 \sim_t C_2$. Therefore, thanks to the congruence theorem, in each connector we can replace the two nodes connected by a Sync channel with a single node without changing the overall behavior. A third bisimilar configuration is in Fig. 11, bottom-right.

6   Modeling the 3-Color Semantics of Reo

As pointed out in [5], the 2-color semantics of Reo fails to fully capture the context-dependent behavior of Reo connectors. Consider in fact the connector in Fig. 12, which is represented by the tile configuration:

Node $\otimes$ Node $\otimes$ Node; $id_\circ \otimes$ LossySync $\otimes$ FIFO1 $\otimes id_\bullet : \varepsilon \rightarrow \circ\bullet$

Fig. 12. Overflow-lossy FIFO1

There are two possible tiles with this initial configuration and with the observation $\langle id_\varepsilon, a \otimes \text{untick} \rangle$ modeling data entering in the connector. The first one loses the data item in the LossySync channel and has the final configuration Node $\otimes$ Node $\otimes$ Node; $id_\circ \otimes$ LossySync $\otimes$ FIFO1 $\otimes id_\bullet$. The second one transports the data item into the buffer of the FIFO1(a) channel and has the final configuration Node $\otimes$ Node $\otimes$ Node; $id_\circ \otimes$ LossySync $\otimes$ FIFO1(a) $\otimes id_\bullet$. The expected behavior corresponds to the second one, since there is no reason for the data to be lost. However, both the 2-color semantics and the tile model we presented above generate both alternatives as permissible behavior for this connector.

The 3-color semantics of Reo discussed in [5] solves this problem by tracking 'reasons to prohibit data flow', and allows LossySync to lose data only if there is a reason for the data not to flow out of the channel (e.g., an attached full buffer or an interface that does not accept data at the other end). The 3-color semantics replaces the □ color by two colors corresponding to 'giving a reason for no data flow' and 'requiring a reason for no data flow.' Briefly, 'giving a reason' is used to model either a choice made by the connector or to capture the absence of data flow on a particular channel end. On the other hand, 'requiring a reason' is used to model that the context determines whether a particular choice is made. Consider the two key tiles for LossySync:

$$\text{LossySync} \xrightarrow[id_\varepsilon]{a \otimes a} \text{LossySync} \qquad \text{LossySync} \xrightarrow[id_\varepsilon]{a \otimes \triangleright} \text{LossySync}$$

The first one simply states that data flow through the LossySync. The second states that data will be lost in the LossySync if a reason for no flow can be provided by the context in which the channel is plugged. If a tile with the label $a \otimes \triangleleft$ was also present, this would say that the LossySync provides a reason for the data to be lost, and thus the LossySync would lose the property that the decision ought to be made by the context.

Composition in the 3-color model includes the additional requirement that at each basic node where there is no data flow, at least one reason for no flow must be present.

We show that tile logic can also easily model this more detailed semantics. To this end, we must refine our untick observation into $\lhd$, which models 'requires a reason for no data flow,' and $\rhd$, which models 'gives a reason for no data flow,' when these symbols occur on the left-hand side of the tile (above the line in the rule format). When these observations occur on the right-hand side of a tile, their meanings are reversed. For instance, one of the rules for Replicator:

$$\text{Replicator} \xrightarrow[\rhd \otimes \rhd]{\rhd} \text{Replicator}$$

means that a reason is required from the channel end on the left of the tile (above the line) and will be given (propagated) to the channel ends on the right of the tile (below the line). This captures that no-input to the Replicator is sufficient to cause no data flow through the Replicator, and that this reason is passed onto the sink ends.

The main tiles for modeling the 3-color semantics of Reo are in Fig. 13. The others are analogous.

$$\gamma \xrightarrow[y \otimes x]{x \otimes y} \gamma \qquad \text{Replicator} \xrightarrow[a \otimes a]{a} \text{Replicator}$$

$$\text{Replicator} \xrightarrow[\rhd \otimes \rhd]{\rhd} \text{Replicator} \qquad \text{Replicator} \xrightarrow[\rhd \otimes \lhd]{\lhd} \text{Replicator} \qquad \text{Replicator} \xrightarrow[\lhd \otimes \rhd]{\lhd} \text{Replicator}$$

$$\text{Node} \xrightarrow[a \otimes a]{id_\varepsilon} \text{Node} \quad \text{Node} \xrightarrow[\lhd \otimes \rhd]{id_\varepsilon} \text{Node} \quad \text{Node} \xrightarrow[\rhd \otimes \lhd]{id_\varepsilon} \text{Node} \quad \text{Node} \xrightarrow[\lhd \otimes \lhd]{id_\varepsilon} \text{Node}$$

$$\text{Merger} \xrightarrow[a \otimes \rhd]{a} \text{Merger} \quad \text{Merger} \xrightarrow[\rhd \otimes a]{a} \text{Merger} \quad \text{Merger} \xrightarrow[\rhd \otimes \rhd]{\rhd} \text{Merger} \quad \text{Merger} \xrightarrow[\lhd \otimes \lhd]{\lhd} \text{Merger}$$

$$\text{Sync} \xrightarrow[id_\varepsilon]{a \otimes a} \text{Sync} \quad \text{Sync} \xrightarrow[id_\varepsilon]{\lhd \otimes \rhd} \text{Sync} \quad \text{Sync} \xrightarrow[id_\varepsilon]{\rhd \otimes \lhd} \text{Sync} \quad \text{Sync} \xrightarrow[id_\varepsilon]{\rhd \otimes \rhd} \text{Sync}$$

$$\text{LossySync} \xrightarrow[id_\varepsilon]{a \otimes a} \text{LossySync} \quad \text{LossySync} \xrightarrow[id_\varepsilon]{a \otimes \rhd} \text{LossySync} \quad \text{LossySync} \xrightarrow[id_\varepsilon]{\rhd \otimes \lhd} \text{LossySync}$$

$$\text{FIFO1} \xrightarrow[id_\varepsilon]{\rhd \otimes \lhd} \text{FIFO1} \quad \text{FIFO1} \xrightarrow[id_\varepsilon]{a \otimes \lhd} \text{FIFO1(a)} \quad \text{FIFO1(a)} \xrightarrow[id_\varepsilon]{\lhd \otimes a} \text{FIFO1} \quad \text{FIFO1(a)} \xrightarrow[id_\varepsilon]{\lhd \otimes \rhd} \text{FIFO1(a)}$$

Fig. 13. Tiles for data-sensitive, 3-color semantics

Note that the tile Node includes a behavior that mimics the so-called *flip rule* in connector coloring [5]. The point of the flip rule is to reduce the size of coloring tables using the fact that nodes need no more than one reason. The fact that nodes can also accept multiple reasons is captured by the tile:

$$\text{Node} \xrightarrow[\lhd \otimes \lhd]{id_\varepsilon} \text{Node}$$

Results analogous to the one in the previous section can be proved, showing that the 3-color tile semantics recovers all the information provided by the standard 3-color semantics of Reo. As for the 2-color semantics, the tile semantics is data-sensitive, and allows to track the state of connectors and model full computations.

Theorem 3 (correspondence between 3-color coloring tables and tiles). *Let T_C be the 3-color coloring table (see [5]) of a data-insensitive Reo connector C with n boundary nodes. T_C contains a coloring c iff for each interface function I_n there exists a tile obtained without using vertical composition with initial configuration $[\![C]\!]_{I_n}$ such that, for each node A:*

- *$c(A)$ is the color for no dataflow with the reason coming into the node (given) iff the observation at the interface element $I_n(A)$ in the tile is $\triangleright$ (this is always below the line);*
- *$c(A)$ is the color for no dataflow with the reason leaving the node (required) iff the observation at the interface element $I_n(A)$ in the tile is $\triangleleft$ (this is always below the line).*

Proof. The proof is similar to the one of Theorem 1.   □

As for the 2-color semantics, tile bisimilarity is a congruence.

Theorem 4 (3-coloring congruence). *Tile bisimilarity is a congruence for the 3-color semantics of Reo connectors.*

7   Reconfiguration of Reo Connectors

Since the tile semantics of a Reo connector includes also the state of the connector after each step, one can model inside the Tile Model also the reconfiguration of Reo connectors triggered by dataflow as presented in [11].

The idea is that some connectors, when suitable conditions concerning their state and the ongoing dataflow are met, can automatically be reconfigured to meet the requirements of the environment. We sketch this approach by demonstrating it through the example of an infinite FIFO buffer [11], and leave a more detailed study of reconfiguration for future work. An infinite FIFO buffer is a FIFO buffer that grows when a new datum arrives to be inserted and its buffer is full, and shrinks when a datum is consumed out of the buffer. To model this we require two new channels: FIFO$_\infty$ is the empty infinite buffer, and FIFOtmp(a) is a temporary buffer, containing value a, that will disappear when the a is consumed.

For simplicity we give semantics to the infinite FIFO buffer using the 2-color semantics, however, the 3-color semantics can be used as well. The necessary basic tiles can be found in Fig. 14. Note that the tile for shrinking the buffer transforms the temporary buffer FIFOtmp(a) into a Sync channel. Thanks to Example 1, up to bisimilarity, this

$$\text{FIFO}_\infty \xrightarrow[id_\varepsilon]{untick \otimes untick} \text{FIFO}_\infty \qquad \text{FIFO}_\infty \xrightarrow[id_\varepsilon]{a \otimes untick} id_\bullet \otimes \text{Node} \otimes id_\circ; \text{FIFO}_\infty \otimes \text{FIFOtmp(a)}$$

$$\text{FIFOtmp(a)} \xrightarrow[id_\varepsilon]{untick \otimes a} \text{Sync} \qquad \text{FIFOtmp(a)} \xrightarrow[id_\varepsilon]{untick \otimes untick} \text{FIFOtmp(a)}$$

Fig. 14. Tiles for 2-color semantics of infinite buffer

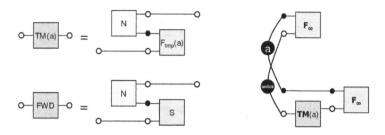

Fig. 15. Some graphical shorthand

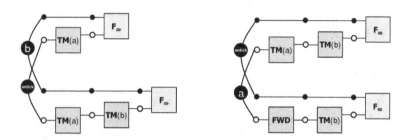

Fig. 16. Datum b arrives (left) and datum a leaves

corresponds to removing the temporary buffer and its nearby node. However, the tile needed to actually do the garbage collection would not satisfy the basic source property, thus we preferr this approach.

To sketch the evolution of infinite buffers, we draw some possible proof steps obtained by horizontal composition of basic tiles. To simplify the graphical notation we introduce some suitable graphical shorthand in Fig. 15 (left) for the composition of a node and a temporary buffer (TM) and for the composition of a node and a synchronous channel (FWD) that basically behaves as a forwarder. Using the shorthand, the tile for inserting a new datum in the infinite buffer can be drawn as in Fig. 15 (right). Figure 16 shows what happens if a new datum b arrives when the buffer already contains a datum a (left) and what happens if a datum is then requested from the buffer (right). Note that it is also allowed for the arrival and departure of data happen at the same time (see Fig. 17).

Proposition 1 (a reconfiguration congruence). *Tile bisimilarity is a congruence for the 2-color semantics of Reo connectors including the infinite FIFO buffer.*

Observe that in this approach reconfiguration and computation are fully integrated (while in [11] and [10,11] the two aspects are dealt with by separate models). Furthermore, reconfigurable connectors and normal connectors can be used together, since reconfiguration is not visible from the outside. However, our tile model currently cannot express more complex reconfigurations that change the interfaces of connectors. Capturing these reconfiguration in such a way as to allow the congruence of bisimilarity to be proved using the basic source property, requires (1) connectors to agree on when and

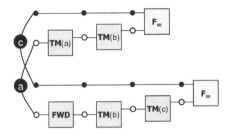

Fig. 17. Datum c arrives while datum a leaves

which reconfiguration to perform, and (2) nodes to propagate this kind of information. We leave an analysis of this approach for future work.

8   Conclusion

We have shown that the Tile Model can be used to describe all main aspects of the semantics of Reo connectors: synchronization, dataflow, context dependency, and reconfiguration. This is the first semantic description of Reo connectors able to present all these aspects natively in a single framework. Furthermore, the semantics is compositional.

As future work we want to consider an alternative approach to the 3-color semantics based on priorities: one can specify that losing data in the LossySync channel has lower priority than data flowing through it. Our goal is to match the expected intuitive semantics of Reo, and solve the problem of causes for data-discard that arises in some cycles in the 3-color semantics, as discussed in [5]. However, further research is necessary to understand how to apply this reasoning to complex connectors. Another long term goal of our work is to understand how to define complex reconfigurations along the lines sketched at the end of Section 7.

References

1. Arbab, F.: Reo: A channel-based coordination model for component composition. Math. Struct. in Comput. Sci. 14(3), 1–38 (2004)
2. Arbab, F., Rutten, J.J.M.M.: A coinductive calculus of component connectors. In: Wirsing, M., Pattinson, D., Hennicker, R. (eds.) WADT 2002. LNCS, vol. 2755, pp. 34–55. Springer, Heidelberg (2003)
3. Baier, C., Sirjani, M., Arbab, F., Rutten, J.J.M.M.: Modeling component connectors in Reo by constraint automata. Sci. Comput. Program 61(2), 75–113 (2006)
4. Bruni, R., Lanese, I., Montanari, U.: A basic algebra of stateless connectors. Theoret. Comput. Sci. 366(1-2), 98–120 (2006)
5. Clarke, D., Costa, D., Arbab, F.: Connector colouring I: Synchronisation and context dependency. Sci. Comput. Program 66(3), 205–225 (2007)
6. Corradini, A., Montanari, U.: An algebraic semantics for structured transition systems and its application to logic programs. Theoret. Comput. Sci. 103, 51–106 (1992)
7. CWI. Reo home page, http://reo.project.cwi.nl

8. CWI. A repository of Reo connectors, http://homepages.cwi.nl/~proenca/webreo/

9. Gadducci, F., Montanar, U.: The tile model. In: Plotkin, G., Stirling, C., Tofte, M. (eds.) Proof, Language and Interaction: Essays in Honour of Robin Milner, pp. 133–166. MIT Press, Cambridge (2000)

10. Koehler, C., Arbab, F., de Vink, E.: Reconfiguring Distributed Reo Connectors. In: Corradini, A., Montanari, U. (eds.) WADT 2008. LNCS, vol. 5486, pp. 221–235. Springer, Heidelberg (2009)

11. Koehler, C., Costa, D., Proença, J., Arbab, F.: Reconfiguration of Reo connectors triggered by dataflow. In: Ermel, C., Heckel, R., de Lara, J. (eds.) Proceedings of GT-VMT 2008. Elect. Communic. of the European Association of Software Science and Technology, vol. 10, pp. 1–13. EASST (2008)

12. Koehler, C., Lazovik, A., Arbab, F.: Connector rewriting with high-level replacement systems. In: Canal, C., Poizat, P., Viroli, M. (eds.) Proceedings of FOCLASA 2007. Elect. Notes in Th. Comput. Sci. Elsevier Science, Amsterdam (2007)

13. Larsen, K.G., Xinxin, L.: Compositionality through an operational semantics of contexts. In: Paterson, M. (ed.) ICALP 1990. LNCS, vol. 443, pp. 526–539. Springer, Heidelberg (1990)

14. MacLane, S.: Categories for the working mathematician. Springer, Heidelberg (1971)

15. Meseguer, J.: Conditional rewriting logic as a unified model of concurrency. Theoret. Comput. Sci. 96, 73–155 (1992)

16. Montanari, U., Rossi, F.: Graph rewriting, constraint solving and tiles for coordinating distributed systems. Applied Categorical Structures 7(4), 333–370 (1999)

17. Plotkin, G.D.: A structural approach to operational semantics. J. Log. Algebr. Program. 60-61, 17–139 (2004)

C-semiring Frameworks for Minimum Spanning Tree Problems

Stefano Bistarelli[1,2,3] and Francesco Santini[3,4]

[1] Dipartimento di Informatica e Matematica, Università di Perugia, Italy
bista@dipmat.unipg.it
[2] Dipartimento di Scienze, Università "G. d'Annunzio" di Chieti-Pescara, Italy
bista@sci.unich.it
[3] Istituto di Informatica e Telematica (CNR), Pisa, Italy
{stefano.bistarelli,francesco.santini}@iit.cnr.it
[4] IMT - Istituto di Studi Avanzati, Lucca, Italy
f.santini@imtlucca.it

Abstract. In this paper we define general algebraic frameworks for the Minimum Spanning Tree problem based on the structure of c-semirings. We propose general algorithms that can compute such trees by following different cost criteria, which must be all specific instantiation of c-semirings. Our algorithms are extensions of well-known procedures, as Prim or Kruskal, and show the expressivity of these algebraic structures. They can deal also with partially-ordered costs on the edges.

1 Introduction

Classical *Minimum Spanning Tree* (MST) problems [1,2] in a weighted directed graph arise in various contexts. One of the most immediate examples is related to the multicast communication scheme in networks with *Quality of Service* (QoS) requirements [3]. For example, we could need to optimize the bandwidth, the delay or a generic cost (for device/link management or to obtain the customer's bill) of the distribution towards several final receivers. Therefore, the aim is to minimize the cost of the tree in order to satisfy the needs of several clients at the same time. Other possible applications may concern other networks in general, as social, electrical/power, pipeline or telecommunication (in a broad sense) ones.

In our study we would like to define a general algebraic framework for the MST problem based on the structure of c-semirings [4,5], that is, a constraint-based semiring; in the following of the paper we will use "c-semiring" and "semiring" as synonyms. We want to give algorithms that work with any semiring covered by our framework, where different semirings are used to model different QoS metrics. Classical MST problems can be generalized to other weight sets, and to other operations. A general algebraic framework for computing these problems has not been already studied, even if a similar work has been already proposed for shortest path problems [6].

More precisely, the algebraic structure that provides the appropriate framework for these problems is a semiring $S = \langle A, +, \times, \mathbf{0}, \mathbf{1} \rangle$. This five-tuple represents the set of preferences/costs (i.e. A), the operation to compose and choose

A. Corradini and U. Montanari (Eds.): WADT 2008, LNCS 5486, pp. 56–70, 2009.

them (i.e. respectively $\times$ and $+$) and the best (i.e. $\mathbf{1}$) and worst (i.e. $\mathbf{0}$) preferences in A. Semirings consist in flexible and parametric algebraic structure that can be simply instantiated to represent different costs, or QoS metrics (e.g. bandwidth) as we mainly suppose in this paper [7,8].

Our goal is to provide a general algebraic framework similar to the one created in [6] for shortest-path algorithms, a work from which we have sensibly taken inspiration for this work. Clearly, our intent is to reach analogous results, but, in this case, for tree structures instead that for plain paths.

The absence of a unifying framework for single-source shortest paths problems was already solved in [6], where the author defines general algebraic frameworks for shortest-distance problems based on the structure of semirings. According to these semiring properties, the author gives also a generic algorithm for finding single-source shortest distances in a weighted directed graph. Moreover, the work in [6] shows some specific instances of this generic algorithm by examining different semirings; the goal is to illustrate their use and compare them with existing methods and algorithms. Notice that, while in [6] the author uses also semirings with a non-idempotent $+$, we would like to focus mainly on c-semirings instead (i.e. even with an idempotent $+$). To further clarify our intents, we would like to say that the ideas in this paper are developed to show the expressivity of semirings, and not to enrich the field of graph theory.

The multi-criteria MST problem has seldom received attention in network optimization. The solution of this problem is a set of Pareto-optimal trees, but their computation is difficult since the problem is NP-hard [9]. One solution, based on a genetic algorithm, has been given in [9]; however, even this solution is not feasible, since a successive work [10] proved that it is not guaranteed that each tree returned by the algorithm in [9] is Pareto optimal. Our goal is to describe this problem from an algebraic point of view.

2 C-semirings

A c-semiring [4,5,11] is a tuple $\langle A, +, \times, \mathbf{0}, \mathbf{1}\rangle$ such that:

1. A is a set and $\mathbf{0}, \mathbf{1} \in A$;
2. $+$ is commutative, associative and $\mathbf{0}$ is its unit element;
3. $\times$ is associative, distributes over $+$, $\mathbf{1}$ is its unit element and $\mathbf{0}$ is its absorbing element.

A *c-semiring* is a semiring $\langle A, +, \times, \mathbf{0}, \mathbf{1}\rangle$ such that $+$ is idempotent, $\mathbf{1}$ is its absorbing element and $\times$ is commutative. Let us consider the relation $\leq_S$ over A such that $a \leq_S b$ iff $a + b = b$. Then it is possible to prove that (see [5]):

1. $\leq_S$ is a partial order;
2. $+$ and $\times$ are monotone on $\leq_S$;
3. $\times$ is intensive on $\leq_S$: $a \times b \leq_S a, b$;
4. $\mathbf{0}$ is its minimum and $\mathbf{1}$ its maximum;
5. $\langle A, \leq_S\rangle$ is a complete lattice and, for all $a, b \in A$, $+$ is the least upper bound operator, that is, $a + b = lub(a, b)$.

Moreover, if $\times$ is idempotent, then: $+$ distributes over $\times$; $\langle A, \leq_S \rangle$ is a complete distributive lattice and $\times$ its glb. Informally, the relation $\leq_S$ gives us a way to compare semiring values and constraints. In fact, when we have $a \leq_S b$, we will say that b *is better than* a. In the following, when the semiring will be clear from the context, $a \leq_S b$ will be often indicated by $a \leq b$. The cartesian product of multiple semirings is still semiring [4]: for instance, $S = \langle \langle [0,1], \mathbb{R}^+ \rangle, \langle max, min \rangle, \langle min, \hat{+} \rangle, \langle 0, +\infty \rangle, \langle 1, 0 \rangle \rangle$ ($\hat{+}$ is the arithmetic sum) corresponds to the cartesian product of a *fuzzy* and a *weighted* semiring, and S is a semiring.

In [12] the authors extended the semiring structure by adding the notion of *division*, i.e. $\div$, as a weak inverse operation of $\times$. An absorptive semiring S is *invertible* if, for all the elements $a, b \in A$ such that $a \leq b$, there exists an element $c \in A$ such that $b \times c = a$ [12]. If S is absorptive and invertible, then, S is *invertible by residuation* if the set $\{x \in A \mid b \times x = a\}$ admits a maximum for all elements $a, b \in A$ such that $a \leq b$ [12]. Moreover, if S is absorptive, then it is *residuated* if the set $\{x \in A \mid b \times x \leq a\}$ admits a maximum for all elements $a, b \in A$, denoted $a \div b$. With an abuse of notation, the maximal element among solutions is denoted $a \div b$. This choice is not ambiguous: if an absorptive semiring is invertible and residuated, then it is also invertible by residuation, and the two definitions yield the same value.

To use these properties, in [12] it is stated that if we have an absorptive and complete semiring[1], then it is residuated. For this reason, since all classical soft constraint instances (i.e. *Classical CSPs*, *Fuzzy CSPs*, *Probabilistic CSPs* and *Weighted CSPs*) are complete and consequently residuated, the notion of semiring division (i.e. $\div$) can be applied to all of them.

The semiring algebraic structure proves to be an appropriate and very expressive cost model to represent QoS metrics. Weighted semirings $\langle \mathbb{R}^+, min, \hat{+}, \infty, 0 \rangle$ ($\hat{+}$ is the arithmetic sum) can be used to find the best MST by optimizing, for instance, the cost of the tree in terms of money, e.g. for link maintenance or billing criteria in order to charge the final user. Fuzzy semirings $\langle [0,1], max, min, 0, 1 \rangle$ represent fuzzy preferences on links, e.g. *low*, *medium* or *high* traffic on the links. Probabilistic semirings $\langle [0,1], max, \hat{\times}, 0, 1 \rangle$ ($\hat{\times}$ is the arithmetic multiplication). As an example, the probabilistic semiring can optimize (i.e. maximize) the probability of successful delivery of packets (due to errors). Classical semirings $\langle \{0,1\}, \vee, \wedge, 0, 1 \rangle$ can be adopted to test the reachability of receivers has to be tested.

3 Algorithms for MST and Totally Ordered Semirings

As a reminder, a MST can be defined as in Def. 1.

---

[1] If S is an absorptive semiring, then S is complete if it is closed with respect to infinite sums, and the distributivity law holds also for an infinite number of summands.

Definition 1. *Given an undirected graph $G \equiv (V, E)$, where each edge $(u, v) \in E$ has a weight $w(u, v)$, the tree $T \subseteq E$ that connects all the vertices V and minimizes the sum $w(t) = \sum\limits_{(u,c) \in T} w(u, v)$ is defined as the MST of G.*

A first sketch of a possible algorithm for a MST problem over a graph $G(V, E)$ is given in Alg. 1. It is obtained by modifying the classical Kruskal algorithm [1] in order to use c-semiring values and operators which are taken as input, i.e. $\langle A, +, \times, \mathbf{0}, \mathbf{1} \rangle$. The algorithm (as Alg. 2) work only with totally ordered edge costs.

In Alg. 1, b corresponds to the best edge in the current iteration of the *repeat* command (line 2) and it is found (in line 3) by applying the $\bigoplus$ operator over all the remaining edges in the set P (i.e. the set of possible edges), instantiated to E at the beginning (line 1); $\bigoplus : E \rightarrow \mathcal{P}(E)$ is a new operator that finds the edge b with the best cost in E, according to the ordering defined by the $+$ operator of the semiring. Then the (partial) solution tree is updated with the $\bigotimes : \mathcal{P}(E) \times \mathcal{P}(E) \rightarrow \mathcal{P}(E)$ operator, which adds the new edge and updates the cost of the tree according to the $\times$ operator of the semiring (line 5). At last, b is removed from P (line 7).

Algorithm 1. Kruskal with semiring structures

INPUT: $G(V, E), \langle A, +, \times, \mathbf{0}, \mathbf{1} \rangle$

1: $T = \emptyset, P = E$
2: **repeat**
3:     *let* $b \in \bigoplus(P)$ \\ Best edge in P
4:     **if** (endpoints of b are disconnected in T) **then**
5:         $T = T \bigotimes \{b\}$ \\ Add the best edge to the solution
6:     **end if**
7:     $P = P \setminus \{b\}$
8: **until** $P == \emptyset$

OUTPUT: $T \equiv MST$ over G

Theorem 1. *To find a Minimum Spanning Tree T, the complexity of the algorithm is $\mathcal{O}(|E| \ln|E|)$ as in the original procedure [1].*

The proof follows the ideas in [1]. Having sorted the edges in $\mathcal{O}(|E| \ln|E|)$, the $\bigoplus$ operator runs in constant time. Consider that here the sorting procedure takes also the $+$ of the chosen semiring as a parameter, in order to select the best values according to the partial ordering defined by $\leq_S$ (see Sec. 2). By using disjoint-set data structures [1], we can check in $\mathcal{O}(\ln|E|)$ time that each of the $\mathcal{O}(|E|)$ edge insertions in T does not create a cycle [1]. This last step is identical to the last check in the classical Kruskal algorithm, and it is used only to keep the structure of a tree.

We can show also that the other best-known algorithm for solving the MST problem can be generalized with semiring structures (see Alg. 2). Step by step, the modified Prim's algorithm [1] adds an edge to the (partial solution) tree

T, instead of joining a forest of trees in a single connected tree, as in Kruskal's algorithm. However, even Prim's procedure proceeds in a greedy way by choosing the best-cost edge (i.e. (v_i, u_j) in line 4) in order to add it to the solution (line 5) through the $\otimes$ operator. The operator $\oplus$ is the same as the one defined for Alg. 1, even if in this case it is applied only to those edges for which one of the endpoints has not been already visited. This set of nodes, i.e. R, is initialized in line 1 with an arbitrary node, and updated at each step (line 5). The algorithm ends when all the nodes of the graph have been visited, that is $R == V$.

Notice also that both Alg. 1 and Alg. 2 properly work only if the set of costs is totally ordered, while they need to be modified for a multicriteria optimization, since the costs of the edges can be partially ordered. In this case, the semiring operators have to deal with multisets of solutions that are Pareto-optimal: in Sec. 4 we modify the $\oplus$ operator in order to select and manage a set of edges (and not only a singleton) with incomparable costs within the same step.

Algorithm 2. `Prim with semiring structures`

INPUT: $G(V, E), \langle A, +, \times, \mathbf{0}, \mathbf{1} \rangle$
1: $T = \emptyset$, $R = \{v_k\}$, v_k is arbitrary
2: **repeat**
3:     *let* $P = \{(v_k, u_z) \in E \mid (v_k \in R) \wedge (u_z \notin R)\}$
4:     *let* $(v_i, u_j) \in \bigoplus(P)$
5:     $T = T \otimes \{(v_i, u_j)\}$
6:     $R = R \cup u_j$
7: **until** $R == V$
OUTPUT: $T \equiv MST$ over G

---

4   Partially Ordered Extensions

As said in Sec. 3, Alg. 1 and Alg. 2 are not able to compute a solution for the MST in case the costs of the edges are partially ordered. The reason is that, since we have a partial order over the chosen semiring S, two costs c_1 and c_2 may possibly be incomparable (i.e. $c_1 <> c_2$). According to this view, the $\oplus$ operators presented in Alg. 1 and Alg. 2 must be extended in order to choose a set of edges within the same step, instead of only a single arc.

In the next paragraph we present the Kruskal algorithm extended to manage partially ordered costs for the edges. Further on, we provide the proof of correctness/soundness and the complexity analysis of its operations (in the second paragraph of this section). Then, in the third paragraph we show an alternative algorithm that incrementally deletes the worst edges from the graph until it reaches the MST; the original version is called *Reverse-delete* algorithm [13].

Kruskal extended with partial order. For a partially ordered set of costs we can use Alg. 3. The most notable difference w.r.t. the totally ordered version of the algorithm (see Alg. 1) is the definition of the $\oplus$ operator (used in line 3 of Alg. 3):

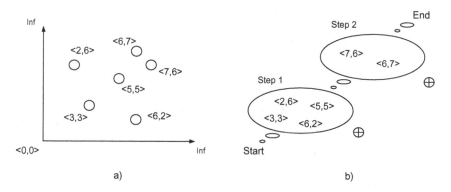

Fig. 1. In *a)* a set of partially ordered costs are represented, and in b) how they are partitioned according to the $\oplus$ operator in Alg. 3

Definition 2. *The $\oplus : \mathcal{P}(E) \to \mathcal{P}(E)$ operator takes a set W of edges and returns a set $W \backslash U = X$, such that $\forall u \in U, x \in X.cost(u) <_S cost(x)$, where $>_S$ (see Sec. 2) depends on the chosen semiring S and the cost function returns the cost of an edge.*

In words, the $\oplus$ operator chooses all the best edges (according to $<_S$) whose costs are incomparable with at least one other cost. To show an example, we consider the cartesian product of two *weighted* semirings, i.e. $S = \langle\langle\mathbb{R}^+, \mathbb{R}^+\rangle,$ $\langle min, min\rangle, \langle\hat{+}, \hat{+}\rangle, \langle\infty, \infty\rangle, \langle 0, 0\rangle\rangle$: given the set W of edges whose costs are represented by $\{\langle 3, 3\rangle, \langle 2, 6\rangle, \langle 6, 2\rangle, \langle 5, 5\rangle, \langle 6, 7\rangle, \langle 7, 6\rangle\}$, $\oplus(W) = X$ whose costs are instead $\{\langle 3, 3\rangle, \langle 2, 6\rangle, \langle 6, 2\rangle, \langle 5, 5\rangle\}$. The partially ordered costs of the edges in W are graphically represented also in the plane of Fig. 1a. Notice that the set X contains also edges whose costs totally dominate the other costs of edges in the same set X: e.g. $\langle 3, 3\rangle >_S \langle 5, 5\rangle$. However, $\langle 5, 5\rangle$ is still selected by $\oplus$ to be in X since it cannot be compared with $\langle 6, 2\rangle$ (and also $\langle 2, 6\rangle$): only $\langle 6, 7\rangle$ and $\langle 7, 6\rangle$ are not chosen, since they are totally dominated by the other costs (they will be chosen by the algorithm in the second step, as shown in Fig. 1b). In other words, the set X is obtained from the Pareto optimal frontier, by adding all the edges with incomparable costs.

The set $X = \oplus(W)$ is then examined in line $4-7$ of Alg. 3, in order to find all its maximal cardinality and best cost subsets of edges (i.e. the R in line 7) that can be added to the solution without introducing cycles. In line 5, $Xset$ collects all the sets of edges in X that do not form a cycle with a partial solution T_i: in this way we enforce the connectivity condition of a tree. Each $T_i \in T$ represents a partial solution, and T collects them all; T_i' represents instead an updated T_i (see line 9). Among all these sets in X, in line 6 we select those subsets with the maximal cardinality, i.e. $Rset$. The reason is that (Lemma 1), in order to minimize the cost of the spanning tree, it is better to connect its components by using as many low cost edges (in X) as possible, having introduced the $\oplus$ operator (see Def. 2).

Algorithm 3. Kruskal extended for partial ordering

INPUT: $G(V, E)$, $\langle A, +, \times, \mathbf{0}, \mathbf{1} \rangle$, A partially ordered

1: *let* $T = \bigcup_i T_i$ where $T_0 = \{\emptyset\}$, $W = E$

2: **repeat**

3:      $X = \bigoplus(W)$

4:      **for all** $T_i \in T$ **do**

5:           $Xset = \{X' | X' \subseteq X, T_i \otimes X' \text{ has not cycles}\}$ \\ No cycles

6:           $Rset = \{X^* | X^* \in Xset, \forall X' \in Xset, |X^*| >= |X'|\}$ \\ Max Cardinality

7:           $R = \{R' | R' \in Rset, \forall R'' \in Rset \text{ s.t. } T_i \otimes R'' \not\succ_S T_i \otimes R'\}$ \\ Best Cost

8:           **for all** $R_i \in R$ **do**

9:                $T_i' = T_i \otimes R_i$

10:          **end for**

11:     **end for**

12:     $W = W \setminus X$

13: **until** $W == \emptyset$

OUTPUT: $T \equiv$ the set of all *MSTs* over G

Therefore, in line 7 we only take the R' subsets in $Rset$ that, composed with the partial solutions T_i (i.e. $T_i \otimes R'$), are not completely dominated by another $R'' \in Rset$. In this way, the algorithm discards the completely dominated partial solutions since they can lead only to a completely dominated final solution (thus, not a MST), as explained in Lemma 1.

In lines $8 - 10$, each $R_i \in R$ is added to the related partial solution T_i, in order to remember all the possible partial solutions that can be obtained within the same step, i.e. the set of all the T_i': they consist in all the best (i.e. dominating) partial trees and need to be stored since they can lead to different MST with an incomparable cost.

At last, the set X of examined edges is removed from W (line 12). This procedure is repeated until all the edges in W have been examined (at the beginning, $W = E$, i.e. the set of edges in the graph). Considering the costs in Fig. 1a, in Fig. 1b it is possible to see the W sets of edges that will be selected at the first and second step of Alg. 3. At the last step, T collects all the MSTs (i.e. T_i) that can be obtained over the graph G. A full example of the algorithm execution is given is Sec. 4.1.

To give a particular example of a single iteration, we suppose that at the first step the algorithm has added the edges (n_k, n_z) and (n_u, n_v) to the solution T, as shown in Fig. 2; thus, the cost of the partial solution is $\langle 2, 3 \rangle$. We still consider the cartesian product of two *weighted* semirings. Then, at the second step $\bigoplus(W) = \{(n_j, n_k), (n_i, n_v), (n_i, n_u), (n_j, n_z)\}$ (represented with dashed lines in Fig. 2), whose costs respectively are $\langle 3, 4 \rangle$, $\langle 4, 3 \rangle$, $\langle 1, 10 \rangle$ and $\langle 10, 1 \rangle$. Following line 5 of Alg. 3, at this step we can add either (n_j, n_k) or (n_j, n_z) to the first component and either (n_i, n_u) or (n_i, n_v) to the second one; otherwise, we would introduce a cycle in the solution. Notice that all these edges are selected within the same step, since their costs are partially ordered (see Def. 2).

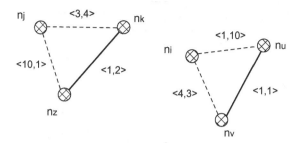

Fig. 2. The graphical intuition of the *mincost* operation in Alg. 3

Therefore, according to lines 5 and 6, $Rset = \{R^1 = \{(n_j, n_z), (n_i, n_v)\}, R^2 = \{(n_j, n_z), (n_i, n_u)\}, R^3 = \{(n_j, n_k), (n_i, n_v)\}, R^4 = \{(n_j, n_k), (n_i, n_u)\}\}$. The costs of these four sets of edges are respectively $\langle 14, 4 \rangle$, $\langle 11, 11 \rangle$, $\langle 7, 7 \rangle$ and $\langle 4, 14 \rangle$. The operation in line 7 of Alg. 3 discards R^2 (whose cost is $\langle 11, 11 \rangle$), since $T \otimes R^2 <_S T \otimes R^3$: $\langle 13, 14 \rangle <_S \langle 9, 10 \rangle$. Therefore, we have that $R = \{\{(n_j, n_z), (n_i, n_v)\}, \{(n_j, n_k), (n_i, n_v)\}, \{(n_j, n_k), (n_i, n_u)\}\}$ (R is obtained at line 7). Then the partial solution T (after the first step $T = \{\{(n_k, n_z), (n_u, n_v)\}\}$) becomes $T = \{T \otimes R^1 = \{(n_k, n_z), (n_u, n_v), (n_j, n_z), (n_i, n_v)\}, T \otimes R^3 = \{(n_k, n_z), (n_u, n_v), (n_j, n_k), (n_i, n_v)\}, T \otimes R^4 = \{(n_k, n_z), (n_u, n_v), (n_j, n_k), (n_i, n_u)\}\}$.

Reverse-delete. In the original version of the *Reverse-delete* algorithm [13], if the graph is disconnected, this algorithm will find a MST for each connected component of the graph. The set of these minimum spanning trees is called a minimum spanning forest, which consists of every vertex in the graph. The Reverse-Delete algorithm starts with the original graph and deletes the worst edges from it, instead of adding solution edges to the empty set, step by step as in Kruskal's algorithm. If the graph is connected, the algorithm is able to find the MST.

Considering Alg. 4, the $\ominus : \mathcal{P}(E) \rightarrow \mathcal{P}(E)$ operator in line 3 selects the set X of the worst completely dominated edges in W, which is the set of edges that still need to be checked; at the beginning $W = E$, and the only one partial solution consists in all the edges in the graph, i.e. $T = \{E\}$. Formally, $\ominus(W) = \{e \in W : \nexists e' \in W, cost(e) \leq_S cost(e')\}$, where S is the chosen semiring and the *cost* function return the weight of an edge. Each $T_i \in T$ represents a partial solution, and T collects them all; T_i' represents instead an updated T_i (see line 9).

Then, like Alg. 3, in lines $5-6$ the algorithm finds $Rset$, i.e. the set of maximal cardinality subsets of X whose removal still keeps the graph connected. Among all these subsets, in line 7 we select R, which is the set of subsets of $Rset$ with the worst possible (incomparable) costs according to the semiring partial order (i.e. $<_S$): to do so we use the $\ominus : \mathcal{P}(E) \times \mathcal{P}(E) \rightarrow \mathcal{P}(E)$ operator, which removes the second set of edges from the first one and then updates the cost of the partial solution according to the $\div$ operator presented in Sec. 2 (i.e. the weak inverse operator of $\times$). We can consider $\ominus$ as the inverse operator of $\otimes$ in Alg. 3. All

---

Algorithm 4. Reverse-delete Kruskal extended for partial ordering

---

INPUT: $G(V, E)$, $\langle A, +, \times, \mathbf{0}, \mathbf{1} \rangle$, A partially ordered

1: $let\ T = \bigcup_i T_i$ where $T_0 = \{E\}, W = E$

2: **repeat**

3:     $X = \ominus(W)$

4:     **for all** $T_i \in T$ **do**

5:         $Xset = \{X'|X' \subseteq X, T_i \ominus X' \text{ is connected}\}$ \\ Connectivity

6:         $Rset = \{X^*|X^* \in Xset, \forall X' \in Xset, |X^*| >= |X'|\}$ \\ Max Cardinality

7:         $R = \{R'|R' \in Rset, \forall R'' \in Rset \text{ s.t. } T_i \ominus R'' \not\succ_S T_i \ominus R'\}$ \\ Worst Cost

8:         **for all** $R_i \in R$ **do**

9:             $T_i' = T_i \ominus R_i$

10:         **end for**

11:     **end for**

12:     $W = W \setminus S$

13: **until** $W == \emptyset$

OUTPUT: $T \equiv$ the set of all MST over G

---

the edge sets in $Rset$ can be removed from the partial solutions T_i (lines $8 - 10$) by still using the $\ominus$ operator.

At last, the procedure updates W by removing the set X of checked edges (line 12). These steps are repeated until all the edges in E have been examined. Following similar steps as for Alg. 3, we can prove Theo. 3.

4.1   Examples

In this section we provide an example to better explain how Alg. 3 and Alg. 4 work in a proper way.

Example on Alg. 3. Concerning Alg. 3 and consequently a non-idempotent semiring, as a reference we consider the graph $G(V, E)$ represented in Fig. 3a, where the edges in E are labeled with partially ordered costs taken from the semiring $S = \langle\langle\mathbb{R}^+, [0, 1]\rangle, \langle\hat{+}, \hat{\times}\rangle, \langle min, max\rangle, \langle\infty, 0\rangle, \langle 0, 1\rangle\rangle$. This semiring is obtained through the cartesian product of the *weighted* and *probabilistic* semirings, and its vectorized $\times$ operator is non-idempotent since at least one of the original $\times$ operators is non-idempotent (in this case, both the operators are non-idempotent). Therefore, the costs are expressed in terms of couples of values, i.e. $\langle c, p \rangle$, and the cost of a tree is obtained by arithmetically summing all the money costs and multiplying all the probability costs of the chosen edges. At the end of the computation, Alg. 3 finds the two best MSTs (i.e. T_1 and T_2) by minimizing c and maximizing p for the entire obtained tree, which are represented in Fig. 3b and Fig. 3c. The first MST has a cost of $\langle 28, 0.6 \rangle$, while the second one, $\langle 29, 0.61 \rangle$: they are not comparable costs and thus they represent two distinct optimal solutions.

Figure 4 reports the steps of the algorithm with the related X, $XSet$ and R sets, as obtained from Alg. 3. At step 1 in Fig. 4, the two edges

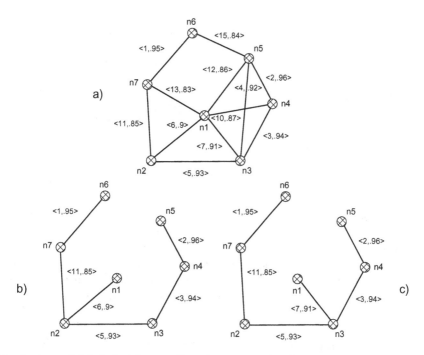

Fig. 3. A graph labeled with partially ordered costs (in a), and the best MST trees (in b) obtained by using $\langle\langle\mathbb{R}^+, [0..1]\rangle, \langle\hat{+}, \hat{\times}\rangle, \langle min, max\rangle, \langle\infty, 0\rangle, \langle 0, 1\rangle\rangle$

Step	X	XSet	R
1	$\{(n_4, n_5), (n_6, n_7)\}$	$\{\{(n_4, n_5), (n_6, n_7)\}, \{(n_4, n_5)\}, \{(n_6, n_7)\}\}$	$\{\{(n_4, n_5), (n_6, n_7)\}\}$
2	$\{(n_3, n_4)\}$	$\{\{(n_3, n_4)\}\}$	$\{\{(n_3, n_4)\}\}$
3	$\{(n_2, n_3), (n_3, n_5)\}$	$\{\{(n_2, n_3)\}\}$	$\{\{(n_2, n_3)\}\}$
4	$\{(n_1, n_2), (n_1, n_3)\}$	$\{\{(n_1, n_2), (n_1, n_3)\}, \{(n_1, n_2)\}, \{(n_1, n_3)\}\}$	$\{\{(n_1, n_2)\}, \{(n_1, n_3)\}\}$
5	$\{(n_1, n_4)\}$	$\emptyset$	$\emptyset$
6	$\{(n_2, n_7), (n_1, n_5)\}$	$\{\{(n_2, n_7)\}\}$	$\{\{(n_2, n_7)\}\}$
7	$\{(n_5, n_6), (n_1, n_7)\}$	$\emptyset$	$\emptyset$

Fig. 4. The steps of Alg. 3 applied on the graph in Fig. 3a

$X = \{(n_4, n_5), (n_6, n_7)\}$ are selected since their costs totally dominate all the other costs (i.e. $\langle 2, 0.96\rangle$ and $\langle 1, 0.95\rangle$) and are partially ordered w.r.t. each other, since the first shows a better (i.e. higher) probability and the second a better (lower) cost. Therefore, since they do not form any cycle, they are both added to the solution, i.e. $R = \{(n_4, n_5)\}, \{(n_6, n_7)\}\}$.

Step 2 works in the same way for the edge (n_3, n_4). At step 3, Alg. 3 chooses $X = \{(n_2, n_3), (n_3, n_5)\}$ with costs $\langle 5, 0.93\rangle$ and $\langle 4, 0.92\rangle$, but only (n_2, n_3) is

Step	X	XSet	R
1	$\{(n_5, n_6), (n_1, n_7)\}$	$\{\{(n_5, n_6), (n_1, n_7)\},$ $\{(n_5, n_6)\}, \{(n_5, n_6)\}$	$\{\{(n_5, n_6), (n_1, n_7)\}\}$
2	$\{(n_2, n_7), (n_1, n_5)\}$	$\{\{(n_1, n_5)\}$	$\{\{(n_1, n_5)\}$
3	$\{(n_1, n_4)\}$	$\{\{(n_1, n_4)\}\}$	$\{\{(n_1, n_4)\}\}$
4	$\{(n_1, n_2), (n_1, n_3)\}$	$\{\{(n_1, n_2), (n_1, n_3)\},$ $\{(n_1, n_2)\}, \{(n_1, n_3)\}\}$	$\{\{(n_1, n_2)\}, \{(n_1, n_3)\}\}$
5	$\{(n_2, n_3), (n_3, n_5)\}$	$\{\{(n_3, n_5)\}\}$	$\{\{(n_3, n_5)\}\}$
6	$\{(n_3, n_4)\}$	$\emptyset$	$\emptyset$
7	$\{(n_4, n_5), (n_6, n_7)\}$	$\emptyset$	$\emptyset$

Fig. 5. The steps of Alg. 4 applied on the graph in Fig. 3a

added to the solution (i.e. $R = \{\{(n_2, n_3)\}\}$), since (n_3, n_5) would create the cycle $n_3 - n_4 - n_5$; for this reason, the operation in line 5 (see Alg. 3) discards it from $X set$.

At step 4, the $\bigoplus$ operator selects $X = \{(n_1, n_2), (n_1, n_3)\}$: in this case, these two edges cannot be added at the same time to the solution, since it would create a cycle among $n_1 - n_2 - n_3$. Therefore, from this bifurcation step, the algorithm remembers and updates two distinct partial solutions T_1 and T_2 (see Alg. 3 at line 9), one given by adding $\{(n_1, n_2)\}$, and one given by adding $\{(n_1, n_3)\}$ (i.e. $R = \{\{(n_1, n_2)\}, \{(n_1, n_3)\}\}$). While steps 5 and 7 cannot respectively add (n_1, n_5) and (n_5, n_6) or (n_1, n_7) since it would create a cycle, at step 6 only (n_2, n_7) can be added because (n_5, n_6) would form a cycle as well.

Example on Alg. 4. Clearly, the two MST solutions in Fig. 3b and Fig. 3c can be obtained also with Alg. 4 as well. The steps of the algorithm are shown in Fig. 5; as a reminder, notice that the sets R of edges are now removed from the set E of graph edges, in order to find a (minimum cost) tree structure: the considered semiring is still $S = \langle\langle\mathbb{R}^+, [0, 1]\rangle, \langle\hat{+}, \hat{\times}\rangle, \langle min, max\rangle, \langle\infty, 0\rangle, \langle 0, 1\rangle\rangle$. In the first step, we can safely remove two edges, i.e. $R = \{\{(n_5, n_6), (n_1, n_7)\}\}$, while at step 2, Alg. 4 can only remove (n_1, n_5) (i.e. $R = \{\{(n_1, n_5)\}\}$), otherwise the resulting graph would be disconnected. At step 3, we can remove $R = \{\{(n_1, n_4)\}\}$, while at step 4 we can remove only one edge between (n_1, n_2) and (n_1, n_3) or graph would be disconnected: from this step we store two different (partially ordered) solutions $T_1 = E \ominus \{(n_5, n_6), (n_1, n_7), (n_1, n_5), (n_1, n_4), (n_1, n_2)\}$ and $T_2 = E \ominus \{(n_5, n_6), (n_1, n_7), (n_1, n_5), (n_1, n_4), (n_1, n_3)\}$.

The two solutions in Fig. 3b and Fig. 3c are then obtained at step 5, which removes $R = \{\{(n_3, n_5)\}\}$ (removing (n_2, n_3) would disconnect the tree). Then the remaining edges are checked (step $6 - 7$) but not removed due to the connectivity property.

4.2   Correctness Considerations

We can show the correctness of Algorithm 3 step-by-step. The following property comes from the definition at line 5 in Alg. 3:

Proposition 1. *Let $X' \in Xset$. Adding even one edge in $X'\backslash X$ to X would produce a cycle in the partial solution T_i (for this reason, at each iteration it is possible to discard X from the edges to check, i.e. $W = W\backslash X$ in Alg. 3).*

Therefore, line 5 maintains a tree structure and avoids cycles. The cost of the R' subsets in Alg. 3 can be obtained by using the $\times$ operator of the considered semiring, and the best subset is the result of applying the $+$ operator in order to choose the best cost according to the ordering defined by $+$, i.e. $\leq_S$ in Sec. 2. Therefore, referring to lines $5 - 6$ in Alg. 3, we can prove that:

Lemma 1. *By adding the maximal cardinality subsets of partially ordered edges $X^* \in Rset$ that do not form any cycle (i.e. $X' \in Xset$), at each step we connect the maximum number of forests possible. Since all the $R' \in R$ are the subsets with the best (incomparable) costs, each $T_i \otimes R_i$ forest is connected with the best possible cost according to the $+$ operator of the semiring.*

As a reminder, a forest is an acyclic undirected graph, while a set of connected forests corresponds to a tree [1]. Lemma 1 extends the *safety* property explained for MST [1]. At each step i we obtain a new forest made of distinct components, which are tree-shaped. For each of these components, the edge that connects them is *light* [1], in the sense that it has the best cost (according to $+$); therefore, all the added edges are *safe*. In words, Alg. 3 extends the classical Kruskal algorithm by connecting more than two components within the same step. This connection shows the best possible cost, since it is characterized by the maximal cardinality (the reason is highlighted in Prop. 2) and the best cost, according to the partial order defined by $+$, among those sets of best cardinality.

We can prove that by replacing an edge chosen with $\oplus$ at one step of Alg. 3 with an edge that will be selected at a successive step, we obtain a worse spanning tree (according to the $+$ operator):

Proposition 2. *Connecting the two same components with an edge (whose cost is c_k) chosen with $\oplus$ at the step $i+n$ (with $n > 0$) instead of an edge selected at step i (with cost c_j) results in a completely dominated cost for the final solution.*

The proof comes from the fact that $c_j >_S c_k$ according to the definition of $\oplus$ (see Def. 2), and the $\times$ operator, used to compose the costs, is monotone. Prop. 2 explains why we need to consider only those $X^* \in Xset$ with the maximal cardinality: otherwise we will need to connect that same component with a completely dominated edge, found at a successive iteration.

Notice that we could have several maximal cardinality subsets $R' \in Rset$ that can be added to the same partial solution T_i, thus obtaining different partially ordered solutions $T_i' = T_i \otimes R_i$ (see line 9 of Alg. 3). These solutions represent the best possible forests that can be obtained at a given step (defined by the $\oplus$ operator). However, some of them can be safely deleted at the successive steps if they become completely dominated by even only one other partial solution T_i, as explained for Fig. 2:

Lemma 2. *Given two sets of edges R', $R'' \in Rset$ such that $T_i \otimes R' >_S T_i \otimes R''$, then R'' is not added to R (line 7 of Alg. 3), and the partial solution $T_i \otimes R''$ is consequently discarded from the possible ones.*

The proof of this Lemma comes from the fact that, if a partial solution $T_i \otimes R''$ is completely dominated ($T_i \otimes R'$) at a given step, it will inevitably create a completely dominated tree at the end of the algorithm (thus, not a MST). The reason is that the $\times$ operator of the semiring is monotone, i.e. if $a \geq_S b$ then $a \times c \geq_S b \times c$ (see Sec. 2).

Theorem 2. *Following the steps of Algorithm 3 over a graph $G = (V, E)$, we find all the Minimum Spanning Trees $T_i \in T$ even if the costs of the edges, taken from a semiring $\langle A, +, \times, \mathbf{0}, \mathbf{1} \rangle$, are partially ordered.*

The proof of Theo. 2 derives from Lemma 1 and Lemma 2. Since Lemma 1 satisfies the safety property at each step, if the graph $G = (V, E)$ is connected, at the end we find a tree spanning all the vertices and satisfying the safety property. The final tree spans all the nodes because we suppose that our graph G is connected, and as stated in Lemma 1, we connect the maximum number of forests possible without adding any cycles. Similar considerations can be proved for Algorithm 4.

Theorem 3. *Following the steps of Algorithm 4 over a graph $G = (V, E)$, we find all the Minimum Spanning Trees $T_i \in T$ even if the costs of the edges, taken from a semiring $\langle A, +, \times, \mathbf{0}, \mathbf{1} \rangle$, are partially ordered.*

4.3   Complexity Considerations

The complexity of Alg. 3 obviously resides in the operations performed at line $5 - 7$, that is in finding the R best-cost subsets among all the possible ones of maximal cardinality. The other operations in Alg. 3 merely delete or add subsets of edges. We suppose the set E of edges as already ordered by according to the cost: this step can be performed in $\mathcal{O}(|E| \, ln|E|)$ [1] and choosing the best edges (with the $\oplus$ operator) can be consequently accomplished in a constant time.

Concerning the space/time complexity, the algorithm is, in the worst case, exponential in the number of the edges with partially ordered costs, since with lines $8 - 10$ we have to store all the possible non-comparable (best) partial solutions, i.e. the number of T_i sets in T can be exponential. This is the case when all the edges in the graph $G = (V, E)$ show incomparable costs, and the number of MSTs can correspond to the number of all the possible spanning trees over G: $|V|^{|V|-2}$ following to Cayley's formula [14]. After having ordered the edges according to their cost, the $\oplus$ operator (see Def. 2) partitions them into k disjoint sets P_i, as represented in Fig. 1; when $k = 1$ all the edges in G are not comparable (i.e. an exponential number of MSTs in the worst case) and when $k = |E|$ all the edge costs are totally ordered and the complexity corresponds to Alg. 1 (i.e. $\mathcal{O}(|E| \, ln|E|)$), as for the original Kruskal procedure.

The complexity of Alg. 3 is then $\mathcal{O}(|E| \, ln|E| + k \, d^{|d|-2})$, where k is the number P_i of the disjoint edge sets and d is the maximum number of nodes which have an incident edge in the P_i set, among all the P_i: for instance if P_1 stores the edges incident in 4 nodes, P_2 in 3 nodes and P_3 in 2, then $d = 4$. When $d << |V|$, i.e. there are few incomparable edges in each P_i, the complexity is linear (i.e. $\mathcal{O}(|E| \ ln|E|)$). Consider that it is possible to estimate the complexity of the algorithm after having ordered the edges (in $\mathcal{O}(|E| \ ln|E|)$), since after that step we know the number k and the respective size of the P_i sets (consequently, we know d). Therefore, we can easily know how the algorithm will perform before executing it.

Notice also that, with an idempotent $\times$ operator (e.g. min), Alg. 3 returns only a subset of the possible MSTs. To find all of them we should keep all the possible spanning trees (deleting line 7 from Alg. 3) until the last iteration, since the cost of the whole tree is flattened on the (not comparable) costs found in the last step. In this case, the number of solutions could not be limited step-by-step. Identical complexity and $\times$-idempotency considerations can be provided for Alg. 4.

5   Conclusions

We have shown that c-semirings are expressive and generic structures that can be used inside slightly modified versions of classical MST algorithms (as Kruskal, Prim and Reverse-delete Kruskal), in order to find the best spanning trees according to different QoS metrics with different features (but still representable with a semiring). Classical algorithms have been extended to deal with semiring structures and partially ordered costs; moreover, an analysis of correctness and complexity has been provided for the extension Kruskal's algorithm for partially ordered costs.

The weight of a tree from a node p to a set of destination nodes D, is obtained by "multiplying" the edge weights along the tree by using the $\times$ semiring operator (see Sec. 2), and the cost of the min-weight tree is the "sum" of the weights of all such trees, obtained by using the $+$ semiring operator. By parametrically varying the semiring, we can represent many different kinds of problems, having features like fuzziness, probability, and optimization [4]. This paper extends other works focused only on semirings and shortest path algorithms [6].

In the future, one ambition could be to merge these frameworks with constraints concerning the considered QoS metrics (e.g. *delay* ≤ 40), since *Soft Constraint Satisfaction Problems* based on c-semirings have been already successfully applied to this field [7,8,15].

References

1. Cormen, T.T., Leiserson, C.E., Rivest, R.L.: Introduction to algorithms. MIT Press, Cambridge (1990)
2. Graham, R., Hell, P.: On the history of the minimum spanning tree problem. IEEE Annals of the History of Computing 07(1), 43–57 (1985)
3. Wang, B., Hou, J.: Multicast routing and its QoS extension: problems, algorithms, and protocols. IEEE Network 14, 22–36 (2000)

4. Bistarelli, S.: Semirings for Soft Constraint Solving and Programming. LNCS, vol. 2962. Springer, London (2004)
5. Bistarelli, S., Montanari, U., Rossi, F.: Semiring-based constraint satisfaction and optimization. J. ACM 44(2), 201–236 (1997)
6. Mohri, M.: Semiring frameworks and algorithms for shortest-distance problems. J. Autom. Lang. Comb. 7(3), 321–350 (2002)
7. Bistarelli, S., Montanari, U., Rossi, F., Santini, F.: Modelling multicast QoS routing by using best-tree search in and-or graphs and soft constraint logic programming. Electr. Notes Theor. Comput. Sci. 190(3), 111–127 (2007)
8. Bistarelli, S., Santini, F.: A formal and practical framework for constraint-based routing. In: ICN, pp. 162–167. IEEE Computer Society, Los Alamitos (2008)
9. Zhou, G., Gen, M.: Genetic algorithm approach on multi-criteria minimum spanning tree problem. European Journal of Operational Research 114, 141–152 (1999)
10. Knowles, J.D., Corne, D.W.: Enumeration of pareto optimal multi-criteria spanning trees - a proof of the incorrectness of Zhou and Gen's proposed algorithm. European Journal of Operational Research 143(3), 543–547 (2002)
11. Kuich, W., Salomaa, A.: Semirings, automata, languages. Springer, London (1986)
12. Bistarelli, S., Gadducci, F.: Enhancing constraints manipulation in semiring-based formalisms. In: Brewka, G., Coradeschi, S., Perini, A., Traverso, P. (eds.) ECAI, pp. 63–67. IOS Press, Amsterdam (2006)
13. Kleinberg, J., Tardos, E.: Algorithm Design. Addison-Wesley Longman Publishing Co. Inc, Boston (2005)
14. Shor, P.W.: A new proof of Cayley's formula for counting labeled trees. J. Comb. Theory Ser. A 71(1), 154–158 (1995)
15. Bistarelli, S., Montanari, U., Rossi, F.: Soft constraint logic programming and generalized shortest path problems. Journal of Heuristics 8(1), 25–41 (2002)

What Is a Multi-modeling Language?

Artur Boronat[1], Alexander Knapp[2], José Meseguer[3], and Martin Wirsing[4]

[1] University of Leicester
aboronat@le.ac.uk
[2] Universität Augsburg
knapp@informatik.uni-augsburg.de
[3] University of Illinois
meseguer@uiuc.edu
[4] Ludwig-Maximilians-Universität München
wirsing@pst.ifi.lmu.de

Abstract. In large software projects often multiple modeling languages are used in order to cover the different domains and views of the application and the language skills of the developers appropriately. Such "multi-modeling" raises many methodological and semantical questions, ranging from semantic consistency of the models written in different sublanguages to the correctness of model transformations between the sublanguages. We provide a first formal basis for answering such questions by proposing semantically well-founded notions of a multi-modeling language and of semantic correctness for model transformations. In our approach, a multi-modeling language consists of a set of sublanguages and correct model transformations between some of the sublanguages. The abstract syntax of the sublanguages is given by MOF meta-models. The semantics of a multi-modeling language is given by associating an institution, i.e., an appropriate logic, to each of its sublanguages. The correctness of model transformations is defined by semantic connections between the institutions.

1 Introduction

In an idealized software engineering world, development teams would follow well-defined processes in which one single modeling language is used for all requirements and design documents; but in practice "multi-modeling" happens: in a large software project entity-relationship diagrams and XML may be used for domain modeling, BPEL for business process orchestration, and UML for design and deployment. In fact, UML itself can be seen as a multi-modeling language comprising several sublanguages such as class diagrams, OCL, and state machines; each sub-modeling language provides a particular view of a software system. Such views have the advantage of complexity reduction: a software engineer can concentrate on a particular aspect of the system such as the domain architecture or dynamic interactions between objects.

On the other hand, multi-modeling raises a host of methodological and semantical questions: are the different modeling sublanguages semantically consistent with each other? How can we correctly transform an abstract model in one modeling language into a more concrete one in another language? How can we detect semantic inconsistencies between heterogeneous models expressed in different modeling sublanguages? More

A. Corradini and U. Montanari (Eds.): WADT 2008, LNCS 5486, pp. 71–87, 2009.

generally, is there a notion of "multi-modeling language" which provides more insight than just an ad-hoc collection of modeling languages put together? Is it possible to give a semantics to multi-modeling languages which allows one to deal with consistency, validation and verification but that retains the advantages of multiple views by providing a local semantics and local reasoning capabilities for each modeling language?

The methodological use of views and viewpoints in software modelling is a long standing research topic [18]. In the literature, there are three main complementary approaches for interrelating modeling notations: the "system model approach", the "model-driven architecture approach", and the "heterogeneous semantics and development approach". In the system model approach the different modeling languages are translated into a common (formally defined) modeling notation called system model [10] which serves as unique semantic basis and for analyzing consistency of software engineering models. In the "model-driven architecture approach" [27] model transformations and consistency issues are typically dealt with at the syntactic level of the modeling notation. In the third approach different modeling languages are interrelated by semantic-preserving mappings [13,24]; a mathematical semantics is given locally for each modeling language and the consistency between different languages is analyzed semantically through the semantic-preserving mappings. All three approaches have been applied to several modeling languages including UML, but to the best of our knowledge, multi-modeling languages in the software engineering sense have never been systematically studied. However, research within the theory of institutions [19] on institution morphisms and comorphims [20], and on "heterogeneous institutions" [24] is directly relevant to this problem.

We combine ideas from model-driven architecture and heterogeneous semantics and propose a new, semantically well-founded notion of a multi-modeling language and a new notion of semantic correctness for model transformations. In particular, our formal definition of a multi-modeling language L: (i) uses the Meta-Object Facility MOF and its algebraic semantics [9] for describing the metamodels and models of the sublanguages of L; (ii) associates an institution to each sublanguage S of L and gives a mathematical semantics to each software engineering model[1] of S by a corresponding (logical) theory in the institution of S; (iii) defines the links between different sublanguages of S by model transformations and provides a notion of semantic correctness for such transformations; and (iv) provides a notion of consistent heterogeneous (software engineering) model in the multi-modeling language L, which is derived from a notion of a class of heterogeneous mathematical models at the institution level.

The approach is illustrated in Fig. 1: There are three sublanguages S_1, S_2, and S_3 of a common multi-modeling language L, software engineering models M_1, M_2, and M_3 conforming to (the meta-model representations of) the sublanguages, and having a formal semantics in the institutions $\mathcal{I}_1, \mathcal{I}_2, \mathcal{I}_3$. The model transformations $trans_{12}$ and $trans_{13}$ between the sublanguages S_1 and S_2, and S_1 and S_3, respectively, are applied

---

[1] For distinguishing semantic models from the models of a modeling language we write "software engineering model (SE-model)" for a (syntactic) model defined in a modeling language such as UML. In contrast to this, "(semantic) models" are part of the mathematical semantics of a modeling language so that a semantic model corresponds to a model of a theory in a suitable logic; here, we will use institutional models (Ins-models).

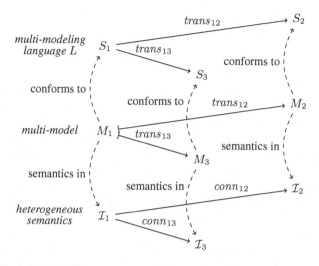

Fig. 1. Relations between metamodels, models, and semantic domains

to M_1 yielding (sub-models of) M_2 and M_3. These model transformations are backed by semantic connections $conn_{12}$ and $conn_{13}$ between $\mathcal{I}_1$ and $\mathcal{I}_2$, $\mathcal{I}_3$ which make it possible to show that these model transformations are correct.

In addition to make these concepts precise, we illustrate them by a case study involving (UML) class diagrams and relational database schema diagrams as modeling languages. Based on earlier work [13] we show that class diagrams and schema diagrams form a multi-modeling language where class diagrams are related to schema diagrams by a semantically correct model transformation.

The paper is organized as follows: In Sect. 2 we briefly recall the necessary background from the theory of institutions. Section 3 shows how MOF metamodels and model transformations are algebraically formalized as membership equational theories. In Sect. 4 we present the institutional semantics of metamodels and in Sect. 5 our formal notions of semantic connections between institutions and of correct model transformations. The notions of multi-modeling languages and consistent multi-models are introduced in Sect. 6. In Sect. 7 we discuss related and future work.

2   Preliminaries: Institutions and Institution (Co-)Morphisms

We briefly recall basic notions on institutions and their morphisms and comorphisms which form the framework for our institutional semantics of multi-modeling languages. We assume familiarity with the most elementary notions of category theory: category, functor, and natural transformation (see, e.g., [21]).

An *institution* [19] $\mathcal{I}$ is a tuple $\mathcal{I} = (Sign_{\mathcal{I}}, Sen_{\mathcal{I}}, Mod_{\mathcal{I}}, \models_{\mathcal{I}})$, with: (i) $Sign_{\mathcal{I}}$ a category whose objects are called *signatures*; (ii) a functor $Sen_{\mathcal{I}} : Sign_{\mathcal{I}} \rightarrow Set$, called the *sentence functor*, from $Sign_{\mathcal{I}}$ to Set, the category of sets; (iii) a contravariant functor $Mod_{\mathcal{I}} : Sign_{\mathcal{I}}^{\mathrm{op}} \rightarrow Cat$, called the *model functor*, from $Sign_{\mathcal{I}}$ to Cat, the category of categories; and (iv) a family $\models_{\mathcal{I}} = \{\models_{\mathcal{I},\Sigma}\}_{\Sigma \in Sign_{\mathcal{I}}}$ of *satisfaction relations* between Σ-models $M \in Mod_{\mathcal{I}}(\Sigma)$ and Σ-sentences $\varphi \in Sen_{\mathcal{I}}(\Sigma)$, such that for each $H : \Sigma \to \Sigma'$ in $Sign_{\mathcal{I}}$, $M' \in Mod_{\mathcal{I}}(\Sigma')$, and $\varphi \in Sen_{\mathcal{I}}(\Sigma)$, we have the equivalence

$$Mod_{\mathcal{I}}(H)(M') \models_{\mathcal{I},\Sigma} \varphi \iff M' \models_{\mathcal{I},\Sigma'} Sen_{\mathcal{I}}(H)(\varphi) .$$

An institution provides a categorical semantics for the model-theoretic aspects of a logic, focusing on the satisfaction relation between models and sentences, and emphasizing that satisfaction is invariant under changes of syntax by signature morphisms. Note that, given an institution $\mathcal{I}$, we can always define an associated category $Th_{\mathcal{I}}$ of *theories* (theory *presentations* to be more exact, see, e.g., [22]), where theories are pairs (Σ, Γ) with $\Gamma \subseteq Sen_{\mathcal{I}}(\Sigma)$, and theory morphisms $H : (\Sigma, \Gamma) \to (\Sigma', \Gamma')$ are signature morphisms $H : \Sigma \to \Sigma'$ such that $\Gamma' \models_{\Sigma'} Sen_{\mathcal{I}}(H)(\Gamma)$, where the satisfaction relation is extended to a semantic consequence relation between sets of sentences in the usual way (see [19]). There is then an obvious functor $sign : Th_{\mathcal{I}} \to Sign_{\mathcal{I}}$ defined on objects by the equation $sign(\Sigma, \Gamma) = \Sigma$.

An *institution morphism* [19] $\mu : \mathcal{I} \twoheadrightarrow \mathcal{I}'$ from an institution $\mathcal{I}$ to another institution $\mathcal{I}'$ is given by: (i) a functor $\mu^{Sign} : Sign_{\mathcal{I}} \to Sign_{\mathcal{I}'}$; (ii) a natural transformation $\mu^{Sen} : \mu^{Sign}; Sen_{\mathcal{I}'} \Rightarrow Sen_{\mathcal{I}}$; and (iii) a natural transformation $\mu^{Mod} : Mod_{\mathcal{I}} \Rightarrow \mu^{Sign^{op}}; Mod_{\mathcal{I}'}$, such that for each $M \in Mod_{\mathcal{I}}(\Sigma)$ and each sentence $\varphi' \in Sen_{\mathcal{I}'}(\mu^{Sign}(\Sigma))$ we have

$$M \models_{\mathcal{I},\Sigma} \mu_{\Sigma}^{Sen}(\varphi') \iff \mu_{\Sigma}^{Mod}(M) \models_{\mathcal{I}',\mu^{Sign}(\Sigma)} \varphi' .$$

Dually, an *institution comorphism* [20] (called a plain map of institutions in [22]) $\rho : \mathcal{I} \to \mathcal{I}'$ is given by: (i) a functor $\rho^{Sign} : Sign_{\mathcal{I}} \to Sign_{\mathcal{I}'}$; (ii) a natural transformation $\rho^{Sen} : Sen_{\mathcal{I}} \Rightarrow \rho^{Sign}; Sen_{\mathcal{I}'}$; and (iii) a natural transformation $\rho^{Mod} : \rho^{Sign^{op}}; Mod_{\mathcal{I}'} \Rightarrow Mod_{\mathcal{I}}$, such that for each $M' \in Mod_{\mathcal{I}'}(\rho^{Sign}(\Sigma))$ and each sentence $\varphi \in Sen_{\mathcal{I}}(\Sigma)$ we have

$$M' \models_{\mathcal{I}',\rho^{Sign}(\Sigma)} \rho_{\Sigma}^{Sen}(\varphi) \iff \rho_{\Sigma}^{Mod}(M') \models_{\mathcal{I},\Sigma} \varphi .$$

Note that, given an institution comorphism $\rho : \mathcal{I} \to \mathcal{I}'$, the functor ρ^{Sign} extends naturally to a functor $\rho^{Th} : Th_{\mathcal{I}} \to Th_{\mathcal{I}'}$ with $\rho^{Th}(\Sigma, \Gamma) = (\rho^{Sign}(\Sigma), \rho_{\Sigma}^{Sen}(\Gamma))$.

3 Algebraic Semantics of MOF and of Model Transformations

We briefly explain how a MOF metamodel defines a modeling language, how it is formalized by means of a membership-equational logic theory, and how model transformations are formalized as equationally-defined functions in MOMENT2.

3.1 MOF

MOF [28] is a semiformal approach to define modeling languages. It provides a four-level hierarchy, with levels M_0, M_1, M_2 and M_3, where level M_{i+1} serves as the *metalevel* for level M_i. The entities populating level M_i are *collections* of a certain *type*, which is defined by means of an entity at level M_{i+1}. Level M_0 contains collections

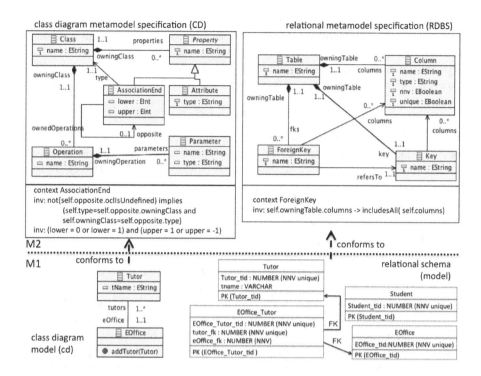

Fig. 2. Levels M_2 and M_1 of the MOF hierarchy: metamodel and model examples

of structured data that are defined by using a specific model in a modeling space, e.g., tuples in a database or class instances of a class diagram. Level M_1 contains *models*, which are used to represent a specific reality by using a well-defined language for computer-based interpretation such as class diagrams or relational schemas. Level M_2 contains *metamodels*. A metamodel is a model specifying the types that can be used in a modeling language, such as the metamodel CD for defining class diagrams and the metamodel RDBS for defining relational schemas, as shown in Fig. 2. An entity at level M_3 is a *meta-metamodel* enabling the definition of metamodels at the level M_2.

For a model M at level M_1 and a metamodel $\mathcal{M}$ at level M_2, we write $M : \mathcal{M}$ to denote the *metamodel conformance* relation. In addition, a metamodel $\mathcal{M}$ can be enriched with a set $\mathcal{C}$ of OCL constraints constituting a *metamodel specification* $(\mathcal{M}, \mathcal{C})$ [8] so that a model M conforms to $(\mathcal{M}, \mathcal{C})$ when it conforms to the metamodel $\mathcal{M}$ *and* satisfies the constraints $\mathcal{C}$. In Fig. 2, the OCL constraints over the CD metamodel defines the concept of opposite association ends and restricts the set of possible cardinalities. The OCL constraint over the RDBS metamodel indicates that the columns of a foreign key should be contained in the same table where the column is defined.

3.2   Algebraic Semantics of Metamodel Specifications and MOMENT2

The goal of the algebraic semantics of metamodel specifications in [8,9] is to give a precise semantics to the conformance relation $M : (\mathcal{M}, \mathcal{C})$ between a model M and a metamodel specification $(\mathcal{M}, \mathcal{C})$ (this subsumes $M : \mathcal{M}$ using $M : (\mathcal{M}, \emptyset)$). This semantics is achieved as follows. First of all, the set of MOF-conformant metamodel specifications $(\mathcal{M}, \mathcal{C})$ is a syntactically well-defined set *MetamodelSpecs*. Second, the set of equational theories in the institution of membership equational logic (MEL [23]) is another well-defined set Th_{MEL}. The algebraic semantics is then defined as a function

$$\mathbb{A} : \textit{MetamodelSpecs} \rightarrow Th_{\text{MEL}} \; : \; (\mathcal{M}, \mathcal{C}) \mapsto \mathbb{A}(\mathcal{M}, \mathcal{C}) \; .$$

The key point of this algebraic semantics is that the set of models M conformant with $(\mathcal{M}, \mathcal{C})$, which we denote $[\![(\mathcal{M}, \mathcal{C})]\!]$, is precisely axiomatized as the carrier of the sort *CModel* in the *initial algebra* $T_{\mathbb{A}(\mathcal{M}, \mathcal{C})}$ of the MEL theory $\mathbb{A}(\mathcal{M}, \mathcal{C})$. That is, we have the definitional equality $[\![(\mathcal{M}, \mathcal{C})]\!] = T_{\mathbb{A}(\mathcal{M}, \mathcal{C}), CModel}$, and hence

$$M : (\mathcal{M}, \mathcal{C}) \iff M \in [\![(\mathcal{M}, \mathcal{C})]\!] \iff M \in T_{\mathbb{A}(\mathcal{M}, \mathcal{C}), CModel} \; .$$

Intuitively, the elements of sort *CModel* are models algebraically represented as sets of objects with an associative, commutative union operation with identity (ACU), corresponding to an algebraic description of graphs. MEL is used in an essential way to impose the OCL constraints $\mathcal{C}$ by means of a conditional membership.

The algebraic semantics supports the notion of *submodel* (see [6] for details). From a graph-theoretic point of view, given $M_1, M_2 \in [\![(\mathcal{M}, \mathcal{C})]\!]$, we say that M_1 is a *submodel* of M_2, written $M_1 \subseteq M_2$ iff it is a *subgraph*, so that all the nodes (objects with attribute values) and edges (association ends) of M_1 are included in M_2 up to name and edge order isomorphism. The submodel relation is a partial order, endowing $[\![(\mathcal{M}, \mathcal{C})]\!]$ with a poset structure $([\![(\mathcal{M}, \mathcal{C})]\!], \subseteq)$. The notion of submodel will be very useful to obtain a flexible notion of multi-model in a multi-modeling language.

These notions are implemented in Maude and integrated within the Eclipse Modeling Framework (EMF) in the MOMENT2 tool [6,8,9].

3.3   Model Transformations

In this work we consider functional model transformations that map input models M, such that $M : (\mathcal{M}, \mathcal{C})$, to output models $\beta(M)$ so that $\beta(M) : (\mathcal{M}', \mathcal{C}')$, where in general $(\mathcal{M}, \mathcal{C}) \neq (\mathcal{M}', \mathcal{C}')$.

Definition 1. *Given metamodel specifications $(\mathcal{M}, \mathcal{C})$ and $(\mathcal{M}', \mathcal{C}')$, a functional model transformation from $(\mathcal{M}, \mathcal{C})$ to $(\mathcal{M}', \mathcal{C}')$ is a function $\beta : [\![(\mathcal{M}, \mathcal{C})]\!] \rightarrow [\![(\mathcal{M}', \mathcal{C}')]\!]$. The transformation β is called* monotonic, *if, in addition, it is a monotonic function $\beta : ([\![(\mathcal{M}, \mathcal{C})]\!], \subseteq) \rightarrow ([\![(\mathcal{M}', \mathcal{C}')]\!], \subseteq)$.*

MEL theories $\mathbb{A}(\mathcal{M}, \mathcal{C})$ associated to MOF metamodel specifications $(\mathcal{M}, \mathcal{C})$ are by construction executable by rewriting in Maude [14]; in fact by confluent and terminating equations modulo ACU. Therefore, the initial algebra $T_{\mathbb{A}(\mathcal{M}, \mathcal{C})}$ is computable [4]. Furthermore, any computable function $\beta : [\![(\mathcal{M}, \mathcal{C})]\!] \rightarrow [\![(\mathcal{M}', \mathcal{C}')]\!]$ can in such a case be specified by a finite set of confluent and terminating equations modulo ACU. This is exactly the approach taken in MOMENT2, where a model transformation β can be specified as a set of recursive model equations, which are automatically translated into ordinary MEL equations, as detailed in [6].

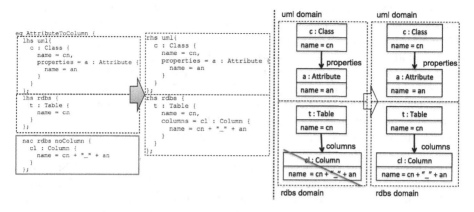

Fig. 3. Model equation: MOMENT2 format (left) and graphical representation (right)

Let us consider a model transformation between the metamodel specifications for class diagrams (CD) and for relational database schemas (RDBS) in Fig. 2: $\beta_{cd2rdbs}$: $[\![(\mathscr{M}_{CD}, \mathscr{C}_{CD})]\!] \rightarrow [\![(\mathscr{M}_{RDBS}, \mathscr{C}_{RDBS})]\!]$. The transformation is defined by several model equations that specify the translation process: classes are transformed into tables with primary keys, class attributes are transformed into table columns, and bidirectional associations are transformed into auxiliar tables that contain foreign keys that point to the tables that correspond to the associated classes. The complete specification of the model transformation is given in [7] using the concrete syntax of MOMENT2. In Fig. 3 we show a simplified version of the model equation that generates columns in a table from attributes of a class. The model equation is specified by a left-hand side (LHS) model pattern, a right-hand side (RHS) model pattern, and a negative application condition (NAC), which is applied over a LHS instance. The NAC ensures that the rule is applied only once for each attribute. MOMENT2 formalizes this model transformation as the function $\beta_{cd2rdbs}$, which is internally defined by equations that are generated from the user defined model equations of the transformation (see [6] for further details).

$\beta_{cd2rdbs}$ maps the class diagram model cd in Fig. 2 to the relational schema $\beta_{cd2rdbs}(cd) \subseteq rs$, where rs is the relational schema shown in Fig. 2. This model transformation is *monotonic* by considering the submodel relation in both source and target metamodel specifications. In particular, we consider the submodel $tutor$ of cd that is constituted only by the class Tutor, i.e., $tutor, cd \in [\![(\mathscr{M}_{CD}, \mathscr{C}_{CD})]\!]$ and $tutor \subseteq cd$. We have then that $\beta_{cd2rdbs}(tutor) \subseteq \beta_{cd2rdbs}(cd)$.

4   Institutional Semantics of Metamodels

In order to capture the semantics of models conforming to a given metamodel, we use the mathematical framework of institutions.

Definition 2. *Given a MOF-compliant metamodel specification* $(\mathscr{M}, \mathscr{C})$, *an institutional semantics for* $(\mathscr{M}, \mathscr{C})$ *is specified by: (i) an institution* $\mathcal{I}$; *and (ii) a functor* $\sigma : ([\![(\mathscr{M}, \mathscr{C})]\!], \subseteq) \rightarrow Th_{\mathcal{I}}$.

Therefore, an SE-model $M \in [\![(\mathcal{M}, \mathcal{C})]\!]$ is interpreted as a *theory* $\sigma(M) \in Th_{\mathcal{I}}$ in the corresponding institution. This definition highlights a crucial difference between "models" in the software engineering sense, which we call *SE-models*, and semantic models in the institution $\mathcal{I}$, which we call *Ins-models*, and in this case $\mathcal{I}$-models. The key point is that an SE-model of a system is only a *partial specification* of such a system, allowing many possible implementations. For example, in a UML class diagram the semantics of the methods involved is typically only partially specified. By contrast, an Ins-model is typically much closer to an actual implementation, and may fully constrain various relevant aspects of such an implementation: for example, the full semantic specification of the methods in a class diagram.

This is captured by the above definition which gives semantics to an SE-model $M \in [\![(\mathcal{M}, \mathcal{C})]\!]$ as a logical theory $\sigma(M) \in Th_{\mathcal{I}}$. That is, $\sigma(M)$ is a "partial" specification describing not a single Ins-model, but a class (actually a category) of Ins-models in the institution $\mathcal{I}$, viz. the class $Mod_{\mathcal{I}}(\sigma(M))$. Such $\mathcal{I}$-models typically fully constrain some relevant aspects of the system partially specified by the SE-model M. For example, if we choose for $\mathcal{I}$ a *computational logic*, some of the $\mathcal{I}$-models associated to M may be *executable* as programs. Therefore, an institutional semantics for a metamodel specification may support program generation methods that are correct by construction. For relational database schemas as SE-models, e.g., the Ins-models may be relational models of actual databases conformant with the given schema.

The functoriality condition in the definition of institutional semantics is very natural. Intuitively, if $M \subseteq M'$, then any implementation of the system partially specified by the SE-model M' should *a fortiori* give us an implementation of the system partially specified by M, essentially by disregarding the implementation of the extra features in $M' \setminus M$. Mathematically, this is captured by the fact that the submodel inclusion $M \subseteq M'$ induces a theory morphism $\sigma(M) \to \sigma(M')$, which, in turn, by the contravariance of the functor $Mod_{\mathcal{I}} : Sign_{\mathcal{I}}^{\mathrm{op}} \to Cat$, induces a forgetful functor $Mod_{\mathcal{I}}(\sigma(M')) \to Mod_{\mathcal{I}}(\sigma(M))$, corresponding to the intuition that from an implementation of M' we can always obtain an implementation of M. This functoriality condition will be useful to arrive at a proper notion of Ins-models for an SE-multi-model.

Let us exemplify the definition by explaining the institutional semantics for our running examples of (simplified) UML class diagrams and relational database schemas. The first is described in more detail in [12], the second builds on the traditional semantics of relational database schemas [15].

Institutional semantics for class diagrams. Signatures of the class diagram institution $\mathcal{I}_{\mathrm{CD}}$ declare class names, typed attributes, operations, and association names with corresponding properties as association ends. On the signature part, the functor $\sigma_{\mathrm{CD}} : ([\![\mathcal{M}_{\mathrm{CD}}, \mathcal{C}_{\mathrm{CD}}]\!], \subseteq) \to Th_{\mathcal{I}_{\mathrm{CD}}}$ maps, for instance, the class diagram in Fig. 2 to the $\mathcal{I}_{\mathrm{CD}}$-signature

$(\{\mathsf{EOffice}, \mathsf{Tutor}, \mathsf{EString}, \mathsf{Void}\},$
$\quad \{\mathsf{tName} : \mathsf{Tutor} \to \mathsf{EString}\},$
$\quad \{\mathsf{addTutor} : \mathsf{EOffice} \times \mathsf{Tutor} \to \mathsf{Void}\},$
$\quad \{\mathsf{tuteof} \subseteq \mathsf{tutors} : \mathsf{Tutor} \times \mathsf{eOffice} : \mathsf{EOffice}\})$.

Sentences associated with a signature of $\mathcal{I}_{\mathrm{CD}}$ declare multiplicities of form $0..1, 0..\star, 1..1, 1..\star$ for associations. Applying σ_{CD} to the class diagram of Fig. 2 yields a theory with a single sentence

$$association(\mathsf{tuteof}, \mathsf{tutors} : \mathsf{Tutor} : 1..\star, \mathsf{eOffice} : \mathsf{EOffice} : 1..1) \,.$$

Models of a class diagram signature are given as sets of object states. Object states are sets of created object identifiers of the declared class names, together with functions that interpret attributes and methods, as well as relations that interpret associations. Moreover, models of a presentation are required to satisfy the constraints put on associations. In our example, for each e-office, there is at least one tutor, and for each tutor there is exactly one e-office, such that if we navigate from a tutor to his e-office, then we can navigate back to the tutor, and vice versa.

A signature morphism between two class diagram signatures consistently maps class names, properties, methods, and association names. For example, there is an embedding signature morphism from the signature induced by our sample model without the method addTutor to the signature of the sample model above. The reduct of any model along a signature morphism simply "forgets" all additional information of the target. Signature morphisms canonically extend to sentences.

Institutional semantics for relational database schemas. Signatures of the relational database schema institution $\mathcal{I}_{\mathrm{RDBS}}$ declare the primitive types, the table names, the columns names, the typing of columns, and the primary keys of tables where each primary key of a table has to be a column of that table. On the signature part, the functor $\sigma_{\mathrm{RDBS}} : (\llbracket \mathscr{M}_{\mathrm{RDBS}}, \mathscr{C}_{\mathrm{RDBS}} \rrbracket, \subseteq) \to Th_{\mathcal{I}_{\mathrm{RDBS}}}$ maps, for instance, the relational schema in Fig. 2 to the $\mathcal{I}_{\mathrm{RDBS}}$-signature

$(\{\mathsf{NUMBER}, \mathsf{VARCHAR}\},$
$\{\mathsf{EOffice}, \mathsf{EOffice\_Tutor}, \mathsf{Tutor}\},$
$\{\mathsf{Tutor\_tid}, \mathsf{tname}, \mathsf{tutor\_fk}, \mathsf{eOffice\_fk}, \mathsf{EOffice\_tid}\},$
$\{(\mathsf{Tutor}, \mathsf{Tutor\_tid}) \mapsto \mathsf{NUMBER}, (\mathsf{Tutor}, \mathsf{tname}) \mapsto \mathsf{VARCHAR},$
$(\mathsf{EOffice\_Tutor}, \mathsf{tutor\_fk}) \mapsto \mathsf{NUMBER}, (\mathsf{EOffice\_Tutor}, \mathsf{eOffice\_fk}) \mapsto \mathsf{NUMBER},$
$(\mathsf{EOffice}, \mathsf{EOffice\_tid}) \mapsto \mathsf{NUMBER}\},$
$\{\mathsf{Tutor} \mapsto \mathsf{Tutor\_tid}, \mathsf{EOffice\_Tutor} \mapsto \mathsf{tutor\_fk}, \mathsf{EOffice} \mapsto \mathsf{EOffice\_tid}\})$

Sentences associated with a signature of $\mathcal{I}_{\mathrm{RDBS}}$ declare constraints on tables and columns: all column entries in a table that do not correspond to either primary keys or foreign keys can be null (not *nnv*) and not *unique*, all columns that correspond to primary keys shall be *nnv* and *unique*, and all columns that correspond to foreign keys (*fk*) encode cardinality constraints as *nnv* and *unique* statements. Applying σ_{RDBS} to the example in Fig. 2 yields a theory with the following sentences:

$nnv(\mathsf{Tutor}, \mathsf{Tutor\_id})$	$unique(\mathsf{Tutor}, \mathsf{Tutor\_id})$
$nnv(\mathsf{EOffice\_Tutor}, \mathsf{tutor\_fk})$	$unique(\mathsf{EOffice\_Tutor}, \mathsf{tutor\_fk})$
$nnv(\mathsf{EOffice\_Tutor}, \mathsf{eOffice\_fk})$	
$fk(\mathsf{EOffice\_Tutor}, \mathsf{tutor\_fk}, \mathsf{Tutor})$	$fk(\mathsf{EOffice\_Tutor}, \mathsf{eOffice\_fk}, \mathsf{EOffice})$
$nnv(\mathsf{EOffice}, \mathsf{EOffice\_id})$	$unique(\mathsf{EOffice}, \mathsf{EOffice\_id})$

Models of a relational database signature are given by relations over interpretations of the primitive types such that the typings of the columns in tables are satisfied. The interpretation of primitive types introduces special null-values. A model satisfies a clause $nnv(t, c)$ if the projection corresponding to the column c of the interpretation of the table t does not contain null values; and it satisfies clause $unique(t, c)$, if the projection of t to c (as a multiset) does not show duplicated entries. A model satisfies a clause $fk(t_1, c_{i_1} \ldots c_{i_k}, t_2)$ if for all tuples in t_1 projected to $c_{i_1} \ldots c_{i_k}$, there exists a tuple in t_2 projected over its primary key columns.

A signature morphism between two relational database signatures consistently maps the primitive types, the table names, and the columns names such that this mapping can be extended to the typing of columns and the primary keys of tables.

5   Semantic Connections and Correct Model Transformations

The institutional semantics of metamodel specifications provides a formal framework without which the following burning question in software engineering cannot be given any precise meaning: *When is a model transformation $\beta : [\![(\mathcal{M}, \mathcal{C})]\!] \rightarrow [\![(\mathcal{M}', \mathcal{C}')]\!]$ correct?*

The point is that, although model transformations can be very useful to leverage model building efforts in one modeling language to be used in another modeling language, we can in principle define many such βs, but some of them may be *disastrous*. Given an SE-model $M \in [\![(\mathcal{M}, \mathcal{C})]\!]$, which gives us a partial specification of a system, we want the transformed model $\beta(M) \in [\![(\mathcal{M}', \mathcal{C}')]\!]$ to be a model of the *same* system from a different perspective. In particular, $\beta(M)$ should *never* have implementations that are *incompatible* with those allowed by M. However, when modeling languages do not have any precise mathematical semantics, this very real problem can be painfully experienced in practice, but there is no way to systematically understand and prevent it.

5.1   Semantic Connections

Institution (co-)morphisms provide relations between different institutions which can be used to reflect model transformations semantically. Intuitively, institution comorphisms map a "poorer" institution into a "richer" one, whereas institution morphisms forget logical structure by mapping a "richer" institution into a "poorer" one. Sometimes, however, we have situations in which two institutions cannot be naturally related by either an institution morphism or an institution comorphism. In the example, $\mathcal{I}_{\mathrm{CD}}$ shows operations which have no counterpart in $\mathcal{I}_{\mathrm{RDBS}}$; on the other hand, $\mathcal{I}_{\mathrm{RDBS}}$ allows the uniqueness constraint to be stated while this cannot be mimicked in $\mathcal{I}_{\mathrm{CD}}$. However, we may choose a "lowest common denominator" institution $\mathcal{I}_{\mathrm{PCD}}$, which is poorer than both $\mathcal{I}_{\mathrm{CD}}$ and $\mathcal{I}_{\mathrm{RDBS}}$: the institution of "poor man's class diagrams", which is defined like $\mathcal{I}_{\mathrm{CD}}$ but does not show operations in its signature. We can then use this $\mathcal{I}_{\mathrm{PCD}}$ to relate $\mathcal{I}_{\mathrm{CD}}$ and $\mathcal{I}_{\mathrm{RDBS}}$ by what we call a *semantic connection*.[2]

---

[2] In recent discussions with A. Tarlecki we have learned that the same idea is also contemplated in his upcoming paper with T. Mossakowski [25].

Definition 3. *A semantic connection between an institution $\mathcal{I}$ and another institution $\mathcal{I}'$ is a pair (μ, ρ) of the form $\mathcal{I} \overset{\mu}{\twoheadrightarrow} \mathcal{I}_0 \overset{\rho}{\to} \mathcal{I}'$, where $\mathcal{I}_0$ is a third institution, μ is an institution morphism, and ρ is an institution comorphism.*

Using $\mathcal{I}_{\mathrm{PCD}}$, we may define a semantic connection $\mathcal{I}_{\mathrm{CD}} \overset{\mu_{\mathrm{C2R}}}{\twoheadrightarrow} \mathcal{I}_{\mathrm{PCD}} \overset{\rho_{\mathrm{C2R}}}{\to} \mathcal{I}_{\mathrm{RDBS}}$ between $\mathcal{I}_{\mathrm{CD}}$ and $\mathcal{I}_{\mathrm{RDBS}}$ as follows: The signature part $\mu_{\mathrm{C2R}}^{Sign}$ of the institution morphism $\mu_{\mathrm{C2R}} : \mathcal{I}_{\mathrm{CD}} \to \mathcal{I}_{\mathrm{PCD}}$ forgets all operations and the sentence part μ_{C2R}^{Sen} is the identity. On the other hand, the institution comorphism ρ_{C2R} is defined along the lines of the model transformation β_{cd2rdbs} in Sect. 3.3, adding primary keys and encoding *association*-clauses as *nnv* and *unique* properties of the columns that are involved in foreign keys; here the model part $\rho_{\mathrm{C2R}}^{Mod}$ is the identity.

A semantic connection $\mathcal{I} \overset{\mu}{\twoheadrightarrow} \mathcal{I}_0 \overset{\rho}{\to} \mathcal{I}'$ also allows us to relate models in $\mathcal{I}$ and $\mathcal{I}'$ by viewing them both as models in the "common semantic ground" $\mathcal{I}_0$:

Definition 4. *Given institutions $\mathcal{I}$, $\mathcal{I}'$, and a semantic connection $\mathcal{I} \overset{\mu}{\twoheadrightarrow} \mathcal{I}_0 \overset{\rho}{\to} \mathcal{I}'$, a pair of models $(\overline{M}, \overline{M}')$, with $\overline{M} \in Mod_{\mathcal{I}}(\Sigma)$, $\overline{M}' \in Mod_{\mathcal{I}'}(\rho^{Sign}(\mu^{Sign}(\Sigma)))$, and $\Sigma \in Sign_{\mathcal{I}}$, is called (μ, ρ)-consistent, if $\mu_{\Sigma}^{Mod}(\overline{M}) = \rho_{\mu^{Sign}(\Sigma)}^{Mod}(\overline{M}')$.*

For the semantic connection $\mathcal{I}_{\mathrm{CD}} \overset{\mu_{\mathrm{C2R}}}{\twoheadrightarrow} \mathcal{I}_{\mathrm{PCD}} \overset{\rho_{\mathrm{C2R}}}{\to} \mathcal{I}_{\mathrm{RDBS}}$, e.g., two models for the class diagram and the relational database schema in Fig. 2 which have a different number of Tutors would be inconsistent.

5.2 Correctness of Model Transformations

Based on semantic connections, we may go on to define a notion of semantic correctness for a model transformation.

Definition 5. *Given metamodel specifications $(\mathcal{M}, \mathcal{C})$ and $(\mathcal{M}', \mathcal{C}')$ with corresponding institutional semantics $(\mathcal{I}, \sigma : ([\![(\mathcal{M}, \mathcal{C})]\!], \subseteq) \to Th_{\mathcal{I}})$ and $(\mathcal{I}', \sigma' : ([\![(\mathcal{M}', \mathcal{C}')]\!], \subseteq) \to Th_{\mathcal{I}'})$, and given a semantic connection $\mathcal{I} \overset{\mu}{\twoheadrightarrow} \mathcal{I}_0 \overset{\rho}{\to} \mathcal{I}'$, a model transformation $\beta : [\![(\mathcal{M}, \mathcal{C})]\!] \to [\![(\mathcal{M}', \mathcal{C}')]\!]$ is called (μ, ρ)-correct, if the following two conditions hold:*

1. *For each $M \in [\![(\mathcal{M}, \mathcal{C})]\!]$ we have*

$$\rho^{Sign}(\mu^{Sign}(sign(\sigma(M)))) = sign'(\sigma'(\beta(M))) \,.$$

This condition can be visualized as the commutativity of the diagram:

$$
\begin{array}{ccc}
[\![(\mathcal{M}, \mathcal{C})]\!] & \overset{\beta}{\longrightarrow} & [\![(\mathcal{M}', \mathcal{C}')]\!] \\
\sigma \downarrow & & \downarrow \sigma' \\
Th_{\mathcal{I}} & & Th_{\mathcal{I}'} \\
sign \downarrow & & \downarrow sign' \\
Sign_{\mathcal{I}} \overset{\mu^{Sign}}{\longrightarrow} & Sign_{\mathcal{I}_0} \overset{\rho^{Sign}}{\longrightarrow} & Sign_{\mathcal{I}'}
\end{array}
$$

where if β is not monotonic this is just a commuting diagram of functions, but if β is monotonic we further require it to be a commuting diagram of functors.

2. *For each $M \in [[(\mathcal{M}, \mathscr{C})]]$ we have the containment:*

$$\rho_{\Sigma_0}^{Mod}(Mod_{\mathcal{I}'}(\sigma'(\beta(M)))) \subseteq \mu_{\Sigma}^{Mod}(Mod_{\mathcal{I}}(\sigma(M)))$$

where $\Sigma = sign(\sigma(M))$, $\Sigma_0 = \mu^{Sign}(\Sigma)$.

Note that condition (1) is a sanity check for the SE-models M and $\beta(M)$ to be *relatable* at the semantic level, since the signatures of their corresponding theories $\sigma(M)$ and $\sigma'(\beta(M))$ should be compatible. Condition (2) assumes condition (1) and adds the further stipulation that each $\mathcal{I}'$-model $\overline{M}'$ of $\beta(M)$, when brought to the common ground $\mathcal{I}_0$, should also be (the downgraded version of) an $\mathcal{I}$-model $\overline{M}$ of M, that is, $(\overline{M}, \overline{M}')$ are (μ, ρ)-consistent. This captures the crucial requirement that an implementation of $\beta(M)$ should never be incompatible with the implementations allowed by M.

The model transformation $\beta_{cd2rdbs}$ is indeed correct w.r.t. the semantic connection $\mathcal{I}_{CD} \overset{\mu_{C2R}}{\rightsquigarrow} \mathcal{I}_{PCD} \overset{\rho_{C2R}}{\rightsquigarrow} \mathcal{I}_{RDBS}$, as, given a class diagram $cd \in (\mathcal{M}_{CD}, \mathscr{C}_{CD})$, the "poor man's"-models of $\sigma_{RDBS}(\beta_{cd2rdbs}(cd))$ in $\mathcal{I}_{PCD}$ still fulfill all cardinality constraints induced by associations.

6   Multi-modeling Languages

At the very least, a multi-modeling language should be a collection of modeling languages supporting different views of a system. But if no interactions of any kind are supported between models in the different modeling sublanguages, a multi-modeling language is not very useful, since there is no way of taking advantage of model building and model analysis efforts in one sublanguage to benefit similar efforts in another sublanguage. Therefore, we assume in what follows that a multi-modeling language supports model transformations between some of its sublanguages.

Definition 6. *A multi-modeling language is specified by*

1. *A family $((\mathcal{M}_i, \mathscr{C}_i))_{i \in I}$ of metamodel specifications.*
2. *An irreflexive relation $K \subseteq I \times I$ where each pair $(i, j) \in K$ is called a connection.*
3. *For each $(i, j) \in K$ a model transformation $\beta_{ij} : [[(\mathcal{M}_i, \mathscr{C}_i)]] \rightarrow [[(\mathcal{M}_j, \mathscr{C}_j)]]$.*

The family $L_{C\&R} = (\mathcal{M}_i, \mathscr{C}_i)_{i \in \{CD, RDBS\}}$ with $K_{C\&R} = \{(CD, RDBS)\}$ and $\beta_{C\&R} = \{\beta_{CD,RDBS}\}$ with $\beta_{CD,RDBS} = \beta_{cd2rdbs}$ may serve as a simple example of a multi-modeling language $C\&R = (L_{C\&R}, K_{C\&R}, \beta_{C\&R})$.

It is assumed that for the purposes of the multi-modeling language there is, given $(i, j) \in K$ a *single* model transformation relating $(\mathscr{C}_i, \mathcal{M}_i)$ to $(\mathscr{C}_j, \mathcal{M}_j)$. This seems reasonable, since such a model transformation is used to provide a systematic "change of viewpoint" from the perspective supported by $(\mathscr{C}_i, \mathcal{M}_i)$ to that of $(\mathscr{C}_j, \mathcal{M}_j)$.

We envision teams of system designers and developers using such a multi-modeling language to design and develop a given system. A useful division of labor is supported by the multi-modeling language, so that some team members may concentrate their efforts on building and validating models mostly in a given sublanguage. If the team is well-coordinated and the multi-modeling language has a good infrastructure, team members working in different sublanguages will benefit from the efforts of their colleagues working in other sublanguages. For example, if a model in sublanguage $(\mathscr{C}_j, \mathscr{M}_j)$ has not yet been developed, a modeler may not have to begin from scratch, but may have available model fragments in $(\mathscr{C}_j, \mathscr{M}_j)$ that have been obtained by transformations from models in other sublanguages $(\mathscr{C}_i, \mathscr{M}_i)$. Or a person responsible for model analysis in sublanguage $(\mathscr{C}_j, \mathscr{M}_j)$ may be asked to verify some properties of a model in $(\mathscr{C}_j, \mathscr{M}_j)$ after it is transformed by $\beta_{i,j}$. This leads to the important question: *What is a multi-model?* for which we provide the following definition.

Definition 7. *Given a multi-modeling language* $(((\mathscr{M}_i, \mathscr{C}_i))_{i \in I}, K, \beta)$ *an I-indexed family* $(M_i)_{i \in I}$ *is called*

1. *a pre-multi-model, if* $M_i \in [\![(\mathscr{M}_i, \mathscr{C}_i)]\!]$ *for all* $i \in I$;
2. *a multi-model, if it is a pre-multi-model and, furthermore, we have* $\beta_{ij}(M_i) \subseteq M_j$ *for all* $(i, j) \in K$.

The notion of pre-multi-model may seem chaotic, but may accurately reflect the real situation of a team at moments when different team members are actively developing different models quite independently of each other. We think of this as a hopefully *transient* but very common situation, reflecting the fact that the software team may be large and geographically distributed, so that it may not be feasible for model changes in different sublanguages to be immediately taken into account across sublanguages.

However, to avoid dangerous and costly design divergences, from time to time team members should try to keep their model building efforts coordinated by freezing the current pre-multi-model $M = (M_i)_{i \in I}$ and asking some hard questions about it. A very natural question to ask is: is it the case that for each $(i, j) \in K$ we have $\beta_{ij}(M_i) \subseteq M_j$? This may not be the case, and then this may perhaps reveal that incompatible design decisions may have been made in different sublanguages.

The idea behind the model inclusion $\beta_{ij}(M_i) \subseteq M_j$ is that the model M_i, even when we transform it, may only account for part of all the information that must be modeled from the $(\mathscr{M}_j, \mathscr{C}_j)$ modeling point of view. Therefore, requiring an equality $\beta_{ij}(M_i) = M_j$ would be too restrictive. The inclusion requirement $\beta_{ij}(M_i) \subseteq M_j$ could perhaps be relaxed to a requirement that $\beta_{ij}(M_i)$ can be "mapped" to a submodel of M_j, but we do not explore this further here. For our example of tutors and e-offices in Fig. 2, the models not only form a multi-model in the multi-modeling language C&R, but also $\beta_{\text{cd2rdbs}}(cd) \subseteq rs$ holds.

Note that a multi-modeling language as defined so far lacks an institutional semantics. This means that the requirements $\beta_{ij}(M_i) \subseteq M_j$, although useful for the coherence of the overall effort, are *primarily syntactic* and do not address the burning issue of the *semantic correctness* of the transformations β_{ij} supported by the multi-modeling language. For this we need an institutional semantics.

Definition 8. *Given a multi-modeling language* $(((\mathscr{M}_i, \mathscr{C}_i))_{i \in I}, K, \beta)$ *an Ins-semantics for it is specified by:*

1. *an Ins-semantics* $(\mathcal{I}_i, \sigma_i)$ *for each* $(\mathcal{M}_i, \mathcal{C}_i)$, $i \in I$;
2. *for each* $(i,j) \in K$ *a semantic connection* $\mathcal{I}_i \overset{\mu_{ij}}{\twoheadrightarrow} \mathcal{I}_{ij} \overset{\rho_{ij}}{\rightarrow} \mathcal{I}_j$ *such that* β_{ij} *is* (μ_{ij}, ρ_{ij})-*correct.*

Applying this definition to C&R we get that $((\mathcal{I}_i, \sigma_i)_{i \in \{CD, RDBS\}}, \mathcal{I}_{CD} \overset{\mu_{C2R}}{\twoheadrightarrow} \mathcal{I}_{PCD} \overset{\rho_{C2R}}{\rightarrow} \mathcal{I}_{RDBS})$ is an Ins-semantics for the multi-modeling language C&R.

The above Ins-semantics for a multi-modeling language can be very useful in several ways. First of all, it can make sure that its model transformations β_{ij} are semantically correct. Since they will be used all the time across many modeling efforts and their incorrectness would be disastrous, this is a very valuable requirement worth verifying. There is, however, a second very useful consequence, namely, that we also obtain a notion of *Ins-model* for a multi-model.

Definition 9. *Let* $M = (M_i)_{i \in I}$ *be a multi-model in a multi-modeling language with an Ins-semantics. Then the class of its Ins-models is defined as the set* $Mod(M)$ *of all families* $(\overline{M}_i)_{i \in I}$ *where*

1. $\overline{M}_i \in Mod_{\mathcal{I}_i}(\sigma_i(M_i))$, *that is, each* $\overline{M}_i$ *is an Ins-model for the model* M_i.
2. *For each* $(i,j) \in K$ *the models* $\overline{M}_i$ *and* $Mod_{\mathcal{I}_j}(\sigma_j(\beta_{ij}(M_i) \subseteq M_j))(\overline{M}_j)$ *are* (μ_{ij}, ρ_{ij})-*consistent.*

The second condition is an "obvious" semantic compatibility condition, but it is somewhat terse in its formulation, so let us unpack it. Since $M = (M_i)_{i \in I}$ is a multi-model, for $(i,j) \in K$ we must have $\beta_{ij}(M_i) \subseteq M_j$. By the functoriality of σ_j this then gives us a theory morphism $\sigma_j(\beta_{ij}(M_i) \subseteq M_j)$, which is also a signature morphism, and which when applying the contravariant functor $Mod_{\mathcal{I}_j}$ to it gives us a reduct of $\overline{M}_j$ to a model in $Mod_{\mathcal{I}_j}(sign(\sigma_j(\beta_{ij}(M_i))))$. This reduct and the model M_i are the ones that must be (μ_{ij}, ρ_{ij})-consistent.

Why are such Ins-models useful from a software engineering point of view? Because they allow us to address another burning practical question: *When is a multi-model inconsistent?* Intuitively, a multi-model is inconsistent when it has no implementation meeting all the requirements imposed by all the models of the multi-model. But since Ins-models are mathematical surrogates for implementations of the different system aspects (and may in fact *be* implementations when the logics are computable), if a multi-model has no Ins-models, then there is no hope for it to have an implementation.

Definition 10. *In a multi-modeling language with an Ins-semantics a multi-model* $M = (M_i)_{i \in I}$ *is called* consistent, *if* $Mod(M) \neq \emptyset$.

The point, therefore, is that if $Mod(M) = \emptyset$, then the whole software design is *inconsistent* and unrealizable: the different models M_i in M place semantic constraints on each other that cannot be simultaneously satisfied.

7   Related Work and Conclusions

Interrelating different modeling notations is a difficult task due to the variety of possible structuring mechanisms and underlying computational paradigms. In the introduction we have already shortly discussed the three main approaches: the "system model approach", the "model-driven architecture approach", and the "heterogeneous semantics and development approach".

Further system model formalisms are for example, stream-based [11], graph grammar [17] and rewrite system models [14], or the integration of different specification formalisms, like CSP and Z [33]. In the model-driven architecture approach, the MOF standard permits the syntactical definition of modeling languages by means of the *meta-model* notion. The formal semantics of the MOF standard and its use for model transformations have been studied in algebraic [9], relational [2], graph grammar [5] and type-theoretic [29,30] settings. OCL-constraints of meta-models have been added in our algebraic setting in [8] and in the relational approach for Alloy in [3]. Most of these model transformation approaches are also well supported by tools such as AGG [1], VIATRA2 [32], and MOMENT2 [6]. The heterogeneous semantics line of research concentrates on the comparison and integration of different specification formalisms, retaining the formalisms most appropriate for expressing parts of the overall problem [34]. The theory of institutions [19] and its subsequent development into a powerful framework for heterogenous specifications [16,24,26,31] provide the mathematical foundations for our approach.

This paper is intended as a first step for developing a consistent and semantically well-founded framework for software development with multiple modeling languages. We have presented a novel notion of multi-modeling language which not only allows the developer to study the consistency of a multi-language design, but makes it also easy to integrate additional modeling languages. In our approach a multi-modeling language consists of a set of sublanguages and correct model transformations between some of the sublanguages. The abstract syntax of the sublanguages is specified by MOF meta-models. The semantics of a multi-modeling language is then given by associating an institution to each of its sublanguages. A further main result of the paper is the notion of semantic correctness of model transformations. It is defined by a so-called semantic connection between the institutions of the source and target meta-model of the transformation. The main correctness condition is given by a model inclusion which expresses the fact that a model transformation is understood as a kind of semantic refinement relation. This definition corresponds well with the use of model transformations in MDA; in other settings one may use other kinds of model transformations such as refactorings and abstraction mappings. For such cases our correctness notion may not be adequate and we may need to distinguish between different notions of correctness such as refinement correctness, abstraction correctness, and structural correctness.

A careful reader may have observed that our algebraic semantics for MOF, which has provided what might be called the "metalevel" at which the Ins-semantics for modeling languages is defined, is itself an *instance* of this Ins-semantics. Specifically, all MOF-compliant metamodels are exactly the SE-*models* of the MOF *meta-metamodel*. Therefore, our algebraic semantics $\mathbb{A}$ for MOF is just an institutional semantics for a modeling language in the general sense we have proposed. Namely, a semantics in which the (meta-)metamodel is MOF itself, and the institution in question is MEL. This suggests several important generalizations of the present work. Why restricting ourselves to MOF? Why not considering similar semantics for multi-modeling languages in other modeling frameworks? More generally, why not considering *multi-framework multi-languages*? Many challenging questions remain open and will be subject of our further studies including verification and tool support for multi-language consistency.

Acknowledgements. We would like to thank María Victoria Cengarle and Andrzej Tarlecki for several fruitful discussions on heterogeneous institutions and their morphisms.

This work has been partially supported by the ONR Grant N00014-02-1-0715, by the NSF Grant IIS-07-20482, by the NAOMI project funded by Lockheed Martin, by the EU project SENSORIA IST-2005-016004 and by the Spanish project META TIN2006-15175-C05-01.

References

1. The AGG website (1997), http://tfs.cs.tu-berlin.de/agg/
2. Akehurst, D.H., Kent, S., Patrascoiu, O.: A Relational Approach to Defining and Implementing Transformations Between Metamodels. Softw. Sys. Model. 2(4), 215–239 (2003)
3. Anastasakis, K., Bordbar, B., Georg, G., Ray, I.: UML2Alloy: A Challenging Model Transformation. In: Engels, G., Opdyke, B., Schmidt, D.C., Weil, F. (eds.) MODELS 2007. LNCS, vol. 4735, pp. 436–450. Springer, Heidelberg (2007)
4. Bergstra, J.A., Tucker, J.V.: A Characterisation of Computable Data Types by Means of a Finite Equational Specification Method. In: Proc. ICALP 1980. LNCS, vol. 85, pp. 76–90. Springer, Heidelberg (1980)
5. Biermann, E., Ermel, C., Taentzer, G.: Precise Semantics of EMF Model Transformations by Graph Transformation. In: Czarnecki, K., Ober, I., Bruel, J.-M., Uhl, A., Völter, M. (eds.) MODELS 2008. LNCS, vol. 5301, pp. 53–67. Springer, Heidelberg (2008)
6. Boronat, A., Heckel, R., Meseguer, J.: Rewriting Logic Semantics and Verification of Model Transformations. Technical Report CS-08-004, University of Leicester (2008)
7. Boronat, A., Knapp, A., Meseguer, J., Wirsing, M.: What is a Multi-Modeling Language? Technical Report UIUCDCS-R-2008-3006, UIUC (2008), http://www.cs.uiuc.edu/research/techreports.php?report=UIUCDCS-R-2008-3006
8. Boronat, A., Meseguer, J.: Algebraic Semantics of OCL-constrained Metamodel Specifications. Technical Report UIUCDCS-R-2008-2995, University of Illinois, Urbana Champaign (2008)
9. Boronat, A., Meseguer, J.: An Algebraic Semantics for MOF. In: Fiadeiro, J.L., Inverardi, P. (eds.) FASE 2008. LNCS, vol. 4961, pp. 377–391. Springer, Heidelberg (2008)
10. Broy, M., Cengarle, M.V., Rumpe, B.: Semantics of UML – Towards a System Model for UML: The Structural Data Model. Technical Report TUM-I0612, Technische Universität München (2006)
11. Broy, M., Stølen, K.: Specification and Development of Interactive Systems: Focus on Streams, Interfaces, and Refinement. Springer, Heidelberg (2001)
12. Cengarle, M.V., Knapp, A.: An Institution for UML 2.0 Static Structures. Technical Report TUM-I0807, Technische Universität München (2008)
13. Cengarle, M.V., Knapp, A., Tarlecki, A., Wirsing, M.: A Heterogeneous Approach to UML Semantics. In: WISTP 2008. LNCS, vol. 5019, pp. 383–402. Springer, Heidelberg (2008)
14. Clavel, M., Durán, F., Eker, S., Meseguer, J., Lincoln, P., Martí-Oliet, N., Talcott, C.: All About Maude. LNCS, vol. 4350. Springer, Heidelberg (2007)
15. Codd, E.F.: A Relational Model of Data for Large Shared Data Banks. Comm. ACM 13(6), 377–387 (1970)

16. Diaconescu, R.: Institution-Independent Model Theory. Birkhäuser, Basel (2008)
17. Engels, G., Heckel, R., Taentzer, G., Ehrig, H.: A Combined Reference Model- and View-Based Approach to System Specification. Int. J. Softw. Knowl. Eng. 7(4), 457–477 (1997)
18. Finkelstein, A., Goedicke, M., Kramer, J., Niskier, C.: Viewpoint Oriented Software Development: Methods and Viewpoints in Requirements Engineering. In: Bergstra, J.A., Feijs, L.M.G. (eds.) Algebraic Methods 1989. LNCS, vol. 490, pp. 29–54. Springer, Heidelberg (1991)
19. Goguen, J.A., Burstall, R.M.: Institutions: Abstract Model Theory for Specification and Programming. J. ACM 39(1), 95–146 (1992)
20. Goguen, J.A., Rosu, G.: Institution Morphisms. Form. Asp. Comp. 13(3–5), 274–307 (2002)
21. MacLane, S.: Categories for the Working Mathematician. Springer, Heidelberg (1971)
22. Meseguer, J.: General Logics. In: Logic Coll. 1987, pp. 275–329. North Holland, Amsterdam (1989)
23. Meseguer, J.: Membership Algebra as a Logical Framework for Equational Specification. In: Parisi-Presicce, F. (ed.) WADT 1997. LNCS, vol. 1376, pp. 18–61. Springer, Heidelberg (1998)
24. Mossakowski, T.: Heterogeneous Specification and the Heterogeneous Tool Set. Habilitationsschrift, Universität Bremen (2005)
25. Mossakowski, T., Tarlecki, A.: Heterogeneous Specification (in preparation)
26. Mossakowski, T., Tarlecki, A.: Heterogeneous Logical Environments for Distributed Specifications. In: WADT 2008. LNCS, vol. 5486, pp. 266–289. Springer, Heidelberg (2009)
27. Object Management Group. MDA Guide Version 1.0.1. Technical report, OMG (2003)
    www.omg.org/docs/omg/03-06-01.pdf
28. Object Management Group. MOF 2.0 Core Specification. Technical report, OMG (2006)
    www.omg.org/cgi-bin/doc?formal/2006-01-01
29. Poernomo, I.: The Meta-Object Facility Typed. In: Proc. SAC 2006, pp. 1845–1849. ACM, New York (2006)
30. Poernomo, I.: Proofs-as-Model-Transformations. In: Vallecillo, A., Gray, J., Pierantonio, A. (eds.) ICMT 2008. LNCS, vol. 5063, pp. 214–228. Springer, Heidelberg (2008)
31. Tarlecki, A.: Moving between Logical Systems. In: Proc. WADT 1995. LNCS, vol. 1130, pp. 478–502. Springer, Heidelberg (1996)
32. Varró, D., Balogh, A.: The Model Transformation Language of the VIATRA2 Framework. Sci. Comp. Prog. 68(3), 187–207 (2007)
33. Wehrheim, H.: Behavioural Subtyping in Object-Oriented Specification Formalisms. Habilitationsschrift, Carl-von-Ossietzky-Universität Oldenburg (2002)
34. Wirsing, M., Knapp, A.: View Consistency in Software Development. In: Wirsing, M., Knapp, A., Balsamo, S. (eds.) RISSEF 2002. LNCS, vol. 2941, pp. 341–357. Springer, Heidelberg (2004)

Generalized Theoroidal Institution Comorphisms

Mihai Codescu

German Research Center for Artificial Inteligence (DFKI GmbH) Bremen, Germany
Mihai.Codescu@dfki.de

Abstract. We propose a generalization of the notion of theoroidal comorphism, motivated by several logic translations of practical importance, encountered in the implementation of Heterogeneous Tool Set HETS. We discuss the impact of this generalization on the level of heterogenous specifications, by presenting the Grothendieck construction over a diagram of institutions and translations modelled as generalized comorphisms. Conditions for heterogeneous proofs are also evaluated.

1 Introduction

Heterogeneous specifications are needed since a large number of logics, like equational logics, description logics, higher-order logics, modal logics etc. are used in formal specification and computer science and when large systems are involved, one may use different formalisms for different parts or aspects. The approach of heterogeneous specification is not to combine the features of each logic into a single logic, but rather to keep them specific and to provide means for translating between formalisms during the specification and verification processes, thus using the logic which suits best the problem to be solved and offers best tool support.

The Heterogeneous tool set (HETS) [17] is an integration tool, providing, (1) at the logic-specific level, parser, static analysis and proof support, via dedicated tools and (2) at the logic-independent level, heterogeneous structuring mechanisms and heterogeneous proof calculus based on the formalism of development graphs. Unlike other specification languages, like UML, HETS is fully formal. The logical notions are formalized using the theory of institutions and the core of HETS is based on a graph of institutions and translations between them. Their combination is obtained via the so-called Grothendieck institution [3], which provides the semantics of heterogeneous specifications. This construction is characterized by the fact that no interaction between logics is made otherwise than via the logic translations.

Several notions of translations between institutions have been proposed, capturing different concepts; among them, institution comorphisms, also called representations or maps of institutions, which usually represent logic encodings or inclusions. This formalization is used in HETS, as the conditions for lifting the properties from the logic-specific level to the logic-independent one are in this case easier to meet in practice; however, other kind of logic translation can also be added [15]. Comorphisms, in their theoroidal variant, are required to map

A. Corradini and U. Montanari (Eds.): WADT 2008, LNCS 5486, pp. 88–101, 2009.

theories of the same signature (i.e. using the same symbols) to theories over the same signature and this mapping must interact well with the translation of sentences. We present several logic translations of practical importance encountered in HETS that do not have this properties. This motivates us to investigate, building on the ideas in [12], in which conditions the notion of comorphism can be generalized by dropping these restrictions and what the impact of this generalization at the heterogeneous level is.

2   Preliminaries

We assume the reader is familiar with category theory notions like functor, natural transformation or colimit. Note that we prefer to use diagrammatic order for composition and denote it with ";".

Institutions [6] formalize in a model-oriented way the notion of logical system, abstracting away the details of signatures, sentences and models by not imposing other restrictions on them than the satisfaction condition, which has the meaning that truth is invariant under change of notation and enlargement of context. The advantage of using the theory of institutions as foundation for specification theory is that the concepts can be defined at the general level, indepedently of the underlying logic.

Definition 1. *An institution $I = (\mathbb{S}ign, \mathbb{S}en, \mathbb{M}od, \models)$ consists of:*

- *a category $\mathbb{S}ign$ of* signatures,
- *a functor $\mathbb{S}en \colon \mathbb{S}ign \longrightarrow \mathbb{S}et$, giving for each signature Σ the set of* sentences *$\mathbb{S}en(\Sigma)$ and for each signature morphism $\varphi \colon \Sigma \longrightarrow \Sigma'$ a sentence translation map $Sen(\varphi) \colon \mathbb{S}en(\Sigma) \longrightarrow \mathbb{S}en(\Sigma')$, where we may write $Sen(\varphi)(e)$ as $\varphi(e)$,*
- *a functor $\mathbb{M}od \colon \mathbb{S}ign^{op} \longrightarrow \mathbb{C}at$ giving for each signature Σ the category of models $\mathbb{M}od(\Sigma)$ and for each signature morphism $\varphi \colon \Sigma \longrightarrow \Sigma'$, the reduct functor $\mathbb{M}od(\varphi) \colon \mathbb{M}od(\Sigma') \longrightarrow \mathbb{M}od(\Sigma)$, where we may write $\mathbb{M}od(\varphi)(M')$ as $M' \!\restriction_{\varphi}$;*
- *a binary relation $\models_{\Sigma} \subseteq |\mathbb{M}od(\Sigma)| \times \mathbb{S}en(\Sigma)$, for each signature Σ, called the* satisfaction relation

such that the following satisfaction condition *holds:*

$$M' \!\restriction_{\varphi} \models_{\Sigma} e \iff M' \models_{\Sigma'} \varphi(e)$$

for each signature morphism $\varphi \colon \Sigma \longrightarrow \Sigma'$, each Σ-sentence e and each Σ'-model M'.

Example 2. **Partial many-sorted first-order logic with equality [14]**
Signatures consist of a set of sorts and sets of total and partial operations and predicates symbols, divided by their profile. Signature morphisms map the sorts and the symbols in a compatible way, and such that the totality of operation symbols is preserved. Models are first-order structures, interpreting sorts as sets,

operation symbols as total/partial functions and predicates as relations. First-order sentences are built from the atomic ones, using the usual first-order logic features (connectives and quantification). Atomic sentences are predications, existential and strong equations and definedness assertions. The satisfaction of formulas is the Tarskian first-order satisfaction. One can check that thus we defined an institution, denoted $PFOL^=$.

Example 3. **Subsorted partial first-order logic [1]**
A subsorted signature extends a many-sorted signature with a pre-order on its set of sorts. A signature morphism is required to preserve the subsort relation (and also overloadings, but because they are not essential for the scope of the paper, they are ommited). Models are defined in terms of an associated many-sorted signature, where embeddings, projections and membership symbols are added to the original subsorted signature. Sentences are also usual many-sorted sentences over the associated signature, and satisfaction is defined as in $PFOL$. Let us denote this logic $SubPFOL^=$.

Example 4. **Subsorted partial constraint first-order logic**
This is the logic of specification language CASL [18], which was designed by *Common Framework Initiative* with the purpose to provide a standard language for algebraic specification of software systems and it combines first-order logic with induction mechanisms for specifying inductive datatypes. We denote this logic CASL or $SubPCFOL^=$. It extends subsorted partial first-order logic with an additional type of sentences, called sort generation constraints, for expressing that all values of a given set of sorts are reachable by some term in the function symbols, possibly containing variables of other sorts. Formally, a sort generation constraint over a signature Σ is represented a a triple (S', F', θ) where $\theta : \Sigma_1 = (S_1, TF_1, PF_1, P_1) \to \Sigma$, $S' \subseteq S_1$ and $F_1 \subseteq TF_1 \cup PF_1$.

A Σ-constraint (S', F', θ) holds in a Σ-model M if the carrier sets of $M \upharpoonright_\theta$ of the sorts in S' are generated by the function symbols in F' i.e. for every sort $s \in S'$ and every value $a \in (M \upharpoonright_\theta)_s$, there is a Σ_1-term t containing only function symbols from F' and variables of sorts not in S' such that $v(t) = a$ for some valuation v into $M \upharpoonright_\theta$ (denoted the same when defined on terms).

A Σ-constraint (S', F', θ) is translated along a signature morphism ϕ to $(S', F', \theta; \phi)$. It can be shown [18] that the satisfaction condition is fulfilled and we get thus an institution.

Definition 5. *A logic is an institution* $(\mathbb{S}ign, \mathbb{S}en, \mathbb{M}od, \models)$ *together with an entailment system, which is a relation* $\vdash_\Sigma \subseteq \mathcal{P}(\mathbb{S}en(\Sigma)) \times \mathbb{S}en(\Sigma)$ *for each signature* Σ, *such that:*

1. *reflexivity: for any* $e \in \mathbb{S}en(\Sigma), \{e\} \vdash_\Sigma e$;
2. *monotonicity: if* $E \vdash_\Sigma e$ *and* $E \subseteq E'$, *then* $E' \vdash_\Sigma e$;
3. *transitivity: if* $E \vdash_\Sigma e_i$, *for* $i \in Ind$ *and* $E \cup \{e_i | i \in Ind\} \vdash_\Sigma e$, *then* $E \vdash_\Sigma e$;
4. *translation: if* $E \vdash_\Sigma e$ *and* $\sigma : \Sigma \to \Sigma'$, *then* $\sigma(E) \vdash_{\sigma(\Sigma)} \sigma(e)$;
5. *soundness: if* $E \vdash_\Sigma e$, *then* $E \models_\Sigma e$.

Moreover, a logic is complete *if* $E \models_\Sigma e$ *implies* $E \vdash_\Sigma e$.

Let us further assume an arbitrary institution $I = (\mathbb{S}ign, \mathbb{S}en, \mathbb{M}od, \models)$.

A *theory* is a pair (Σ, E) where Σ is a signature and E is a set of Σ-sentences. A model of a theory (Σ, E) is a Σ-model which satisfies E.

A theory morphism $\sigma : (\Sigma, E) \longrightarrow (\Sigma', E')$ is a signature morphism $\sigma : \Sigma \longrightarrow \Sigma'$ such that $E' \models_{\Sigma'} \sigma(E)^1$.

Definition 6. *A theory morphism* $\phi : (\Sigma, E) \to (\Sigma', E')$ *is called* model-theoretically conservative *if any* (Σ, E)-*model* M *has (at least) an expansion along* ϕ *to a* (Σ', E')-*model, i.e. a model* M' *that satisfies* E' *such that* $M' \upharpoonright_\phi = M$. *Moreover, a theory morphism* $\phi : (\Sigma, E) \to (\Sigma', E')$ *is called* proof-theoretically conservative *if* $E' \models \phi(e)$ *implies* $E \models e$, *for any* Σ-*sentence* e.

It is known that the model-theoretic implies proof-theoretic conservativity, but the converse is not true in general (see [10] for an example involving description logics).

The *institution of theories* I^{th} has as signature category the category of theories of I. The remaining components are inherited from I, but with models of a theory restricted to those actually satisfying its axioms.

Given a diagram $D : J \to \mathbb{S}ign$, a cocone $(\Sigma, (\mu_j)_{j \in |J|})$ is called weakly amalgamable if for any family of models $(M_j)_{j \in |J|}$ such that $M_k \upharpoonright_{D(\sigma)} = M_j$ for any $\sigma : j \to k \in J$, there exists a Σ-model M with $M \upharpoonright_{\mu_j} = M_j$ for each j in $|J|$. If this model is unique, the cocone is called amalgamable. If such cocone exists whenever D is a span, we say that I is quasi-semi-exact, and if it is a colimiting cocone, we say that I is semi-exact.

Several types of translations between institutions have been introduced. Among them, institution comorphisms typically express that an institution is included or encoded into another one.

Definition 7. *Given two institutions* $I_1 = (\mathbb{S}ign_1, \mathbb{S}en_1, \mathbb{M}od_1, \models_1)$ *and* $I_2 = (\mathbb{S}ign_2, \mathbb{S}en_2, \mathbb{M}od_2, \models_2)$, *an institution comorphism consists of a functor* $\phi : \mathbb{S}ign_1 \to \mathbb{S}ign_2$, *a natural transformation* $\beta : \phi; \mathbb{M}od_2 \Rightarrow \mathbb{M}od_1$ *and a natural transformation* $\alpha : \mathbb{S}en_1 \to \phi; \mathbb{S}en_2$ *such that the following satisfaction condition holds for each* $\Sigma \in |\mathbb{S}ign_1|$, $M' \in |\mathbb{M}od_2(\phi(\Sigma))|$ *and* $e \in \mathbb{S}en_1(\Sigma)$

$$\beta_\Sigma(M') \models_\Sigma e \Leftrightarrow M' \models_{\phi(\Sigma)} \alpha_\Sigma(e)$$

As noticed in [11], it is often a natural case that the signatures of the source logic are translated to theories of the target one rather than just signatures. This leads to a generalization of the concept of institution comorphism, called theoroidal comorphism in [5]. A *theoroidal comorphism* between two institutions is defined as a regular comorphism between their corresponding institutions of theories such that the theory translation is (1) signature preserving, i.e. denoting ϕ the theory translation functor of the comorphism and $sign_i : \mathbb{T}h_i \to \mathbb{S}ign_i$ the forgetful functor, there is a functor ϕ' translating signatures such that $\phi; sign_2 =$

---

^{1} Given two set of Σ-sentences, E and E', we say that $E \models E'$ if for any Σ-model M such that $M \models E$, we have $M \models E'$.

$sign_1; \phi'$ (notice that two theories over the same signature are mapped by ϕ to theories over the same signature in the target logic) and (2) α-sensible, i.e. given a theory T in the source logic, the consequences of its translated theory along the comorphism are exactly the consequences of the theory obtained by first translating the signature of T along the comorphism, paired with the empty sets of axioms and goals (notice that axioms may occur in the resulting theory) and then adding as axioms the translations of the axioms of T along the sentence translation component of the comorphism.

Given an institution comorphism $\mu = (\phi, \alpha, \beta) : I_1 \to I_2$, we say that μ has (weak) amalgamation property if for any signature morphism $\sigma : \Sigma_1 \to \Sigma_2$ the diagram:

$$
\begin{array}{ccc}
\mathbb{M}od_2(\phi(\Sigma_2)) & \xrightarrow{\ \beta_{\Sigma_2}\ } & \mathbb{M}od_1(\Sigma_2) \\
{\scriptstyle \mathbb{M}od_2(\phi(\sigma))} \big\downarrow & & \big\downarrow {\scriptstyle \mathbb{M}od_1(\sigma)} \\
\mathbb{M}od_2(\phi(\Sigma_1)) & \xrightarrow[\ \beta_{\Sigma_1}\ ]{} & \mathbb{M}od_1(\Sigma_1)
\end{array}
$$

admits weak amalgamation, in the sense that for any two models $M_2 \in |\mathbb{M}od_1(\Sigma_2)|$ and $M_1' \in |\mathbb{M}od_2(\phi(\Sigma_1))|$ such that $M_2 \restriction_\sigma = \beta_{\Sigma_1}(M_1')$, there is (at least) a model $M_2' \in |\mathbb{M}od_2(\phi(\Sigma_2))|$ such that $\beta_{\Sigma_2}(M_2') = M_2$ and $M_2' \restriction_{\phi(\sigma)} = M_1'$.

3   Generalized Comorphisms

We will now present several translations between logics whose theory translation part is not signature preserving. For space limitation reasons, we decided not to give definitions of the institutions and translations involved, but rather to present them in an intuitive fashion, using examples.

Example 8. The comorphism $CASL2SubCFOL$ encoding partiality with the help of bottom or 'undefined' elements, described as translation (5a') in [14] acts at the level of signature by introducing, for each sort s such that there is a term of sort s using a partial function or a projection, a bottom element bot_s, a definedness predicate $defined_s$, total function symbols for projections to subsorts (when the case) and by turning the partial functions into total ones. Moreover, axioms are introduced for expressing the undefinedness of the bottom element, the non-emptiness of each sort, injectivity of projection and that projection maps elements identically from the supersort to the subsort.

However, the resulting theory may still contain many symbols that are not actually needed for the proofs, so we try to optimize this translation by mapping an entire theory and introduce symbols for encoding partiality depending on its sentences. Namely, the set of sorts for which there exists a partial term is computed considering only the subsort projections on those subsorts for which there is a sentence in the theory with a membership or a cast on the subsort, and then adding all their supersorts. The motivation for making this simplification is that provers are more efficient on smaller theories.

spec SP =
    sorts *Nat < Int*; *Car < Vehicle*
    ops   *speed_limit* : *Vehicle* → *Int*;
          car_speed_limit : *Int*
    • ∀ *v* : *Vehicle*
      • *v* ∈ *Car* ⇒ *speed_limit*(*v*) = *car_speed_limit*
end

Fig. 1. First-order specification of vehicles

spec SPEC =
    sorts *Nat < Int*; *Car < Vehicle*
    op   *car_speed_limit* : *Int*
    op   *gn_bottom_Car* : *Car*
    op   *gn_bottom_Int* : *Int*
    op   *gn_bottom_Vehicle* : *Vehicle*
    op   *gn_proj_Vehicle_Car* : *Vehicle* ↠ *Car*
    op   *speed_limit* : *Vehicle* → *Int*
    pred *gn_defined* : *Car*; **pred** *gn_defined* : *Int*; **pred** *gn_defined* : *Vehicle*
    ∀ *x*, *y* : *Vehicle*
    • *gn_defined*(*gn_proj_Vehicle_Car*(*x*))
      ∧ *gn_defined*(*gn_proj_Vehicle_Car*(*y*))
      ∧ *gn_proj_Vehicle_Car*(*x*) = *gn_proj_Vehicle_Car*(*y*)
      ⇒ *x* = *y*
    ∀ *x* : *Car*
    • *gn_defined*(*x*) ⇒ *gn_proj_Vehicle_Car*(*x*) = *x*
    • ∃ *x* : *Car* • *gn_defined*(*x*)
    ∀ *x* : *Car*
    • ¬ *gn_defined*(*x*) ⇔ *x* = *gn_bottom_Car*
    ∀ *v* : *Vehicle*
    • *gn_defined*(*v*)
      ⇒ *gn_defined*(*gn_proj_Vehicle_Car*(*v*))
        ⇒ *speed_limit*(*v*) = *car_speed_limit*

Fig. 2. Translated signature and axioms illustrating the effect of the translation on sort Car

Let us consider the specification from Fig. 1, where the axiom tests whether a *Vehicle* is a *Car*. Then the projection from *Vehicle* to *Car* is considered for determining the partial terms. Note that *speed_limit*(*v*) can be undefined if *v* is the bottom element on sort *Vehicle*, so *Int* gets a bottom as well. However, no membership or cast involves the sort *Nat*, so it shall not get a bottom element. Fig. 2 presents the resulting signature and also the axioms introduced by the translation on a sort, for exemplification purposes.

Notice that one can also decompose this translation as the composition of the original comorphism with an endo-translation on *SubCFOL*, making the

spec NAT =
    free type $Nat ::= 0 \mid suc(Nat)$
end

spec COMMPLUS =
    NAT
then op $\_\_ + \_\_ : Nat \times Nat \to Nat$
    vars $x, y : Nat$
    • $0 + y = y$
    • $suc(x) + y = suc(x + y)$
    • $0 + x = x + 0$ **%implied**
    • $suc(x) + y = x + suc(y)$ **%implied**
    • $x + y = y + x$ **%implied**
end

Fig. 3. Specification of natural numbers, with goals marked as implied

simplification. As this endo-translation maps theories in a formula-dependent way, it fails to be a comorphism.

Example 9. In [9] a comorphism from CASL to SoftFOL (an untyped variant of first-order logic with sort generation constraints, details ommited) is introduced with the purpose of connecting CASL to theorem provers. The resulting SoftFOL theory is translated to provers' input format; however, existing provers like SPASS do not provide support for inductive datatypes. Recovering induction proofs can be done, as explained also in [9], by instanciating the induction principles corresponding to sort generation constraints for each given proof goal. Note that lemmas still have to be provided by the user.

For example, consider the specification of natural numbers as a free type generated by 0 and successor and assume we want to prove commutativity of +, using two lemmas, as in Fig. 3. Then Fig. 4 presents the axioms introduced by translation CASL2SoftFOLInduction, which extends CASL2SoftFOL as described above.

This translation is preserving signatures when mapping theories, but, since new axioms are introduced, the α-sensibility condition of the comorphism does not hold.

Example 10. This example actually contains two translations, CASL2HasCASL [16] and CASL2Isabelle (translation (7) in [14]), which are similar in the sense that the same type of problem is encountered when translating theories.

HasCASL [20] is a higher-order extension of CASL allowing polymorphic datatypes and functions. It is closely related to the functional programming language Haskell [7]. Isabelle [19] is the logic of the interactive theorem prover Isabelle. Both in Haskell and in Isabelle, it is essential to know the constructors of a datatype when doing the analysis of program blocks, because pattern-matching is allowed only against the constructors. Therefore, this information has to be stored in the signature (see Fig. 5 with the translations of the CASL theory

- $(0 + 0 = 0 + 0$
  $\land \; \forall \; y : Nat \bullet 0 + y = y + 0 \Rightarrow 0 + succ(y) = succ(y) + 0)$
  $\Rightarrow \forall \; x : Nat \bullet 0 + x = x + 0$

%(Ax4)%

- $((\forall \; y : Nat \bullet succ(0) + y = 0 + succ(y))$
  $\land \; \forall \; y1 : Nat$
    - $(\forall \; y : Nat \bullet succ(y1) + y = y1 + succ(y))$
    $\Rightarrow \forall \; y : Nat$
      - $succ(succ(y1)) + y = succ(y1) + succ(y))$
  $\Rightarrow \forall \; x, y : Nat \bullet succ(x) + y = y + succ(x)$

%(Ax5)%

- $((\forall \; y : Nat \bullet 0 + y = y + 0)$
  $\land \; \forall \; y1 : Nat$
    - $(\forall \; y : Nat \bullet y1 + y = y + y1)$
    $\Rightarrow \forall \; y : Nat \bullet succ(y1) + y = y1 + succ(y))$
  $\Rightarrow \forall \; x, y : Nat \bullet x + y = y + x$

%(Ax6)%

Fig. 4. Axioms added by the translation CASL2SoftFOLInduction

logic HASCASL.PPOLYHOL=

spec SPEC =
    type *Nat*
    op $0 : Nat$ %(constructor)%
    op $suc : Nat \rightarrow Nat$ %(constructor)%
    $\forall \; X1 : Nat; Y1 : Nat$
    - $suc \; X1 = suc \; Y1 \Leftrightarrow X1 = Y1;$
    $\forall \; Y1 : Nat \bullet \neg \; 0 = suc \; Y1;$
    free type $Nat ::= 0 \; | \; suc \; Nat$
end

Fig. 5. Translation of spec *Nat* to HasCASL with 0 and *suc* marked as constructors

NAT from Fig. 3 to HasCASL, where 0 and *suc* are constructors in the resulting HasCASL theory and displayed as such), unlike the case of CASL, where datatypes are sentences. This causes the theory mapping of the comorphism not to be signature-preserving, as it depends on the presence of sort generation constraints in the source CASL theory.

Notice that it is not a solution to keep the constructors of datatypes in the CASL signatures as well: if we would restrict signature morphisms to map datatypes to datatypes, we would lose many views which are now correct in CASL and if we allow signature morphisms to map datatypes to ordinary sorts, a comorphism from this new logic to CASL which introduces a sort generation constraint axiom for each datatype loses functoriality of the signature translation.

The previous examples show that there are cases when logic translations can not be formalized as (theoroidal) comorphisms, as one of the two requirements on the theory mapping components, namely signature preserving and α-sensibility, fails to hold. We have therefore to generalize the notion to a concept that does not have these restrictions anymore.

Specification frames [4] formalize abstract specifications and models of specifications, while there is no notions of sentence and satisfaction.

Definition 11. *A specification frame $\mathcal{F} = (Th, Mod)$ consists of*

- *a category Th whose objects are called theories and*
- *a functor $Mod : (Th)^{op} \to CAT$ giving the category of models of a theory.*

Translation between specification frames provide the generality that they make no restriction on the way the objects of Th are mapped.

Definition 12. *A specification frame comorphism (or representation) $\mu : \mathcal{F} \to \mathcal{F}'$ consists of*

- *a functor $\phi : Th \to Th'$ and*
- *a natural transformation $\beta : \phi^{op}; Mod' \to Mod$.*

For generalizing the concept of theoroidal comorphism, we proceed in two steps. First, given an institution I, we can associate with it a specification frame in a natural way, by defining the category Th to be the category of the theories of I and the functor Mod of the specification frame assigns to each theory its category of models. Let us denote $SF(I)$ the specification frame assigned to I [2]. Notice that the sentences are not completely lost, but rather stored in the theories. We can then define generalized theoroidal comorphisms.

Definition 13. *A generalized theoroidal institution comorphism $\mu : I \to I'$ is just a specification frame comorphism $\mu : SF(I) \to SF(J)$.*

Let us denote $genIns$ the category of institutions and generalized theoroidal institution comorphisms. We can define a functor, denoted $SF : genIns->$ $Spec$, (where $Spec$ is the category of specification frames and specification frame comorphisms) which assigns to each institution its associated specification frame and maps generalized theoroidal institution comorphisms identically.

4   Heterogeneous Specifications

Heterogeneous specification is based on some graph of logics and logic translations. The so-called *Grothendieck institution* [3,13] is a technical device for giving a semantics to heterogeneous specifications. This institution is basically a flattening, or disjoint union, of the logic graph, but with Grothendieck signature morphisms comprising both ordinary (intra-institution) signature morphisms as well as (inter-institution) translations. Notice that we have not chosen some formalization for translations in this intuitive description; in HETS, the comorphism-based Grothendieck construction is preferred.

Generalized comorphisms do not come with an explicit sentence translation component, but this can be in some sense recovered, because the sentences of the logics are stored in the theories. Given a theory $T = (\Sigma, E)$ and a Σ-sentence e, we can add e to the set of sentences to obtain a *theory extension*. This leads us to consider all theory morphisms of source T as sentences of the theory T. We can also define a notion of satisfaction for morphisms, in terms of expansions: a T-model M satisfies a theory morphism $\phi : T \to T'$ if there exists at least a ϕ-expansion of M to a T'-model. The idea of theory extensions as sentences originates from [12].

To make the Grothendieck construction over a diagram $I : Ind^{op} \to genIns$ of institutions and generalized theoroidal comorphisms, we first compose I with the functor SF to obtain a diagram of specification frames and specification frame comorphisms. We will then investigate the hypotheses under which specification frames and their comorphisms can be extended to institutions/institution comorphisms, using morphisms as sentences, as described above. Thus, we will have defined a functor INS to $coIns$ and, by further composing $I; SF$ with INS, we get a diagram to $coIns$ for which we can build the known comorphism-based Grothendieck institution.

Proposition 14. *Let $Spec^{amalg}$ be the subcategory of $Spec$ such that:*

- *each object S of $Spec^{amalg}$ has pushouts of theories, with a canonical selection of pushouts such that selected pushouts compose and is weakly semi-exact and,*
- *each morphism of $Spec^{amalg}$ has weak amalgamation property and preserves selected pushouts.*

We can then define a canonical functor $INS : Spec^{amalg} \to coIns$ using theory morphisms as sentences.

Proof:
We define the functor on objects: let $S = (Th, Mod)$ be an object of $Spec^{amalg}$ and denote the institution that we define $INS(S) = (\mathbb{S}ign^I, \mathbb{S}en^I, \mathbb{M}od^I, \models)$.

The signatures and the models of $INS(S)$ are inherited from S (that is, $\mathbb{S}ign^I = Th$ and $\mathbb{M}od^I = Mod$) and we only have to define sentences and the satisfaction relation.

For any object T of Th, we define the sentences of T to be all morphisms of source T in Th, i.e. $\mathbb{S}en^I(T) = Th(T, \bullet)$.[2] For any objects T, T' of Th and any morphism $f \in Th(T, T')$, $\mathbb{S}en^I(f) : \mathbb{S}en^I(T) \to \mathbb{S}en^I(T')$ is the function that maps each morphism $e : T \to T_1$ to the morphism of source T' of the selected pushout of the span formed by e and f.

---

[2] Note that morphisms of source T form rather a class than a set. This problem can be overcomed if we consider institutions with small signature categories, which suffice in practical situations.

$$T \xrightarrow{\ f\ } T'$$
$$\downarrow e \qquad\qquad \downarrow \theta_1 = f(e)$$
$$T_1 \xrightarrow{\ \theta_2\ } T_1'$$

The functoriality of $\mathbb{S}en^I$ is ensured by the fact that selected pushouts compose.

Let T be an object of Th, M a T-model and $e : T \to T'$ be a T-sentence in $INS(S)$. Then $M \models e$ if there exists a T'-expansion of M.

The satisfaction condition follows easily from weak semi-exactness of S. Thus we have defined the institution $INS(S)$.

Given a specification frame comorphism in $Spec^{amalg}$, $\mu = (\phi, \beta) : S \to S'$, we need to define an institution comorphism $\rho : INS(S) \to INS(S')$. As expected, the action of ρ on signatures and models is inherited from μ and we only need to define the sentence translation componenent.

Let T be an object of S. Then we define $\alpha_T : Sen(T) \to Sen'(\phi(T))$ to be the function that maps each morphism e of source T to his image $\phi(e)$ along the signature morphism translation of μ. The naturality of α is ensured by the fact that translations preserve selected pushouts, while the satisfaction condition of the comorphism follows immediately from weak amalgamability property of μ. □

Note that the hypotheses about the objects and morphisms of $Spec^{amalg}$ have to hold for the institutions in the image of I, for the composition $(I; SF); INS$ to be well defined.

Let us compare our resulting institution $(I; SF; INS)^{\#}$ with the Grothendieck logic obtained by flattening a diagram $D : Ind^{op} \to coIns$ which only involves institutions existing in the logic graph. The differences are at the levels of sentences and satisfaction relation. In the case of morphisms $\iota : (i, (\Sigma, E)) \to (i, (\Sigma, E \cup \{e\}))$, where e is a sentence in I^i and ι is the identity morphism, notice that the satisfaction of ι in the Grothendieck institution $I^{\#}$ coincides with the 'local' satisfaction of the sentence e in institution I^i. From a practical point of view, this is important because it allows us to write specifications using the sentences of the logics and to obtain sentences as morphisms only when translating along a generalized theoroidal comorphism. Moreover, when we translate along a theoroidal comorphism that is α-simple in the sense of [11], i.e. the functor ϕ between theories is the α-extension to theories of a functor taking signatures to theories, then by translation along the corresponding generalized comorphism, we still obtain $\alpha_\Sigma(e)$ as the translation of ι.

5   Heterogeneous Proofs

Heterogeneous proving is done in the Grothendieck institution using the formalism of development graphs. For the construction based on a graph of institutions and non-generalized comorphisms, conditions needed for completeness of development graph calculus have been investigated in [13], [16].

When using generalized comorphisms, the sentences of the Grothendieck institution are, as defined in the previous section, theory morphisms of the original institution. We would like to obtain an entailment system on sentences of $INS(SF(I))$ which extends or at least approximates the entailment system of the original logic I. We start by noticing that the semantic entailment of $INS(SF(I))$ can be expressed in terms of model-theoretic conservativity.

Remark 15. Let I be an institution with pushouts of signatures and weakly semi-exact. Then for any theory T of I, for any set of T-sentences $E \cup \{e\}$ in $INS(SF(I))$, we have $E \models_T \phi$ in $INS(SF(I))$ if and only if the unique morphism from the colimit of the diagram formed by the theory morphisms in E to the colimit of the diagram formed by the theory morphisms in $E \cup \{e\}$, denoted $\chi_{E,e}$ is model-theoretically conservative.

Unfortunately, there is no known logic-independent characterization or approximation of model theoretical conservativity based on proof theoretical one, that could be employed in defining entailment in $INS(SF(I))$. We can instead assume the existence of a "proof-theoretical conservativity" predicate on theory morphisms of I, logic specific, with the property that whenever the predicate holds for a morphism ϕ, ϕ is model-theoretically conservative. This would allow us to define entailment in $INS(SF(I))$ based on it as follows:

$$E \vdash e \Leftrightarrow \text{ the proof } - \text{theoretical conservativity predicate holds for } \chi_{E,e},$$

where $\chi_{E,e}$ is as denoted above. The property of the predicate ensures that entailment thus defined is sound. To prove that the relation $\vdash$ is indeed an entailment system, one could attempt to make an analysis of the properties that the predicate should fulfill, but it is the case that they can not be completely derived in a logic-independent manner i.e. some of the properties expected for entailment system rely on the particular choice of the predicate.

In practical situations, proof-theoretical conservativity needs to be studied for the logics involved. For the case of CASL, we refer the reader to [8].

6   Conclusions

We have introduced the notion of generalized theoroidal institution comorphism, which eliminates the restrictions on the the way theories are translated. This new notion broadens the class of logic encodings formalized as comorphisms. We also describe a framework for heterogeneous specifications based on a graph of institutions and generalized comorphism and briefly discuss conditions in which we can equip the resulting Grothendieck institution with an entailment system.

Comparing our resulting framework with the comorphism-based Grothendieck institution of [13], notice that at the heterogeneous specification level the differences are almost invisible for the user, since sentences of the logics can still be used and logic translations that can be formalized as comorphisms do not map sentences in a different way when they are represented as generalized theoroidal comorphisms. Moreover, notice that at this level pushouts of signatures are not actually needed and therefore we can use approximations of colimits i.e. weakly amalgamable cocones, so the hypotheses turn to be equivalent. However, the changes are semnificative when it comes to heterogeneous proving. The assumption that pushouts should exists is, on one side, mandatory and, on the other side, too strong to hold in all practical situations. In particular, this is also the case for some institutions in the logic graph of HETS.

Future work should concern the study of interaction of logic-specific tools with the heterogenous sentences and the implementation of this interaction in HETS.

Acknowledgment. I would like to thank Till Mossakowski, Christian Maeder and Dominik Lücke for discussions and suggestions. I am grateful to the anonymous referees for suggesting a series of improvements of the presentation and for constructive criticism. This work has been supported by the German Federal Ministry of Education and Research (Project 01 IW 07002 FormalSafe).

References

1. Cerioli, M., Haxthausen, A., Krieg-Brückner, B., Mossakowski, T.: Permissive subsorted partial logic in CASL. In: Johnson, M. (ed.) AMAST 1997. LNCS, vol. 1349, pp. 91–107. Springer, Heidelberg (1997)
2. Cornelius, F., Baldamus, M., Ehrig, H., Orejas, F.: Abstract and behaviour module specifications. Mathematical Structures in Computer Science 9(1), 21–62 (1999)
3. Diaconescu, R.: Grothendieck institutions. Applied categorical structures 10, 383–402 (2002)
4. Ehrig, H., Pepper, P., Orejas, F.: On recent trends in algebraic specification. In: Ronchi Della Rocca, S., Ausiello, G., Dezani-Ciancaglini, M. (eds.) ICALP 1989, vol. 372, pp. 263–288. Springer, Heidelberg (1989)
5. Goguen, J., Roşu, G.: Institution morphisms. Formal aspects of computing 13, 274–307 (2002)
6. Goguen, J.A., Burstall, R.M.: Institutions: Abstract model theory for specification and programming. Journal of the Association for Computing Machinery 39, 95–146 (1992)
7. Jones, S.P.: Haskell 98 Language and Libraries: The Revised Report. Cambridge University Press, Cambridge (2003)
8. Lüth, C., Roggenbach, M., Schröder, L.: CCC - the CASL consistency checker. In: Fiadeiro, J.L., Mosses, P.D., Orejas, F. (eds.) WADT 2004. LNCS, vol. 3423, pp. 94–105. Springer, Heidelberg (2005)
9. Lüttich, K., Mossakowski, T.: Reasoning Support for CASL with Automated Theorem Proving Systems. In: Fiadeiro, J.L., Schobbens, P.-Y. (eds.) WADT 2006. LNCS, vol. 4409, pp. 74–91. Springer, Heidelberg (2007)
10. Lutz, C., Walther, D., Wolter, F.: Conservative extensions in expressive description logics. In: Proc. of IJCAI 2007, pp. 453–459. AAAI Press, Menlo Park (2007)
11. Meseguer, J.: General logics. In: Logic Colloquium 1987, pp. 275–329. North Holland, Amsterdam (1989)
12. Mossakowski, T.: Representations, hierarchies and graphs of institutions. PhD thesis, Universität Bremen, Also appeared as book in Logos Verlag (1996)

13. Mossakowski, T.: Comorphism-based Grothendieck logics. In: Diks, K., Rytter, W. (eds.) MFCS 2002. LNCS, vol. 2420, pp. 593–604. Springer, Heidelberg (2002)
14. Mossakowski, T.: Relating CASL with other specification languages: the institution level. Theoretical Computer Science 286, 367–475 (2002)
15. Mossakowski, T.: Foundations of heterogeneous specification. In: Wirsing, M., Pattinson, D., Hennicker, R. (eds.) WADT 2003. LNCS, vol. 2755, pp. 359–375. Springer, Heidelberg (2003)
16. Mossakowski, T.: Heterogeneous specification and the heterogeneous tool set. Technical report, Universitaet Bremen, Habilitation thesis (2005)
17. Mossakowski, T., Maeder, C., Lüttich, K.: The Heterogeneous Tool Set. In: Grumberg, O., Huth, M. (eds.) TACAS 2007. LNCS, vol. 4424, pp. 519–522. Springer, Heidelberg (2007)
18. Mosses, P.D. (ed.): Casl Reference Manual: The Complete Documentation of The Common Algebraic Specification Language. LNCS, vol. 2960. Springer, Heidelberg (2004)
19. Nipkow, T., Paulson, L.C., Wenzel, M.: Isabelle/HOL — A Proof Assistant for Higher-Order Logic. LNCS, vol. 2283. Springer, Heidelberg (2002)
20. Schröder, L., Mossakowski, T.: HasCASL: towards integrated specification and development of functional programs. In: Kirchner, H., Ringeissen, C. (eds.) AMAST 2002. LNCS, vol. 2422, pp. 99–116. Springer, Heidelberg (2002)

Graph Transformation with Dependencies for the Specification of Interactive Systems

Andrea Corradini[1], Luciana Foss[2], and Leila Ribeiro[2,3]

[1] Dipartimento di Informatica, Università di Pisa, Italy
[2] Instituto de Informática, Universidade Federal do Rio Grande do Sul, Brazil
[3] Computer Science Department, University of York, England

Abstract. We propose a notion of Graph Transformation Systems (GTS) with dependency relation, more expressive than a previously proposed one, and suitable for the specification of interactions. We show how a specification using GTS with dependencies can be implemented, at a lower level of abstraction, by a transactional GTS, that is, a GTS equipped with the notion of observable (stable) items in which computations that correspond to "complete" interactions are characterized as transactions.

1 Introduction

Reactive systems, in contrast to transformational systems, are characterized by continuously having to react to stimuli from the environment [1]. This kind of system is naturally data-driven, since reactions are triggered by the presence of signals or messages. The basic behavior of a reactive system is that the arrival of a signal from the environment, requiring that some task should be performed, triggers a reaction, that is a computation typically involving a series of interactions between the environment and the system until the task is completed. Therefore, a suitable abstract description of a reaction is an interaction pattern, rather than a relation between inputs and outputs, which is the usual abstraction for transformational systems. Consequently, a method for the specification of reactive systems should provide a way to describe abstractly a whole computation (a reaction) in which the interaction pattern is highlighted, because what the users (and also the developers) of a reactive system must know is in which series of interaction patterns the system may engage.

Specification of Reactive Systems
One of the techniques that is widely used in practice to define possible interaction patterns of a system is the language of Message Sequence Charts (MSC) and its variants (like Sequence Diagrams). However, although very useful in the early stages of the software development process, they do not provide sufficient information for driving the implementation of a system [2], since they represent a set of desired interaction patterns a system may engage in, but offer no information about forbidden or required runs, nor about independence of events within one interaction (that could lead to the existence of interaction patterns

A. Corradini and U. Montanari (Eds.): WADT 2008, LNCS 5486, pp. 102–118, 2009.

which differ only because independent events occur in different orders). Therefore, most approaches for the specification of reactive systems do not consider them as starting points.

Several methods for design and analysis of reactive systems propose synchronous languages as specification formalism [3,4,5], where the time of reaction to an event is null, i.e., all actions involved in the reaction occur simultaneously. This characteristic is useful to simplify the model, but frequently it does not correspond to the reality, like in the case of distributed systems where the communication between components may take some time. Other methods [6,7] propose to use asynchronous languages to specify communication between components and define mechanisms to model activities that are performed atomically.

These formalisms propose different ways to abstract computations and to represent interactions. For example, in [5], the interaction pattern of the system is defined by a discrete sequence of instants: at each instant several input and output signals can occur simultaneously forming an instantaneous computation. Instead in [6] an abstract computation consists of a sequence of steps: at each step the enabled actions are executed by first modifying the local variables, and then updating simultaneously all the shared global variables. However, these methods do not provide a clear relationship between the abstract level, which shows the interactions a system may engage in the *big-steps* of execution, and the lower level, in which these interactions are implemented by *small-steps*. This relation is really needed if one wants to validate the implementation with respect to a (set of) message sequence charts.

There are methods for specifying reactive systems that are extensions of MSCs. Live sequence charts (LSCs) [8] describe the set of interactions a system may engage in using a modal logics, by defining the properties the runs of a system must obey. In this way, it is clear how to check whether an implementation is suitable for the specification. However, generating such an implementation automatically is not straightforward (a user-assisted way is provided by a tool). Another interesting approach are causal MSCs [9], in which the linear order of messages within a MSC is replaced by a partial order describing the causality relationships among the exchanged messages.

Using Graph Transformation to Specify Reactive Systems
Graph transformation systems (GTSs) are a flexible formalism for the specification of complex systems, that may take into account aspects such as object-orientedness, concurrency, mobility and distribution [10]. The states are modeled as graphs and the events occurring in the system, which are responsible for the evolution from one state into another, are modeled by the application of suitable transformation rules (or *productions*). This specification formalism has a *data-driven* nature: the productions of a system can be (non-deterministically) applied to the graph representing the current state whenever it contains a copy of the left-hand side of the production. Due to the use of productions to specify the behavior of the systems, this specification formalism is well-suited for reactive systems: the left-hand side of a production describes the incoming signals and the right-hand side defines the reactions to these signals.

Transactional graph transformation systems (T-GTS) are an extension of GTSs, providing a notion of transaction. A graph transaction is an abstraction of a suitable kind of computation using the productions of a GTS. This approach is data driven: when a transaction starts, some items are created in the state graph and marked as *unstable* items, and the transaction is completed when all unstable items created during its execution are deleted. The intuitive idea is that, to perform a task, the system creates resources that are not observable to the environment, and only when these internal resources are consumed, the task is completed. We can have two different views on transactions. At a lower level of abstraction, we can see them as computations that start and end in states without unstable items. But at a more abstract level, the unstable items can be forgotten and a transaction can be seen as an atomic production. Using this latter view, a whole computation of a reactive system can be considered as an atomic step, taking a null time to produce a reaction.

Since an interaction may consists of several signals exchanged in both directions with the environment, it could be represented, naively, by a production which includes in the left-hand side the signals that are needed (received), and in the right-hand side the signals that are generated. However such a representation would not be adequate, because it would abstract out from the relationships between input and output signals. In [11] an extension of GTSs, called GTS *with dependency relations* was introduced, equipping productions with a relation defining the dependencies between consumed (input) and generated (output) items. However, this relation was not rich enough to enable the description of complex interaction patterns like the ones occurring in (causal) MSCs.

In this paper we extend this framework, by allowing a more expressive dependency relationship to be associated with each production. This relationship allows to describe more faithfully the possible interactions a component may engage in – one production with dependencies specifies a set of *big-steps* of a system, that is a set of possible interactions, equivalent with respect to concurrency. Moreover, we will define a way to implement a production with dependencies by a set of *small-step* productions (a transactional graph transformation system), and show that the provided construction is really a suitable implementation.

This paper is organized as follows: in Section 2 we introduce the main concepts of graph transformation systems and the new notion of productions with dependencies; in Section 3 we review transactional graph transformation system and in Section 4 we define how to obtain an implementation for a d-GTS.

2   Graph Transformation

The basic definitions of typed graph transformation systems (GTS) in the algebraic (double-pushout) approach will be reviewed in Section 2.1. Then we will define the notion of graph transformation systems with dependency relation (Section 2.2), that is a kind of GTS in which productions specify interactions. This definition is an extension of the corresponding notion presented in [11].

2.1 Graph Transformation Systems

We present here the basics of the DPO-approach to GTSs [12]. Intuitively, a GTS describes the evolution of a system in terms of productions that change the state, where states are represented as graphs. We use a typing mechanism for graphs (see [13] for more details), that can be seen as a labeling technique which allows to label each graph over a structure that is itself a graph (called the *type graph*). The basis of this approach is the category of graphs and graph morphisms.

Definition 1 ((typed) graphs and graph morphisms). *A graph is a tuple* $G = \langle V_G, E_G, s^G, t^G \rangle$, *where* V_G *and* E_G *are the sets of nodes and edges with* $V_G \cap E_G = \emptyset$, *and* $s^G, t^G : E_G \to V_G$ *are the source and target functions. We denote by* $|G|$ *the set of items of G, i.e.,* $|G| = V_G \cup E_G$. *A (total) graph morphism* $f : G \to G'$ *is a pair of functions* $(f_V : V_G \to V_{G'}, f_E : E_G \to E_{G'})$ *such that* $f_V \circ s^G = s^{G'} \circ f_E$ *and* $f_V \circ t^G = t^{G'} \circ f_E$. *A graph morphism is an inclusion, denoted* $f : G \hookrightarrow G'$, *if* f_E *and* f_V *are both inclusions. The category of graphs and total graph morphisms is called* **Graph**.

Let $T \in$ **Graph** *be a fixed graph, called the* type graph. *A T-typed graph* G^T *is given by a graph G and a graph morphism* $t_G : G \to T$. *When the type graph is clear from the context we will write G instead of* G^T. *A morphism of T-typed graphs* $f : G^T \to G'^T$ *is a graph morphism* $f : G \to G'$ *that satisfies* $t_{G'} \circ f = t_G$. *A typed graph* G^T *is called injective if the typing morphism* t_G *is injective. The category of T-typed graphs and T-typed graph morphisms is called T-***Graph**.[1]

The behavior of a graph transformation system is determined by the application of rewriting rules, also called graph productions.

Definition 2 (productions and graph transformation systems). *A (T-typed graph) production is a tuple* $q : L_q \xleftarrow{l_q} K_q \xhookrightarrow{r_q} R_q$, *where q is the name of the production,* L_q, K_q *and* R_q *are T-typed graphs (called the* left-hand side, *the* interface *and the* right-hand side *of p, respectively), and* l_q *and* r_q *are inclusions. Additionally we require that* l_q *is not an isomorphism and it is surjective on nodes. Without loss of generality, we assume that* K_q *is the intersection, componentwise, of* L_q *and* R_q, *i.e.,* $|K_q| = |L_q| \cap |R_q|$. *The class of all T-typed graph productions is denoted by T-***Prod**. *We define the following sets of items:* **consumed**$(q) = |L_q| - |K_q|$, **created**$(q) = |R_q| - |K_q|$ *and* **preserved**$(q) = |K_q|$, *where "$-$" denotes set difference.*

A (typed) graph transformation system (GTS) is a tuple $\mathcal{G} = \langle T, P, \pi \rangle$, *where T is a type graph, P is a set of production names, and* π *is a function mapping production names to productions in T-***Prod**.

If there exists a homomorphic image of the left-hand side of a production in a given graph, this production can be applied if a diagram involving two pushouts in the category T-**Graph** can be constructed [12], obtaining a *direct derivation*. The first pushout has the effect of removing from the given graph all items in

---

[1] T-**Graph** can be defined equivalently as the comma category **Graph**$\downarrow T$.

the image of **consumed**(q), while the second one adds to it a fresh copy of all items in **created**(q).

Definition 3 ((direct) derivations). *Given a T-typed graph G, a T-typed graph production $q = L_q \overset{l}{\hookleftarrow} K_q \overset{r}{\hookrightarrow} R_q$ and a **match** (an injective T-typed graph morphism) $m : L_q \to G$, a **direct derivation** from G to H using q (based on m), denoted by $\delta : G \overset{q,m}{\Rightarrow} H$, exists if and only if both squares in the diagram below are pushouts in T-**Graph**.*

$$
\begin{array}{ccccc}
L_q & \overset{l}{\longleftarrow} & K_q & \overset{r}{\hookrightarrow} & R_q \\
{\scriptstyle m}\downarrow & (1) & {\scriptstyle k}\downarrow & (2) & \downarrow{\scriptstyle m^*} \\
G & \underset{l^*}{\longleftarrow} & D & \underset{r^*}{\longrightarrow} & H
\end{array}
$$

*Given a GTS $\mathcal{G} = \langle T, P, \pi \rangle$, a **derivation** $\rho : G_0 \overset{p_1,m_1}{\Rightarrow} G_1 \overset{p_2,m_2}{\Rightarrow} G_2 \cdots G_{n-1} \overset{p_n,m_n}{\Rightarrow} G_n$ of $\mathcal{G}$ is a finite sequence of direct derivations $\delta_i : G_i \overset{p_{i+1},m_{i+1}}{\Rightarrow} G_{i+1}$ for $0 \le i < n$.*

Since we consider *injective* matches only and productions do not delete nodes, a rule is always applicable at a given match, because the *gluing conditions* [12] are always satisfied.

2.2   Graph Transformations Systems with Dependencies

In our approach, reactions of a system are modeled by computations (derivations), but we are also interested in a more abstract specification, which is conceptually at the same abstraction level of, for example, Message Sequence Charts. If we consider reactions where the input signals present in the beginning are consumed in an arbitrary order, and the output signals are produced in any order and are all available at the end, then we can use productions to abstractly represent such reactions: the left-hand side is the initial graph of the derivation modeling the reaction, and the right-hand side is its final graph. Then the application of a single "abstract" production simulates the execution of a whole interaction, hiding the internal processing. However, when a reaction involves a more complex interaction pattern, where the input signals cannot be consumed in an arbitrary order and, similarly, the output signals cannot be produced in any order, then a production (interpreted in the traditional way) is not a good abstraction, because we should be able to specify the relation between the elements of left- and right-hand sides in a more sophisticated way.

  To this aim, we equip productions with a *dependency relation*, which describes some extra relationships between the deleted and created elements. A production with dependencies describes a set of interactions: the set of all concrete traces in which items are created/deleted according to the dependency relation. Of course, dependency relations must satisfy suitable constraints, which are listed in the next definition and are explained intuitively after. Based on these constraints, we will show (in Section 4) that it is always possible to find an implementation (in terms of the small steps that build an interaction) for a production with dependencies.

  In the following, for a set of items $C \subseteq |G|$ we will denote by $\text{LEAST}_G(C)$, the least subgraph of G containing all elements of C. Furthermore, we denote by $\mathcal{MC}(p)$ the set of all *maximal created connected components* of a production p. More precisely, $M \in \mathcal{MC}(p)$ if $M \subseteq \mathbf{created}(p)$, M is connected ($x, y \in M \Rightarrow \exists \langle x = x_0, x_1, \ldots, x_n = y \rangle \subseteq M . \forall 0 \le i < n . (x_i = s(x_{i+1}) \vee x_i = t(x_{i+1}) \vee s(x_i) = x_{i+1} \vee t(x_i) = x_{i+1}))$, and it is maximal ($x \in M, y \in \mathbf{created}(p) \wedge (s(y) = x \vee t(y) = x \vee y = s(x) \vee y = t(x)) \Rightarrow y \in M$).

Definition 4 (dependency relation and *dep*-production). *Given a T-typed production $p : L_p \hookleftarrow K_p \rightarrow R_p$. A **dependency relation** $\prec_p$ for p is a relation over $(L_p \cup R_p)$ satisfying the following conditions:*

1. *$\prec_p$ is acyclic and transitive;*
2. *$min(\prec_p) \subseteq L_p$;*
3. *$max(\prec_p) \subseteq R_p$;*
4. *for each element $b \in \mathbf{created}(p)$, there exists at least one element $a \in \mathbf{consumed}(p)$, such that $a \prec_p b$;*
5. *for each element $a \in L_p$, there exists at least one element $b \in \mathbf{created}(p)$, such that $a \prec_p b$;*
6. *two items a and b in $\mathbf{consumed}(p)$ (resp. $\mathbf{created}(p)$) can only be related if there exists an item $c \in \mathbf{created}(p)$ (resp. $\mathbf{consumed}(p)$) such that $a \prec_p c \prec_p b$.*
7. *each element $a \in \mathbf{preserved}(p)$ can be related directly via $\prec_p$ only to elements of $\mathbf{created}(p)$;*
8. *there is at least one element in $max(\prec_p)$ that is related to all elements of $min(\prec_p)$;*
9. *for each $M \in \mathcal{MC}(p)$, all elements of M must have the same dependencies.*

A dep-production is a tuple $\langle L_p \hookleftarrow K_p \hookrightarrow R_p, \prec_p \rangle$, where $L_p \hookleftarrow K_p \hookrightarrow R_p$ is the span of production p and $\prec_p$ is a dependency relation for p. The class of all T-typed dep-production is denoted by T-$\mathbf{DProd}$.

The definition above is more expressive than the one we proposed earlier in [11], because dependencies between items of the left-hand side (or the right-hand side) are possible. This allows for the definition of much more complex interaction patterns within a rule. The restrictions imposed for a dependency relation are (1) the relation is acyclic, to enable the existence of a computation implementing it; (2, 3) the minimal and maximal elements with respect to the dependency relation must be contained in the left- and right-hand sides of the rule, respectively; (4, 5) each element of the production is related to at least another element – a production with dependencies models an interaction, and should not include items that are neither a cause nor a reaction within the interaction; (6) two items both consumed or both created by a production may only be related by transitivity – if two input signals s_1 and s_2 are related, then there should be an output signal that is sent (produced) before s_2 is received (consumed), and analogously for the relations between output signals; (7) only items created by a production can depend directly on elements preserved by this production; (8) there is at least one element that is caused by all elements that are minimal of the relation – this characterizes the interaction as a whole, as a unit that can not

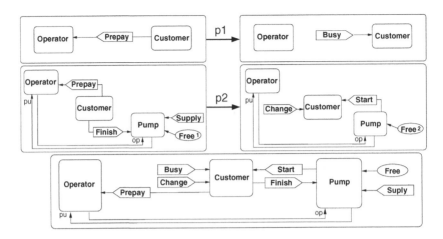

Fig. 1. Productions and type graph of a d-GTS modeling a gas station system

be split, as one "complete" task; (9) the last condition formalizes the fact that the output signals produced by the interaction modeled by the production (the created items) cannot be used again in the same interaction, not even to connect a new output signal, thus connected signals have to be created simultaneously. A GTS with *dep*-productions is defined as a GTS where the π function maps each production name into a *dep*-production.

Example 1 (GTS with dependency relation). Figure 1 shows an example of GTS with dependency relation. It models the behavior of a gas station OPERATOR when a customer arrives. There are two possible interactions, described by productions P1 and P2, which are equipped with the dependency relations depicted below. In the first case, when the Operator receives a request for pre-payment from a Customer, it declines it generating a Busy signal. The other possible behavior is to engage in an interaction modeled by production P2: when receiving a pre-payment (consuming the input signal Prepay), the Operator generates an output signal Start for the Customer, after which the other available input signals can be consumed: Supply, Finish and Free. No specific order is prescribed by relation $\prec_{p2}$ among these last signals, but only after all of them are consumed the output signals Free and Change can be produced. In representing the dependencies, curly brackets group items with the same dependencies. Items above the dashed line are created or deleted, while items below (if any) are preserved by the production.

$$\left.\begin{array}{c} Prepay \\ \text{.............} \\ Operator \\ Customer \end{array}\right\} \prec_{\text{P1}} Busy \qquad \left.\begin{array}{c} Prepay \\ \text{.............} \\ Operator \\ Pump \\ Customer \\ pu \end{array}\right\} \prec_{p2} Start \prec_{p2} \left\{\begin{array}{c} Free^1 \\ Supply \\ Finish \\ \text{.............} \\ op \end{array}\right\} \prec_{p2} \left\{\begin{array}{c} Free^2 \\ Change \end{array}\right.$$

Given the intuitive meaning of a *dep*-production as an abstraction of an interaction, using it to transform a given state graph with a direct derivation would not be sound. In fact, this would require that all the items of the left-hand side are present in the given state, but some of them could depend on produced items in the right-hand side, and thus should not be available in the given state. Therefore a production of a d-GTS does not describe a derivation step, but can be seen as a *big-step*, an abstraction for a set of computations that implement this interaction via *small-steps*. In the next section we review the notion of transactions for GTSs, which are a good description for synchronous reactions: a computation that occurs atomically at an abstract level. We will use transactions to provide small-step implementations for d-GTS.

3   Transactional Graph Transformation Systems

In *transactional graph transformation systems* (T-GTSs), introduced in [14] and inspired by *zero-safe Petri nets* [15], the basic tool is a typing mechanism for graphs which induces a distinction between *stable* and *unstable* graph items. Transactions are "minimal" computations starting from a completely stable graph, evolving through graphs with unstable items and eventually ending up in a new stable state. We will use T-GTSs in the next section to provide an implementation for the productions with dependency relations introduced before.

Definition 5 (transactional graph transformation systems). *A transactional GTS is a pair $\langle \mathcal{G}, T_s \rangle$, where $\mathcal{G} = \langle T, P, \pi \rangle$ is a T-typed GTS and $T_s \subseteq T$ is a subgraph of the type graph of $\mathcal{G}$, called the stable type graph.*

Example 2 (transactional GTS). Figure 2 shows a transactional graph transformation system. The dashed items in the type graph and productions are the unstable ones. The idea is that these productions may be used as *small-steps* to construct the *big-step* productions of Figure 1. Here we see how the interaction

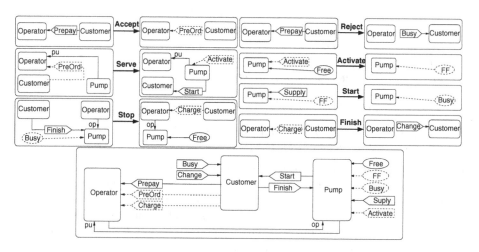

Fig. 2. A T-GTS specification for the gas station system

can actually be implemented. For example, in production ACCEPT, when the
Prepay signal is received, an internal signal is sent (PreOrd), and this will later
be used to trigger the activation of the Pump, while sending a Start signal to
the Customer (production SERVE).

Given a T-typed graph $t_G\colon G \to T$, we denote by $\mathcal{S}(G)$ its *stable subgraph*, i.e.,
the T_s-typed graph obtained as the (codomain) restriction of t_G with respect to
$T_s \subseteq T$. A T-typed graph G is *stable* if $G \approx \mathcal{S}(G)$, otherwise it is *unstable*.

As anticipated above, the transactions of a T-GTS are defined in [14] as
"minimal" derivations starting from and ending into stable graphs, where all
intermediate graphs are unstable. Actually, since the order in which independent productions are applied is considered irrelevant, transactions are defined
as shift-equivalence classes of derivations. In [16], by exploiting the one-to-one
correspondence between shift-equivalence classes of derivations and *graph processes*, it was shown that transactions could be characterized more conveniently
as graph processes satisfying suitable properties.

A graph process is a special kind of T-GTS which can be obtained via a colimit
construction applied to a derivation of $\mathcal{Z}$. This construction essentially builds
the type graph of the process as a copy of the start graph and of all the items
created during the derivation. Furthermore, the productions of the process are
the occurrences of production applications of the original derivation. The process
is also equipped with a mapping to the original T-GTS which is a morphism in a
suitable category, but this structure will not be needed in the present framework.

Definition 6 (process from a derivation). *Let $\mathcal{Z} = \langle\langle T, P, \pi\rangle, T_s\rangle$ be a T-
GTS, and let $\rho = G_0 \overset{q_1,m_1}{\Rightarrow} G_1 \overset{q_2,m_2}{\Rightarrow} \ldots \overset{q_n,m_n}{\Rightarrow} G_n$ be a derivation in $\mathcal{Z}$. A process
ϕ associated with ρ is a T-GTS $\mathcal{O}_\phi = \langle\langle T_\phi, P_\phi, \pi_\phi\rangle, T_{\phi_s}\rangle$ obtained as follows*

- *$T_\phi^T = \langle T_\phi, t_{T_\phi}\rangle$ is a colimit object (in
  T-**Graph**) of the diagram representing
  derivation ρ, as depicted (for a single
  derivation step) in the diagram on the
  right, where $c_{X_i}\colon X_i^T \to T_\phi^T$ is the induced
  injection for $X \in \{D, G, L, K, R\}$;*

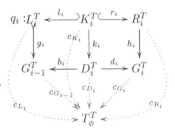

- *Graph $T_{\phi_s} \hookrightarrow T_\phi$ and morphism $T_{\phi_s} \to T_s$
  are obtained as the codomain restriction of
  $t_{T_\phi}\colon T_\phi \to T$ along $T_s \hookrightarrow T$;*
- *$P_\phi = \{\langle q_i, i\rangle \mid i \in \{1, \ldots, n\}\}$;*
- *$\pi_\phi(\langle q_i, i\rangle) = \langle L_i, c_{L_i}\rangle \overset{l_i}{\hookleftarrow} \langle K_i, c_{K_i}\rangle \overset{r_i}{\hookrightarrow}
  \langle R_i, c_{R_i}\rangle$ (see the diagram on the right);*

In order to characterize the processes that are transactions, we need to introduce some auxiliary notions. Since in a derivation all matches are injective,
in the associated process all productions are typed injectively, i.e., morphisms
$c_{X_i}\colon X_i^T \to T_\phi^T$ are injective for $X \in \{L, K, R\}$. Therefore we can say unam-
biguously that a production *consumes*, *preserves* or *creates* elements of the
type graph. Using the notation introduced in Definition 2, for each produc-
tion $q = \langle q_i, i\rangle$ of the process we define the following sets of items of T_ϕ^T:

$\mathbf{consumed}_T(q) = c_{L_i}(\mathbf{consumed}(q))$, $\mathbf{created}_T(q) = c_{R_i}(\mathbf{created}(q))$ and $\mathbf{preserved}_T(q) = c_{K_i}(\mathbf{preserved}(q))$. Furthermore, for each item $x \in |T^T_\phi|$ we define $\mathbf{pre}_T(x) = \{q \mid x \in \mathbf{created}_T(q)\}$, $\mathbf{post}_T(x) = \{q \mid x \in \mathbf{consumed}_T(q)\}$, and $\mathbf{preserve}_T(x) = \{q \mid x \in \mathbf{preserved}_T(q)\}$. Exploiting these definitions, we can define the *minimal* and *maximal* subgraphs of the type graph of a process, as well as a causal relation among the items of the type graph and the productions of a process.

Definition 7 (minimal and maximal graphs, causal relation). *Let $\mathcal{O}_\phi = \langle\langle T_\phi, P_\phi, \pi_\phi\rangle, T_{\phi_s}\rangle$ be a process. The* minimal graph *of $\mathcal{O}_\phi$, denoted by $Min(\mathcal{O}_\phi)$, is the subgraph of T_ϕ consisting of the items x such that $\mathbf{pre}_T(x) = \emptyset$. The* maximal graph *of $\mathcal{O}_\phi$, denoted by $Max(\mathcal{O}_\phi)$, is the subgraph of T_ϕ consisting of the items x such that $\mathbf{post}_T(x) = \emptyset$.*

The causal relation *of a process $\mathcal{O}_\phi$ is the least transitive and reflexive relation $\leq_\phi$ over $|T_\phi| \uplus P_\phi$ such that for all $x \in |T_\phi|$ and $q, q' \in P_\phi$:*

$$x \leq_\phi q \quad \text{if } x \in \mathbf{consumed}_T(q),$$
$$q \leq_\phi x \quad \text{if } x \in \mathbf{created}_T(q), \text{ and}$$
$$q \leq_\phi q' \quad \text{if } ((\mathbf{created}_T(q) \cap \mathbf{preserved}_T(q')) \cup$$
$$(\mathbf{preserved}_T(q) \cap \mathbf{consumed}_T(q'))) \neq \emptyset$$

For a $\leq_\phi$-left-closed $P' \subseteq P_\phi$, the reachable set *associated with P' is $S_{P'} \subseteq |T_\phi|$, defined as $x \in S_{P'}$ iff $\forall q \in P_\phi . (x \leq_\phi q \Rightarrow q \notin P') \wedge (q \leq_\phi x \Rightarrow q \in P')$.*

The reachable sets of a process are subsets of items of the type graph of the process. It can be shown that each of them represents a graph reachable from the minimal graph, applying a subset of productions of the process. A transaction is a special process.

Definition 8 (Transaction). *Let $\mathcal{Z} = \langle\langle T, P, \pi\rangle, T_s\rangle$ be a T-GTS. A transaction (t-process) is a process $\mathcal{O}_\phi$ of $\mathcal{Z}$ such that*

1. *for any stable item $x \in T_{\phi_s}$, at most one of the sets $\mathbf{pre}_T(x)$, $\mathbf{post}_T(x)$, $\mathbf{preserve}_T(x)$ is not empty;*
2. *for any $x \in Min(\mathcal{O}_\phi)$, there exists $q \in P_\phi$ such that either $x \in \mathbf{consumed}_T(q)$ or $x \in \mathbf{preserved}_T(q)$;*
3. *for any reachable set $S_{P'}$ associated with a non-empty $P' \subset P_\phi$, there exists $x \in S_{P'}$ such that $x \notin T_{\phi_s}$;*
4. *$Min(\mathcal{O}_\phi) \cup Max(\mathcal{O}_\phi) \subseteq T_{\phi_s}$.*

The family of t-processes of $\mathcal{Z}$ is denoted by $\mathbf{tProc}(\mathcal{Z})$.

Condition 1 implies that each stable item is either in the source or in the target graph of the process. Additionally, each stable item that is preserved by at least one production cannot be generated nor consumed in the process itself. By condition 2, any item in the source state is used in the computation. Condition 3 ensures that the process is not decomposable into "smaller pieces": by executing only an initial, non-empty subset P' of the productions of the process, we end up in a graph $S_{P'}$ which contains at least one unstable item. Finally, in a transaction the source and target states are required to be stable.

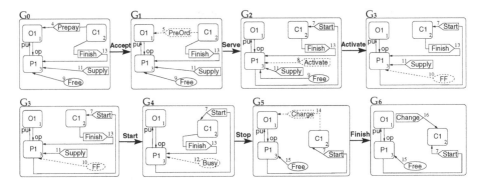

Fig. 3. An example of transaction

Example 3. Figure 3 depicts a derivation of the T-GTS of Example 2, whose process is a transaction. The rules involved in this derivation are just indicated by their names, the matches being obvious). Note that the initial and final states are totally stable, whereas in any intermediate state there is at least one unstable item.

4 Implementing GTS with Dependencies

In this section we will show that, given a d-GTS, it is possible to find a suitable implementation for it, that is, a GTS that has the observable behavior according to the dependencies specified in the d-GTS. We will call this a *small-step* GTS *corresponding to a d-GTS*, since the idea is that this graph transformation system contains the small steps needed to implement the *big-steps* specified by the productions of the d-GTS. The existence of such an implementation will be shown using T-GTS, in which the stable part is the type graph of the original d-GTS, and in which the transactions are the desired observable behavior (i.e., they implement the productions of the d-GTS).

First we show that, given the restriction on the dependency relation of the productions, it is possible to define a set of (small-step) productions that implements the (big-step) *dep*-production. This set of productions is such that, if all elements of the left-hand side of the *dep*-production are given, it will produce a graph containing all elements of the right-hand side of the production. This output graph will be constructed stepwise applying the small-step productions, and the dependency induced by this derivation is coherent with the dependency relation of the *dep*-production. The small-step productions are typed over the original type of the *dep*-production, with additional unstable types that are needed to recast the dependencies relation of the *dep*-production.

If we order all elements of a *dep*-production, based on its dependencies, the functions **before**() and **after**() give us the elements that are related directly (not by transitivity). Given a *dep*-production p, for all $a \in |L_p| \cup |R_p|$, we define

$$\mathbf{before}(a) = \{b \mid b \prec_p a \wedge \nexists c \in |L_p| \cup |R_p| : b \prec_p c \wedge c \prec_p a\}$$

$$\mathbf{after}(a) = \{b \mid a \prec_p b \wedge \nexists c \in |L_p| \cup |R_p| : a \prec_p c \wedge c \prec_p b\}$$

For a graph G, we will write $\mathbf{before}(G)$ (resp. $\mathbf{after}(G)$) to denote $\bigcup_{x \in |G|} \mathbf{before}(x)$ (resp. $\bigcup_{x \in |G|} \mathbf{after}(x)$).

The small-step GTS implementing a *dep*-production will be a T-GTS having one production for each element deleted by the production and one production for each maximal connected component created by it. In the following, for $a, b \in |G|$ we write $Kind_G(a, b) = \mathtt{true}$ if and only if a and b are both nodes, or they are edges with the same source and target nodes. Formally,

$$Kind_G(a, b) \iff (a, b \in V_G) \vee (a, b \in E_G \wedge s^G(a) = s^G(b) \wedge t^G(a) = t^G(b))$$

Definition 9 (Small-step GTS of a *dep*-production). *Given a dep-production* $r : \langle L_p \hookleftarrow K_p \hookrightarrow R_p, \prec_p \rangle$ *over type graph* T^d, *the* **small-step GTS of** r, *denoted by smallStep(r), is the* T-GTS $\langle \mathcal{G}, T^d \rangle$ *where* $\mathcal{G} = \langle T, P, \pi \rangle$, *and the productions and type graph of* $\mathcal{G}$ *are obtained as follows:*

- $T = T^d \cup \{\langle a, x \rangle \mid x \in \mathbf{consumed}_T(p) \wedge a \in \mathbf{before}(x)\} \cup \{\langle x, b \rangle \mid x \in \mathbf{created}_T(p) \wedge b \in \mathbf{after}(x)\}$;
- $P = \{delete_x \mid x \in \mathbf{consumed}_T(p)\} \cup \{create_G \mid G \in \mathcal{MC}(p)\}$
- $\forall\, delete_x \in P : \pi(delete_x) = L_q \hookleftarrow K_q \hookrightarrow R_q$, *where*
  - $L_q = K_q \cup \{x\} \cup \{\langle a, x \rangle \mid a \in \mathbf{before}(x)$, *with* $Kind(\langle a, x \rangle, x) = \mathtt{true}\}$,
  - $K_q = D_x \cup C_x$ *and*
  - $R_q = K_q \cup \{\langle x, a \rangle \mid a \in \mathbf{after}(x)$, *with* $Kind(\langle x, a \rangle, a) = \mathtt{true}\}$

  with $D_x = (\mathrm{LEAST}_{L_p}(\{x\}) - \{x\})$ *and* $C_x = \mathrm{LEAST}_{R_p}(\mathbf{after}(x)) - \mathbf{after}(x)$.
- $\forall\, create_G \in P : \pi(create_G) = L_q \hookleftarrow K_q \hookrightarrow R_q$, *where* G *is a maximal created connected component of* p *and*
  - $L_q = K_p \cup \{\langle a, x \rangle \mid x \in G \wedge a \in (\mathbf{before}(x) \cap \mathbf{consumed}_T(p))\}$, *with* $Kind(\langle a, x \rangle, x) = \mathtt{true}\}$,
  - $K_q = D_G \cup C_G$ *and*
  - $R_q = K_p \cup G \cup \{\langle x, a \rangle \mid x \in G \wedge a \in \mathbf{after}(x)$, *with* $Kind(\langle x, a \rangle, a) = \mathtt{true}\}$

  with $D_G = \mathrm{LEAST}_{L_p}(\mathbf{before}(G)) - (\mathbf{before}(G) \cap \mathbf{consumed}_T(p))$ *and* $C_G = (G \cap \mathbf{preserved}_T(p)) \cup (\mathrm{LEAST}_{L_p}(\mathbf{after}(G)) - \mathbf{after}(G))$.

In the previous definition, each $delete_x$ production consumes the stable element x and one unstable element $\langle a, x \rangle$ for each element $a \in \mathbf{before}(x)$; preserves the least set of elements needed to $\{x\}$ be a graph[2]; and creates one unstable element $\langle x, a \rangle$ for each element $a \in \mathbf{after}(x)$. Each $create_G$ production consumes all unstable created elements $\langle a, x \rangle$, where x is an element of G (a maximal created connected component of the *dep*-production) and a is a deleted element in $\mathbf{before}(x)$[3]. Furthermore, for each x in G, it preserves the items in $\mathbf{before}(x)$

---

[2] When x is an edge, we need its source and target, too.

[3] Note that, by definition of $\prec_p$, $\mathbf{before}(x)$ contains also the preserved elements if x is created by p.

that are preserved by the production, x (if it is preserved by the production), and the least sets of elements needed to have a graph containing the elements of **after**(x) and the preserved elements of **before**(x). Finally, they create each element x in G and all unstable element $\langle x, a \rangle$, where $a \in$ **after**(x).

Example 4 (Small-step GTS of a dep-production). The *small-step* GTS corresponding to the production P2 of Figure 1 is shown in Figure 4 (the type graph was omitted, since it is just the type graph of the original d-GTS including all non-observable items that appear in Figure 4).

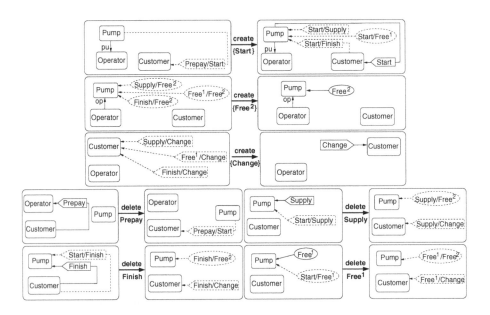

Fig. 4. Small-step of production P2 of Example 1

Now, given a d-GTS, we can define an implementation for it by constructing suitable implementations for each of its productions.

Definition 10 (Small-step GTS of a d-GTS). *Let be given a d-GTS* $d\mathcal{G} = \langle T^d, P^d, \pi^d \rangle$ *and, for all* $p_i \in P^d$, $smallStep(p_i) = \langle \langle T_{p_i}, P_{p_i}, \pi_{p_i} \rangle, T^d \rangle$. *The* **small-step GTS** *of* $d\mathcal{G}$, *denoted by* $smallStep(d\mathcal{G})$, *is the* T-GTS $\langle \langle T, P, \pi \rangle, T^d \rangle$, *where*

- *T is the colimit along all $T^d \hookrightarrow T_{p_i}$;*
- *$P = \biguplus_{p_i \in P^d} P_{p_i}$ and $\pi = \biguplus_{p_i \in P^d} \pi_{p_i}$.*

To be able to check whether the small-step T-GTS is really a suitable implementation for a d-GTS, we need to define the notion of implementation formally. Intuitively, we will require that (i) the observable part of the small-step GTS is the type graph of the d-GTS, and (ii) the big-step productions of the d-GTS correspond to transactions of the small step GTS, that model the "complete" computations (or complete interactions). A transaction is actually a (special equivalence class of) derivation, and the causal dependency relation of the corresponding process is the one obtained by the actual production application within this derivation. Just requiring that the dependency relation of a *dep*-production is a subset of this causal relation is not adequate because the dependency relation may relate items within the left-hand side (or right-hand side) of a production, and this relation will never be present in a causal order generated by a derivation. Nevertheless, it is possible to enrich the causal relation of a process adding dependencies to describe the fact that some elements will always be consumed/produced in a certain order within any derivation described by the process. We will call this new relation *dependency relation of a process*.

Definition 11 (dependency relation of a process). *Let R^* be the transitive closure of R. Given a process $\mathcal{O}_\phi$ of a T-GTS $\mathcal{Z}$, its dependency relation is the relation $\prec_\phi$ over $Min(\mathcal{O}_\phi) \cup Max(\mathcal{O}_\phi)$, defined by $\prec_\phi = (R_1 \cup R_2)^* \cap \{(a,b) \mid a,b \in Min(\mathcal{O}_\phi) \cup Max(\mathcal{O}_\phi)\}$, where $R_1 = \bigcup_{p \in P_\phi} \{(a,b) \mid a \in L_p$ and $b \in \mathbf{created}_T(p)\}$ and R_2 is defined by*

$$R_2 = \{(a,b) \mid \exists q_1, q_2 \in P_\phi \,.\, q_1 \leq_\phi q_2 \wedge a \in (\mathbf{created}_T(q_1) \cap Max(\mathcal{O}_\phi)) \wedge$$

$$b \in (\mathbf{consumed}_T(q_2) \cap Min(\mathcal{O}_\phi))\}$$

The relation R_1 contains the obvious dependencies between created and consumed/preserved elements. R_2 describes instead the dependencies between the productions, i.e, if one production depends on another one, all consumed (in the minimal graph) elements of the second production depends on all created (in the maximal graph) elements of the first one. By transitivity, the relation R_2 captures the order in which the elements must be consumed/created in any derivation that corresponds to this process.

We can obtain a *dep*-production associated with a transaction, considering its minimal and maximal graphs as left- and right-hand sides, and their intersection as interface. The dependency relation of this production is given by the dependency relation associated with the transactional process. Thus, we obtain an abstract description (*dep*-production) of a transaction, and we can see the transaction as an implementation of this abstract description.

Definition 12 (*dep*-production associated with a t-process). *Given a process $\mathcal{O}_\phi$ of a T-GTS $\mathcal{Z}$, we have $\Pi_\mathcal{Z}(\mathcal{O}_\phi) = \langle Min(\mathcal{O}_\phi) \hookleftarrow Min(\mathcal{O}_\phi) \cap Max(\mathcal{O}_\phi) \hookrightarrow Max(\mathcal{O}_\phi), \prec_\phi \rangle$, where the intersection $Min(\mathcal{O}_\phi) \cap Max(\mathcal{O}_\phi)$ is taken componentwise and $\prec_\phi$ is the dependency relation associated with ϕ.*

Now we can formalize the notion of implementation of a d-GTS via a T-GTS. This is done in the following definition. An implementation is given by a pair of mappings: (*i*) a morphism between the type graphs and (*ii*) a function mapping each production of the source d-GTS into a transaction of the target one.

Definition 13 (d-GTS implementation relationship)

Given a d-GTS $\mathcal{G} = \langle T, P, \pi \rangle$ and a T-GTS $\mathcal{Z} = \langle \langle T', P', \pi' \rangle, T_s \rangle$, an (d-GTS) implementation relationship from $\mathcal{G}$ to $\mathcal{Z}$ is a pair $\langle f_T, f_P \rangle$, where $f_T : T \to T'_s$ is a graph morphism from the type graph of $\mathcal{G}$ to the stable type graph of $\mathcal{Z}$, and $f_P : P \to \mathbf{tProc}(\mathcal{Z})$ maps each dep-production of $\mathcal{G}$ to a transaction of $\mathcal{Z}$, preserving (modulo morphism f_T) the left- and right-hand sides and the interface, as well as the dependency relation. More formally, for each $q \in P$ there must exist isomorphisms (in T-$\mathbf{Graph}$) $f_\iota^L : Min(f_P(q)) \cong t_{L_q}; f_T$ and $f_\iota^R : Max(f_P(q)) \cong t_{R_q}; f_T$ which agree on $K_q = L_q \cap R_q$, and such that for all $x, y \in |L_q| \cup |R_q|$, $x \prec_q y \iff f_\iota(x) \prec_{f_P(q)} f_\iota(y)$.

Finally, we can prove that the standard implementation defined above is actually an implementation of the original d-GTS. By proving this theorem we (i) prove that there is always at least one possible implementation of a d-GTS, that is, all its rules are feasible; and (ii) give an automatic construction method to obtain an implementation for a the give d-GTS.

Theorem 1. *Given a d-GTS $\mathcal{G} = \langle T^d, P^d, \pi^d \rangle$ and its small-step GTS $\mathcal{Z} = \langle \langle T, P, \pi \rangle, T^d \rangle$, there exists an d-GTS implementation relationship $f : \mathcal{G} \to \mathcal{Z}$.*

Proof. (Sketch) Let us define $f = \langle f_T, f_P \rangle$ as follows:

- f_T is the identity of T^d in **Graph**;
- for each $p \in P^d$, $f_P(p)$ is the transaction of $\mathcal{Z}$ obtained from $smallStep(p)$.

In order to show that this definition is well-given, we need to prove that each $smallStep(p)$ has a transaction compatible with p (with the same associated dep-production) and that all these transactions are in $\mathbf{tProc}(\mathcal{Z})$. Since $\mathcal{Z}$ is obtained by gluing $smallStep(p)$, for all $p \in P^d$, all transactions of each $smallStep(p)$ are in $\mathbf{tProc}(\mathcal{Z})$. Then it remains to prove that for each $p \in P^d$:

1. $smallStep(p)$ **has a transaction** $\mathcal{O}_\phi$ **of** $\mathcal{Z}$. Let's consider a process containing all productions in $smallStep(p)$, obtained from the derivation applying the productions from L_p. The productions are constructed considering the dependency relation of p, then, since $\prec_p$ is acyclic, it is possible to apply all productions because: the elements needed to apply $delete_x$ productions are in L_p (stable consumed or preserved elements) or are created by $create_G$ productions (unstable elements); and the elements needed to apply $create_G$ productions are created by $delete_x$ productions. Therefore, there exists such a derivation and it corresponds to a transaction if it matches the following restrictions, that can be proved to hold:
   (a) *for any stable element it is either consumed, or deleted, or preserved;*
   (b) *for any element in $Min(\mathcal{O}_\phi)$ there exists a production that consumes or preserves it;*
   (c) *there is no intermediate stable graph;*
   (d) *all elements in the minimal and maximal graphs are stable.*

2. **the *dep*-production associated with $\mathcal{O}_\phi$ is isomorphic to p:** as stated in (*b*) above, the minimal graph of $\mathcal{O}_\phi$ corresponds to L_p. Since all elements in R_p are created by some *create*$_G$ production and all unstable created items are consumed by some production, the maximal graph of $\mathcal{O}_\phi$ corresponds to R_p. Therefore, by definition of *dep*-production associated with a t-process and because the dependency relation of $\mathcal{O}_\phi$ is the same of p by construction, the production associated with $\mathcal{O}_\phi$ and p are the same.

5   Conclusion

In this paper we presented an extension of Graph Transformation Systems (GTS) to enable the specification of complex interaction patterns using productions, called GTS *with dependencies* (d-GTS). Moreover, we showed by providing an explicit construction that it is always possible to find an implementation for the productions of a d-GTS in terms of transactions of a suitable, more refined *transactional* GTS (T-GTS). This is a relevant contribution for software development, since the user may start by specifying the possible interactions a system may engage in (like it is usually done with Message Sequence Charts, but here we propose a more expressive formalism to describe interactions), providing then an implementation (as a T-GTS), and finally checking whether its implementation is really compatible with the interaction patterns he defined. This can be done by generating all possible transactions of the T-GTS (an algorithm that can construct this set of transactions for T-GTS with certain restrictions was proposed in [17]), and then comparing the corresponding abstractions (productions associated to transactions) with the productions with dependencies of the original d-GTS. The interest in comparing the set of transactions of the implementation of a d-GTS with its original set of rules is to find out whether unexpected behavior was introduced in the small-step GTS. Note that, since any implementation of a d-GTS is compatible with the dependencies specified in the *dep*-productions, this situation reveals that there are dependencies that were not explicitly specified in productions, but may arise from interference among different interaction patterns (different productions deleting/creating the same signals). This might be a hint that the specification of interaction patterns of the system, given by the d-GTS, should be revised.

The new notions presented here may serve as a starting point for defining restrictions on sets of interaction patterns (on d-GTS) that may assure that they define completely the possible behaviors of a system (in the sense that no unexpected behavior can arise from implementations respecting the dependencies given by the productions with dependencies). We are also interested in investigating the relationships between different implementations of the same system, and whether it is possible to define a "universal" implementation (at least for d-GTS with some restrictions, like the one discussed above). Other research topics include the development of a graphical representation of production with dependencies that is closer to the MSC notation, since this would make it easier for developers already familiar with that notation to use the proposed framework.

References

1. Manna, Z., Pnueli, A.: The Temporal Logic of Reactive and Concurrent Systems: Specification. Springer, Berlin (1992)
2. Harel, D., Thiagarajan, P.S.: Message sequence charts. In: Lavagno, L., Martin, G., Selic, B. (eds.) UML for Real – Design of Embedded Real-Time Systems, pp. 77–105. Kluwer Academic Publishers, Dordrecht (2003)
3. Berry, G., Couronne, P., Gonthier, G.: Synchronous programming of reactive systems: an introduction to Esterel. In: 1st. Franco-Japanese Symposium on Programming of Future Generation Computers, pp. 35–56. Elsevier, Amsterdam (1988)
4. Halbwachs, N., Caspi, P., Raymond, P., Pilaud, D.: The synchronous data-flow programming language Lustre. Proceedings of the IEEE 79(9), 1305–1320 (1991)
5. LeGuernic, P., Gautier, T., Le Borgne, M., Le Maire, C.: Programming real-time applications with SIGNAL. Proceedings of the IEEE 79(9), 1321–1336 (1991)
6. Seceleanu, C.C., Seceleanu, T.: Synchronization can improve reactive systems control and modularity. Universal Computer Science 10(10), 1429–1468 (2004)
7. Riesco, M., Tuya, J.: Synchronous Estelle: Just another synchronous language? In: 2nd. Synchronous Languages, Applications and Programming. ENTCS, vol. 88, pp. 71–86. Elsevier, Amsterdam (2004)
8. Damm, W., Harel, D.: Breathing live into message sequence charts. Formal Methods in System Design 19(1), 48–80 (2001)
9. Gazagnaire, T., Genest, B., Hélouët, L., Thiagarajan, P.S., Yang, S.: Causal message sequence charts. In: Caires, L., Vasconcelos, V.T. (eds.) CONCUR 2007. LNCS, vol. 4703, pp. 166–180. Springer, Heidelberg (2007)
10. Ehrig, H., Engels, G., Kreowski, H.J., Rozenberg, G. (eds.): Handbook of Graph Grammars and Computing by Graph Transformation: Applications, Languages and Tools. World Scientific, River Edge (1999)
11. Foss, L., Machado, R., Ribeiro, L.: Graph productions with dependencies. In: 10th. Brazilian Symposium on Formal Methods, pp. 128–143. SBC, Ouro Preto (2007)
12. Ehrig, H., Pfender, M., Schneider, H.J.: Graph-grammars: An algebraic approach. In: 14th Annual Symposium on Foundations of Computer Science, pp. 167–180. IEEE Computer Society, Washington (1973)
13. Corradini, A., Montanari, U., Rossi, F.: Graph processes. Fundamenta Informaticae 26(3/4), 241–265 (1996)
14. Baldan, P., Corradini, A., Dotti, F.L., Foss, L., Gadducci, F., Ribeiro, L.: Towards a notion of transaction in graph rewriting. In: 5th International Workshop on Graph Transformation and Visual Modeling Techniques. ENTCS, vol. 211C, pp. 39–50. Elsevier, Amsterdam (2008)
15. Bruni, R., Montanari, U.: Zero-safe nets: Comparing the collective and individual token approaches. Information and Computation 156(1-2), 46–89 (2000)
16. Baldan, P., Corradini, A., Foss, L., Gadducci, F.: Graph transactions as processes. In: Corradini, A., Ehrig, H., Montanari, U., Ribeiro, L., Rozenberg, G. (eds.) ICGT 2006. LNCS, vol. 4178, pp. 199–214. Springer, Heidelberg (2006)
17. Foss, L.: Transactional Graph Transformation Systems. PhD Thesis, Federal University of Rio Grande do Sul - UFRGS (2008)

Finitely Branching Labelled Transition Systems from Reaction Semantics for Process Calculi*

Pietro Di Gianantonio, Furio Honsell, and Marina Lenisa

Dipartimento di Matematica e Informatica, Università di Udine
via delle Scienze 206, 33100 Udine, Italy
{digianantonio,honsell,lenisa}@dimi.uniud.it

Abstract. We investigate Leifer-Milner *RPO approach* for CCS and π-calculus. The basic category in which we carry out the construction is the category of term contexts. Several issues and problems emerge from this experiment; for them we propose some new solutions.

Introduction

Recently, much attention has been devoted to derive *labelled transition systems* and *bisimilarity congruences* from *reactive systems*, in the context of process languages and graph rewriting, [1,2,3,4,5,6]. In the theory of process algebras, the operational semantics of CCS was originally given by a labelled transition system (LTS), while more recent process calculi have been presented by means of reactive systems plus structural rules. Reactive systems naturally induce behavioral equivalences which are *congruences* w.r.t. contexts, while LTS's naturally induce *bisimilarity equivalences* with coinductive characterizations. However, such equivalences are not congruences in general, or else it is an heavy, ad-hoc task to prove that they are congruences.

Leifer and Milner [1] presented a general categorical method, based on the notion of Relative Pushout (RPO), for deriving a transition system from a reactive system, in such a way that the induced *bisimilarity* is a *congruence*. The labels in Leifer-Milner's transition system are those contexts which are *minimal* for a given reaction to fire.

In the literature, some case studies have been carried out in the setting of process calculi, for testing the expressivity of Leifer-Milner's approach. Some difficulties have arisen in applying the approach directly to such languages, viewed as Lawvere theories, because of structural rules. Thus more complex categorical constructions have been introduced by Sassone and Sobocinski in [6].

In this work, we apply the RPO technique to the prototypical examples of CCS and π-calculus.

Aims and basic choices are the following.

(i) To consider simple and quite fundamental case studies in which to experiment the RPO approach.

---

* Work partially supported by ART PRIN Project prot. 2005015824 and by the FIRB Project RBIN04M8S8 (both funded by MIUR).

A. Corradini and U. Montanari (Eds.): WADT 2008, LNCS 5486, pp. 119–134, 2009.

(ii) To apply the RPO approach in the category of term contexts. In this category, arrows represent syntactic terms or contexts. The use of a category so strictly related to the original syntax has the advantage that the generated LTS has a quite direct and intuitive interpretation.

In carrying out the simpler case study given by CCS, we have found the following problems. For all of them we propose some new solutions.

– *Structural rules.* In [6], Sassone and Sobocinski proposed the use of G-categories to deal with reduction systems like CCS, where, beside the reduction rules, there is a series of structural rules. However, so far G-categories have been used to treat just tiny fragments of CCS, while in other treatments of CCS [2], structural rules are avoided through a graph encoding; namely there is a single graph representation for each class of structurally equivalent terms. In this work, we show how, using a suitably defined G-category, one can directly apply the RPO approach to the full CCS calculus.

– *Names.* Another issue is given by names, and name biding. In this work we propose *de Brujin indexes* as a suitable instrument to deal with the issues that name manipulation poses. We found out that de Brujin indexes can be used also for π-calculus, where name manipulation is more sophisticated than in CCS.

– *Pruning the LTS.* The simple application of the RPO approach generates LTS's that are quite redundant, in the sense that most of the transitions can be eliminated from the LTS without affecting the induced bisimilarity. From a practical point of view, having such large trees makes the proofs of bisimilarity unnecessarily complex. In this work, we propose a general technique that can be used in order to identify sets of transitions that can be eliminated from the LTS, without modifying the induced bisimilarity. In detail, we introduce a notion of *definability* of a transition in terms of a set of other transitions T. We prove that, given a LTS constructed by the RPO technique, if the class T of transitions is such that any other transition in the original LTS is definable from T, then the restricted LTS, obtained by considering only transitions in T, induces the same bisimilarity of the original LTS.

The result of the above technique is a LTS for CCS that coincides with the original LTS proposed by Milner. The above construction, applied to the more sophisticated case of the π-calculus, gives us a notion of bisimilarity which turns out to coincide with the *syntactical bisimilarity* of [4], and it is strictly included in Sangiorgi's *open bisimilarity*. In the π-calculus case, the LTS that we obtain, although quite reduced, is not directly finitely branching. However, it can be turned into a finite one, by working in the setting of *categories of second-order term contexts* of [7], where *parametric rules* can be represented.

Summary. In Section 1, we present CCS syntax and reaction semantics with de Brujin indexes. In Section 2, we summarize the theory of G-categories, and Leifer-Milner theory of reactive systems in a G-category setting. In Section 3, we present a construction allowing to prune the LTS obtained by applying the previous theory of reactive systems. In Sections 4 and 5, we study LTS's and bisimilarities obtained by applying the above general constructions to CCS, and π-calculus, respectively. Final remarks and directions for future work appear in Section 6.

1   CCS Processes with de Brujin Indexes

In this section, we present a version of Milner's CCS with *de Brujin indexes*, together with the reactive system. Such presentation allows us to deal smoothly with binding operators, and it is needed for extending in a natural way the structural congruence on processes to contexts.

In our presentation, CCS *names* $a_0, a_1, \ldots$ are replaced by de Brujin indexes $r_0, r_1, \ldots$, which are *name references*. The intuition about indexes is that

- the index r_i refers to the free name a_j if $j = i - n \geq 0$ and r_i appears under the scope of n ν's;
- otherwise, if $i < n$, then r_i is bound by the $i + 1$-th ν on its left;
- binding operators ν do not contain any name.

E.g. in $\nu\nu\bar{r}_0.r_2.0$, $\bar{r}_0$ is bound by the internal ν, while r_2 refers to the free name a_0. Formally:

Definition 1 (CCS Processes). *Let* $r_0, r_1, \ldots \in \mathcal{NR}$ *be a set of* name references, *let* $\alpha \in Act = \{r_i, \bar{r}_i | r_i \in \mathcal{N}\} \cup \{\tau\}$ *be a set of* actions, *and let* $x, y, z, \ldots \in \mathcal{X}$ *be a set of process variables, then we define*

$$(\mathcal{G} \ni) \ M ::= 0 \ | \ \alpha.P \ | \ M_1 + M_2 \ | \ \alpha.x \qquad \text{guarded processes}$$
$$(\mathcal{P} \ni) \ P ::= M \ | \ \nu P \ | \ P_1|P_2 \ | \ rec \ x.P \ | \ \varphi P \qquad \text{processes}$$

where φ *is a (injective)* index transformation, *obtained as a finite composition of the transformations* $\{\delta_i\}_{i \geq 0} \cup \{s_i\}_{i \geq 0}$, *where* δ_i, s_i *represent the i-th* shifting *and the i-th* swapping, *respectively, defined by*

$$\delta_i(r_j) = \begin{cases} r_{j+1} & \text{if } j \geq i \\ r_j & \text{if } j < i \end{cases} \qquad s_i(r_j) = \begin{cases} r_j & \text{if } j \neq i, i+1 \\ r_{i+1} & \text{if } j = i \\ r_i & \text{if } j = i+1 \end{cases}$$

A closed process *is a process in which each occurrence of a variable is in the scope of a rec operator.*

The index transformations φ in Definition 1 above are needed for dealing with α-rule explicitly.

In order to apply the GRPO technique to CCS, it is convenient to extend the structural congruence, which is usually defined only on processes, to all contexts. Here is where the syntax presentation à la de Brujin plays an important rôle. Namely the CCS rule

$$(\nu a P) \ | \ Q \equiv \nu a(P \ | \ Q) \ , \ \text{if } a \text{ not free in } Q$$

is problematic to extend to contexts with the usual syntax, since, if Q is a context, we have to avoid captures, by the ν-operator, of the free variables of the processes, that will appear in the holes of Q. Using de Brujin indexes (and index transformations), the above rule can be naturally extended to contexts as:

$$C[\ ]|(\nu C'[\ ]) \equiv \nu((\delta_0 C[\ ])|C'[\ ])$$

where the shifting operator δ_0 avoids the capture of free name references.

The complete definition of structural congruence is as follows:

Definition 2 (Structural Congruence). *Let $C[\ ], C'[\ ], C''[\ ]$ denote 0-holed or 1-holed contexts. The structural congruence is the relation $\equiv$, closed under process constructors, inductively generated by the following set of axioms:*

(**par**) $\quad C[\ ]|0 \equiv C[\ ] \quad C[\ ]|C'[\ ] \equiv C'[\ ]|C[\ ]$
$\qquad\quad\ C[\ ]|(C'[\ ]|C''[\ ]) \equiv (C[\ ]|C'[\ ])|C''[\ ]$

(**plus**) $\quad C[\ ]+0 \equiv C[\ ] \quad C[\ ]+C'[\ ] \equiv C'[\ ]+C[\ ]$
$\qquad\quad\ C[\ ]+(C'[\ ]+C''[\ ]) \equiv (C[\ ]+C'[\ ])+C''[\ ]$

(**rec**) $\quad\ rec\ x.C[\ ] \equiv C[\ ][rec\ x.C[\ ]/x]$

(**nu**) $\quad\ \nu 0 \equiv 0 \quad C[\ ]|(\nu C'[\ ]) \equiv \nu((\delta_0 C[\ ])|C'[\ ]) \quad \nu\nu C[\ ] \equiv \nu\nu s_0 C[\ ]$

(**phi**) $\quad\ \varphi(\nu C[\ ]) \equiv \nu(\varphi_{+1}C[\ ]) \quad \varphi 0 \equiv 0 \quad \varphi(\alpha.C[\ ]) \equiv \varphi(\alpha).\varphi(C[\ ])$
$\qquad\quad\ \varphi(C[\ ]|C'[\ ]) \equiv \varphi(C[\ ])|\varphi(C'[\ ]) \quad \varphi(rec\ x.C[\ ]) \equiv rec\ x.(\varphi C[\ ])$
$\qquad\quad\ \varphi(C[\ ]+C'[\ ]) \equiv \varphi(C[\ ])+\varphi(C'[\ ])$
$\qquad\quad\ \varphi_1 \ldots \varphi_m[\ ] \equiv \varphi'_1 \ldots \varphi'_n[\ ] , \quad if\ \varphi_1 \circ \ldots \circ \varphi_m = \varphi'_1 \circ \ldots \circ \varphi'_n$

$$where\ \varphi_{+1}(r_i) = \begin{cases} r_0 & if\ i = 0 \\ (\varphi(r_{i-1}))_{+1} & otherwise \end{cases} \qquad \varphi(\alpha) = \begin{cases} \overline{\varphi}(r) & if\ \alpha = \overline{r} \\ \varphi(r) & if\ \alpha = r \\ \tau & if\ \alpha = \tau \end{cases}$$

The last (**phi**)-rule in the above definition is useful for dealing with structural congruence of contexts (but of course is not necessary when dealing only with processes). Notice that there is an effective procedure to determine whether $\varphi_1 \circ \ldots \circ \varphi_m = \varphi'_1 \circ \ldots \circ \varphi'_n$. Namely, the two compositions are equal if and only if they contain the same number of transformations in the forms δ_i and their behavior coincides on an initial segment of indexes (whose length can be calculated from the δ_i's and the s_i's involved.)

As in the standard presentation, one can easily show that each CCS process is structurally congruent to a process in *normal form*, i.e. a process of the shape $\nu^k(\Sigma_{j=1}^{m_1}\alpha_{1,j}.P_{1,j} \mid \ldots \mid \Sigma_{j=1}^{m_n}\alpha_{n,j}.P_{n,j})$, where all unguarded restrictions are at the top level, and index transformations do not appear at the top level. A similar normal form can be defined also for contexts. Reaction semantics, defined up-to structural congruence, is as follows:

Definition 3 (Reaction Semantics). *The reaction relation $\rightarrow$ is the least relation (on closed processes) closed under the following reaction rules and reactive contexts:*

Reaction rules. $\qquad r.P + M \mid \overline{r}.Q + N \rightarrow P|Q \qquad\qquad \tau.P + M \rightarrow P$

Reaction contexts. $\quad D[\ ] ::= [\ ] \mid \nu D[\ ] \mid P|D[\ ] \mid D[\ ]|P$

Of course, one can easily define a mapping from standard CCS syntax into our de Brujin presentation, in such a way that reaction semantics is preserved. We omit the details.

2 Reactive Systems in the G-category Setting

In this section, we summarize the categorical notions necessary in the remaining of the article. These are the theories of G-categories, and the reactive systems formulated in a G-category setting [1,8].

The basic idea is to formulate the notion of *reactive system*, in a setting whereby *contexts* are modeled as arrows of a category, *terms* are arrows having as source a special object 0, and reaction rules are pairs of terms.

For our purpose it is necessary to consider a more involved formulation of the theory where G-categories are used. G-categories are a particular form of 2-categories where morphisms between arrows are all isomorphisms. G-categories are useful in dealing with calculi where there are structural equivalence relations on terms, CCS and π-calculus are typical examples. For these calculi, two cells isomorphisms represent equivalence relations on contexts. The extra complexity of using G-categories is motivated by the fact that the simpler approach of using categories with arrows representing equivalence classes of contexts (or terms) induces an incorrect bisimilarity, [8].

Definition 4. *A 2-category $\mathcal{C}$ consists of*

- *A set of objects: $X, Y, Z, ...$*
- *For any pair of objects $X, Y \in \mathcal{C}$, a category $C(X, Y)$. Objects in $C(X, Y)$ are called 1-cells morphisms, and denoted by $f : X \to Y$. Arrows in $C(X, Y)$ are*

  called 2-cells isomorphisms and represented by $\alpha : f \Rightarrow g$ or by $X \underset{g}{\overset{f}{\Longrightarrow}} Y$.

  Composition in $C(X, Y)$, called vertical composition, *is denoted by $\bullet$.*
- *For all objects X, Y and Z, there is a functor $\circ : \mathcal{C}(Y, Z) \times \mathcal{C}(X, Y) \to \mathcal{C}(X, Y)$, called* horizontal composition, *which is associative and admits the identity 2-cells of id_X as identities.*

A G-category is a 2-category whose 2-cells morphisms are all isomorphisms.

We define here the G-category formed by the *finite* (*i.e.* without the *rec* operator) CCS terms and contexts, with terms (and contexts) equipped with a structural equivalence. Since the CCS grammar needs to distinguish between guarded and generic terms, the category needs to contain two distinct objects. Formally:

- Objects are $0, \mathcal{G}, \mathcal{P}$.
- Arrows from 0 to $\mathcal{G}$ ($\mathcal{P}$) are guarded processes (generic processes). Arrows from $\mathcal{G}$ ($\mathcal{P}$) are contexts that take a guarded term (a term). More formally, the arrows $\mathcal{A} \to \mathcal{B}$ are the contexts $C_{\mathcal{A}}^{\mathcal{B}}[\ ]$ generated by the grammar:

$$C^{\mathcal{G}}_{\mathcal{G}}[\,] ::= [\,] \mid \alpha.C^{\mathcal{P}}_{\mathcal{G}}[\,] \mid C^{\mathcal{G}}_{\mathcal{G}}[\,] + M \mid M + C^{\mathcal{G}}_{\mathcal{G}}[\,]$$
$$C^{\mathcal{G}}_{\mathcal{P}}[\,] ::= \quad \alpha.C^{\mathcal{P}}_{\mathcal{P}}[\,] \mid C^{\mathcal{G}}_{\mathcal{P}}[\,] + M \mid M + C^{\mathcal{G}}_{\mathcal{P}}[\,]$$
$$C^{\mathcal{P}}_{\mathcal{G}}[\,] ::= \quad C^{\mathcal{G}}_{\mathcal{G}}[\,] \mid \nu C^{\mathcal{P}}_{\mathcal{G}}[\,] \mid C^{\mathcal{P}}_{\mathcal{G}}[\,]\|P \mid P|C^{\mathcal{P}}_{\mathcal{G}}[\,] \mid \delta C^{\mathcal{P}}_{\mathcal{G}}[\,]$$
$$C^{\mathcal{P}}_{\mathcal{P}}[\,] ::= [\,] \mid C^{\mathcal{G}}_{\mathcal{P}}[\,] \mid \nu C^{\mathcal{P}}_{\mathcal{P}}[\,] \mid C^{\mathcal{P}}_{\mathcal{P}}[\,]\|P \mid P|C^{\mathcal{P}}_{\mathcal{P}}[\,] \mid \delta C^{\mathcal{P}}_{\mathcal{P}}[\,]$$

For simplicity, in what follows we will omit the tag $\mathcal{P}, \mathcal{G}$ from the contexts.

- 2-cell isomorphisms between $C[\,]$ and $C'[\,]$ are the one-to-one maps between the instances of actions in $C[\,]$ and $C'[\,]$ induced by the proof of structural congruence. By structural induction on the proof of structural congruence, it is possible to show that two structurally congruent finite terms have the same number of instances for each action, and each proof of congruence induces a one to one map between actions in an obvious way.

Here we restrict the G-category to contain only finite processes because we need the 2-cell morphisms to be isomorphisms. When CCS processes contain the *rec* operator, two congruent processes can contain two different numbers of actions, so there cannot exists a one-to-one map between the sets of actions. However, it is possible to recover a LTS for the whole CCS processes by defining the LTS associated to an infinite process P (a term containing *rec*) as the supremum of the LTS associated to the approximants of P. For lack of space we omit the details.

Definition 5 (G-Reactive System). *A* G-reactive system **C** *consists of:*

- *a G-category $\mathcal{C}$;*
- *a distinguished object $0 \in |\mathcal{C}|$;*
- *a collection $\mathcal{D}$ of 1-cells morphisms, in $\mathcal{C}$. $\mathcal{D}$ is referred as the set of* reactive contexts, *it is required to be closed under 2-cells, and to reflect composition.*
- *a set of pairs $\mathbf{R} \subseteq \bigcup_{I \in |\mathcal{C}|} \mathcal{C}[0, I] \times \mathcal{C}[0, I]$ of reaction rules.*

The reactive contexts are those in which a reaction can occur. By composition-reflecting we mean that $d \circ d' \in \mathcal{D}$ implies $d, d' \in \mathcal{D}$, while by closure under 2-cells we mean that if $d \in \mathcal{D}, \alpha : d \Rightarrow d'$ then $d' \in \mathcal{D}$.

In our leading example a G-reactive system for CCS is obtained by taking as reaction rules and reactive contexts the ones given in Definition 3. It is immediate to check that this definition is correct.

A G-reactive system induces a *reaction relation* $\rightarrow$ on 1-cells, defined by: $t \rightarrow u$ if there exist $\langle l, r \rangle \in \mathbf{R}$, $\alpha : dl \Rightarrow t$ and $\beta : u \Rightarrow dr$. Observe that the reaction relation is closed by 2-cell isomorphisms. In the CCS example, the above reaction relation coincides with the canonical one given in Definition 3.

The behavior of a reactive system is expressed as an unlabelled transition system. On the other hand, many useful behavioral equivalences are only defined for LTS's.

From a reactive system it is possible to derive a LTS by taking as labels the contexts that transform a term into a term for which a reduction rule applies. In [1], the authors formalize these ideas and propose to take as labels the "minimal contexts allowing for a reaction". A categorical criterion for identifying the smallest contexts is given by the *relative pushouts* construction. In [8] this categorical construction is extended to G-categories.

Definition 6 (GRPO/GIPO)

(i) *Let $\mathcal{C}$ be a G-category and let us consider the commutative diagram in Fig. 1(i). Any tuple $\langle I_5, e, f, g, \beta, \gamma, \delta \rangle$ which makes diagram in Fig. 1(ii) commute and such that $\delta l \bullet g\beta \bullet \gamma t = \alpha$ is called a* candidate *for (i).*

(ii) *A G relative pushout (RPO) is the smallest such candidate, i.e. it satisfies the universal property that given any other candidate $\langle I_6, e', f', g', \beta', \gamma', \delta' \rangle$, there exists a mediating morphism given by a tuple $\langle h, \varphi, \psi, \tau \rangle$, with $\tau : g'h \Rightarrow g$, such that diagrams in Fig. 1(iii) commute. Moreover, the following identities on two cells need to be satisfied: $\gamma = \tau e \bullet g'\varphi \bullet \gamma'$, $\delta = \delta' \bullet g'\psi \bullet \tau^{-1}f$, $\beta' = \psi l \bullet h\beta \bullet \varphi t$. Such a mediating morphism must be unique, up to 2-cell isomorphisms.*

(iii) *A commuting square such as diagram in Fig 1(i) is a* G-idem pushout *(GIPO) if $\langle I_4, c, d, id_{I_4}, \alpha, 1_c, 1_d \rangle$ is its GRPO.*

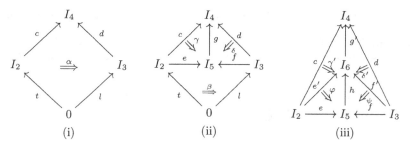

(i)   (ii)   (iii)

Fig. 1. Redex Square and Relative Pushout

Definition 7 (GIPO Transition System)

– *States: equivalence classes of arrows $[t] : 0 \to I$ in $\mathcal{C}$, for any I; two arrows are in the same equivalence class if there exists a 2-cell isomorphism between them;*

– *Transitions: $[t] \xrightarrow{[c]}_I [dr]$ iff $d \in \mathcal{D}$, $\langle l, r \rangle \in \mathbf{R}$ and the diagram in Fig. 1(i) is a GIPO.*

An important property of GIPO squares is that they are preserved by the substitution of one edge with a two 2-cell isomorphic one, [8]. It follows that the transition relation is independent from the chosen representative of an equivalence class. Let $\sim_I$ denote the bisimilarity induced by the GIPO LTS.

Another important property is the pasting property for GIPO squares.

Lemma 1 (GIPO pasting, [8]). *Suppose that the square in Fig. 2(i) has an GRPO and that both squares in Fig. 2(ii) commute.*

(i) *If the two squares of Fig. 2(ii) are GIPOs so is the outer rectangle.*

(ii) *It the outer rectangle and the left square of Fig. 2(ii) are GIPOs so is the right square.*

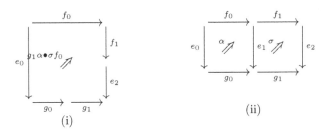

Fig. 2. IPO pasting

Definition 8 (Redex GRPO). *Let* **C** *be a G-reactive system and* $t : 0 \to I_2$ *an arrow in* $\mathcal{C}$. *A* redex square *is a diagram in the form of Fig. 1(i), with l the left-hand side of a reaction rule and d a reactive context. A G-reactive system* **C** *is said to* have redex GRPOs *if every redex square has a GRPO.*

The following fundamental theorem is provable using the GIPO pasting lemma:

Theorem 1. *Let* **C** *be a G-reactive system having redex GRPOs. Then the GIPO bisimilarity* $\sim_I$ *is a congruence w.r.t. all contexts, i.e. if* $a \sim_I b$ *then for all c of the appropriate type,* $ca \sim_I cb$.

3   Pruning the GIPO LTS

In this section we present a construction allowing to prune the LTS obtained by the GIPO construction. In this way it is possible to derive simpler and more usable LTS's. The key notion is that of *definability*. We will prove that in a GIPO LTS, the GIPO transitions that are "definable" in some suitable sense can be removed without affecting the bisimilarity induced by the LTS.

Definition 9. *Given a G-reactive system* $\mathbb{C}$, *having redex GRPOs, let T be a subset of the whole set of GIPO transitions,*

(i) *we say that T is* closed under bisimulation *if for any* $[t_1], [t'_1], [t_2], [t'_2], [f]$, *such that* $[t_1] \sim_I [t'_1]$, $[t_2] \sim_I [t'_2]$, $[t_1] \xrightarrow{[f]}_I [t_2]$, $[t'_1] \xrightarrow{[f]}_I [t'_2]$, *we have that:*
$[t_1] \xrightarrow{[f]}_I [t_2] \in T$ *iff* $[t'_1] \xrightarrow{[f]}_I [t'_2] \in T$

(ii) *we say that the whole GIPO LTS is* definable *from T if there exists a set of tuples* $\{ \langle [f_k], [f'_k], P_k, e_k \rangle | k \in K \}$ *of the following form:*
  - *$[f_k]$ GIPO label, $[f'_k]$ GIPO label or $f'_k = \epsilon$ with $f_k : I_k \to I'_k$, $f'_k : I_k \to J_k$ (where we set $\epsilon : I_k \to I_k$)*
  - *P_k is a Hennessy-Milner proposition with modal operators labeled by GIPO labels*
  - *$e_k : J_k \to I_k$ (with J_k possibly 0)*

and such that, in the whole GIPO LTS, there is a transition $[t] \xrightarrow{[f]}_I [t']$ *if and only if there exist* $k \in K$, $t'' : 0 \to J_k$ *satisfying:*
  - *$[f] = [f_k]$,*

- $([t] \xrightarrow{[f_k']}_I [t''] \in T)$ or $(t'' = t \wedge f_k' = \epsilon)$
- in the T LTS, the state $[t'']$ satisfies the proposition P_k
- $([t'] = [e_k(t'')] \wedge J_k \neq 0)$ or $([t'] = [e_k] \wedge J_k = 0)$

Remark 1. Intuitively a tuple $\langle [f_k], [f_k'], P_k, e_k \rangle$ says that some of the transitions with label $[f_k]$ can be simulated by transitions with label $[f_k']$ and contexts e_k. We allow the extra case of $f_k' = \epsilon$ to deal with those transitions that can be simulated by just inserting the original term in a contexts e_k, following [9] we can call *not engaged* these kind of transitions. The Hennessy-Milner propositions P_k are a sort of guard conditions. In the present work these propositions have no use: in all cases considered P_k is the proposition true. However, there are examples of calculi where the extra expressivity given by the Hennessy-Milner propositions is useful, and so we prefer to present the proposition below in its full strength.

Proposition 1. *Given a reactive system $\mathbb{C}$, and a subset T of transition that is closed under GIPO bisimulation and such that the whole GIPO LTS is definable from T, then $\sim_I = \sim_T$, i.e. the two GIPO LTS induce the same bisimilarity.*

Proof. Consider the relation $S = \{\langle [ct], [cu] \rangle \mid [t] \sim_T [u],\ c \text{ context}\}$. It is easy to prove that $\sim_I \subseteq \sim_T \subseteq S$. If we prove that S is an GIPO bisimilarity (*i.e.* $S \subseteq \sim_I$), then the three relations are equal. So we prove that, for any $\langle [ct], [cu] \rangle \in S$, if $[ct] \xrightarrow{[f]}_I [t']$, then there exists u' s.t. $[cu] \xrightarrow{[f]}_I [u']$ with $[t']S[u']$.

Consider the following diagram:

$$
\begin{array}{ccccc}
0 & \xrightarrow{\ t\ } & I_0 & \xrightarrow{\ c\ } & I_2 \\
{\scriptstyle l}\downarrow & {\scriptstyle \alpha}\nearrow\!\!\!\!\nearrow & \downarrow{\scriptstyle f'} & {\scriptstyle \beta}\nearrow\!\!\!\!\nearrow & \downarrow{\scriptstyle f} \\
I_3 & \xrightarrow{\ d\ } & I_1 & \xrightarrow{\ d'\ } & I_4
\end{array}
$$

where the outer rectangle is the GIPO inducing the transition $[ct] \xrightarrow{[f]}_I [t']$, namely $[t'] = [d'dr]$ with $\langle l, r \rangle$ reaction rule, and the left square is obtained from an GIPO construction starting from l and t. There are two cases to consider:

- If the transition labeled by $[f']$ is in T, then, since $[t] \xrightarrow{[f']}_I [dr]$, there exists u'', $[u] \xrightarrow{[f']}_I [u'']$, $[u''] \sim_T [dr]$. By composition of GRPO squares, $[cu] \xrightarrow{[f]}_I [d'u'']$, from which the thesis.
- If the transition labeled by $[f']$ is not in T, then it is definable by T, and since $[t] \xrightarrow{[f']}_I [dr]$, there exists a tuple $\langle [f'], [f_k], P_k, e_k \rangle$ and a term t'' such that $[t] \xrightarrow{[f_k]}_I [t'']$, $P_k([t'])$, and $[dr] = [e_k t'']$ (or $[dr] = [e_k]$). From the last equality it follows $[t'] = [d'dr] = [d'e_k t'']$ $(= [d'e_k])$. Since $[t] \sim_T [u]$, there exists u'', $[u] \xrightarrow{[f_k]}_I [u'']$, $[t''] \sim_T [u'']$ and so $P_k([u''])$ (Hennessy-Milner propositions cannot separate bisimilar elements). From this, $[u] \xrightarrow{[f']}_I [e_k u'']$ $(\xrightarrow{[f']}_I = [e_k])$. By composition of GRPO squares, $[cu] \xrightarrow{[f]}_I [d'e_k u'']$ $(\xrightarrow{[f]}_I [d'e_k])$, from which the thesis. $\qquad\square$

In using the above proposition for the CCS and π-calculus cases, we are not going to use the extra expressivity given by the Hennessy-Milner propositions P_k; in all the tuples $\langle [f_k], [f'_k], P_k, e_k \rangle$ defined in the following, the propositions P_k will be set equal to *true*. Nevertheless, we prefer here to present this general version of the proposition.

4 Applying the GRPO Technique to CCS

In this section, we study LTS's obtained by applying the GRPO technique to CCS. First, we consider the LTS obtained by applying Leifer-Milner theorem in the GRPO setting (Theorem 1 of Section 2). This turns out to be still infinitely branching. However, by applying our general pruning technique of Section 3, we are able to get a *finitely branching LTS* and *GIPO bisimilarity*, which coincide with the original Milner's LTS and strong bisimilarity, respectively.

The property that allows to apply the GRPO construction is the following:

Proposition 2. *The G-reactive system of finite CCS processes has redex GRPO.*

Proof. There are several cases to consider depending also on the reaction rule involved, here we consider only the reaction rule $r.P + M \mid \bar{r}.Q + N \rightarrow P|Q$. Given the commuting redex square $\alpha : C[\ ] \circ P \Rightarrow D[\ ] \circ L$, by structural rules, the reactive contexts $D[\ ]$ can be written in the form $\nu^n([\ ]|P_1|\ldots|P_h)$, while the context $C[\ ]$ can be written as $\nu^m C'[\varphi[\ ]]$, with $C'[\ ]$ not containing any name transformation φ', or hiding operator ν.

If the redex L is all contained (or better mapped by α^{-1}) in the process P, the GRPO has form $\alpha' : \varphi[\ ] \circ P \Rightarrow (\nu^m[\ ]|P_{i_1}|\ldots|P_{i_k}) \circ L$. Notice that the transformation $\varphi[\ ]$, cannot be factorized by the GRPO construction.

If the process P contains only one side of the redex L, the GRPO has form $\alpha' : (\varphi[\ ]|P') \circ P \Rightarrow (\nu^m[\ ]|P_{i_1}|\ldots|P_{i_k}) \circ L$, with the process P' giving the other side of the redex.

If the redex L is all contained in the context $C[\ ]$, the GRPO has form $\alpha' : C''[\varphi[\ ]] \circ P \Rightarrow (\nu^m[\ ]|P') \circ L$, with φP contained in P'.

In the remaining cases, where the main connectives of the redex L are contained in $C'[\ ]$, the GRPO has the form $\alpha' : C''[\varphi[\ ]] \circ P \Rightarrow (\nu^m[\ ]) \circ L$ □

Table 4 summarizes the set of GIPO contexts (up-to structural congruence) obtained by applying Theorem 1. For simplicity, we denote an equivalence class $[C[\ ]]$ by a special representative.

The second raw in Table 4 corresponds to the case where an internal transition of the process P is considered. In such case the GIPO context is empty. If the process P exposes a non-τ action (third raw), then a communication with a context exposing a complementary action can arise: the formula $\delta_0^k(\alpha) = \overline{\varphi_{+k}(\alpha_{i,j})}$ expresses that the actions are complementary. Finally, the last raw shows all GIPO contexts where the reduction is "all inside the context" (and the process plays a passive rôle).

Table 1. CCS GIPO contexts

Process $P \equiv \nu^k(\Sigma_{j=1}^{m_1}\alpha_{1,j}.P_{1,j} \mid \ldots \mid \Sigma_{j=1}^{m_n}\alpha_{n,j}.P_{n,j})$	**GIPO contexts**
$\exists i,j.\ \alpha_{i,j} = \tau \ \vee \ (\exists i,j,i',j'.\ \alpha_{i,j} = r_l \ \wedge \ \alpha_{i',j'} = \overline{r}_l)$	$\varphi[\,]$
$\exists i,j.\ \alpha_{i,j} \neq \tau$	$\varphi[\,] \ + \ M \mid \alpha.Q + N$
	with $\delta_0^k(\alpha) = \overline{\varphi_{+k}(\alpha_{i,j})}$
	$C[\,] \mid \alpha.Q + M \mid \overline{\alpha}.R + N$
	$C[\,] + \alpha.Q \mid \overline{\alpha}.R + N$
	$C[\,] + \tau.Q$
	$C[\,] \mid \tau.Q + N$
	$\alpha.C[\,] + M \mid \overline{\alpha}.Q + N$
	$\tau.C[\,] + N$

Notice that if we do not use G-categories and work on the category where arrows are equivalence classes of processes, the process $\alpha.0 \mid \overline{\alpha}.0$ has as only GIPO contexts the contexts $\varphi[\,]$ and the GIPO contexts $\varphi[\,] + M \mid \alpha.Q + N$ are missing.

The GIPO LTS described above is still infinitely branching. However, there are many GIPO contexts which are intuitively redundant, *e.g.* all the contexts in the last raw. These are not engaged, *i.e.* the reduction is all inside the context (and this is why they are not considered in the LTS of [2]). Also the class of contexts in the third raw is redundant; namely, the contexts of the shape $[\,]\mid\alpha.0$ are sufficient to define the whole class, in the sense of Definition 9 of Section 3. More precisely, we have:

Proposition 3
i) The GIPO LTS is definable from set of GIPO transitions labeled by $\{[\,]\} \cup \{[\,]\mid\alpha.0 \mid \alpha \in \mathcal{A}\}$.
ii) The bisimilarity induced by the LTS defined by such GIPO contexts (see Table 4) is a congruence.

Proof. i) Transitions corresponding to GIPO contexts of the shape $\varphi[\,]$ (second raw in Table 4) are definable by the tuple $\langle\varphi[\,],[\,],true,\varphi[\,]\rangle$. Transitions corresponding to GIPO contexts of the shape $\varphi[\,] + M \mid \alpha.Q + N$ (third raw in Table 4) are definable by the tuple $\langle\varphi[\,] + M \mid \alpha.Q + N,[\,]\mid\alpha'.0,true,\varphi[\,]\mid Q\rangle$, where $\delta_0^k(\alpha') = \overline{\alpha}_{ij}$. Transitions corresponding to the contexts $C[\,]$ in the last raw of Table 4 are definable by tuples of the shape $\langle C[\,],\epsilon,true,E[\,]\rangle$, where $E[\,]$ is a 0 or 1-holed context defined according to the following table:

GIPO context	**E**$[\,]$
$C[\,] \mid \alpha.Q + M \mid \overline{\alpha}.R + N$	$C[\,] \mid Q \mid R$
$C[\,] + \alpha.Q \mid \overline{\alpha}.R + N$	$Q \mid R$
$C[\,] + \tau.Q$	Q
$C[\,] \mid \tau.Q + N$	$C[\,] \mid Q$
$\alpha.C[\,] + M \mid \overline{\alpha}.Q + N$	$C[\,] \mid Q$
$\tau.C[\,] + N$	$C[\,]$

Table 2. CCS reduced GIPO contexts

Process $P \equiv \nu^k(\Sigma_{j=1}^{m_1}\alpha_{1,j}.P_{1,j} \mid \ldots \mid \Sigma_{j=1}^{m_n}\alpha_{n,j}.P_{n,j})$	GIPO contexts
$\exists i,j.\ \alpha_{i,j} = \tau\ \vee\ (\exists i,j,i',j'.\ \alpha_{i,j} = r_l\ \wedge\ \alpha_{i',j'} = \overline{r}_l)$	$[\,]$
$\exists i,j.\ \alpha_{i,j} \neq \tau$	$[\,] \mid \overline{\alpha}_{i,j}.0$

ii) The proof follows from Proposition 1. $\square$

Now, it is immediate to see that the above GIPO LTS coincides with the standard LTS; namely the GIPO context $[\,]$ corresponds to the τ-transition, while the GIPO context $[\,] \mid \alpha.0$ corresponds to a $\overline{\alpha}$-transition.

Summarizing, we have:

Proposition 4. *The reduced GIPO LTS coincides with the original LTS for CCS, and the GIPO bisimilarity coincides with CCS strong bisimilarity.*

5   The π-calculus Case

In this section, we apply the above machinery to π-calculus. The latter is significantly more difficult to deal with than CCS, because of name substitutions, which arise in the reaction semantics. We will show that the reduced GIPO LTS for π-calculus induces the *syntactical bisimilarity* of [4], which is finer than Sangiorgi's *open bisimilarity*. Our pruning technique does not give us directly a finitely branching LTS, however we will briefly discuss how a finitary GIPO LTS can be obtained by working in the setting of *categories of second-order term contexts* of [7].

We start by introducing the π-calculus syntax with de Brujin indexes.

Definition 10 (π-calculus Processes). *Let* $r_0, r_1, \ldots, s_0, s_1, \ldots \in \mathcal{NR}$ *be* name references, *and let* $x, y, z, \ldots \in \mathcal{X}$ *be process variables. We define*

$$(\mathcal{A}ct \ni) \ \alpha ::= \tau \mid r() \mid \overline{r}s \qquad \text{actions}$$
$$(\mathcal{G} \ni) \ M ::= 0 \mid \alpha.P \mid M_1 + M_2 \mid \alpha.x \qquad \text{guarded processes}$$
$$(\mathcal{P} \ni) \ P ::= M \mid \nu P \mid P_1|P_2 \mid rec\ x.P \mid \sigma P \qquad \text{processes}$$

where σ *is a substitution obtained as a finite composition of shifting operators* δ_i*'s, swapping operators* s_i*'s, and singleton substitutions* $t_{i,j}$*'s, defined by:*

$$t_{i,j}(r_k) = \begin{cases} r_k & \text{if } k \neq i \\ r_j & \text{if } k = i \end{cases}$$

A closed process *is a process in which each occurrence of a variable is in the scope of a rec operator.*

We denote by $dom(\sigma)$ the set of name references on which σ is not the identity, i.e. $\{r_i \mid \sigma(r_i) \neq r_i\}$. π-calculus contexts are defined similarly to CCS contexts. The structural congruence extended to contexts is defined as follows:

Definition 11 (Structural Congruence). *Let* $C[\,], C'[\,], C''[\,]$ *denote 0-holed or 1-holed contexts. The* structural congruence *is the relation* $\equiv$*, closed under process constructors, inductively generated by the following set of axioms:*

(**par**) $C[\,]|0 \equiv C[\,]$ $C[\,]|C'[\,] \equiv C'[\,]|C[\,]$
$C[\,]|(C'[\,]|C''[\,]) \equiv (C[\,]|C'[\,])|C''[\,]$

(**plus**) $C[\,]+0 \equiv C[\,]$ $C[\,]+C'[\,] \equiv C'[\,]+C[\,]$
$C[\,]+(C'[\,]+C''[\,]) \equiv (C[\,]+C'[\,])+C''[\,]$

(**rec**) $rec\ x.C[\,] \equiv C[\,][rec\ x.C[\,]/x]$

(**nu**) $\nu 0 \equiv 0$ $C[\,]|(\nu C'[\,]) \equiv \nu((\delta_0 C[\,])|C'[\,])$ $\nu\nu C[\,] \equiv \nu\nu s_0 C[\,]$

(**sigma**) $\sigma 0 \equiv 0$ $\sigma(\bar{r}s.C[\,]) \equiv \overline{\sigma(r)}\sigma(s).\sigma(C[\,])$
$\sigma(\tau.C[\,]) \equiv \tau.\sigma(C[\,])$ $\sigma(r().C[\,]) \equiv \sigma(r)().\sigma_{+1}C[\,]$
$\sigma(C[\,]|C'[\,]) \equiv \sigma(C[\,])|\sigma(C'[\,])$ $\sigma(rec\ x.C[\,]) \equiv rec\ x.(\sigma C[\,])$
$\sigma(C[\,]+C'[\,]) \equiv \sigma(C[\,])+\sigma(C'[\,])$ $\sigma(\nu C[\,]) \equiv \nu(\sigma_{+1}C[\,])$
$\sigma_1 \ldots \sigma_m[\,] \equiv \sigma'_1 \ldots \sigma'_n[\,]$, *if* $\sigma_1 \circ \ldots \circ \sigma_m = \sigma'_1 \circ \ldots \circ \sigma'_n$

Notice that, similarly to the CCS case, the last (**sigma**)-rule is effective, by definition of the substitutions σ_i, σ'_i.

As in the standard presentation, one can easily show that each π-calculus process P is structurally congruent to a process in *normal form*, *i.e.* a process of the shape $\nu^k(\Sigma_{j=1}^{m_1}\alpha_{1,j}.P_{1,j} \mid \ldots \mid \Sigma_{j=1}^{m_n}\alpha_{n,j}.P_{n,j})$, where all unguarded restrictions are at the top level, and substitutions do not appear at the top level.

Definition 12 (Reaction Semantics). *The* reaction relation $\rightarrow$ *is the least relation closed under the following reaction rules* and *reactive contexts:*

Reaction rules. $r().P + M \mid \bar{r}r_j.Q + N \rightarrow (\nu(t_{0,j+1}P))|Q$ $\tau.P + M \rightarrow P$

Reaction contexts. $D[\,] ::= [\,] \mid \nu D[\,] \mid P|D[\,] \mid D[\,]|P$

The above rule for communication may seem strange, but one can easily check that it is equivalent to the original one. It is motivated by the fact that, by using a ν operator in the resulting process, we avoid the introduction of operators for index decrementing, which would be problematic for the GRPO construction.

Table 3 summarizes the set of GIPO contexts (up-to structural congruence) obtained by applying Theorem 1. Table 4 summarizes the set of reduced GIPO contexts, which define the whole LTS, according to Definition 9 of Section 3.

Notice that, when the process exposes an output action and the context an input one, *i.e.* $C[\,] = \sigma[\,] + M \mid r'().Q+N$ (fifth raw in Table 3), we cannot get rid of Q in the reduced context (last raw of Table 4). This is because the transition provides a substitution for Q, depending on the process P (and hence the context $e()$ required in Definition 9 would not be uniform on all processes). Moreover, if σ acts also on $fr(\nu Q)$, then we cannot get rid of it, since otherwise it would appear in the context $e()$ and it would act also on names in Q, which we do not want. Therefore, the reduced GIPO LTS that we obtain, although significantly simpler than the original one, is still infinitely branching, since a process P,

Table 3. π-calculus GIPO contexts

Process $P \equiv \nu^k(\Sigma_{j=1}^{m_1}\alpha_{1,j}.P_{1,j} \mid \ldots \mid \Sigma_{j=1}^{m_n}\alpha_{n,j}.P_{n,j})$	GIPO contexts
$\exists i,j.\ \alpha_{i,j} = \tau$	$\sigma[\,]$
$\exists i,j,i',j'.\ i \neq i' \ \wedge\ \alpha_{i,j} = r_h() \ \wedge\ \alpha_{i',j'} = \overline{r}_l s$	$\sigma[\,]$ with $\sigma_{+k}(r_h) = \sigma_{+k}(r_l)$
$\exists i,j.\ \alpha_{i,j} = r()$	$\sigma[\,] + M \mid \overline{r}'s.Q + N$ with $\delta_0^k(r') = \sigma_{+k}(r)$
$\exists i,j.\ \alpha_{i,j} = \overline{r}s$	$\sigma[\,] + M \mid r'().Q + N$ with $\delta_0^k(r') = \sigma_{+k}(r)$
	$C[\,] \mid r().Q + M \mid \overline{r}s.R + N$ $C[\,] + r().Q \mid \overline{r}s.R + N$ $C[\,] + \tau.Q$ $C[\,] \mid \tau.Q + N$ $r().C[\,] + M \mid \overline{r}s.Q + N$ $\tau.C[\,] + N$

Table 4. π-calculus reduced GIPO contexts

Process $P \equiv \nu^k(\Sigma_{j=1}^{m_1}\alpha_{1,j}.P_{1,j} \mid \ldots \mid \Sigma_{j=1}^{m_n}\alpha_{n,j}.P_{n,j})$	Reduced GIPO Contexts
$\exists i,j.\ \alpha_{i,j} = \tau$	$[\,]$
$\exists i,j,i',j'.\ \alpha_{i,j} = r_h() \ \wedge\ \alpha_{i',j'} = \overline{r}_l s$	$\sigma[\,]$ with $\cdot\ \sigma$ identity, if $h = l$ $\cdot\ \sigma$ singleton, $\sigma_{+k}(r_h) = \sigma_{+k}(r_l)$, if $h \neq l$
$\exists i,j.\ \alpha_{i,j} = r()$	$[\,] \mid \overline{r}'s.0$ with $\delta_0^k(r') = r$
$\exists i,j.\ \alpha_{i,j} = \overline{r}s$	$\sigma[\,] \mid r'().Q$ with $dom(\sigma) \subseteq fr(\nu Q)$, $\delta_0^k(r') = \sigma_{+k}(r)$

which exposes an output action, makes infinitely many transitions $P \xrightarrow{\ \sigma[\,]\|r'().Q\ }$, for any Q. In Section 5.1, we will sketch how to overcome this problem getting a finitely branching characterization of the GIPO LTS and bisimilarity.

5.1   Finitely Branching LTS's for π-calculus

First of all, notice that in the context $[\,] \mid \overline{r}'s.0$ (Table 4, fourth raw), it is sufficient for the name s to range over the names in P together with a fresh name. Moreover, the substitution σ appearing in the GIPO context $\sigma[\,]\|r'().Q$ in the last raw of Table 4 is actually redundant, even if it cannot be eliminated using Proposition 1. However, by a direct reasoning, one can show that contexts of the shape $[\ ]|r'().Q$ are sufficient.

Moreover, one can show that the GIPO bisimilarity that we have obtained coincides with the extension to the whole π-calculus of the *syntactical bisimilarity* introduced in [4] for the open π-calculus. In the syntactical bisimilarity one essentially observes input/output actions and name fusions allowing for a communication. The prefix and the communication rules of the symbolic LTS in [4] are represented as follows, in our setting:

$$\text{(pre)} \quad \frac{P \xrightarrow{\bar{r}r_j} P' \quad Q \xrightarrow{r'()} Q'}{P \mid Q \xrightarrow{r=r'} P' \mid \nu(t_{0,j+1}Q')}$$

The notion of syntactical bisimilarity is as defined by:

Definition 13 (Syntactical Bisimilarity). *A symmetric relation $\mathcal{R}$ is a syntactical bisimulation if whenever $P\mathcal{R}Q$ it holds that:*

- *if $P \xrightarrow{\alpha} P'$ then $Q \xrightarrow{\alpha} Q'$,*
- *if $P \xrightarrow{r=r'} P'$ then $Q \xrightarrow{r=r'} Q'$ and $(\sigma P')\mathcal{R}(\sigma Q')$,*

where σ is a fusion that fuses r to r'.
The union of all syntactical bisimulations is syntactical bisimilarity.

Intuitively, our (reduced) GIPO LTS corresponds to the one of [4]. Notice in particular that, when P exposes an output action, in the LTS of [4] we have a transition $P \xrightarrow{\bar{r}s} P'$, where we recall both the output channel $\bar{r}$ and the object s in the label, while in our LTS we have $P \xrightarrow{[\ ]\mid r().Q} P' \mid \nu(\sigma Q)$, where we keep track of the object s not in the label, but in the substitution applied to Q. Summarizing, one can prove:

Theorem 2. *The GIPO bisimilarity on π-calculus coincides with the syntactical bisimilarity.*

As it has been observed in [4], the above bisimilarity is strictly included in Sangiorgi's open bisimilarity (where there is the extra freedom of matching a fusion transition of a process with a τ-transition of the other).

The LTS of [4] provides a finitely branching characterization of our GIPO bisimilarity. However, it is possible to get a more direct finitary characterization of the GIPO equivalence, by working in the setting of *categories of second-order term contexts*, introduced in [7]. In this setting one can represent *parametric transitions* such as $P \xrightarrow{[\ ]\mid r().X} P' \mid \nu(\sigma X)$, where X is a second-order variable representing a generic term, which will be possibly instantiated in the future (with the most general substitution allowing for a reaction). In this way we can avoid to have infinitely many ground transitions. We aim to present the whole construction in a further work.

6   Conclusions and Directions for Further Work

In this paper, we have refined Leifer-Milner construction for deriving LTS's from reactive systems, by studying general conditions under which we can prune the

GIPO LTS. Then we have carried out two fundamental case studies in process calculi, by working in categories of term contexts. In order to deal properly with structural rules, we had to work in the setting of G-categories. For CCS the result is quite satisfactory, since we have obtained as GIPO LTS exactly Milner's original LTS together with strong bisimilarity. There are other works in the literature, where the case study of CCS has been considered, but often a graph encoding is used and furthermore the original LTS is not directly obtained from the general construction, but it is recovered only a posteriori, using an ad hoc reasoning. A similar observation applies also to π-calculus, to which the RPO approach has not been previously applied directly to its reaction semantics, but to a (often ad hoc) enriched semantics. Under this perspective, it would be interesting to develop in all details also the π-calculus case study in the second-order setting, as hinted at the end of Section 5.1.

Finally, it would be interesting to compare our work with [10], where a LTS for the π-calculus is presented, whose labels are taken to be contexts on a higher order syntax. However, that work does not apply the Leifer-Milner technique to derive the LTS, and the higher-order syntax does not coincide with the one proposed in [7] for second-order contexts. It is not clear how the proposed LTS is related to the one obtained by the GIPO technique in the second-order setting.

References

1. Leifer, J.J., Milner, R.: Deriving bisimulation congruences for reactive systems. In: Palamidessi, C. (ed.) CONCUR 2000. LNCS, vol. 1877, pp. 243–258. Springer, Heidelberg (2000)
2. Bonchi, F., Gadducci, F., König, B.: Process bisimulation via a graphical encoding. In: Corradini, A., Ehrig, H., Montanari, U., Ribeiro, L., Rozenberg, G. (eds.) ICGT 2006. LNCS, vol. 4178, pp. 168–183. Springer, Heidelberg (2006)
3. Bonchi, F., König, B., Montanari, U.: Saturated semantics for reactive systems. In: LICS, pp. 69–80. IEEE Computer Society, Los Alamitos (2006)
4. Ferrari, G.L., Montanari, U., Tuosto, E.: Model checking for nominal calculi. In: Sassone, V. (ed.) FOSSACS 2005. LNCS, vol. 3441, pp. 1–24. Springer, Heidelberg (2005)
5. Gadducci, F., Montanari, U.: Observing reductions in nominal calculi via a graphical encoding of processes. In: Middeldorp, A., van Oostrom, V., van Raamsdonk, F., de Vrijer, R. (eds.) Processes, Terms and Cycles: Steps on the Road to Infinity. LNCS, vol. 3838, pp. 106–126. Springer, Heidelberg (2005)
6. Sassone, V., Sobocinski, P.: Deriving bisimulation congruences using 2-categories. Nord. J. Comput. 10(2), 163–190 (2003)
7. Di Gianantonio, P., Honsell, F., Lenisa, M.: RPO, second-order contexts, and lambda-calculus. In: Amadio, R.M. (ed.) FOSSACS 2008. LNCS, vol. 4962, pp. 334–349. Springer, Heidelberg (2008),
   http://www.dimi.uniud.it/pietro/papers/
8. Sobocinski, P.: Deriving process congruences from reduction rules. PhD thesis, University of Aarhus (2004)
9. Jensen, O.H., Milner, R.: Bigraphs and transitions. In: POPL, pp. 38–49. ACM, New York (2003)
10. Sobocinski, P.: A well-behaved lts for the pi-calculus (abstract). Electr. Notes Theor. Comput. Sci. 192(1), 5–11 (2007)

A Rewriting Logic Approach to Type Inference[*]

Chucky Ellison, Traian Florin Şerbănuţă, and Grigore Roşu

Department of Computer Science, University of Illinois at Urbana-Champaign
{celliso2,tserban2,grosu}@cs.uiuc.edu

Abstract. Meseguer and Roşu proposed rewriting logic semantics (RLS) as a programing language definitional framework that unifies operational and algebraic denotational semantics. RLS has already been used to define a series of didactic and real languages, but its benefits in connection with defining and reasoning about type systems have not been fully investigated. This paper shows how the same RLS style employed for giving formal definitions of languages can be used to define type systems. The same term-rewriting mechanism used to execute RLS language definitions can now be used to execute type systems, giving type checkers or type inferencers. The proposed approach is exemplified by defining the Hindley-Milner polymorphic type inferencer $\mathcal{W}$ as a rewrite logic theory and using this definition to obtain a type inferencer by executing it in a rewriting logic engine. The inferencer obtained this way compares favorably with other definitions or implementations of $\mathcal{W}$. The performance of the executable definition is within an order of magnitude of that of highly optimized implementations of type inferencers, such as that of OCaml.

1 Introduction

Meseguer and Roşu proposed rewriting logic as a semantic foundation for the definition and analysis of languages [1,2], as well as type systems and policy checkers for languages [2]. More precisely, they proposed rewriting integer values to their types and incrementally rewriting a program until it becomes a type or other desired abstract value. That idea was further explored by Roşu [3], but not used to define polymorphic type systems. Also, no implementation, no proofs, and no empirical evaluation of the idea were provided. A similar idea has been recently proposed by Kuan, MacQueen, and Findler [4] in the context of Felleisen et al.'s reduction semantics with evaluation contexts [5,6] and Matthews et al.'s PLT Redex system [7].

In this paper we show how the same rewriting logic semantics (RLS) framework and definitional style employed in giving formal semantics to languages can be used to also define type systems as rewrite logic theories. This way, both the language and its type system(s) can be defined using the same formalism, facilitating reasoning about programs, languages, and type systems.

---

[*] Supported in part by NSF grants CCF-0448501, CNS-0509321 and CNS-0720512, by NASA contract NNL08AA23C, by the Microsoft/Intel funded Universal Parallel Computing Research Center at UIUC, and by several Microsoft gifts.

A. Corradini and U. Montanari (Eds.): WADT 2008, LNCS 5486, pp. 135–151, 2009.
© Springer-Verlag Berlin Heidelberg 2009

We use the Hindley-Milner polymorphic type inferencer $\mathcal{W}$ [8] for Milner's Exp language to exemplify our technique. We give one rewrite logic theory for $\mathcal{W}$ and use it to obtain an efficient, executable type-inferencer.

Our definitional style gains modularity by specifying the minimum amount of information needed for a transition to occur, and compositionality by using *strictness* attributes associated with the language constructs. These allow us, for example, to have the rule for function application corresponding to the one in $\mathcal{W}$ look as follows (assuming application was declared strict in both arguments):

$$\frac{\langle\!\langle t_1\ t_2 \rangle\!\rangle_k\ \langle\!\langle \underline{\qquad\cdot\qquad} \rangle\!\rangle\, eqns}{tvar\qquad t_1 = t_2 \rightarrow tvar} \qquad \text{where } tvar \text{ is a fresh type variable}$$

which reads as follows: once all constraints for both sides of an application construct are gathered, the application of t_1 to t_2 will have a new type, $tvar$, with the additional constraint that t_1 is the function type $t_2 \rightarrow tvar$.

This paper makes two novel contributions:

1. It shows how non-trivial type systems are defined as RLS theories, following the same style used for defining languages and other formal analyses;
2. It shows that RLS definitions of type systems, when executed on existing rewrite engines, yield competitive type inferencers.

Related work. In addition to the work mentioned above, there has been other previous work combining term rewriting with type systems. For example, the Stratego reference manual [9] describes a method of using rewriting to add type-check notations to a program. Also, pure type systems, which are a generalization of the λ-cube [10], have been represented in membership equational logic [11], a subset of rewriting logic. There is a large body on term graph rewriting [12,13] and its applications to type systems [14,15]. There are similarities with our work, such as using a similar syntax for both types and terms, and a process of reduction or normalization to reduce programs to their types. A collection of theoretical papers on type theory and term rewriting can be found in [16]. Adding rewrite rules as annotations to a particular language in order to assist a separate algorithm with type checking has been explored [17], as well as adding type annotations to rewrite rules that define program transformations [18]. Much work has been done on defining type systems modularly [19,20,21,22,23]. The style we propose in this paper is different from previous approaches combining term rewriting with type systems. Specifically, we use an executable definitional style within rewriting logic semantics, called K [3,24]. The use of K makes the defined type inferencers easy to read and understand, as well as efficient when executed.

Section 2 introduces RLS and the K definitional style, and gives an RLS definition of Milner's Exp language. Section 3 defines the Hindley-Milner $\mathcal{W}$ algorithm as an RLS theory and reports on some experiments. Section 4 shows that our RLS definition faithfully captures the Hindley-Milner algorithm, and gives a summary of preservation results. Section 5 concludes the paper.

2   Rewriting Semantics

This section recalls the RLS project, then presents the K technique for designing programming languages. We briefly discuss these, and show how Milner's Exp language [8] can be defined as an RLS theory using K. The rest of the paper employs the same technique to define type systems as RLS theories.

2.1   Rewriting Logic

Term rewriting is a standard computational model supported by many systems. Meseguer's rewriting logic [25], not to be confused with term rewriting, organizes term rewriting modulo equations as a logic with a complete proof system and initial model semantics. Meseguer and Roşu's RLS [1,2] seeks to make rewriting logic a foundation for programming language semantics and analysis that unifies operational and algebraic denotational semantics.

In contrast to term rewriting, which is just a method of computation, rewriting logic is a computational logic built upon equational logic, proposed by Meseguer [25] as a logic for true concurrency. In equational logic, a number of *sorts* (types) and *equations* are defined, specifying which terms are to be considered equivalent. Rewriting logic adds *rules* to equational logic, thought of as irreversible transitions: a rewrite theory is an equational theory extended with rewrite rules. Rewriting logic admits a complete proof system and an initial model semantics [25] that makes inductive proofs valid. Rewriting logic is connected to term rewriting in that all the equations $l = r$ can be transformed into term rewriting rules $l \rightarrow r$. This provides a means of taking a rewriting logic theory, together with an initial term, and "executing" it. Any of the existing rewrite engines can be used for this purpose. Some of the engines, e.g., Maude [26], provide even richer support than execution, such as an inductive theorem prover, a state space exploration tool, a model checker, and more.

RLS builds upon the observation that programming languages can be defined as rewrite logic theories. By doing so, one gets "for free" not only an interpreter and an initial model semantics for the defined language, but also a series of formal analysis tools obtained as instances of existing tools for rewriting logic. Operationally speaking, the major difference between conventional reduction semantics, with [5] or without [27] evaluation contexts, and RLS is that the former typically impose contextual restrictions on applications of reduction steps and the reduction steps happen one at a time, while the latter imposes no such restrictions. To avoid undesired applications of rewrite steps, one has to obey certain methodologies when using rewriting logic. In particular, one can capture the conventional definitional styles by appropriate uses of conditional rules. Consequently, one can define a language many different ways in rewriting logic. In this paper, we use Roşu's K technique [3], which is inspired by abstract state machines [28] and continuations [29], and which glosses over many rewriting logic details that are irrelevant for programming languages. Roşu's K language definitional style optimizes the use of RLS by means of a definitional technique and a specialized notation.

2.2 K

The idea underlying K is to represent the program configuration as a nested "soup" (multiset) of *configuration item* terms, also called *configuration cells*, representing the current infrastructure needed to process the remaining program or fragment of program; these may include the current computation (a continuation-like structure), environment, store, remaining input, output, analysis results, bookkeeping information, etc. The set of configuration cells is not fixed and is typically different from definition to definition. K assumes lists, sets and multisets over any sort whenever needed; for a sort S, List$[S]$ denotes comma-separated lists of terms of sort S, and Set$[S]$ denotes white space separated sets of terms of sort S. For both lists and sets, we use "·" as unit (nil, empty, etc.). To use a particular list- or set-separator, one writes it as an index; for example, List$_\curvearrowright[S]$ stands for $\curvearrowright$-separated lists of terms of sort S. Lists and sets admit straightforward equational definitions in rewriting logic (a list is an associative binary operator, while a set is an associative, commutative, and idempotent binary operator). Formally, configurations have the following structure:

$$ConfigLabel ::= \top \mid k \mid env \mid store \mid ...$$
$$\qquad\qquad \text{(descriptive names; first two common, rest differ with language)}$$
$$Config ::= [\![K]\!] \mid ... \mid (\![S]\!)_{ConfigLabel}$$
$$\qquad\qquad (S \text{ can be any sort, including Set}[Config])$$

The advantage of representing configurations as nested "soups" is that language rules only need to mention applicable configuration cells. This is one aspect of K's modularity. We can add or remove elements from the configuration set as we like, only impacting rules that use those particular items. Rules do not need to be changed to match what the new configuration looks like.

Almost all definitions share the configuration labels $\top$ (which stands for "*top*") and k (which stands for "*current computation*"). The remaining configuration labels typically differ with the language or analysis technique to be defined in K. A configuration $(\![c]\!)_l$ may also be called a *configuration item* (or *cell*) *named* (or *labeled*) l; interesting configuration cells are the nested ones, namely those where $c \in$ Set$[Config]$. One is also allowed to define some language-specific configuration constructs, to more elegantly intialize and terminate computations. A common such additional configuration construct is $[\![p]\!]$, which takes the given program to an initial configuration. An equation therefore needs to be given, taking such special initializing configuration into an actual configuration cell; in most definitions, an equation identifies a term of the form $[\![p]\!]$ with one of the form $(\![(\![p]\!)_k...]\!)_\top$, for some appropriate configuration items replacing the dots. If one's goal is to give a dynamic semantics of a language and if p is some terminating program, then $[\![p]\!]$ eventually produces (after a series of rewrites) the result of evaluating p; if one's goal is to analyze p, for example to type check it, then $[\![p]\!]$ eventually rewrites to the result of the analysis, for example a type when p is well-typed or an error term when p is not well-typed.

The most important configuration item, present in all K definitions and "wrapped" with the *ConfigLabel k*, is the *computation*, denoted by K. Computations generalize abstract syntax by adding a special list construct (associative operator) $\_ \curvearrowright \_$:

$$K ::= KResult \mid KLabel(\mathsf{List}[K]) \mid \mathsf{List}_{\curvearrowright}[K]$$
$$KResult ::= \text{(finished computations, e.g., values, or types, etc.)}$$
$$KLabel ::= \text{(one per language construct)}$$

The construct $KLabel(\mathsf{List}[K])$ captures any programming language syntax, under the form of an abstract syntax tree. If one wants more K syntactic categories, then one can do that, too, but we prefer to keep only one here. In our Maude implementation, thanks to Maude's mixfix notation for syntax, we write, e.g., "if b then s_1 else s_2" in our definitions instead of "if_then_else_(b, s_1, s_2)".

The distinctive K feature is $\_ \curvearrowright \_$. Intuitively, $k_1 \curvearrowright k_2$ says "process k_1 then k_2". How this is used and what the meaning of "process" is left open and depends upon the particular definition. For example, in a concrete semantic language definition it can mean "evaluate k_1 then k_2", while in a type inferencer definition it can mean "type and accumulate type constraints in k_1 then in k_2". A K definition consists of two types of sentences: structural equations and rewrite rules. Structural equations carry no computational meaning; they only say which terms should be viewed as identical and their role is to transparently modify the term so that rewrite rules can apply. Rewrite rules are seen as irreversible computational steps and can happen concurrently on a match-and-apply basis. The following are examples of structural equations:

$$a_1 + a_2 = a_1 \curvearrowright \square + a_2$$
$$a_1 + a_2 = a_2 \curvearrowright a_1 + \square$$
$$\text{if } b \text{ then } s_1 \text{ else } s_2 = b \curvearrowright \text{if } \square \text{ then } s_1 \text{ else } s_2$$

Note that, unlike in evaluation contexts, $\square$ is not a "hole," but rather part of a *KLabel*, carrying the obvious "plug" intuition; e.g., the *KLabels* involving $\square$ above are $\square + \_$, $\_ + \square$, and if $\square$ then_else_, respectively. To avoid writing such obvious, distracting, and mechanical structural equations, the language syntax can be annotated with *strict* attributes when defining language constructs: a *strict* construct is associated with an equation as above for each of its subexpressions. If an operator is intended to be strict in only some of its arguments, then the positions of the strict arguments are listed as arguments of the *strict* attribute; for example, the above three equations correspond to the attributes *strict* for $\_ + \_$ and *strict*(1) for if_then_else_. All these structural equations are automatically generated from strictness attributes in our implementation.

Structural equations can be applied back and forth; for example, the first equation for $\_ + \_$ can be applied left-to-right to "schedule" a_1 for processing; once evaluated to i_1, the equation is applied in reverse to "plug" the result back in context, then a_2 is scheduled with the second equation left-to-right, then its result i_2 plugged back into context, and then finally the rewrite rule can apply the irreversible computational step. Special care must be taken so that side effects are propagated appropriately: they are only generated at the leftmost side of the computation.

The following are examples of rewrite rules:

$$i_1 + i_2 \rightarrow i, \text{ where } i \text{ is the sum of } i_1 \text{ and } i_2$$
$$\text{if true then } s_1 \text{ else } s_2 \rightarrow s_1$$
$$\text{if false then } s_1 \text{ else } s_2 \rightarrow s_2$$

In contrast to structural equations, rewrite rules can only be applied left to right.

2.3 A Concrete Example: Milner's Exp Language

Milner proved the correctness of his $\mathcal{W}$ type inferencer in the context of a simple higher-order language that he called Exp [8]. Recall that $\mathcal{W}$ is the basis for the type checkers of all statically typed functional languages.

Exp is a simple expression language containing lambda abstraction and application, conditional, fix point, and "let" and "letrec" binders. To exemplify K and also to remind the reader of Milner's Exp language, we next define it using K. Figure 1 shows its K annotated syntax and Figure 2 shows its K semantics. We also use this to point out some other K notational conventions. Note that application is strict in both its arguments (call-by-value) and that let and letrec are desugared. Additionally, syntactic constructs may be annotated with desugaring equations. In Figure 2, we see that λ-abstractions are defined as values, which are also *KResults* in this definition; *KResults* are not further "scheduled" for processing in structural equations. Since Exp is simple, there is only one *ConfigItem* needed, wrapped by *ConfigLabel k*. The first two equations initialize

$$
\begin{aligned}
Var ::=&\ \text{standard identifiers} \\
Exp ::=&\ Var \mid \dots \text{ add basic values (Bools, ints, etc.)} \\
\mid&\ \lambda\, Var.\, Exp \\
\mid&\ Exp\ Exp && [strict] \\
\mid&\ \mu\, Var.\, Exp \\
\mid&\ \text{if } Exp \text{ then } Exp \text{ else } Exp && [strict(1)] \\
\mid&\ \text{let } Var = Exp \text{ in } Exp && [(\text{let } x = e \text{ in } e') = ((\lambda x.e')\, e)] \\
\mid&\ \text{letrec } Var\ Var = Exp \text{ in } Exp && [(\text{letrec } f\ x = e \text{ in } e') = (\text{let } f = \mu f.(\lambda x.e) \text{ in } e')]
\end{aligned}
$$

Fig. 1. K-Annotated Syntax of Exp

$$Val ::= \lambda\, Var.\, Exp \mid \dots (\text{Bools, ints, etc.})$$
$$KResult ::= Val$$
$$Config ::= Val \mid [\![K]\!] \mid (\!|K|\!)_k$$

$$[\![e]\!] = (\!|e|\!)_k$$
$$(\!|v|\!)_k = v$$
$$\frac{(\!|\ (\lambda x.e)\ v\ |\!)_k}{e[x \leftarrow v]}$$
$$\frac{(\!|\quad \mu\ x.e\quad |\!)_k}{e[x \leftarrow \mu\ x.e]}$$
$$\text{if true then } e_1 \text{ else } e_2 \rightarrow e_1$$
$$\text{if false then } e_1 \text{ else } e_2 \rightarrow e_2$$

Fig. 2. K Configuration and Semantics of Exp

and terminate the computation process. The third applies the β-reduction when $(\lambda x.e)\, v$ is the first item in the computation; we here use two other pieces of K notation: list/set fragment matching and the two-dimensional writing for rules. The first allows us to use angle brackets for unimportant fragments of lists or sets; for example, $\langle\!(T\rangle$ matches a list whose prefix is T, $\langle T\rangle\!)$ matches a list whose suffix is T, and $\langle T\rangle$ matches a list containing a contiguous fragment T; same for sets, but the three have the same meaning there. Therefore, special parentheses $\langle\!($ and $\rangle\!)$ represent respective ends of a list/set, while angled variants mean "the rest." The second allows us to avoid repetition of contexts; for example, instead of writing a rule of the form $C[t_1, t_2, ..., t_n] \rightarrow C[t'_1, t'_2, ..., t'_n]$ (rewriting the above-mentioned subterms in context C) listing the context (which can be quite large) C twice, we can write it $C[\underset{t'_1}{t_1}, \underset{t'_2}{t_2}, ..., \underset{t'_n}{t_n}]$ with the obvious meaning, mentioning the context only once. The remaining Exp semantics is straightforward. Note that we used the conventional substitution, which is also provided in our Maude implementation.

The Exp syntax and semantics defined in Figures 1 and 2 is all we need to write in our implementation of K. To test the semantics, one can now execute programs against the obtained interpreter.

3   Defining Milner's $\mathcal{W}$ Type Inferencer

We next define Milner's $\mathcal{W}$ type inferencer [8] using the same K approach. Figure 3 shows the new K annotated syntax for $\mathcal{W}$; it changes the conditional to be strict in all arguments, makes let strict in its second argument, and desugars letrec to let (because let is typed differently than its Exp desugaring). Unification over type expressions is needed and defined in Figure 4 (with $t_v \in \mathit{TypeVar}$). Fortunately, unification is straightforward to define equationally using set matching; we define it using rewrite rules, though, to emphasize that it is executable. Our definition is equivalent to the nondeterministic unification algorithm by Martelli and Montanari [30, Algorithm 1], instantiated to types and type variables, and with a particular rule evaluation order. [30, Theorem 2.3] provides a proof of correctness of the strategy. We implement their multi-set of equations by collecting equations and using associative and commutative (AC) matching. Finally, we should note that our substitution is kept canonical and is calculated as we go.

$Var ::=$ standard identifiers
$Exp ::= Var \mid ...$ add basic values (Bools, ints, etc.)
$\quad\mid \ \lambda\, Var.\, Exp$
$\quad\mid \ Exp\; Exp$ $[strict]$
$\quad\mid \ \mu\, Var.\, Exp$
$\quad\mid \ \mathsf{if}\; Exp \;\mathsf{then}\; Exp \;\mathsf{else}\; Exp$ $[strict]$
$\quad\mid \ \mathsf{let}\; Var = Exp \;\mathsf{in}\; Exp$ $[strict(2)]$
$\quad\mid \ \mathsf{letrec}\; Var\; Var = Exp \;\mathsf{in}\; Exp \;\; [(\mathsf{letrec}\; f\; x = e \;\mathsf{in}\; e') = (\mathsf{let}\; f = \mu f.(\lambda x.e) \;\mathsf{in}\; e')]$

Fig. 3. K-Annotated Syntax of Exp for $\mathcal{W}$

$$Type ::= \ldots \mid int \mid bool \mid Type \mapsto Type \mid Type\,Var$$
$$Eqn ::= Type = Type$$
$$Eqns ::= \mathsf{Set}[Eqn]$$

$$(t = t) \to \cdot$$
$$(t_1 \mapsto t_2 = t_1' \mapsto t_2') \to (t_1 = t_1'),\ (t_2 = t_2')$$
$$(t = t_v) \to (t_v = t) \quad \text{when } t \notin Type\,Var$$
$$t_v = t,\ t_v = t' \to t_v = t,\ t = t' \quad \text{when } t, t' \neq t_v$$
$$t_v = t,\ t_v' = t' \to t_v = t,\ t_v' = t'[t_v \leftarrow t]$$
$$\quad \text{when } t_v \neq t_v',\ t_v \neq t,\ t_v' \neq t',\ \text{and } t_v \in vars(t')$$

Fig. 4. Unification

The first rule in Figure 4 eliminates non-informative type equalities. The second distributes equalities over function types to equalities over their sources and their targets; the third swaps type equalities for convenience (to always have type variables as lhs's of equalities); the fourth ensures that, eventually, no two type equalities have the same lhs variable; finally, the fifth rule canonizes the substitution. As expected, these rules take a set of type equalities and eventually produce a most general unifier for them:

Theorem 1. *Let $\gamma \in Eqns$ be a set of type equations. Then:*

- *The five-rule rewrite system above terminates (modulo AC); let $\theta \in Eqns$ be the normal form of γ.*
- *γ is unifiable iff θ contains only pairs of the form $t_v = t$, where $t_v \notin vars(t)$; if that is the case, then we identify θ with the implicit substitution that it comprises, that is, $\theta(t_v) = t$ when there is some type equality $t_v = t$ in θ, and $\theta(t_v) = t_v$ when there is no type equality of the form $t_v = t$ in θ.*
- *If γ is unifiable then θ is idempotent (i.e., $\theta \circ \theta = \theta$) and is a most general unifier of γ.*

Therefore, the five rules above give us a rewriting procedure for unification. The structure of θ in the second item above may be expensive to check every time the unification procedure is invoked; in our Maude implementation of the rules above, we sub-sort (once and for all) each equality of the form $t_v = t$ with $t_v \notin vars(t)$ to a "proper" equality, and then allow only proper equalities in the sort $Eqns$ (the improper ones remain part of the "kind" $[Eqns]$). If $\gamma \in Eqns$ is a set of type equations and $t \in Type$ is some type expression, then we let $\gamma[t]$ denote $\theta(t)$; if γ is not unifiable, then $\gamma[t]$ is some error term (in the kind $[Type]$).

Figure 5 shows the K definition of $\mathcal{W}$. The configuration has four items: the computation, the type environment, the set of type equations (constraints), and a counter for generating fresh type variables. Due to the strictness attributes, we can assume that the corresponding arguments of the language constructs (in which these constructs were defined strict) have already been "evaluated" to their

$$KResult ::= Type$$
$$TEnv ::= \mathsf{Map}[Name, Type]$$
$$Type ::= \ldots \mid let(\,Type)$$
$$ConfigLabel ::= k \mid tenv \mid eqns \mid nextType$$
$$Config ::= Type \mid [\![K]\!] \mid (\!\langle S \rangle\!)_{ConfigLabel} \mid (\!\langle \mathsf{Set}[ConfigItem]\rangle\!)_\top$$
$$K ::= \ldots \mid Type \to K \quad [strict(2)]$$

$$[\![e]\!] = (\!\langle (\!\langle e \rangle\!)_k\ (\!\langle \cdot \rangle\!)_{tenv}\ (\!\langle \cdot \rangle\!)_{eqns}\ (\!\langle t_0 \rangle\!)_{nextType} \rangle\!)_\top$$
$$(\!\langle (\!\langle t \rangle\!)_k\ (\!\langle \gamma \rangle\!)_{eqns} \rangle\!)_\top = \gamma[t]$$
$$i \to int, \quad true \to bool, \quad false \to bool$$

$$(\!\langle t_1 + t_2 \rangle\!)_k\ (\!\langle \underline{\quad \cdot \quad} \rangle\!)_{eqns}$$
$$\overline{int}\overline{t_1 = int,\ t_2 = int}$$

$$(\!\langle \underline{\quad x \quad} \rangle\!)_k\ (\!\langle \eta \rangle\!)_{tenv}\ (\!\langle \gamma \rangle\!)_{eqns}\ (\!\langle \underline{\ t_v\ } \rangle\!)_{nextType} \quad \text{when } \eta[x] = let(t),$$
$$\overline{(\gamma[t])[tl \leftarrow tl']}\overline{t_v + |tl|} \qquad\qquad\quad tl = vars(\gamma[t]) - vars(\eta),$$
$$\text{and } tl' = t_v \ldots (t_v + |tl| - 1)$$

$$(\!\langle \underline{\ x\ } \rangle\!)_k\ (\!\langle \eta \rangle\!)_{tenv} \quad \text{when } \eta[x] \neq let(t)$$
$$\overline{\eta[x]}$$

$$(\!\langle \underline{\qquad \lambda x.e \qquad} \rangle\!)_k\ (\!\langle \underline{\quad \eta \quad} \rangle\!)_{tenv}\ (\!\langle \underline{\ t_v\ } \rangle\!)_{nextType}$$
$$\overline{(t_v \to e) \curvearrowright restore(\eta)}\overline{\eta[x \leftarrow t_v]}\overline{t_v + 1}$$

$$(\!\langle t_1\ t_2 \rangle\!)_k\ (\!\langle \underline{\quad \cdot \quad} \rangle\!)_{eqns}\ (\!\langle \underline{\ t_v\ } \rangle\!)_{nextType}$$
$$\overline{t_v}\overline{t_1 = t_2 \to t_v}\overline{t_v + 1}$$

$$(\!\langle \underline{\qquad \mu x.e \qquad} \rangle\!)_k\ (\!\langle \underline{\quad \eta \quad} \rangle\!)_{tenv}\ (\!\langle \underline{\ t_v\ } \rangle\!)_{nextType}$$
$$\overline{e \curvearrowright ?_{=}(t_v) \curvearrowright restore(\eta)}\overline{\eta[x \leftarrow t_v]}\overline{t_v + 1}$$

$$(\!\langle \underline{t \to ?_{=}t_v} \rangle\!)_k\ (\!\langle \underline{\quad \cdot \quad} \rangle\!)_{eqns}$$
$$\overline{\cdot}\overline{t_v = t}$$

$$(\!\langle \underline{\ \mathsf{let}\ x = t\ \mathsf{in}\ e\ } \rangle\!)_k\ (\!\langle \underline{\quad \eta \quad} \rangle\!)_{tenv}$$
$$\overline{e \curvearrowright restore(\eta)}\overline{\eta[x \leftarrow let(t)]}$$

$$(\!\langle \mathsf{if}\ t\ \mathsf{then}\ t_1\ \mathsf{else}\ t_2 \rangle\!)_k\ (\!\langle \underline{\quad \cdot \quad} \rangle\!)_{eqns}$$
$$\overline{t_1}\overline{t = bool,\ t_1 = t_2}$$

$$(\!\langle \underline{restore(\eta)} \rangle\!)_k\ (\!\langle \eta' \rangle\!)_{tenv}$$
$$\overline{\cdot}\overline{\eta}$$

Fig. 5. K Configuration and Semantics of $\mathcal{W}$

types and the corresponding type constraints have been propagated. Lambda and fix-point abstractions perform normal bindings in the type environment, while the let performs a special binding, namely one to a type wrapped with a new "*let*" type construct. When names are looked up in the type environment, the "*let*" types are instantiated with fresh type variables for their "universal" type variables, namely those that do not occur in the type environment.

We believe that the K definition of $\mathcal{W}$ is as simple to understand as the original $\mathcal{W}$ procedure proposed [8] by Milner, once the reader has an understanding of the K notation. However, note that Milner's procedure is an *algorithm*, rather than a formal definition. The K definition above is an ordinary rewriting logic theory—the same as the definition of Exp itself. That does not mean that our K definition, when executed, must be slower than an actual implementation of

Table 1. Speed of various $\mathscr{W}$ implementations

System	Average Speed	Stress test				
		n = 10	n = 11	n = 12	n = 13	n = 14
OCaml	0.6s	0.6s	2.1s	7.9s	30.6s	120.5s
Haskell	1.2s	0.5s	0.9s	1.5s	2.5s	5.8s
SML/NJ	4.0s	5.1s	14.6s	110.2s	721.9s	-
$\mathscr{W}$ in K	1.1s	2.6s	7.8s	26.9s	103.1s	373.2s
!$\mathscr{W}$ in K	2.0s	2.6s	7.7s	26.1s	96.4s	360.4s
$\mathscr{W}$ in PLT/Redex	134.8s	>1h	-	-	-	-
$\mathscr{W}$ in OCaml	49.8s	105.9s	9m14	>1h	-	-

$\mathscr{W}$. Experiments using Maude (see [31] for the complete Maude definition) show that our K definition of $\mathscr{W}$ is comparable to state of the art implementations of type inferencers in conventional functional languages: in our experiments, it was only about twice slower on average than that of OCaml, and had average times comparable, or even better than Haskell ghci and SML/NJ.

We have tested type inferencers both under normal operating conditions and under stressful conditions. For normal operating conditions, we have used small programs such as factorial computation together with small erroneous programs such as $\lambda x.(xx)$. We have built a collection of 10 such small programs and type-checked each of them 1,000 times. The results in Table 1 (average speed column) show the average time in seconds in which the type inferencer can check the type of a program. For the "stress test" we have used a program for which type inferencing is known to be exponential in the size of the input. Concretely, the program (which is polymorphic in $2^n + 1$ type variables!):

$$\text{let } f_0 = \lambda x.\lambda y.x \text{ in}$$
$$\text{let } f_1 = \lambda x.f_0(f_0 x) \text{ in}$$
$$\text{let } f_2 = \lambda x.f_1(f_1 x) \text{ in}$$
$$\cdots$$
$$\text{let } f_n = \lambda x.f_{n-1}(f_{n-1} x) \text{ in } f_n$$

takes the time shown in columns 3–7 to be type checked using OCaml (version 3.10.1), Haskell (ghci version 6.8.2), SML/NJ (version 110.67), our K definition executed in Maude (version 2.3), the PLT-Redex definition of the $\mathscr{W}$ procedure [4], and an "off-the-shelf" implementation of $\mathscr{W}$ using OCaml [32].

These experiments have been conducted on a 3.4GHz/2GB Linux machine. All the tests performed were already in normal form, so no evaluation was necessary (other than type checking and compiling). For OCaml we have used the type mode of the Enhanced OCaml Toplevel [33] to only enable type checking. For Haskell we have used the type directive ":t". For SML the table presents the entire compilation time since we did not find a satisfactory way to only obtain typing time. Only the user time has been recorded. Except for SML, the user time was very close to the real time; for SML, the real time was 30% larger than the user time. Moreover, an extension to $\mathcal{W}$, which we call $!\mathcal{W}$ (see Figures 6 and 7), containing lists, products, side effects (through referencing, dereferencing, and assignment) and weak polymorphism did not add any significant slowdown. Therefore, our K *definitions* yield quite realistic *implementations* of type checkers/inferencers when executed on an efficient rewrite engine.

$$Exp ::= ... \mid \mathsf{ref}\ Exp \mid \&\ Var \mid !\ Exp \mid Exp := Exp \mid [ExpList]$$
$$\mid\ \mathsf{car}\ Exp \mid \mathsf{cdr}\ Exp \mid \mathsf{null?}\ Exp \mid \mathsf{cons}\ Exp\ Exp \mid Exp\ ;\ Exp$$

$$KResult ::= Type$$
$$TEnv ::= \mathsf{Map}[Name, Type]$$
$$ConfigLabel ::= k \mid tenv \mid eqns \mid nextType \mid results \mid mkLet$$
$$Config ::= Type \mid [\![K]\!] \mid (\!(S)\!)_{ConfigLabel} \mid (\!(\mathsf{Set}[Config])\!)_\top$$

Fig. 6. The $!\mathcal{W}$ type inferencer, Syntax & Configuration

Our Maude "implementation" of an extension[1] to the K definition of $\mathcal{W}$ has about 30 lines of code. How is it possible that a formal definition of a type system, written in 30 lines of code, can be executed *as is* with comparable efficiency to well-engineered implementations of the same type system in widely used programming languages? We think that the answer to this question involves at least two aspects. On one hand, Maude, despite its generality, is a well-engineered rewrite engine implementing state-of-the-art AC matching and term indexing algorithms [34]. On the other hand, our K definition makes intensive use of what Maude is very good at, namely AC matching. For example, note the fourth rule in Figure 4: the type variable t_v appears twice in the lhs of the rule, once in each of the two type equalities involved. Maude will therefore need to search and then index for two type equalities in the set of type constraints which share the same type variable. Similarly, the fifth rule involves two type equalities, the second containing in its t' some occurrence of the type variable t_v that appears in the first. Without appropriate indexing to avoid rematching of rules, which is what Maude does well, such operations can be very expensive. Moreover, note that our type constraints can be "solved" incrementally (by applying the five unification rewrite rules), as generated, into a most general substitution; incremental solving of the type constraints can have a significant impact on the complexity of unification as we defined it, and Maude indeed does that.

---

[1] With conventional arithmetic and boolean operators added for writing and testing our definition on meaningful programs.

Rules from Figure 5, with modifications or additions as follows:

$$\cfrac{\langle\ \underline{x}\ \rangle_k\ \langle\!\langle\eta\rangle\!\rangle_{tenv}\ \langle\!\langle\gamma\rangle\!\rangle_{eqns}\ \langle\ \underline{t_v}\ \rangle_{nextType}}{t'[tl\leftarrow tl']\qquad\qquad\qquad\qquad t_v+|tl|}\quad\begin{array}{l}\text{when }\eta[x]=let(t),\\ t'=\gamma[t],\ t':RefType,\\ tl=vars(\gamma[t])-vars(\eta),\\ \text{and }tl'=t_v\ldots(t_v+|tl|-1)\end{array}$$

$$\cfrac{\langle\ \underline{\lambda xl.e}\ \rangle_k\ \langle\!\langle\eta\rangle\!\rangle_{tenv}}{bind(xl)\curvearrowright e\curvearrowright mkFunType(xl)\curvearrowright restore(\eta)}$$

$$t\curvearrowright mkFunType(tl)=tl\to t$$

$$\cfrac{\langle t\curvearrowright mkFunType(xl)\rangle_k\ \langle\!\langle\eta\rangle\!\rangle_{tenv}}{\eta[xl]\to t}$$

$$\cfrac{\langle\ \underline{let\ xl=el\ in\ e}\ \rangle_k\ \langle\!\langle\eta\rangle\!\rangle_{tenv}}{el\curvearrowright mkLet(\cdot)\curvearrowright bindTo(xl)\curvearrowright e\curvearrowright restore(\eta)}$$

$$\cfrac{\langle\underline{t}\rangle_{results}\curvearrowright\langle\ \underline{\cdot}\ \rangle_{mkLet}}{\cdot\qquad\qquad let\ (t)}$$

$$\cfrac{\langle\ \cdot\ \rangle_{results}\curvearrowright\langle\underline{tl}\rangle_{mkLet}}{tl\qquad\qquad\cdot}$$

$$\cfrac{\langle\ \underline{letrec\ xl=el\ in\ e}\ \rangle_k\ \langle\!\langle\eta\rangle\!\rangle_{tenv}}{bind(xl)\curvearrowright el\curvearrowright addEqns(xl)\curvearrowright mkLet(\cdot)\curvearrowright bindTo(xl)\curvearrowright e\curvearrowright restore(\eta)}$$

$$\cfrac{\langle results(tl')\curvearrowright\underline{addEqns(tl)}\rangle_k\ \langle\ \underline{\cdot}\ \rangle_{eqns}}{\cdot\qquad\qquad\qquad tl=tl'}$$

$$results(tl)\curvearrowright addEqns(xl)=xl\curvearrowright addEqns(tl)$$

$$\cfrac{\langle\ \underline{[tl]}\ \rangle_k\ \langle\!\langle\Gamma,\underline{\cdot}\ \rangle\!\rangle_{eqns}\ \langle\ \underline{t_v}\ \rangle_{nextType}}{list\ t_v\qquad t_v*=tl\qquad t_v+1}\qquad\text{where }t*=\cdot\text{ and }\cfrac{\cdot}{\cdot}\ ,t_v*=\langle\underline{t}\rangle\atop t_v=t$$

$$\cfrac{\langle\ \underline{cdr\ t}\ \rangle_k\ \langle\!\langle\Gamma,\underline{\cdot}\ \rangle\!\rangle_{eqns}\ \langle\ \underline{t_v}\ \rangle_{nextType}}{list\ t_v\qquad t=list\ t_v\qquad t_v+1}$$

$$\cfrac{\langle\underline{car\ t}\rangle_k\ \langle\!\langle\Gamma,\underline{\cdot}\ \rangle\!\rangle_{eqns}\ \langle\ \underline{t_v}\ \rangle_{nextType}}{t_v\qquad t=list\ t_v\qquad t_v+1}$$

$$\cfrac{\langle cons\ t_1\ t_2\rangle_k\ \langle\!\langle\Gamma,\underline{\cdot}\ \rangle\!\rangle_{eqns}}{t_2\qquad t_2=list\ t_1}$$

$$\cfrac{\langle null?\ t\rangle_k\ \langle\!\langle\Gamma,\underline{\cdot}\ \rangle\!\rangle_{eqns}\ \langle\ \underline{t_v}\ \rangle_{nextType}}{bool\qquad t=list\ t_v\qquad t_v+1}$$

$$\cfrac{\langle!\ t\rangle_k\ \langle\!\langle\Gamma,\underline{\cdot}\ \rangle\!\rangle_{eqns}\ \langle\ \underline{t_v}\ \rangle_{nextType}}{t_v\qquad t=ref\ t_v\qquad t_v+1}$$

$$\&\ t\to ref\ t$$

$$\cfrac{\langle t_1:=t_2\rangle_k\ \langle\!\langle\Gamma,\underline{\cdot}\ \rangle\!\rangle_{eqns}}{unit\qquad t_1=ref\ t_2}$$

$$\cfrac{\langle t_1\ ;\ t_2\rangle_k\ \langle\!\langle\Gamma,\underline{\cdot}\ \rangle\!\rangle_{eqns}}{t_2\qquad t_1=unit}$$

Fig. 7. The !$\mathscr{W}$ type inferencer, Semantics

4 Analysis and Proof Technique

Here, we sketch an argument that although different in form, our definition of $\mathscr{W}$ is equivalent to Milner's. We additionally give a brief summary of our work in proving type system soundness using our formalism.

4.1 Equivalence of Milner's and Our $\mathscr{W}$

We would like to make explicit how our rewriting definition is effectively equivalent to Milner's $\mathscr{W}$, up to the additions of some explicit fundamental data types and operators. To do this, it is easiest to look at $\mathscr{J}$, Milner's simplified algorithm, which he proved equivalent to $\mathscr{W}$. Milner's definition of $\mathscr{J}$ is given in Figure 8 as a convenience for the reader. The main questions of equivalence center around recursive calls and their environments, as well as the substitution.

$\mathscr{J}$ is a recursive algorithm. It calls itself on subexpressions throughout the computation. We achieve the same effect through the use of strictness attributes and the saving and restoring of environments. Our strictness attributes cause subexpressions to be moved to the front of the computation structure, effectively disabling rules that would apply to the "context," and enabling rules applying to the subexpression itself.

To each call of $\mathscr{J}$, a type environment (also called a typed prefix in Milner's notation) is passed. Because we have only one global type environment, it is not immediately obvious that changes to the type environment when evaluating subexpressions cannot affect the remaining computation. In Milner's algorithm, this is handled by virtue of passing the environments by value. We ensure this by always placing, at the end of the local computation, a *restore* marker and a copy of the current environment before affecting the environment. Thus, when

$\mathscr{J}(\bar{p}, f) = \tau$

1. If f is x then:
   If λx_σ is active in $\bar{p}$, $\tau := \sigma$.
   If $let\ x_\sigma$ is active in $\bar{p}$, $\tau = [\beta_i/\alpha_i]E_\sigma$, where α_i are the generic type variables of $let\ x_{E\sigma}$ in $E\bar{p}$, and β_i are new variables.
2. If f is de then:
   $\rho := \mathscr{J}(\bar{p}, d)$; $\sigma := \mathscr{J}(\bar{p}, e)$;
   UNIFY$(\rho, \sigma \rightarrow \beta)$; $\tau := \beta$; (β new)
3. If f is $(if\ d\ then\ e\ else\ e')$, then:
   $\rho := \mathscr{J}(\bar{p}, d)$; UNIFY$(\rho, bool)$;
   $\sigma := \mathscr{J}(\bar{p}, e)$; $\sigma' := \mathscr{J}(\bar{p}, e')$;
   UNIFY(σ, σ'); $\tau := \sigma$
4. If f is $(\lambda x \cdot d)$ then:
   $\rho := \mathscr{J}(\bar{p} \cdot \lambda x_\beta, d)$; $\tau := \beta \rightarrow \rho$; ($\beta$ new)
5. If f is $(fix\ x \cdot d)$, then:
   $\rho := \mathscr{J}(\bar{p} \cdot fix\ x_\beta, d)$; ($\beta$ new)
   UNIFY(β, ρ); $\tau = \beta$;
6. If f is $(let\ x = d\ in\ e)$ then:
   $\rho := \mathscr{J}(\bar{p}, d)$; $\sigma := \mathscr{J}(\bar{p} \cdot\ let\ x_\rho, e)$; $\tau := \sigma$.

UNIFY is a procedure that delivers no result, but has a side effect on a global substitution E. If UNIFY(σ, τ) changes E to E', and if $\mathscr{U}(E\sigma, E\tau) = U$, then $E' = UE$, where $\mathscr{U}$ is a unification generator.

Fig. 8. Milner's $\mathscr{J}$ Algorithm

the local computation is complete, the environment is restored to what it was before starting the subcomputation.

Both definitions keep a single, global substitution, to which restrictions are continually added as side effects. In addition, both only apply the substitution when doing variable lookup. The calls to UNIFY in the application, if/then/else, and fix cases are reflected in our rules by the additional formulas added to the *eqns* configuration item. Indeed, for the rules of Exp (disregarding our extensions with integers), these are the only times we affect the unifier. As an example, let us look at the application de in an environment $\bar{p}$. In $\mathscr{J}$, two recursive calls to $\mathscr{J}$ are made: $\mathscr{J}(\bar{p}, d)$ and $\mathscr{J}(\bar{p}, e)$, whose results are called ρ and σ respectively. Then a restriction to the global unifier is made, equating ρ with $\sigma \to \beta$, with β being a new type variable, and finally β is returned as the type of the expression.

We do a very similar thing. The strictness attributes of the application operator force evaluation of the arguments d and e first. These eventually transform into $\rho\sigma$. We can then apply a rewrite rule where we end up with a new type β, and add an equation $\rho = \sigma \to \beta$ to the *eqns* configuration item. The evaluations of d and e are guaranteed not to change the environment because we always restore environments upon returning types.

4.2   Type Preservation

We attempted to prove the preservation property of $\mathscr{W}$ using our methodology. We briefly outline the approach below. For more details of the partial proofs of soundness for this and other type systems defined in K, see [35]. We use a few conventions to shorten statements. The variables V, E, and K stand for values, expressions, and computation items respectively. Additionally, we add $\mathscr{E}$ and $\mathscr{W}$ subscripts on constructs that are shared between both the Exp language and the $\mathscr{W}$ algorithm. We then only mention the system in which reductions are taking place if it is not immediately clear from context. Finally, a statement like $\mathscr{W} \models R \xrightarrow{*} R'$ means that R reduces to R' under the rewrite rules for $\mathscr{W}$.

A distinguishing feature of our proof technique is that we use an abstraction function, α, to enable us to convert between a configuration in the language domain to a corresponding configuration in the typing domain. Using an abstraction function in proving soundness is a technique used frequently in the domain of processor construction, as introduced in [36], or compiler optimization [37,38].

Lemma 1. *Any reachable configuration in the language domain can be transformed using structural equations into a unique expression.*

Proof. This follows from two key ideas. One, you cannot use the structural equations to transform an expression into any other expression, and two, each structural equation can be applied backwards even after the rules have applied.

Because of Lemma 1, we can use a simple definition for α: $\alpha(\llbracket E \rrbracket_{\mathscr{E}}) = \llbracket E \rrbracket_{\mathscr{W}}$. By the lemma, this definition is well-defined for all reachable configurations, and homomorphic with respect to structural rules. While this function is effectively the identity function, we have experimented with much more complicated abstraction functions, which lead us to believe the technique scales [35].

Lemma 2. *If $\mathscr{W} \models \alpha(V) \xrightarrow{*} \tau$ then $[\![V]\!]_{\mathscr{W}} \xrightarrow{*} \tau$.*

Proof. This follows directly from the $\mathscr{W}$ rewrite rules for values.

Lemma 3 (Preservation 1). *If $[\![E]\!]_{\mathscr{W}} \xrightarrow{*} \tau$ and $[\![E]\!]_{\mathscr{E}} \xrightarrow{*} R$ for some type τ and configuration R, then $\mathscr{W} \models \alpha(R) \xrightarrow{*} \tau'$ for some τ' unifiable with τ.*

Lemma 4 (Preservation 2). *If $[\![E]\!]_{\mathscr{W}} \xrightarrow{*} \tau$ and $[\![E]\!]_{\mathscr{E}} \xrightarrow{*} V$ for some type τ and value V, then $[\![V]\!]_{\mathscr{W}} \xrightarrow{*} \tau'$ for some type τ'.*

Proof. This follows directly from Lemmas 2 and 3.

In comparison, the definition of (strong) preservation as given by Wright and Felleisen [6, Lemma 4.3] states: "If $\Gamma \rhd e_1 : \tau$ and $e_1 \longrightarrow e_2$ then $\Gamma \rhd e_2 : \tau$." We cannot define preservation in the same way, because our terms do not necessarily remain terms as they evaluate. If one accepts the idea of our abstraction function, then subject reduction is actually closer in spirit to the above Lemma 3. We were able to verify Lemma 3 for many of the cases, but were unable to show the necessary correspondence between the environment and replacement.

5   Conclusions and Further Work

We have shown that rewriting logic, through K, is amenable for defining feasible type inferencers for programming languages. Evaluation suggests that these equationally defined type inferencers are comparable in speed with "off-the-shelf" ones used by real implementations of programming languages. Since both the language and its type system are defined uniformly as theories in the same logic, one can use the standard RLS proof theory to prove properties about languages and type systems for those languages. These preliminary results lead us to believe our approach is a good candidate for the POPLMARK Challenge [39].

References

1. Meseguer, J., Roşu, G.: Rewriting logic semantics: From language specifications to formal analysis tools. In: Basin, D., Rusinowitch, M. (eds.) IJCAR 2004. LNCS, vol. 3097, pp. 1–44. Springer, Heidelberg (2004)
2. Meseguer, J., Roşu, G.: The rewriting logic semantics project. J. TCS 373(3), 213–237 (2007)
3. Roşu, G.: K: A rewrite-based framework for modular language design, semantics, analysis and implementation. Technical Report UIUCDCS-R-2006-2802, Computer Science Department, University of Illinois at Urbana-Champaign (2006)
4. Kuan, G., MacQueen, D., Findler, R.B.: A rewriting semantics for type inference. In: De Nicola, R. (ed.) ESOP 2007. LNCS, vol. 4421, pp. 426–440. Springer, Heidelberg (2007)
5. Felleisen, M., Hieb, R.: A revised report on the syntactic theories of sequential control and state. J. TCS 103(2), 235–271 (1992)

6. Wright, A.K., Felleisen, M.: A syntactic approach to type soundness. Information and Computation 115(1), 38–94 (1994)
7. Matthews, J., Findler, R.B., Flatt, M., Felleisen, M.: A visual environment for developing context-sensitive term rewriting systems. In: van Oostrom, V. (ed.) RTA 2004. LNCS, vol. 3091, pp. 301–311. Springer, Heidelberg (2004)
8. Milner, R.: A theory of type polymorphism in programming. J. Computer and System Sciences 17(3), 348–375 (1978)
9. Bravenboer, M., Kalleberg, K.T., Vermaas, R., Visser, E.: Stratego/XT Tutorial, Examples, and Reference Manual. Department of Information and Computing Sciences, Universiteit Utrecht (August 2005) (Draft)
10. Barendregt, H.: Introduction to generalized type systems. J. Functional Programming 1(2), 125–154 (1991)
11. Stehr, M.-O., Meseguer, J.: Pure type systems in rewriting logic: Specifying typed higher-order languages in a first-order logical framework. In: Owe, O., Krogdahl, S., Lyche, T. (eds.) From Object-Orientation to Formal Methods. LNCS, vol. 2635, pp. 334–375. Springer, Heidelberg (2004)
12. Barendregt, H.P., van Eekelen, M.C.J.D., Glauert, J.R.W., Kennaway, R., Plasmeijer, M.J., Sleep, M.R.: Term graph rewriting. In: de Bakker, J.W., Nijman, A.J., Treleaven, P.C. (eds.) PARLE 1987. LNCS, vol. 259, pp. 141–158. Springer, Heidelberg (1987)
13. Plump, D.: Term graph rewriting. In: Handbook of Graph Grammars and Computing by Graph Transformation, vol. 2. World Scientific, Singapore (1998)
14. Banach, R.: Simple type inference for term graph rewriting systems. In: Rusinowitch, M., Remy, J.-L. (eds.) CTRS 1992. LNCS, vol. 656, pp. 51–66. Springer, Heidelberg (1993)
15. Fogarty, S., Pasalic, E., Siek, J., Taha, W.: Concoqtion: Indexed types now! In: PEPM 2007, pp. 112–121. ACM, New York (2007)
16. Kamareddine, F., Klop, J.W.(eds.): Special Issue on Type Theory and Term Rewriting: A Collection of Papers. Journal of Logic and Computation 10(3). Oxford University Press, Oxford (2000)
17. Hünke, Y., de Moor, O.: Aiding dependent type checking with rewrite rules (2001) (unpublished), http://citeseer.ist.psu.edu/huencke01aiding.html
18. Mametjanov, A.: Types and program transformations. In: OOPSLA 2007 Companion, pp. 937–938. ACM, New York (2007)
19. Levin, M.Y., Pierce, B.C.: TinkerType: A language for playing with formal systems. J. Functional Programing 13(2), 295–316 (2003)
20. Lee, D.K., Crary, K., Harper, R.: Towards a mechanized metatheory of standard ML. In: POPL 2007, pp. 173–184. ACM, New York (2007)
21. Klein, G., Nipkow, T.: A machine-checked model for a Java-like language, virtual machine and compiler. TOPLAS 28(4), 619–695 (2006)
22. Sewell, P., Nardelli, F.Z., Owens, S., Peskine, G., Ridge, T., Sarkar, S., Strniša, R.: Ott: Effective tool support for the working semanticist. In: ICFP 2007: Proceedings of the 2007 ACM SIGPLAN international conference on Functional programming, pp. 1–12. ACM, New York (2007)
23. van den Brand, M., et al.: The ASF+SDF meta-environment: A component-based language development environment. In: Wilhelm, R. (ed.) CC 2001. LNCS, vol. 2027, p. 365. Springer, Heidelberg (2001)
24. Roşu, G.: K: A rewriting-based framework for computations—an informal guide. Technical Report UIUCDCS-R-2007-2926, University of Illinois at Urbana-Champaign (2007)

25. Meseguer, J.: Conditional rewriting logic as a unified model of concurrency. J. TCS 96(1), 73–155 (1992)
26. Clavel, M., Durán, F., Eker, S., Lincoln, P., Martí-Oliet, N., Meseguer, J., Quesada, J.F.: Maude: Specification and programming in rewriting logic. Theor. Comput. Sci. 285(2), 187–243 (2002)
27. Plotkin, G.D.: A structural approach to operational semantics. Journal of Logic and Algebraic Programming 60-61, 17–139 (2004)
28. Gurevich, Y.: Evolving algebras 1993: Lipari guide. In: Specification and validation methods, pp. 9–36. Oxford University Press Inc., New York (1995)
29. Strachey, C., Wadsworth, C.P.: Continuations: A mathematical semantics for handling full jumps. Higher-Order and Symb. Computation 13(1/2), 135–152 (2000)
30. Martelli, A., Montanari, U.: An efficient unification algorithm. ACM Trans. Program. Lang. Syst. 4(2), 258–282 (1982)
31. Roşu, G.: K-style Maude definition of the W type inferencer (2007),
    http://fsl.cs.uiuc.edu/index.php/Special:WOnline
32. Kothari, S., Caldwell, J.: Algorithm W for lambda calculus extended with Milnerlet Implementation used for Type Reconstruction Algorithms—A Survey. Technical Report, University of Wyoming (2007)
33. Li, Z.: Enhtop: A patch for an enhanced OCaml toplevel (2007),
    http://www.pps.jussieu.fr/~li/software/index.html
34. Eker, S.: Fast matching in combinations of regular equational theories. In: WRLA 1996. ENTCS, vol. 4, pp. 90–109 (1996)
35. Ellison, C.: A rewriting logic approach to defining type systems. Master's thesis, University of Illinois at Urbana-Champaign (2008),
    http://fsl.cs.uiuc.edu/pubs/ellison-2008-mastersthesis.pdf
36. Hosabettu, R., Srivas, M.K., Gopalakrishnan, G.: Decomposing the proof of correctness of pipelined microprocessors. In: Y. Vardi, M. (ed.) CAV 1998. LNCS, vol. 1427, pp. 122–134. Springer, Heidelberg (1998)
37. Kanade, A., Sanyal, A., Khedker, U.P.: A PVS based framework for validating compiler optimizations. In: SEFM 2006, pp. 108–117. IEEE Computer Society, Los Alamitos (2006)
38. Kanade, A., Sanyal, A., Khedker, U.P.: Structuring optimizing transformations and proving them sound. In: COCV 2006. ENTCS, vol. 176(3), pp. 79–95. Elsevier, Amsterdam (2007)
39. Aydemir, B.E., Bohannon, A., Fairbairn, M., Foster, J.N., Pierce, B.C., Sewell, P., Vytiniotis, D., Washburn, G., Weirich, S., Zdancewic, S.: Mechanized metatheory for the masses: The PoplMark challenge. In: Hurd, J., Melham, T. (eds.) TPHOLs 2005. LNCS, vol. 3603, pp. 50–65. Springer, Heidelberg (2005)

A Term-Graph Syntax
for Algebras over Multisets*

Fabio Gadducci

Dipartimento di Informatica, Università di Pisa
Polo Scientifico "Guglielmo Marconi", via dei Colli 90, La Spezia

Abstract. Earlier papers argued that term graphs play for the specification of relation-based algebras the same role that standard terms play for total algebras. The present contribution enforces the claim by showing that term graphs are a sound and complete representation for *multiset algebras*, i.e., algebras whose operators are interpreted over multisets.

1 Introduction

Cartesian categories (i.e., with binary products and terminal object) offer the right tool for interpreting equational logic: objects are tuples of sorts, and arrows are tuples of terms, typed accordingly. This is confirmed by the presentation of the category of (total) algebras for a signature Σ as the category of product-preserving functors from the cartesian category $\mathbf{Th}(\Sigma)$, the *algebraic theory* of Σ, to the category $\mathbf{Set}$ of sets and functions. Two arrows in $\mathbf{Th}(\Sigma)$ coincide if and only if they denote the same function for every functor. Thus, equational signatures (Σ, E) and their categories of algebras are recast in the framework by quotienting the arrows corresponding to pairs of terms $(s, t) \in E$ (see e.g. [1]).

The remarks above may be summarized by the following sentence: for ordinary algebras we have a syntactical presentation for terms and term substitutions, basically based on their correspondence with trees and tree replacement; and a complete semantics, based on the interpretation of terms as functions on sets, and of term substitution as function composition. The two sides are reconciled by the notion of cartesian theory, which underlines the syntax (the free cartesian category $\mathbf{Th}(\Sigma)$) and the semantics (the cartesian functors to $\mathbf{Set}$).

Such a neat characterization proved elusive for more complex algebraic formalisms, such as partial algebras and especially multi-algebras [2], where operators are interpreted as partial functions and as additive relations, respectively. Indeed, there is a long tradition, mostly arising from theoretical computer science, looking for suitable syntactical presentations of equational laws for such algebras. Roughly, after quite some work by German algebraists [3], this search boiled down to identifying suitable algebraic characterizations, similar to Lawvere theories, of various tree-like structures (as e.g. data-flow networks). We refer here to the work on flow graphs by Ştefănescu and others (see e.g. the survey [4],

---

* Research supported by the EU FET integrated project Sensoria, IST-2005-016004.

and specifically [5] for the algebraic presentation of multiset relations), even if mostly relevant in recent years have been the graphical characterization of traced monoidal categories [6], and their duality with compact closed categories.

These categorical characterizations of graphical formalisms have been exploited early on: see e.g. [7,8] for a recollection of the main players and of some of the main results, and an application to (term) graph rewriting. However, such characterizations should have a counterpart in the development of the model-theoretical side. Namely, in the identification of the semantical domain for which a chosen algebraic formalism is complete. Our starting point is [9], introducing categories for a signature Σ (based on the gs-monoidal theory $\mathbf{GS\text{-}Th}(\Sigma)$ of Σ, generalizing the algebraic theory $\mathbf{Th}(\Sigma)$) for obtaining a functorial representation of the categories of partial and multi-algebras for Σ.[1]

The solution proved satisfactory for partial algebras, since the arrows of a quotient category of $\mathbf{GS\text{-}Th}(\Sigma)$ (namely, the g-monoidal theory $\mathbf{G\text{-}Th}(\Sigma)$) are in bijective correspondence with conditioned terms, i.e., with pairs $s \mid D$, where s is the principal term and D is a set of terms used for restricting the domain of definition of s. Most importantly, two arrows in $\mathbf{G\text{-}Th}(\Sigma)$ coincide if and only if they always denote the same partial function for every possible functor to $\mathbf{PFun}$ (the category of sets and partial functions), thus allowing the development of a sound and complete deduction system for equational signatures based on so-called conditioned Kleene equations [11].

Things went less smoothly for multi-algebras. The functorial presentation still holds, and indeed, the arrows of the gs-monoidal theory $\mathbf{GS\text{-}Th}(\Sigma)$ are in bijective correspondence with (acyclic) term graphs, i.e, trees with possibly shared nodes. However, as shown in [12], only term graphs up-to *garbage equivalence* are identified by every functor to $\mathbf{Rel}$ (sets and additive relations). Such an equivalence is defined set-theoretically, and an axiomatic presentation is missing.

The present paper further investigates the latter issue. In particular, taking hints from the solution for Frobenius algebras, as recollected in [13] (also echoed in the solution for traced monoidal categories [14]), our work shows that the gs-monoidal theory $\mathbf{GS\text{-}Th}(\Sigma)$ allows for a functorial presentation of what we called *multiset algebras*, that is, algebras whose operators are interpreted as (additive) multiset relations. Most importantly, we prove that two term graphs denote the same multiset relation if and only if they are isomorphic, thus laying the base for a simple deduction system for such algebras.

The paper is structured as follows. Section 2 recalls the basic definitions concerning (term) graphs with interfaces, our chosen graphical formalism, while Section 3 presents a few notions and facts concerning gs-monoidal theories, lifted from [9]. Section 4 discusses instead a few alternative presentations for multi-set relations, partly original and partly drawn from classical mathematical literature, that are needed later on, in Section 5, for proving our main completeness theorem. Finally, Section 6 recasts our results against the algebraic background, and highlights some possible directions for future works.

---

[1] The acronym *gs* stands for *graph substitution*: it was originally introduced in [10] for highlighting the correspondence between arrow composition and graph replacement.

2 Graphs and Graphs with Interfaces

This section presents some definitions concerning (hyper-)graphs, typed graphs and graphs with interfaces. It also introduces two operators on graphs with interfaces. We refer to [15] and [10] for a detailed introduction.

Definition 1 (Graphs). *A (hyper-)graph is a four-tuple* $\langle V, E, s, t \rangle$ *where* V *is the set of nodes,* E *is the set of edges and* $s, t : E \to V^*$ *are the source and target functions (for* V^* *the free monoid over the set of nodes).*

From now on we denote the components of a graph G by V_G, E_G, s_G and t_G.

Definition 2 (Graph morphisms). *Let* G, G' *be graphs. A (hyper-)graph morphism* $f : G \to G'$ *is a pair of functions* $\langle f_V, f_E \rangle$*, such that* $f_V : V_G \to V_{G'}$ *and* $f_E : E_G \to E_{G'}$ *and source and target functions are preserved, i.e.* $f_V^* \circ s_G = s_{G'} \circ f_E$ *and* $f_V^* \circ t_G = t_{G'} \circ f_E$ *(for* f_V^* *extension of the node function).*

The category of graphs is denoted by **Graph**. We now give the definition of typed graph [16], i.e., a graph labelled over a structure that is itself a graph.

Definition 3 (Typed graphs). *Let* T *be a graph. A* typed graph G *over* T *is a graph* $|G|$ *with a graph morphism* $\tau_G : |G| \to T$.

The related class of morphisms is obviously defined.

Definition 4 (Typed graph morphisms). *Let* G, G' *be typed graphs over* T. *A typed graph morphism* $f : G \to G'$ *is a graph morphism* $f : |G| \to |G'|$ *consistent with the typing, i.e., such that* $\tau_G = \tau_{G'} \circ f$.

The category of graphs typed over T is denoted by T-**Graph**. In the following, we assume a chosen type graph T.

To define a notion of substitution, we need operations for composing graphs. So, we equip typed graphs with suitable "handles" for interacting with an environment. To this aim, we consider the class of *discrete* graphs, i.e., containing only nodes (thus, often depicted just as a set).

Definition 5 (Graphs with interfaces). *Let* J, K *be typed, discrete graphs. A graph with input interface* J *and output interface* K *is a triple* $\mathbb{G} = \langle j, G, k \rangle$*, where* G *is a typed graph,* $j : J \to G$ *and* $k : K \to G$ *are typed graph morphisms, and they are called input and output morphisms, respectively.*

A graph with input interface J and output interface K is denoted $J \xrightarrow{j} G \xleftarrow{k} K$. In the following, we consider graphs with *ordered* interfaces, i.e., such that the set of nodes is totally ordered. Moreover, we also assume that the morphism j is injective. Abusing notation, we refer to the nodes belonging to the image of the input morphism as inputs, and similarly for the nodes belonging to the image of the output morphism as outputs. We often refer implicitly to a graph with interfaces as the representative of its isomorphism class. Moreover, we sometimes denote the class of isomorphic graphs and its components by the same symbol.

Note that graphs with interfaces are arrows in the category of (left-linear) cospans $Cospan(T\text{-}\mathbf{Graph})$ [17]. This characterization suggests us the definition of the following two binary operators on graphs with discrete interfaces.

Definition 6 (Sequential composition). *Let* $\mathbb{G} = J \xrightarrow{j} G \xleftarrow{k} K$ *and* $\mathbb{G}' = K \xrightarrow{j'} G' \xleftarrow{k'} I$ *be graphs with discrete interfaces. Their sequential composition is the graph with discrete interfaces* $\mathbb{G} \circ \mathbb{G}' = J \xrightarrow{j''} G'' \xleftarrow{k''} I$, *where* G'' *is the disjoint union* $G \uplus G'$, *modulo the equivalence on nodes induced by* $k(x) = j'(x)$ *for all* $x \in V_K$, *and* j'' *and* k'' *are the uniquely induced arrows.*

Now, we define *IT*-**Graph** the category of *ordered* interfaces, i.e., such that the set of nodes is totally ordered, and typed graphs with (ordered) interfaces. Most importantly, we collapse all "isomorphic" typed graphs with interfaces: two typed graphs $\mathbb{G}, \mathbb{G}'$ with the same interfaces are isomorphic if there exists a typed graph isomorphism between the underlying graphs G, G' that preserves the input and output morphisms.

Definition 7 (Parallel composition). *Let* $\mathbb{G} = J \xrightarrow{j} G \xleftarrow{k} K$ *and* $\mathbb{G}' = J' \xrightarrow{j'} G' \xleftarrow{k'} K'$ *be graphs with discrete interfaces. Their parallel composition is the graph with discrete interfaces* $\mathbb{G} \otimes \mathbb{G}' = (J \uplus J') \xrightarrow{j''} G \uplus G' \xleftarrow{k''} (K \uplus K')$, *where* $\uplus$ *is the disjoint union operator, and* j'', k'' *are the uniquely induced arrows.*

Intuitively, the sequential composition $\mathbb{G} \circ \mathbb{G}'$ is obtained just by taking the disjoint union of the graphs underlying $\mathbb{G}$ and $\mathbb{G}'$, and gluing the outputs of $\mathbb{G}$ with the corresponding inputs of $\mathbb{G}'$. The parallel composition $\mathbb{G} \otimes \mathbb{G}'$ is instead obtained by taking the disjoint union of the graphs underlying $\mathbb{G}$ and $\mathbb{G}'$. Note that both operations are defined on "concrete" graphs. However, their results do not depend on the choice of the representatives of their isomorphism classes.

Note that the sequential composition introduced in Definition 6 corresponds to the standard composition in the category $Cospan(T\text{-}\mathbf{Graph})$, and the parallel composition in Definition 7 corresponds to a monoidal operator in that category.

A *graph expression* is a term over the syntax containing all graphs with discrete interfaces as constants, and parallel and sequential composition as binary operators. An expression is *well-formed* if all the occurrences of both sequential and parallel composition are defined for the interfaces of their arguments, according to Definitions 6 and 7. The interfaces of a well-formed graph expression are computed inductively from the interfaces of the graphs occurring in it; the value of the expression is the graph obtained by evaluating all its operators.

Definition 8 (Term graphs). *Let* G *be an acyclic graph. It is a term graph if* $s(e) \in V_G$ *for all its edges, and moreover* $s(e) = s(e')$ *implies* $e = e'$.

In other terms, a signature Σ is a graph satisfying the first condition, interpreting each operator $f \in \Sigma_{\sigma_x \dots \sigma_n, \sigma}$ an edge with source σ and target the tuple $\langle \sigma_1, \dots, \sigma_n \rangle$; while a term graph over Σ is an acyclic graph typed over Σ such that each node is in the image of the source function of at most one edge.

Note that the binary operators of sequential and parallel composition preserved the properties of being a term graphs. So, in the following we denote as TIT-**Graph** the category of term graphs with interfaces, typed over T.

3   Categories with a GS-Monoidal Structure

This section recalls the definition of gs-monoidal category, and states the correspondence between arrows of a gs-monoidal category and term graphs [10].

Definition 9 (GS-Monoidal categories). *A gs-monoidal category* $\mathbf{C}$ *is a six-tuple* $\langle \mathbf{C}_0, \otimes, e, \rho, \nabla, ! \rangle$, *where* $\langle \mathbf{C}_0, \otimes, e, \rho \rangle$ *is a symmetric strict monoidal category (see [18]) and* $! : Id \Rightarrow e : \mathbf{C}_0 \rightarrow \mathbf{C}_0$, $\nabla : Id \Rightarrow \otimes \circ D : \mathbf{C}_0 \rightarrow \mathbf{C}_0$ *are two transformations (D is the diagonal functor), such that* $!_e = \nabla_e = id_e$ *and they satisfy the* coherence *axioms*

$$\nabla; (\nabla_a \otimes id_a) = \nabla; (id_a \otimes \nabla_a) \qquad \nabla_a^*; (id_a \otimes !_a) = id_a \qquad \nabla_a; \rho_{a,a} = \nabla_a$$

and the monoidality *axioms*

$$\nabla_{a \otimes b}; (id_a \otimes \rho_{b,a} \otimes id_b) = \nabla_a \otimes \nabla_b \qquad !_a \otimes !_b = !_{a \otimes b}$$

A gs-monoidal functor $\langle F, \phi, \phi_e \rangle : \mathbf{C} \rightarrow \mathbf{C}'$ *is a symmetric monoidal functor (that is, a functor F equipped with two natural isomorphisms* $\phi_e : F(e) \rightarrow e'$ *and* $\phi : F(a \otimes b) \rightarrow F(a) \otimes' F(b)$) *such that* $F(!_a); \phi_e =!'_{F(a)}$ *and* $F(\nabla_a); \phi = \nabla'_{F(a)}$; *it is strict if* ϕ *and* ϕ_e *are identities. The category of gs-monoidal categories and their strict functors is denoted by* **GSM-Cat**.

Mimicking the correspondence between terms and trees, morphisms of a gs-monoidal category correspond to term graphs [10], in the same way *terms* over a signature are represented by arrows of its algebraic theory. In particular, the lack of naturality of morphisms ∇ (i.e., of axioms $s; \nabla = \nabla; (s \otimes s)$) allows the distinction between the sharing of a term and the occurrence of two copies of it.

Now, let us consider again the category of term graphs with (discrete and ordered) interfaces: it is actually gs-monoidal. Indeed, for each interface X, the morphism ∇_X is represented by the triple $\langle X, X, X \uplus X \rangle$, and the obvious arrows; while the morphism $!_X$ is represented by the triple $\langle X, \emptyset, \emptyset \rangle$. Now, by abuse of notation let TIT-**Graph** denote also the gs-monoidal category of term graphs, typed over T, with the additional, gs-monoidal structure outlined above.

Proposition 1. *Let Σ be a signature, let T_Σ be the associated graph, and let* **GS-Th**(Σ) *be the free gs-monoidal category over Σ. Then, there exists a full and faithful, strict gs-monoidal functor from* **GS-Th**(Σ) *to* TIT_Σ-**Graph**.

The result above is a multi-sorted version of [10, Theorem 23]. It implies that arrows of **GS-Th**(Σ) are in bijective correspondence with term graphs with (discrete and ordered) interfaces, typed over Σ. Most importantly, the theorem states that the isomorphism of these term graphs can be recast in equational terms, and it can be verified by using the laws holding for gs-monoidal categories.

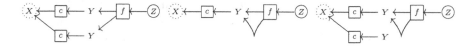

Fig. 1. Term graphs F_1, F_2, and F_3

Example 1. In order to help visualising the correspondence stated in Proposition 1 above, consider the term graphs depicted in Fig. 1: they are taken from [9, Fig. 2], adopting a slightly more agile presentation. In particular, nodes in the input (output) interface are denoted by circling them with a dotted (solid) circle; while edges are boxes with an entering tentacle (from the source) and possibly many leaving tentacles (to the target).[2]

The signature under consideration contains three sorts, represented as X, Y, and Z. The nodes are simply indicated by their sort. For example, term graph F_1 contains four nodes and three edges. The node labelled X is an input node, while Z is an output node. The signature also contains a unary operator $c : X \to Y$ and a binary operator $f : Y \times Y \to Z$: also edges are labelled by boxing their type. The edge in F_1 labelled f has an incoming tentacle from the node labelled Z, and two outgoing tentacles to two different nodes, both labelled Y.

Now, term graphs F_1 and F_2 are the graphical counterpart of the arrows $f \circ (c \otimes c) \circ \nabla_X$ and $\alpha = f \circ \nabla_Y \circ c$, respectively, belonging to hom-set **GS-Th**$(\Sigma)[X, Z]$. These two arrows differ because of the lack of sharing of the term c in F_1, and this is mirrored by the fact that ∇_Y is not natural.

We must now remark that it would be impossible to make the distinction if the standard term presentation is adopted, since both arrows would just represent the term $f(c(x), c(x))$: it is necessary to introduce a suitable *let*-notation, after the seminal work on computational λ-calculus, so that the let-term associated to F_2 would look like *let y be c(x) in f(y, y)*.

Similarly, the term graph F_3 is the counterpart of the $(\alpha \otimes (!_Y \circ c)) \circ \nabla_X$, and it differs from F_2 because $!_Y$ is not natural. In let-notation, F_3 would be represented by an expression such as *let [y be c(x) and y_1 be c(x)] in f(y, y)*.

4   Some Characterizations of Multi-relations

By interpreting the operators of a signature as relations, instead of as functions, one easily models in an algebraic framework some sort of nondeterministic behavior [2]. For *multi-algebras*, the relation associated with an operator is regarded as a function mapping each input value to a *set* of (possible) output values. Quite obviously, this is just an alternative definition of relation, equivalent to the standard one as subset of the direct product.

Definition 10 (Relations). *Let A and B be two sets. A relation $R : A \leftrightarrow B$ is a function $R : A \to \mathcal{P}(B)$, where $\mathcal{P}$ is the finite power-set operator. A relation*

---

[2] Thus, the chosen direction for arrows follows the term graph tradition, adopting a "control flow" view, instead of a (possibly better suited here) "data flow" one.

is total *if* $|R(a)| \geq 1$ *for all* $a \in A$*; it is* functional (univalent) *if* $|R(a)| \leq 1$ *for all* $a \in A$*. Given two relations* $P : A \leftrightarrow B$, $R : B \leftrightarrow C$*, their* composition $P; R : A \leftrightarrow C$ *denotes the relation obtained by composing* P *with the function* $\mathcal{P}(R) : \mathcal{P}(B) \to \mathcal{P}(C)$*, the additive extension of* R.

A total relation is a function $R : A \to \mathcal{P}^+(B)$; while a functional relation is a partial function $R : A \rightharpoonup B$. A functional and total relation is then just a function $R : A \to B$. In the following, we denote with **Rel** and **Pfn** the categories having sets as objects, and relations or functional relations, respectively, as arrows; clearly, there are obvious inclusion functors **Set** $\hookrightarrow$ **Pfn** and **Pfn** $\hookrightarrow$ **Rel**.

4.1   Multiset Relations and Semi-modules

We now plan to further generalize the previous definition.

Definition 11 (Multiset relations). *Let* X *and* Y *be two sets. A* multiset relation *is a function from* X *to* $[Y \to \mathbb{N}]_f$*, i.e., associating to each element* $x \in X$ *a function* h *(with finite support) from* Y *to the natural numbers.*

Being of "finite support" for a function $h : Y \to \mathbb{N}$ means that $h(y) \neq 0$ only for a finite set of elements y. Thus, a multiset relation associates to an element $x \in X$ a set of "weighted" elements in Y. A partial function requires that $h(y) = 1$ for at most one element y, while $h(z) = 0$ for all the others. An additive relation is obtained by replacing $\mathbb{N}$ with the boolean algebra $\{0, 1\}$.

An element of $[Y \to \mathbb{N}]_f$ is going to be denoted as a *(finite) multiset* over Y. It is a standard result that finite multisets (with coefficient in $\mathbb{N}$) form a bi-semi-module [19, p.102]. For the rest of the paper, we adopt a polynomial-like presentation for finitely supported functions, considered as elements of a semi-module $\mathcal{L}[X]$. More explicitly, an expression as $n_1 \cdot y_1 \oplus \ldots \oplus n_k \cdot y_k$ represents the function associating a coefficient $n_i \in \mathbb{N}$ to y_i, and 0 to the other elements (assuming the y_i's to be pairwise different). Moreover, the sum of two polynomials is defined as the component-wise sum of the coefficients. In more abstract terms, the situation boils down to the proposition below.

Proposition 2 (Multisets and semi-modules). *Let* X *be a set. A (finite) multiset over a set* X *is an element of the semi-module* $\mathcal{L}[X]$ *of (finite) linear polynomial with coefficients on the semiring* $\mathbb{N}$ *of natural number: products and sums on the scalar components of a polynomial are defined in the obvious way, assuming they satisfy the laws*

  – $\forall n \in \mathbb{N}.x, y \in \mathcal{L}[X].n \cdot (x \oplus y) = n \cdot x \oplus n \cdot y$,
  – $\forall n, m \in \mathbb{N}.x \in \mathcal{L}[X].(n + m) \cdot x = n \cdot x \oplus m \cdot x$,
  – $\forall n, m \in \mathbb{N}.x \in \mathcal{L}[X].n \cdot (m \cdot x) = nm \cdot x$.

and requiring furthermore $1 \cdot x = x$ *and* $0 \cdot x = \emptyset$ *(the empty polynomial).*

Furthermore, a multiset relation from X *to* Y *consists of a semi-module homomorphism from* $\mathcal{L}[X]$ *to* $\mathcal{L}[Y]$.

A semi-module homomorphism is uniquely induced by a function from X to $\mathcal{L}[Y]$. The one-to-one correspondence between multiset relations and homomorphisms puts forward the definition below, accounting for relation composition.

Definition 12 (Composition). *Let $h : X \to Y$ and $k : Y \to Z$ be multiset relations. Then, their composition $k \circ h$ is defined as*

- $\forall x \in X. \, k \circ h(x) = \bigoplus_{n_y \cdot y \in h(x)} n_y \cdot k(y);$

*We denote with **MRel** the category having sets as objects, and multiset relations as arrows.*

The definition is well-given: since for each $x \in X$ the number of elements $y \in Y$, such that their coefficient n_y in $h(x)$ is different from 0, is finite, the resulting multiset relation has finite support.

Clearly, there is an inclusion functor from **Pfn** to **MRel**. However, the same fact does not hold for **Rel**, since composition needs not to be preserved by the obvious inclusion relation. Intuitively, each partial function $X \to Y$ can be interpreted as an homomorphism between the semi-modules $\mathcal{L}[X]$ to $\mathcal{L}[Y]$, mapping each $x \in X$ to either 0 (the empty polynomial) or 1_y (the polynomial where only the coefficient of y is 1), and such property is preserved by homomorphism composition. Also a relation $X \to Y$ can be in the same way interpreted as a semi-module over the free (idempotent) boolean algebra $\mathbb{B}$ with elements $\{0, 1\}$, but the multiset composition of (the embedding of) two relations does not necessarily result in a relation, but it may describe a multiset relation.

4.2   Multiset Relations and Kleisli Categories

We need to equip the category of multi-relation with a suitable gs-monoidal structure. To this end, we first present yet another characterization of **MRel**.

First of all, let us consider the endofunctor L on **Set**, associating to each set X the set of finitely supported functions $[X \to \mathbb{N}]_f$. Given a morphism $f : X \to Y$, the function $F^f : [X \to \mathbb{N}]_f \to [Y \to \mathbb{N}]_f$ is defined as

$$[F^f(h)](y) = \bigoplus_{x \in f^{-1}(y)} h(x)$$

Proposition 3 (Multisets and monads). *The endofunctor L on **Set** induces a monad (see [18]), with natural morphisms $\eta_X : X \to [X \to \mathbb{N}]_f$ and $\mu_X : [[X \to \mathbb{N}]_f \to \mathbb{N}]_f \to [X \to \mathbb{N}]_f$ defined in the following way*

- $\eta_X(x) = 1_x$, *and*
- $[\mu_X(\lambda)](x) = \bigoplus_{h:X \to \mathbb{N}} \lambda(h)h(x)$

for any $\lambda \in [[X \to \mathbb{N}]_f \to \mathbb{N}]_f$ and with 1_x the usual polynomial.

The cumbersome task of checking the coherence axioms is left to the reader: we just note that the key requirement about the finite support of $[\mu_X(\lambda)]$ is verified, as it is standard practice, via König's lemma.

Since we have a monad construction, it is relevant to look at the Kleisli category, since it provides a representative for free algebras. It is noteworthy that the Kleisli category $K_{\mathcal{F}}$ associated to $\mathcal{F}$ is yet another description for **MRel**. Thus, we can now lift the cartesian product on **Set** to the Kleisli category $K_{\mathcal{F}}$.

Definition 13 (Product). *Let $h : X \rightarrow Y$ and $k : W \rightarrow Z$ be multiset relations. Then, their product $h \otimes k$ is defined as*

$$- \forall xw \in X \times W. h \otimes k(xw) = \bigoplus_{n \cdot y \in h(x), m \cdot z \in k(w)} nm \cdot yz.$$

The category is indeed gs-monoidal, for $!_X$ associating to each element $x \in X$ the empty polynomial $\emptyset$; and for ∇_X associating to each $x \in X$ the polynomial 1_{xx} over $X \times X$. This is far from surprising, since it is a general property holding for the Kleisli category of any monad over a cartesian category. From now on, **MRel** denotes also the gs-monoidal category of multiset relations with the additional, gs-monoidal structure outlined above.

The choice is obvious999ly not unique: we might as well have lifted the cocartesian product of **Set**. However, as argued in [9, Fact 5] and the subsequent paragraph, the point is that this interpretation would be "in contrast with the intuition that binary operators, even if non-deterministic, should have as domain the direct product of the carrier with itself;" and it would be, more explicitly, inconsistent with the established notions for partial and multi-algebras.

5   Syntax and Semantics for Multiset Algebras

It is now the time to turn to our main theorem, establishing the completeness of multiset relations for term graphs. However, since we want to spell out the result in standard algebraic terms, we start by introducing a suitable formalism.

Definition 14 (Multiset algebras). *Let Σ be a signature. A multiset algebra A over Σ consists of the following items*

- *a set A_σ for each sort $\sigma \in \Sigma$;*
- *a multiset relation $A_{\sigma_1} \times \ldots \times A_{\sigma_n} \rightarrow A_\sigma$ for each operator $f \in \Sigma_{\sigma_1 \times \ldots \times \sigma_n, \sigma}$.*

A multiset homomorphism $\rho : A \rightarrow B$ is a family of functions $\rho_\sigma : A_\sigma \rightarrow B_\sigma$, one for each sort $\sigma \in \Sigma$, such that $f(\rho_{\sigma_1}(t_1), \ldots, \rho_{\sigma_n}(t_n)) = [\mathcal{F}(\rho_\sigma)](f(t_1, \ldots t_n))$ for each operator $f \in \Sigma_{\sigma_1 \times \ldots \times \sigma_n, \sigma}$, and for $\mathcal{F}(\rho_\sigma)$ the polynomial lifting of ρ_σ.

These homomorphisms are called tight, point-to-point homorphisms in the literature on multi-algebras. In multiset jargon, each element in the carrier is mapped into a single term, and the weight of each element is preserved. Indeed, $\mathcal{F}$ is the endofunctor on **Set** used for our definition of the Kleisli category.

Proposition 4 (Functorial semantics). *Let Σ be a signature. Then there exists an equivalence of categories between the category **MSAlg** of multiset algebras and (tight, point-to-point) homomorphisms and the category of gs-monoidal functors $[\textbf{GS-Th}(\Sigma), \textbf{MRel}]$ and symmetric monoidal natural transformations.*

As it is standard in functorial semantics, the functors are not strict, that is, as stated in Definition 9, a monoidal functor F may map a product $a \otimes b$ into an element that is just isomorphic to $F(a) \otimes F(b)$, via a natural isomorphism ϕ. Moreover, requiring a natural transformation to be symmetric monoidal just means that such ϕ's are preserved. Furthermore, recall that gs-monoidal functors must preserve the transformations ∇ and !.

The functor category [**GS-Th**(Σ), **MRel**] thus offers a faithful description for multiset algebras and (tight, point-to-point) homomorphisms. The result is actually obvious at the object level, i.e., the correspondence between algebras and functors is easily established. At the morphism level, note that the components of any natural transformation between gs-monoidal functors (i.e., preserving the transformations ∇ and !) has to be just a function. Indeed, the same property holds for multi-algebras [9, Theorem 17].

As we already noted, in [9] theorems that are equivalent to the above were established for partial and multi-algebras. However, as it happens for partial algebras and conditioned terms, a stronger property holds for multiset algebras.

Theorem 1 (Syntax completeness). *Let Σ be a signature, and let s, t be arrows in* **GS-Th**(Σ). *Then, s and t represent isomorphic term graphs if and only if they are mapped to the same multiset relation for each gs-monoidal functor from* **GS-Th**(Σ) *to* **MRel**.

The correspondence established by the above theorem suggests that term graphs are an adequate syntax on which to build an equational deduction system, and possibly an inequational one, for multiset algebras. The expressiveness of such calculi, of course, should of course be properly assessed.

5.1   A Remark on Partial Functions and Relations

First of all, let us note that one of the main results in [9] was the proof that the functorial category [**GS-Th**(Σ), **Rel**] faithfully represents multi-algebras. As we argue in the concluding remarks, this is now a consequence of Theorem 1 above. Moreover, functorial equivalence was not captured axiomatically, as it is shown by the result below (lifted from [20, Theorem 6]).

Proposition 5 (Garbage completeness). *Let Σ be a signature, and let s, t be arrows in* **GS-Th**(Σ). *Then, s and t are mapped to the same relation for each gs-monoidal functor from* **GS-Th**(Σ) *to* **Rel** *if and only if they represent garbage equivalent term graphs.*

Two typed term graphs with interfaces $\mathbb{G}$, $\mathbb{H}$ are garbage equivalent if there exist graph morphisms $j : G \rightarrow H$ and $i : H \rightarrow G$ between the underlying graphs, such that inputs and outputs are preserved.

Intuitively, it means that $\mathbb{G}$ and $\mathbb{H}$ may only differ for the sub-graph which is not reached by the outputs. The main drawback was that such negative result forbade the development of an equational deduction system (only partly solved by the inequational one presented in [12]).

Instead, the arrows of **G-Th**(Σ) (the category obtained by quotienting **GS-Th**(Σ) with respect to axioms $\nabla_Y \circ f = (f \otimes f) \circ \nabla_X$ for all $f \in \Sigma$) represent conditioned terms, and $[\textbf{G-Th}(\Sigma), \textbf{PFun}]$ and $[\textbf{GS-Th}(\Sigma), \textbf{PFun}]$ turn out to be equivalent: this allows the development of a sound and complete deduction system for partial algebras based on conditioned Kleene equations.

Example 2. Let us now look at the examples. As arrows of the g-monoidal category **G-Th**(Σ), the three term graphs of Fig. 1 represent the same (possibly) partial function, corresponding to the conditioned term $f(c(x), c(x)) \mid \emptyset$. However, they differ from the term graph F_4 in Figure 2, which corresponds instead to conditioned term $f(c(x), c(x)) \mid \{b(x)\}$ (in let notation, we have *let [y be c(x) and y_1 be b(x)] in f(y, y)*). In other words, the sub-term $b(x)$ acts only as a possible restriction over the domain of definition of $f(c(x), c(x))$.

The degree of sharing is instead relevant for relations. Indeed, note that F_2 and F_3 coincide as additive relations, since garbage equivalent: F_2 is a subgraph of F_3, while F_3 can be embedded in F_2 by collapsing the edges labelled by c.

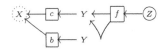

Fig. 2. Term graph F_4

Let us now consider a suitable interpretation for the signature with sorts X, Y and Z, with unary operators $b, c \in \Sigma_{X,Y}$ and the binary operator $f \in \Sigma_{Y \times Y, Z}$. More specifically, we will consider the multiset algebra $\mathcal{E}$ with sort interpretation $X = \{x\}$, $Y = \{y_1, y_2\}$ and $Z = \{z_{ij} \mid i = 1, 2\}$, and operator interpretation $b(x) = n_1 \cdot y_1$, $c(x) = \bigoplus_{i=1,2} n_i \cdot y_i$ and $f(y_i y_j) = z_{ij}$ for $i, j = 1, 2$.

Note now that the derived operation $c \otimes c$ evaluates xx to $\bigoplus_{i,j=1,2} n_i n_j \cdot y_i y_j$, and the derived operation $c \otimes b$ instead evaluates xx to $\bigoplus_{i=1,2} n_i n_1 \cdot y_i y_1$. We leave to the reader to check, by exploiting their arrow representation, that

- $F_1^{\mathcal{E}}$ maps x to $\bigoplus_{i,j=1,2} n_i n_j \cdot z_{ij}$;
- $F_2^{\mathcal{E}}$ maps x to $\bigoplus_{i=1,2} n_i \cdot z_{ii}$;
- $F_3^{\mathcal{E}}$ maps x to $\bigoplus_{i,j=1,2} n_i n_j \cdot z_{ii}$; and
- $F_4^{\mathcal{E}}$ maps x to $\bigoplus_{i=1,2} n_i n_1 \cdot z_{ii}$.

So, the four term graphs differ as multisets. We can see instead that the first three graphs coincide as partial functions: if at most one of the coefficients n_1 and n_2 is 1, and the other has to be 0, then the expressions coincide and evaluate to either $\emptyset$, z_{11} or z_{22}. If $n_1 = 0$, F_4 evaluates to $\emptyset$, regardless of the value of n_2.

The situation is more complex when interpreting the expressions as additive relations. It means to replace natural numbers with the boolean algebra $\{0, 1\}$, thus interpreting $\oplus$ as $\cup$. Thus, F_1 is evaluated to either $\emptyset$ (if $n_1 = n_2 = 0$) or to the set Z. Instead, F_2 and F_3 do coincide, but they may be evaluated to $\emptyset$, the singletons $\{z_{11}\}$ and $\{z_{22}\}$, or to the set $\{z_{11}, z_{22}\}$. As for F_4, it may take the same values of F_2, with the exception of the set $\{z_{22}\}$.

6   Conclusions and Further Works

We presented a functorial description for multiset algebras, as well as stated a completeness property which would allow for a sound and complete (in)equational deduction system for such algebras based on terms graphs. Such a proof calculus would be based on the manipulation of graph expressions, as allowed by the free construction of g- as well as gs-monoidal categories, as e.g. found in [8,9]; or it could be instead directly instantiated on the set-theoretical presentation of term graphs, as suggested in [12].

Previous work on the use of (free) semi-modules over $\mathbb{N}$ for modeling non-determinism appeared, see among others [21]. However, we believe that the completeness property we showed, and the underlying connection between the graphical and the axiomatic presentation of the syntax for multiset algebras, is new. The relevance of the resulting algebraic formalism is still to be assessed, even if to a certain extent subsumes multi-algebras. In fact, it is born out of the attempt to provide an equational presentation for multi-algebras, and we hope it may help out to find a sound and complete set of axioms for garbage equivalence. This should hopefully turn out to be a rather simple system, favorably compared with the solutions presented in the literature (see e.g. the recent [22]).

For the sake of readability we focused on the semirings of natural numbers. Nevertheless, we could have chosen any other semiring, and all the calculations would have carried through. Hence, term graphs represent a sound formalism for the description of any algebraic structure whose operators take value in linear polynomials, with coefficients on a semiring S. Indeed, this fact suggests a way of distilling a functorial semantics for such algebras, simply considering the gs-monoidal functors from $\mathbf{GS\text{-}Th}(\Sigma)$ to the category of semi-modules over S, as it happened for multi-algebras in [9].

One of the future aims where we want to exploit our completeness result is in the definition of a suitable institution for graph rewriting. Indeed, we already know that we can present term graph rewriting by equipping (traced) gs-monoidal categories with a suitable ordering between arrows [8]. Intuitively, the rewriting of a term $f(c(x), b(x))$ into the term $g(c(x))$ can be described by an ordering between the respective representations, as in Figure 3: all "internal" nodes has to become roots, and the rewriting may possibly leave some garbage. This way, Proposition 1 of this paper can be lifted to term graph rewriting.

Indeed, it is known that for term rewriting the model-theoretic side of the associated institution is obtained by considering pre-ordered algebras (see e.g. [23]). We plan to seek a suitable notion of pre-ordered multiset algebras. Such notion should exploit the obvious ordering on multiset relations, as induced by the

Fig. 3. A rewriting rule

ordering on its coefficients: indeed, the binary relation $a \leq b$ if there exists a c such that $a + c = b$ is a pre-order for each semiring, hence for any semi-module.

Finally, note that our completeness result may induce a denotational semantics for those specification formalisms whose components may be described as graphs: indeed, this is the case for most nominal calculi, whose process congruence and operational semantics can be faithfully described by means of graph isomorphism and rewriting, respectively (see e.g. [24]).

References

1. Pitts, A.M.: Categorical Logic. In: Abramsky, S., Gabbay, D.M., Maibaum, T.S.E. (eds.) Handbook of Logic in Computer Science, vol. 5, pp. 39–123. Oxford University Press, Oxford (2001)
2. Walicki, M., Meldal, S.: Algebraic approaches to nondeterminism: An overview. ACM Computing Surveys 29, 30–81 (1997)
3. Hoenke, H.J.: On partial algebras. In: Csákány, B., Fried, E., Schmidt, E. (eds.) Universal Algebra. Colloquia Mathematica Societatis János Bolyai, vol. 29, pp. 373–412. North Holland, Amsterdam (1977)
4. Ştefănescu, G.: Network Algebra. Springer, Heidelberg (2000)
5. Căzănescu, V.E., Ştefănescu, G.: Classes of finite relations as initial abstract data types II. Discrete Mathematics 126, 47–65 (1994)
6. Joyal, A., Street, R., Verity, D.: Traced monoidal categories. Mathematical Proceedings of the Cambridge Philosophical Society 119, 425–446 (1996)
7. Corradini, A., Gadducci, F.: Categorical rewriting of term-like structures. In: Bauderon, M., Corradini, A. (eds.) GETGRATS Closing Workshop. Electr. Notes in Theor. Comp. Sci, vol. 51, pp. 108–121. Elsevier, Amsterdam (2002)
8. Corradini, A., Gadducci, F.: Rewriting on cyclic structures: Equivalence between the operational and the categorical description. Informatique Théorique et Applications/Theoretical Informatics and Applications 33, 467–493 (1999)
9. Corradini, A., Gadducci, F.: A functorial semantics for multi-algebras and partial algebras, with applications to syntax. Theor. Comp. Sci. 286, 293–322 (2002)
10. Corradini, A., Gadducci, F.: An algebraic presentation of term graphs, via gs-monoidal categories. Applied Categorical Structures 7, 299–331 (1999)
11. Burmeister, P.: Partial algebras - An introductory survey. In: Rosenberg, I.G., Sabidussi, G. (eds.) Algebras and Orders. NATO ASI Series C, pp. 1–70. Kluwer Academic, Dordrecht (1993)
12. Corradini, A., Gadducci, F., Kahl, W., König, B.: Inequational deduction as term graph rewriting. In: Mackie, I., Plump, D. (eds.) Term Graph Rewriting. Electr. Notes in Theor. Comp. Sci, vol. 72.1, pp. 31–44. Elsevier, Amsterdam (2002)
13. Kock, J.: Frobenius Algebras and 2-D Topological Quantum Field Theories. London Mathematical Society Student Texts, vol. 59. Cambridge University Press, Cambridge (2003)
14. Hasegawa, M., Hofmann, M., Plotkin, G.D.: Finite dimensional vector spaces are complete for traced symmetric monoidal categories. In: Avron, A., Dershowitz, N., Rabinovich, A. (eds.) Pillars of Computer Science. LNCS, vol. 4800, pp. 367–385. Springer, Heidelberg (2008)
15. Bruni, R., Gadducci, F., Montanari, U.: Normal forms for algebras of connections. Theor. Comp. Sci. 286, 247–292 (2002)

16. Corradini, A., Montanari, U., Rossi, F.: Graph processes. Fundamenta Informaticae 26, 241–265 (1996)
17. Sassone, V., Sobociński, P.: Reactive systems over cospans. In: Logic in Computer Science, pp. 311–320. IEEE Computer Society Press, Los Alamitos (2005)
18. Mac Lane, S.: Categories for the Working Mathematician. Springer, Heidelberg (1971)
19. Golan, J.: Semirings and Affine Equations over Them. Kluwer, Dordrecht (2003)
20. Corradini, A., Gadducci, F., Kahl, W.: Term graph syntax for multi-algebras. Technical Report TR-00-04, Università di Pisa, Department of Informatics (2000)
21. Benson, D.: Counting paths: nondeterminism as linear algebra. IEEE Transactions on Software Engineering 10, 785–794 (1984)
22. Lamo, Y., Walicki, M.: Quantifier-free logic for nondeterministic theories. Theor. Comp. Sci. 355(2), 215–227 (2006)
23. Diaconescu, R.: Institution-independent Model Theory. Birkhäuser, Basel (2008)
24. Gadducci, F.: Graph rewriting for the π-calculus. Mathematical Structures in Computer Science 17(3), 407–437 (2008)

Transformations of Conditional Rewrite Systems Revisited

Karl Gmeiner and Bernhard Gramlich

TU Wien, Austria
{gmeiner,gramlich}@logic.at

Abstract. We revisit known transformations of conditional rewrite systems to unconditional ones in a systematic way. We present a unified framework for describing, analyzing and classifying such transformations, discuss the major problems arising, and finally present a new transformation which has some advantages as compared to the approach of [6]. The key feature of our new approach (for left-linear confluent normal 1-CTRSs) is that it is backtracking-free due to an appropriate encoding of the conditions.

1 Background and Overview

Conditional term rewrite systems (CTRSs) and conditional equational specifications are very important in algebraic specification, prototyping, implementation and programming. They naturally occur in most practical applications. Yet, compared to unconditional term rewrite systems (TRSs), CTRSs are much more complicated, both in theory (especially concerning criteria and proof techniques for major properties of such systems like confluence and termination) and practice (implementing conditional rewriting in a clever way is far from being obvious, due to the inherent recursion when evaluating conditions). For these (theoretical and practical) reasons, transforming CTRSs into (unconditional) TRSs in an adequate way has been studied for a long time cf. e.g. [4, 9, 18, 12, 15, 5, 2, 6, 13, 17, 10]. In many other early papers (like [1, 8]) the issue of transforming conditional into unconditional TRSs is not studied in depth, but at least touched from a programming language point of view.

Roughly, all transformations work by translating the original syntax (signature and terms) into an extended or modified one using auxiliary function symbols, and by translating the rules in a corresponding way such that the evaluation of conditions and some control structure is (appropriately) encoded within the resulting unconditional TRS (in which in some cases reduction is additionally restricted, see below).

In the papers mentioned above certain of these issues have been investigated for particular (quite different) transformations and with different terminology. In order to better understand and relate the different approaches together with their results, we will propose a kind of unified terminology for such transformations and their properties.

A. Corradini and U. Montanari (Eds.): WADT 2008, LNCS 5486, pp. 166–186, 2009.

In the second main part of the paper we will deal with the issue of backtracking and the question whether a transformed system is computationally adequate for simulating the original one. Here we will propose a new approach whose characteristic feature is "backtracking-freeness". The underlying goal here is as follows: If, for some given conditional system, we start a simulation (a reduction in the transformed TRS) from an "initial" term and obtain a normal form in the transformed system, then the latter should correspond to a normal form of the initial term in the original CTRS (this property, together with a few other requirements, is called *computational equivalence* in [6]). Otherwise, some form of backtracking would be needed, because then we are stuck with a failed attempt of verifying conditions, and may need to try another conditional rule.

The rest of the paper is structured as follows. In Section 2 we introduce the necessary background about (conditional) term rewriting. Then, in Section 3 we present and discuss a unifying framework for describing transformations from CTRSs to TRSs. Furthermore known and new unsoundness phenomena are dealt with briefly. Then, in Section 4 we investigate how to design a transformation that avoids explicit backtracking during simulation in such a way that the transformed system still enjoys most desired preservation properties. We motivate the approach by a careful analysis, give a formal definition and present the main results. Finally, in Section 5 we report on some first experiments with our new transformation, briefly discuss related work, sketch possible optimizations, refinements and alternatives, and mention a few interesting perspectives. Due to lack of space, proofs are omitted in the paper.[1]

2   Preliminaries

We assume familiarity with the basic notations and terminology in rewriting, cf. e.g. [3]. We denote by $\mathcal{O}(t)$ the set of all subterm positions of a term t, that is partitioned into all variable positions $\mathcal{O}_{\mathcal{X}}(t) = \{p \in \mathcal{O}(t) \mid t|_p \text{ is a variable}\}$ and all non-variable positions $\overline{\mathcal{O}}(t) = \mathcal{O}(t) \setminus \mathcal{O}_{\mathcal{X}}(t)$. By $\mathcal{V}ars(t)$ we denote the set of all variables occurring in a term t. This notion is extended in the obvious way to rules and conditions. The set of normal forms of a rewrite system $\mathcal{R}$ is denoted by $\mathrm{NF}(\mathcal{R})$. Left- and right-hand sides of rewrite rules are also abbreviated as lhs and rhs, respectively. Slightly abusing notation, we sometimes confuse a rewrite system $\mathcal{R} = (\mathcal{F}, R)$ and its set R of rewrite rules.

Definition 1 (Conditional rewrite system, conditional rewrite relation, depth of reductions). *A conditional term rewriting system (CTRS) $\mathcal{R}$ (over some signature $\mathcal{F}$) consists of rules $l \to r \Leftarrow c$ where c is a conjunction of equations $s_i = t_i$. Equality in the conditions may be interpreted (recursively) e.g. as $\leftrightarrow^*$ (semi-equational case), as $\downarrow$ (join case), or as $\to^*$ (oriented case). In the latter case, if all right-hand sides of conditions are ground terms that are irreducible*

---

[1] More theoretical results, complete proofs and more details about experiments, related work and possible optimizations and refinements can be found in the full version of this paper (forthcoming).

w.r.t. the unconditional version $\mathcal{R}_u = \{l \to r \mid l \to r \Leftarrow c \in \mathcal{R}\}$ of $\mathcal{R}$, the system is said to be a normal one. Furthermore, according to the distribution of variables, a conditional rule $l \to r \Leftarrow c$ may satisfy (1) $\mathcal{V}ars(r) \cup \mathcal{V}ars(c) \subseteq \mathcal{V}ars(l)$, (2) $\mathcal{V}ars(r) \subseteq \mathcal{V}ars(l)$, (3) $\mathcal{V}ars(r) \subseteq \mathcal{V}ars(l) \cup \mathcal{V}ars(c)$, or (4) no variable constraints. If all rules of a CTRS $\mathcal{R}$ are of type (i), $1 \le i \le 4$, respectively, we say that $\mathcal{R}$ is an i-CTRS. The rewrite relation of an oriented CTRS $\mathcal{R}$ is recursively defined as follows: $R_0 \overset{\text{def}}{=} \emptyset$, $R_{n+1} \overset{\text{def}}{=} \{l\sigma \to r\sigma \mid l \to r \Leftarrow s_1 \to^* t_1, \ldots, s_k \to^* t_k \in \mathcal{R}, s_i\sigma \to^*_{R_i} t_i\sigma \text{ for all } 1 \le i \le k\}$, $\to_R \overset{\text{def}}{=} \bigcup_{n \ge 0} R_n$.

In the rest of the paper we will mainly deal with **normal 1-CTRSs**.

3   A Unifying Approach to Transformations

3.1   Basic Transformation Approaches

Basically, two different lines of approaches can be distinguished, according to the way in which the conditions and the intermediate condition evaluation process are encoded. Consider a conditional rule of a given normal 1-CTRS and a term $s = s[l\sigma]$ to be reduced. Obviously, the actual reduction of $s = s[l\sigma]$ into $s' = s[r\sigma]$ has to be delayed until the conditions $s_i\sigma \to^* t_i\sigma = t_i$ have been verified. To this end, the condition evaluation needs to be initiated and performed, while keeping the relevant context, i.e., about the current rule, in order to be finally able to produce $r\sigma$. In one line of approaches (historically the earlier one), the latter information is encoded in an abstract way that hides any concrete structure of l, but keeps the variable bindings of the matching substitution σ. Using the latter, after successful verification of the conditions the corresponding instantiated right-hand side $r\sigma$ can be produced. This means, we need two rules, an *introduction* or *initialization* rule $\rho' : l \to U_\rho(s_1, \ldots, s_n, \mathcal{V}ars(l))$ where $\mathcal{V}ars(s)$ denotes the sequence of the set of all variables occurring in s (in an arbitrary, but fixed order) and the fresh function symbol U_ρ (of appropriate arity) stands for rule ρ, and an *elimination* (or *reducing*) rule $\rho'' : U_\rho(t_1, \ldots, t_n, \mathcal{V}ars(l)) \to r$ that completes the successful rule application after the instantiated conditions $s_i\sigma \to^* t_i\sigma = t_i$ have been verified (by other rewrite steps in between). The most prominent representative of this type of approach are Marchiori's *unravelings* [12]. Early forerunners (with some restrictions/modifications or special cases) and successors of *unraveling* approaches in the literature are among others [4, 8, 15].

In the other main line of approaches, when trying to apply a conditional rule, the left-hand side is not completely abstracted away during verification of the conditions, but instead is kept in a modified form such that the conditions become additional arguments of some function symbol(s) in l, typically of the root function symbol. That means, the arity of this function symbol is increased appropriately by *conditional (argument) positions* which are used to represent the *conditional arguments*. Suppose $l = f(u_1, \ldots, u_k)$. Then f is modified into f' by increasing its arity to $k+n$. For example, for the rule $\rho' : f(x) \to x \Leftarrow x \to^* 0$ the *introduction* and *elimination* rules become $f'(x, \perp) \to f'(x, x)$ and $f'(x, 0) \to x$, respectively. Here, the fresh constant $\perp$ stands for an uninitialized condition. In order to prevent trivial cases of non-preservation of termination[2] we will wrap conditional arguments in some fresh syntactic structure, e.g. as follows: $f'(x, \perp) \to f'(x, \langle x \rangle)$, $f'(x, \langle 0 \rangle) \to x$.[3] Now, increasing the arity of some function symbols in general for storing conditional arguments there requires a more sophisticated construction of the transformed system, since for every occurrence of such function symbols in left- and right hand sides as well in the conditions one has to specify how these conditional arguments should be filled and dealt with during rewriting. And the basic transformation step has to be done for every conditional rule! The basic idea underlying this approach goes back at least till [1]. Yet, the work that inspired many later approaches in this direction is by Viry [18].[4] Other more recent transformation approaches along this line of reasoning include [2, 6, 16].

Intuitively, in both lines of approaches certain reductions in the transformed system during the evaluation of conditions do not correspond to what is done in the conditional system, e.g., reduction in the variable bindings $\mathcal{V}ars(l)$ of $U_\rho(s_1, \ldots, s_n, \mathcal{V}ars(l))$ for unravelings, reduction in the "original arguments" of $f'(x, \langle x \rangle)$, i.e., outside of $\langle x \rangle$, in the conditional argument approach, and reduction above "non-completed" conditional arguments (in both approaches). This phenomenon which may have (and indeed has) problematic consequences for transformations is well-known for a long time. For that reason several approaches in the literature impose some form of (context-sensitivity or strategy or order-sortedness) restrictions on rewriting in the transformed system, e.g. [10, 14, 16, 17, 18], which may lead to better results in theory and/or practice. We will keep this issue in mind, but not deepen it here due to lack of space.

What makes papers and results about transforming conditional systems sometimes hard to read and to compare, is the diversity of the terminology used to reason about their properties. In particular, *soundness* and *completeness* notions are usually defined in different ways. We will instead provide now a proposal for a unified description of such transformations including the relevant terminology.

3.2   A Unified Parameterized Description of Transformations

In view of the existing transformations and since one wants to simulate conditional rewriting in the original system by unconditional rewriting in the transformed one, extending the syntax appears to be unavoidable. So, generally instead of *original terms* from $\mathcal{T} \overset{\text{def}}{=} \mathcal{T}(\mathcal{F}, \mathcal{V})$ the simulation will happen with

---

[2] Note that the introduction rule is obviously non-terminating, whereas the original conditional rule terminates (and is even decreasing cf. [7]).

[3] Strictly speaking, the symbols $\perp$ and $\langle \ldots \rangle$ here are variadic, since they have as many arguments as there are conditions in the respective rule. In a fixed-arity setting one would have to use k-adic symbols $\perp_k$ and $\langle \ldots \rangle_k$ instead, for appropriate arities k.

[4] Even though several main results (and proofs) in [18] are flawed, the ideas and the concrete approach developed there have been very influential.

terms from $T' \stackrel{\text{def}}{=} T(\mathcal{F}', \mathcal{V})$ over an extended or modified signature $\mathcal{F}'$. Moreover, it may be necessary to initially explicitly translate original terms into the new syntax and associate results obtained in the transformed system to original terms. For unravelings this would not be absolutely necessary, but still yields a generalized point of view that turns out to be beneficial for the analysis. For approaches with encoding conditional arguments at new conditional argument positions these mappings are essential, though.

Let us start with the general form of a transformation.[5]

Definition 2 (Transformations of CTRSs). *A* transformation *from a class of CTRSs into a class of TRSs is a total mapping T that associates to every conditional system $\mathcal{R} = (\mathcal{F}, R)$ from the class a triple $((\mathcal{F}', R', \rightarrow_{\mathcal{R}'}), \phi, \psi)$, where $(\mathcal{F}', R')$ is a TRS and $\rightarrow_{\mathcal{R}'}$ is a subset of the rewrite relation induced by $R' = (\mathcal{F}', R')$.[6] We call $\phi: T \rightarrow T'$ the* initialization mapping *(or* encoding*) and $\psi: T' \rightarrow T$ the* backtranslation *(or* decoding*). Furthermore T has to satisfy the following requirements:*

(1) *If $\mathcal{R} = (\mathcal{F}, R)$ is finite (i.e., both $\mathcal{F}$ and R are finite), then $R' = (\mathcal{F}', R')$ is also finite.*

(2) *The restriction of T to finite systems (from the considered class of CTRSs) is effectively constructible.*

(3) *The initialization mapping $\phi: T \rightarrow T'$ is an injective total function.*

(4) *The backtranslation $\psi: T' \rightarrow T$ is a (partial) function that is defined at least on T'_r, the set of all reachable terms[7] which is given by $T'_r \stackrel{\text{def}}{=} \{t' \in T' \mid \phi(s) \rightarrow^*_{\mathcal{R}'} t'$ for some $s \in T\}$.*

(5) *The backtranslation $\psi: T' \rightarrow T$ satisfies $\psi(\phi(s)) = s$ for all $s \in T$, i.e., on all initialized original terms it acts as inverse function w.r.t ϕ.*

Discussion of requirements: Let us briefly discuss this abstract definition of transformation and in particular the requirements mentioned.

First, we parameterize transformations by the class of CTRSs that we want to transform, because this reflects the fact that for different types of CTRSs transformations are typically defined in different ways.

The transformation of $\mathcal{R}$ into $((\mathcal{F}', R', \rightarrow_{\mathcal{R}'}), \phi, \psi)$ allows to impose particular restrictions on $\rightarrow_{\mathcal{R}'}$ like innermost rewriting or context-sensitivity constraints. This ability is crucial in some existing transformation approaches. Intuitively, this happens there in order to simulate more accurately the evaluation of conditions in the transformed setting and to exclude computations that have no analogue in the original system. If such a restriction of ordinary reduction is involved in $\rightarrow_{\mathcal{R}'}$, we will mention this explicitly. Otherwise, again abusing notation, we will simply omit the third component in $(\mathcal{F}', R', \rightarrow_{\mathcal{R}'})$.

---

[5] Alternatively, instead of *transformation* also *encoding, embedding* or *simulation* are used in the literature.

[6] Actually, this is an abuse of notation. By writing $\rightarrow_{\mathcal{R}'}$ we simply want to cover the case, too, where the rewrite relation induced by $\mathcal{R}'$ is somehow restricted.

[7] The notion *reachable term* stems from [6]. In [2], *term of interest* was used instead.

Next, for a given CTRS of the respective class, the transformation does not only yield a transformed signature and rewrite system, but also an *initialization mapping* ϕ and a *backtranslation* ψ. For practical reasons, here the need for requirement (2) is obvious.

Requirement (1) actually is a precondition for (2): When starting with a finite system, we clearly don't want to end up with an infinite one after transformation which usually would lead to non-computability and undecidability problems. Without (1), the ideal and natural candidate for the transformed unconditional system would be $\mathcal{R}' = (\mathcal{F}, R')$ where $R' = \bigcup_{i \geq 0} R_i$ with R_i as in Definition 1 (over the same signature), and with ϕ and ψ the identity function. However, then in general $\mathcal{R}'$ would be infinite and its rewrite relation undecidable! Actually, from a logical point of view this choice of $\mathcal{R}'$ would be ideal, since virtually all typical problems with transformations (like unsoundness phenomena) would simply disappear, because the rewrite relations (of the original and the transformed system) are exactly the same in this case. Yet, for the mentioned reasons, this kind of transformation is practically useless.

Since in general the transformation may work on a modified or extended syntax, the original terms possibly need to be translated (via ϕ) before computation starts. Hence, ϕ should be total and also injective (3), since one clearly wants to be able to distinguish different original terms also after transformation. The other way round is more subtle. Namely, what should one be able to infer from derivations in the transformed system, in terms of the original one? First of all, the backtranslation ψ need not necessarily be total, because there may be terms in $\mathcal{T}'$ which do not correspond to intermediate results of computations in the transformed system. For such *garbage terms* nothing should be required. In particular, there need not exist corresponding original terms for them. However, for every reachable term in the transformed system we do require that there exists indeed some original term to which the former corresponds (4). In a sense, this condition is quite strong.[8] Yet, on an abstract level in general we do not know how the treatment of conditions (of $\mathcal{R}$) in $\mathcal{R}'$ actually works. The intuition behind (4) is that ψ should deliver those "original parts" of an intermediate result $t' \in \mathcal{T}'$ which can obviously be obtained by translating back (i.e., which directly correspond to original syntax). For the other "conditional parts" which correspond to started attempts of applying conditional rules, ψ should go back to the "beginning" of this attempt and recursively only translate back (conditional) rule applications that have been entirely completed. Since the former aspect requires solving reachability problems which are undecidable in general, due to (2) reasonable computable versions of ψ can only approximate "precise backtranslations". Injectivity of ψ would be too strong a requirement, because typically there exist many terms in $\mathcal{T}'$ corresponding in a natural way to a single original term. For initialized original terms $\phi(s), s \in \mathcal{T}$, it should be obvious and

---

[8] In this sense, (4) is perhaps the most debatable requirement in our abstract definition of transformation. Though this definition covers all the major approaches from the literature, it is conceivable that there exist other transformational approaches for which a step-by-step backtranslation need not make sense.

intuitive that via ψ we should get back the original term s (5), simply because the initialization ϕ is the standard or canonical way of translating original terms into the transformed setting.

Based on Definition 2 we will now define various important properties of transformations that are crucial not only in theory, but also in practical applications. Note that the resulting terminology differs from the existing literature.

Definition 3 (Properties of transformations). *Let T be a transformation from a class of CTRSs into the class of TRSs according to Definition 2 and let $\mathcal{R} = (\mathcal{F}, R)$ range over the former class of CTRSs. We define (two versions of) soundness and completeness properties relating (properties $\mathcal{P}$ of) the original and the transformed system (the rectangular brackets indicate the second versions; $\mathcal{P}$ may be e.g. confluence or termination):*

(a) *T is said to be* sound (for reduction) *(or* simulation sound*) [w.r.t. reachable terms] if for every $\mathcal{R} = (\mathcal{F}, R)$ with $T(\mathcal{R}) = ((\mathcal{F}', R', \to_{\mathcal{R}'}), \phi, \psi)$ we have: $\forall s, t \in \mathcal{T}: \phi(s) \to_{\mathcal{R}'}^* \phi(t) \Longrightarrow s \to_{\mathcal{R}}^* t$ [$\forall s', t' \in T_r': s' \to_{\mathcal{R}'}^* t' \Longrightarrow \psi(s') \to_{\mathcal{R}}^* \psi(t'))$].*

(b) *T is said to be* complete (for reduction) *(or* simulation complete*) [w.r.t. reachable terms] if for every $\mathcal{R} = (\mathcal{F}, R)$ with $T(\mathcal{R}) = ((\mathcal{F}', R', \to_{\mathcal{R}'}), \phi, \psi)$ we have: $\forall s, t \in \mathcal{T}: s \to_{\mathcal{R}}^* t \Longrightarrow \phi(s) \to_{\mathcal{R}'}^* \phi(t)$ [$\forall s' \in T_r', t \in \mathcal{T}: \psi(s') \to_{\mathcal{R}}^* t \Longrightarrow \exists t' \in T_r': s' \to_{\mathcal{R}'}^* t', \psi(t') = t$].*

(c) *T is said to be* sound for convertibility *[w.r.t. reachable terms] if for every $\mathcal{R} = (\mathcal{F}, R)$ with $T(\mathcal{R}) = ((\mathcal{F}', R', \to_{\mathcal{R}'}), \phi, \psi)$ we have: $\forall s, t \in \mathcal{T}: \phi(s) \leftrightarrow_{\mathcal{R}'}^* \phi(t) \Longrightarrow s \leftrightarrow_{\mathcal{R}}^* t$ [$\forall s', t' \in T_r': s' \leftrightarrow_{\mathcal{R}'}^* t' \Longrightarrow \psi(s') \leftrightarrow_{\mathcal{R}}^* \psi(t'))$].*

(d) *T is said to be* sound for preserving normal forms *[w.r.t. reachable terms] if for every $\mathcal{R} = (\mathcal{F}, R)$ with $T(\mathcal{R}) = ((\mathcal{F}', R', \to_{\mathcal{R}'}), \phi, \psi)$ we have: $\forall s, t \in \mathcal{T}: \phi(s) \to_{\mathcal{R}'}^* \phi(t) \in NF(\mathcal{R}') \Longrightarrow t \in NF(\mathcal{R})$ [$\forall s' \in T_r': s' \to_{\mathcal{R}'}^* t' \in NF(\mathcal{R}') \Longrightarrow \psi(t') \in NF(\mathcal{R})$[9]].*

(e) *T is said to be* sound for $\mathcal{P}$ *[w.r.t. reachable terms] if $\mathcal{P}(\mathcal{R}')$ implies $\mathcal{P}(\mathcal{R})$ [if $\mathcal{P}(\mathcal{R}')$ on reachable terms implies $\mathcal{P}(\mathcal{R})$].*

(f) *T is said to be* complete for $\mathcal{P}$ *[w.r.t. reachable terms] if $\mathcal{P}(\mathcal{R})$ implies $\mathcal{P}(\mathcal{R}')$ [on reachable terms].*

The above preservation properties of T are "localized" for particular $\mathcal{R}$, $\mathcal{R}'$ and T in the obvious way.[10]

Discussion of terminology: Let us briefly discuss some aspects of these definitions. First, in general the two variants of "T sound / complete for $\mathcal{P}$" and "T sound / complete for $\mathcal{P}$ w.r.t. reachable terms" are not equivalent. We will see an example below. One main goal for the design of any transformation should be *soundness for reduction*. Technically, the stronger *soundness for reduction w.r.t. reachable terms* may be preferable due to proof-technical reasons. Concerning

---

[9] This is equivalent to: $\forall s \in \mathcal{T} \forall t' \in \mathcal{T}': \phi(s) \to_{\mathcal{R}'}^* t' \in NF(\mathcal{R}') \Longrightarrow \psi(t') \in NF(\mathcal{R})$.

[10] Observe that, regarding termination, in practice one is typically interested in slightly different preservation properties, namely of the shape "$\mathcal{R}$ operationally terminating (on s) if (only if, iff) $\mathcal{R}'$ terminating (on $\phi(s)$))".

the computation of normal forms, *soundness for preserving normal forms* is in general strictly weaker than *soundness for preserving normal forms w.r.t. reachable terms*, because it is unrealistic to expect that the (or a) normal form of $\phi(s)$ (for $s \in \mathcal{T}$) has the shape $\phi(t)$ (for some $t \in \mathcal{T}$). Hence, in practice we need here the latter notion. It is a specialized version of *soundness for reduction*, but strengthened by the requirement that the property of being a normal form is preserved backward.

Now, which properties of a transformation T would one expect to have for a computationally feasible simulation of a given system $\mathcal{R}$. Assuming that $\mathcal{R}$ is confluent (a) and operationally terminating (b) (or, equivalently, decreasing, if we assume only one condition per rule), computing the final unique result $t \in \mathrm{NF}(\mathcal{R})$ for some initial $s \in \mathcal{T}$ could be done via $\mathcal{R}'$ as follows: Initialize s into $\phi(s)$, normalize $\phi(s)$ in $\mathcal{R}'$ yielding some t', and finally translate back into $\psi(t')$. For this to work properly, one needs (c) completeness for termination w.r.t. reachable terms, (d) soundness for preserving normal forms w.r.t. reachable terms, and (e) soundness for reduction w.r.t. reachable terms. Then we get $\phi(s) \to_{\mathcal{R}'}^* u' \in \mathrm{NF}(\mathcal{R}')$ for some u' by (b) and (c). Then, by (d) and (e) we obtain $s \to_{\mathcal{R}}^* \psi(u') \in \mathrm{NF}(\mathcal{R})$. This together with $s \to_{\mathcal{R}}^* t \in \mathrm{NF}(\mathcal{R})$ and (a) implies $\psi(u') = t$, i.e., the desired final result.

In fact, it is not difficult to view most transformations from the literature, in particular [12], [2] and [6], as instances of Definition 2 above (with corresponding ϕ and ψ), together with the appropriately adapted terminology according to Definition 3. Due to lack of space, we refrain from describing this in detail. Let us instead briefly discuss a few aspects of unsoundness.

3.3   On-Unsoundness Phenomena

All transformation approaches that do not strongly restrict rewriting in the transformed system (like [10, 14, 16, 17]) such that the simulation of conditional rule application corresponds very closely to what is done in the original CTRS, exhibit in general unsoundness phenomena. More precisely, soundness for reduction (simulation soundness) is violated in general. This has been shown for unravelings by Marchiori in the technical report version [11] of [12] via a tricky counterexample. We present here a slightly simplified variant of this counterexample.[11]

Example 1 (unsoundness (for reduction) in general). The normal 1-CTRS $\mathcal{R}$ consisting of the rules $a \to c$, $a \to d$, $b \to c$, $b \to d$, $c \to e$, $c \to k$, $d \to k$, $h(x, x) \to g(x, x, f(d))$, $g(d, x, x) \to A$, $f(x) \to x \Leftarrow x \to^* e$ is unraveled according to [12] into $\mathcal{R}'$ which is $\mathcal{R}$ with the conditional rule replaced by $f(x) \to U(x, x)$ and $U(e, x) \to x$. In $\mathcal{R}'$ we have the following derivation between original terms: $h(f(a), f(b)) \to_{\mathcal{R}'}^* h(U(c, d), U(c, d)) \to g(U(c, d), U(c, d), f(d)) \to_{\mathcal{R}'}^* g(d, U(k, k), U(k, k)) \to_{\mathcal{R}'} A$. In the original system, however, $h(f(a), f(b)) \to_{\mathcal{R}}^* A$ does not hold as is easily seen!

---

[11] Compared to our version, [11, Example 4.3] has two more rules and two more constants.

Note that Example 1 also constitutes a counterexample to soundness (in general) for most other transformation approaches including [2, 6]. We will not go into details about sufficient conditions that nevertheless ensure soundness for particular transformations (some of them are well-known to exist, cf. e.g. [11], [15], [2], [6]). Instead, let us ask – concerning our proposed terminology – whether soundness w.r.t reachable terms is strictly stronger than ordinary soundness. In fact, the answer is Yes, again for most approaches!

Example 2 (soundness is weaker than soundness w.r.t. reachable terms). Consider the normal 1-CTRS $\mathcal{R} = \{a \rightarrow b, a \rightarrow c, f(x) \rightarrow x \Leftarrow x \rightarrow^* c\}$ that, via [2], is transformed into $\mathcal{R}' = \{a \rightarrow b, a \rightarrow c, f'(x, \bot) \rightarrow f'(x, \langle x \rangle), f'(x, \langle c \rangle) \rightarrow x\}$. In $\mathcal{R}'$ we have $f'(a, \bot) \rightarrow_{\mathcal{R}'} f'(a, \langle a \rangle) \rightarrow^*_{\mathcal{R}'} f'(b, \langle c \rangle) \rightarrow b$. However, back-translation of the last two reachable terms here yields $\psi(f'(b, \langle c \rangle)) = f(b)$ and $\psi(b) = b$. Yet, in $\mathcal{R}$, we clearly do not have $f(b) \rightarrow^*_{\mathcal{R}} b$. Hence the transformation is not sound w.r.t. reachable terms. Note, however, that we do not obtain ordinary unsoundness here, because $\phi(f(a)) = f'(a, \bot) \rightarrow^*_{\mathcal{R}'} \phi(b) = b$ implies indeed $f(a) \rightarrow^*_{\mathcal{R}} b$.

Furthermore let us just mention that unravelings are in general also unsound for convertibility (a counterexample is obtained by extending Example 1 above.

And finally, we observe that some transformations in the literature (e.g. [4] and the technical report version [11] of [12]) are inherently unsound, because they do not transmit all variable bindings in the encoding process (in the introduction rule). Counterexamples for these cases can easily be constructed by adding a non-left-linear rule.

4   The New Transformation

4.1   Motivation, Goal and Basic Idea

Our transformation will be based on the approach of [2]. The transformation in [2] is complete, sound w.r.t. reachable terms and sound for preserving normal forms w.r.t. reachable terms for "constructor-based" (normal 1-)CTRSs that are left-linear and confluent. However, for other CTRSs it may be unsound or incomplete for confluence.

Example 3 (Incompleteness for confluence cf. [2, Example 4]). Consider the confluent CTRS $\mathcal{R} = \{g(s(x)) \rightarrow g(x) , f(g(x)) \rightarrow x \Leftarrow x \rightarrow^* 0\}$. The transformation of [2] returns the TRS $\mathcal{R}' = \{g(s(x)) \rightarrow g(x), f'(g(x), \bot) \rightarrow f'(g(x), \langle x \rangle), f'(g(x), \langle 0 \rangle) \rightarrow x\}$. In $\mathcal{R}'$ we have the derivation $f'(g(s(0)), \bot) \xrightarrow{(1)}_{\mathcal{R}'} f'(g(s(0)), \langle s(0) \rangle) \xrightarrow{(2)}_{\mathcal{R}'} f'(g(0), \langle s(0) \rangle)$. The latter term is irreducible and does not rewrite to 0 although $f(g(0)) \rightarrow_{\mathcal{R}} 0$. $f'(g(s(0)), \bot)$ corresponds to $f(g(s(0)))$ in $\mathcal{R}$ that is matched by the lhs of the conditional rule l with the matcher $\tau = \{x \mapsto s(0)\}$. After introducing the introduction step (1) the unconditional g-rule is applied to the g-subterm (2). $f'(g(0), \langle s(0) \rangle)$ corresponds to $f(g(0))$ in the original CTRS that is matched by l with the matcher $\sigma = \{x \mapsto 0\}$. Since there are no derivations $x\tau \to_{\mathcal{R}}^* x\sigma$ $(x \in \mathit{Vars}(l))$ the conditional argument is "outdated" and "inconsistent" with the original argument.

Such inconsistencies may block further derivations as in Example 3 or lead to unsound derivations as in [2, Example 6]. We will refer to derivations that reproduce a term that is matched by the lhs of a conditional rule as being *pattern preserving*. Our goal is to provide a transformation that conservatively extends the transformation of [2] and does not require explicit propagation of reset information such that for confluent normal 1-CTRSs no explicit backtracking is needed. This means that in particular "critical" pattern preserving derivations are dealt with correctly.

In the derivation of Example 3, the unconditional rule $g(s(x)) \to g(x)$, that is applied in (2) and leads to the "outdated" conditional argument, should eliminate (and re-initialize) the conditional argument that has been introduced in (1). Yet the conditional argument is "out of reach" for the unconditional rule. By encoding the conditional argument in the g-subterm of the conditional rule, however, we can eliminate the conditional argument in the unconditional g-rule. This way $\mathcal{R}$ is transformed into $\mathcal{R}' = \{g'(s(x), z) \to g'(x, \bot),\ f(g'(x, \bot)) \to f(g'(x, \langle x \rangle)),\ f(g'(x, \langle 0 \rangle)) \to x\}$. Now the conditional argument can be reintroduced: $f(g'(s(0), \bot)) \to f(g'(s(0), \langle s(0) \rangle)) \to f(g'(0, \bot)) \to^* 0$.

Following this example our strategy is to encode the conditions in all subterms of the lhs of a conditional rule that otherwise would give rise to "inconsistencies" of conditional and original arguments. To avoid confusion we will refer to subterms that directly (i.e., as additional argument of the root function symbol) contain a conditional argument as subterms encoding a conditional argument.

In certain conditional rules several subterms subterm of the lhs may lead to pattern preserving derivations. Then we have to encode the conditions multiple times:

Example 4 (Multiple conditional arguments). Consider the CTRS

$$\mathcal{R} = \left\{ \begin{array}{cc} g(s(x)) \to g(x) & h(s(x)) \to h(x) \\ f(g(x), h(y)) \to i(x, y) \Leftarrow x \to^* 0, y \to^* 0 \end{array} \right\}$$

Both subterms $g(x)$ and $h(y)$ of the lhs of the conditional rule lead to a critical pattern preserving derivation. Therefore we need to add a conditional argument to the g-subterm and the h-subterm.

Whenever a conditional argument was eliminated and reinitialized, hence has become $\bot$, we have to reintroduce both conditional arguments via an introduction step. Therefore we need one introduction rule for each conditional argument. Only if both conditional arguments "satisfy" the conditions we may reproduce the corresponding rhs. Hence the transformed TRS $\mathcal{R}'$ here should be

$$\mathcal{R}' = \left\{ \begin{array}{c} g'(s(x), z) \to g'(x, \bot) \qquad h'(s(x), z) \to h'(x, \bot) \\ f(g'(x, \bot), h'(y, z_2)) \to f(g'(x, \langle x, y \rangle), h'(y, \langle x, y \rangle)) \\ f(g'(x, z_1), h'(y, \bot)) \to f(g'(x, \langle x, y \rangle), h'(y, \langle x, y \rangle)) \\ f(g'(x, \langle 0, 0 \rangle), z_2), h'(y, z_1, \langle 0, 0 \rangle)) \to i(x, y) \end{array} \right\}$$

Consider a conditional rule $\rho : l \to r \Leftarrow s_1 \to^* t_1, \ldots, s_n \to^* t_n$ of a CTRS $\mathcal{R}$. Subterms where possibly "critical" pattern preserving derivations start are always (instances of) non-variable subterms $l|_p$ $(p \in \overline{\mathcal{O}}(l))$ of l, because inconsistencies of conditional arguments and original arguments in variables can be resolved by rewrite steps in the variable bindings (for confluent CTRSs). However, in order to detect all possibilities of overlaps (also after rewriting in 'variable parts'), we must linearize l into l^{lin}. Then we can identify possible lhs's that overlap into l^{lin} by systematic unification of all non-variable subterms of l with lhs's of $\mathcal{R}$ and encode the conditions in such subterms of l.

It is not necessary to encode conditions in all subterms of l^{lin} that are unifiable with some lhs. If a subterm does not contain any variables that occur in the conditions, rewrite steps in this subterm do not influence the satisfiability of the conditions and therefore it is not necessary to introduce the conditions here. Yet, we must consider the case that (only) after some rewrite steps in such subterms a rule may become applicable that did not overlap into l^{lin} initially.

Example 5 (Iterative abstraction of overlapping subterms). Consider the CTRS

$$\mathcal{R} = \left\{ \begin{array}{ll} a \to b & g(s(x), k(b)) \to g(x, h(a)) \\ h(b) \to k(b) & f(g(x, h(a))) \to x \Leftarrow x \to^* 0 \end{array} \right\}$$

The only subterm of the linearized lhs of the conditional rule l^{lin} into which an lhs of some rule overlaps is the constant a which does not contain any variable of the condition. Since there are no other overlaps, we would just encode the conditions in the root symbol of the conditional rule:

$$\mathcal{R}' = \left\{ \begin{array}{ccc} a \to b & g(s(x), k(b)) \to g(x, h(a)) & h(b) \to k(b) \\ f'(g(x, h(a)), \bot) \to f'(g(x, h(a)), \langle x \rangle) & & f'(g(x, h(a)), \langle s(0) \rangle) \to x \end{array} \right\}$$

In $\mathcal{R}'$ we have the unsound derivation

$$f'(g(s(0), h(a)), \bot) \to f'(g(s(0), h(a)), \langle s(0) \rangle) \to f'(g(s(0), h(b)), \langle s(0) \rangle)$$
$$\to f'(g(s(0), k(b)), \langle s(0) \rangle) \to f'(g(0, h(a)), \langle s(0) \rangle) \to 0 .$$

Although (the lhs of) the g-rule does not overlap into l^{lin}, it is applicable after some rewrite steps. Therefore, we abstract all non-variable subterms of l^{lin}, that are unifiable with some lhs of $\mathcal{R}$, into fresh variables iteratively and try to unify the non-variable subterms of the resulting terms with lhs's of $\mathcal{R}$. In the example, the a-rule overlaps into $l_0 = f(g(x, h(a)))$ so that we abstract it into $l_1 = f(g(x, h(y)))$. Because of the overlap with the h-rule this term then is abstracted into $l_2 = f(g(x, z))$. Now the g-rule overlaps into the g-subterm of l_2 that contains the variable x that also occurs in the conditions. We therefore encode a conditional argument in the g-subterm instead of f and obtain the transformed TRS

$$\mathcal{R}' = \left\{ \begin{array}{ccc} a \to b & g'(s(x), k(b), z) \to g'(x, h(a), \bot) & h(b) \to k(b) \\ f(g'(x, h(a), \bot)) \to f(g'(x, h(a), \langle x \rangle)) & & f(g'(x, h(a), \langle 0 \rangle)) \to x \end{array} \right\}$$

In CTRSs that give rise to multiple conditional arguments, it may be possible to "recombine" them in an inconsistent way, if we do not iterate the sketched construction:

Example 6 (Recombination of conditional arguments). Consider the CTRS

$$\mathcal{R} = \left\{ \begin{array}{lll} i(0, s(0)) \to 0 & i(s(0), 0) \to 0 & f(s(x), y) \to s(x) \\ f(x, s(y)) \to s(y) & g(s(x)) \to g(x) & h(s(x)) \to h(x) \\ \multicolumn{3}{c}{f(s(g(x)), s(h(y))) \to i(x, y) \ \Leftarrow\ i(x, y) \to^* 0} \end{array} \right\}$$

The g-rule and the h-rule overlap into the lhs of the conditional rule, therefore we would encode the condition in both subterms:

$$\mathcal{R}' = \left\{ \begin{array}{lll} i(0, s(0)) \to 0 & i(s(0), 0) \to 0 & f(s(x), y) \to s(x) \\ f(x, s(y)) \to s(y) & g'(s(x), z) \to g'(x, \perp) & h'(s(x), z) \to h'(x, \perp) \\ \multicolumn{3}{l}{f(s(g'(x, \perp)), s(h'(y, z_2))) \to f(s(g'(x, \langle i(x, y) \rangle)), s(h'(y, \langle i(x, y) \rangle)))} \\ \multicolumn{3}{l}{f(s(g'(x, z_1)), s(h'(y, \perp))) \to f(s(g'(x, \langle i(x, y) \rangle)), s(h'(y, \langle i(x, y) \rangle)))} \\ \multicolumn{3}{l}{f(s(g'(x, \langle 0 \rangle)), s(h'(y, \langle 0 \rangle))) \to i(x, y)} \end{array} \right\}$$

In $\mathcal{R}'$, we now have the following derivation:

$$f(f(s(g'(0, \perp)), s(h'(s(0), \perp))), f(s(g'(s(0), \perp)), s(h'(0, \perp))))$$
$$\to^* f(f(s(g'(0, t_1)), s(h'(s(0), t_1))), f(s(g'(s(0), t_2)), s(h'(0, t_2))))$$
$$\to^* f(s(g'(0, t_1)), s(h'(0, t_2))) \to^* f(s(g'(0, \langle 0 \rangle)), s(h'(0, \langle 0 \rangle))) \to i(0, 0)$$

with $t_1 = \langle i(0, s(0)) \rangle$ and $t_2 = \langle i(s(0), 0) \rangle$, whereas in $\mathcal{R}$ we have

$$f(f(s(g(0)), s(h(s(0)))), f(s(g(s(0))), s(h(0)))) \not\to_{\mathcal{R}}^* i(0, 0)\,.$$

In order to avoid that "fragments" of introduction steps can be inconsistently rearranged, we will iteratively abstract (in parallel) all non-variable subterms of the lhs of the conditional rule, that (after transformation) contain conditional arguments, into new variables. This way the lhs of the conditional rule $f(s(g(x)), h(y)))$ is abstracted into $f(s(z_1), s(z_2))$. But then we have an overlap with the unconditional f-rules at root position. Hence, we will also encode the conditional argument at root position. Thus the above problem disappears.

4.2   Definition of the Transformation

In our transformation we iteratively abstract "overlapping" subterms of lhs's of conditional rules into new variables and append conditional arguments to such subterms provided they contain variables that also occur in the conditions. Additionally we have to take into account that also rules into which the lhs of a conditional rule overlaps may lead to inconsistencies. Before defining our transformation we define some mappings to increase (decrease) the arity of function symbols:

Definition 4 (Initialization mapping and backtranslation)

$$\phi_f^{\perp}(t) = \begin{cases} f'(\phi_f^{\perp}(t_1), \dots, \phi_f^{\perp}(t_n), \perp) & \text{if } t = f(t_1, \dots, t_n) \\ g(\phi_f^{\perp}(t_1), \dots, \phi_f^{\perp}(t_m)) & \text{if } t = g(t_1, \dots, t_m), \ g \neq f \\ t & \text{if } t \text{ is a variable} \end{cases}$$

$$\psi_f(t') = \begin{cases} f(\psi_f(t_1), \dots, \psi_f(t_n)) & \text{if } t' = f'(t_1, \dots, t_n, u_1) \\ g(\psi_f(t_1), \dots, \psi_f(t_m)) & \text{if } t' = g(t_1, \dots, t_m), \ g \neq f' \\ t' & \text{if } t' \text{ is a variable} \end{cases}$$

$$\phi_f^{\mathcal{X}}(t) = \begin{cases} f'(\phi_f^{\mathcal{X}_1}(t_1), \dots, \phi_f^{\mathcal{X}_n}(t_n), z) & \text{if } t = f(t_1, \dots, t_n) \\ g(\phi_f^{\mathcal{X}_1}(t_1), \dots, \phi_f^{\mathcal{X}_m}(t_m)) & \text{if } t = g(t_1, \dots, t_m), \ g \neq f \\ t & \text{if } t \text{ is a variable} \end{cases}$$

where $\mathcal{X}$ is an infinite set of new variables, $z \in \mathcal{X}$ and $\mathcal{X}_i$ ($i \geq 1$) are infinite disjoint subsets of $\mathcal{X}$ such that $z \notin \mathcal{X}_i$. We abbreviate multiple applications of these mappings: $\phi_{f_1, \dots, f_n}^{\perp}(t) = \phi_{f_2, \dots, f_n}^{\perp}(\phi_{f_1}^{\perp}(t))$, $\psi_{f_1, \dots, f_n}(t) = \psi_{f_2, \dots, f_n}(\psi_{f_1}(t))$ and $\phi_{f_1, \dots, f_n}^{\mathcal{X}}(t) = \phi_{f_2, \dots, f_n}^{\mathcal{X} \setminus \mathcal{V}ars(\phi_{f_1}^{\mathcal{X}}(t))}(\phi_{f_1}^{\mathcal{X}}(t))$.

By abuse of notation we assume in the following that, if $\phi^{\mathcal{X}}$ is used multiple times in a term, then always only mutually distinct variables are inserted.

Definition 5 (Definition of the transformation T). Let $\mathcal{R}$ be a normal 1-CTRS so that the rules are arranged in some arbitrary but fixed total order $<$. Let $\rho : l_\rho \to r_\rho \Leftarrow s_{\rho,1} \to^* t_{\rho,1}, \dots, s_{\rho,n_\rho} \to^* t_{\rho,n_\rho}$ be a conditional rule of $\mathcal{R}$. Let l_i and P_i be the following

$$l_0 = l_\rho^{lin} \qquad l_{i+1} = l_i[z_1]_{q_1} \dots [z_m]_{q_m}$$

$$P_0 = \emptyset \qquad P_{i+1} = P_i \cup \{q \in \overline{Q} \mid \mathcal{V}ars(l_\rho|_q) \cap \mathcal{V}ars(s_{\rho,1}, \dots, s_{\rho,n_\rho}) \neq \emptyset\}$$

where $Q = \{q \in \overline{\mathcal{O}}(l_i) \mid l_i\sigma = l_i\sigma[l_{\rho'}\sigma]_q, \rho' \in \mathcal{R}, \rho' \neq \rho \ \vee \ q \neq \epsilon\}$ are all positions of l_i that are unifiable with some lhs $l_{\rho'}$ (except $\rho' = \rho$ at root position), $\{q_1, \dots, q_m\} = \overline{Q} = \{q \in Q \mid \nexists q' \in Q : q < q'\}$ are the innermost positions of Q and $z_1, \dots, z_m$ are fresh new variables.

Let $\overline{l}_\rho = l_j$ be the first l_j such that $l_j = l_{j+1}$. Then the set of conditional positions P_ρ is

$$P_\rho = \begin{cases} P_j \cup \{\epsilon\} & \text{if } \exists \rho', l_{\rho'}\sigma = l_{\rho'}\sigma[\overline{l}_\rho\sigma]_q \text{ with } q \in \overline{\mathcal{O}}(l_{\rho'}) \text{ and } \rho' \neq \rho \vee q \neq \epsilon, \\ & \text{or } P_j = \emptyset \text{ and } \rho \text{ is a conditional rule} \\ P_j & \text{otherwise} \end{cases}$$

Let $\{p_{\rho,1}, \dots, p_{\rho,k_\rho}\} = P_\rho$ and $f_{\rho,j} = \text{root}(l|_{p_{\rho,j}})$. Then the position of the conditional argument encoded in $l_\rho|_{p_{\rho,j}}$ is

$$i_{\rho,j} = \text{arity}(f_{\rho,j}) + 1 + |\{\langle \rho', j' \rangle \mid \langle \rho', j' \rangle <_{lex} \langle \rho, j \rangle, f_{\rho',j'} = f_{\rho,j}\}|$$

Let $\mathcal{R} = \{\rho_1, \ldots, \rho_m\}$. The initialization mapping ϕ is $\phi^{\perp}_{f_{\rho_1},1,\ldots,f_{\rho_m},k_{\rho_m}}$, $\phi^{\mathcal{X}}$ is $\phi^{\mathcal{X}}_{f_{\rho_1},1,\ldots,f_{\rho_m},k_{\rho_m}}$ and the backtranslation ψ is $\psi_{f_{\rho_1},1,\ldots,f_{\rho_m},k_{\rho_m}}$.

A rule $\rho \in \mathcal{R}$ is transformed into the rules

$$\rho'_{\rho,j} : \phi^{\mathcal{X}}(l_\rho)[\perp]_{p_{\rho,j} \cdot i_{\rho,j}} \to \phi^{\mathcal{X}}(l_\rho)[\langle \phi^{\perp}(s_{\rho,1}), \ldots, \phi^{\perp}(s_{\rho,n_\rho}) \rangle]_{p_{\rho,1} \cdot i_{\rho,1}, \ldots, p_{\rho,k_\rho} \cdot i_{\rho,k_\rho}}$$

$$\rho'_{\rho,k_\rho+1} : \phi^{\mathcal{X}}(l_\rho)[\langle \phi^{\mathcal{X}}(t_{\rho,1}), \ldots, \phi^{\mathcal{X}}(t_{\rho,n_\rho}) \rangle]_{p_{\rho,1} \cdot i_{\rho,1}, \ldots, p_{\rho,k_\rho} \cdot i_{\rho,k_\rho}} \to \phi^{\perp}(r_\rho)$$

$\rho'_{\rho,1}, \ldots, \rho'_{\rho,k_\rho}$ are the introduction rules and $\rho'_{k_\rho+1}$ is the elimination rule of ρ. The transformed TRS $T(\mathcal{R})$ then is $(\mathcal{R}', \phi, \psi)$ where $\mathcal{R}' = \{\rho'_{\rho_0,1}, \ldots, \rho'_{\rho_m,k_{\rho_m}+1}\}$.

In constructor-based normal 1-CTRSs, P_ρ is $\{\epsilon\}$ for all conditional rules ρ so that in this case our transformation coincides with the transformation of [2], except for the additional wrapping $\langle \ldots \rangle$ of the conditional arguments. In unconditional rules ρ, $P_\rho = \emptyset$.

The following example of [6] can be interpreted as a "self-sorting" list structure:

Example 7 (Sorting CTRS of [6]). Consider the CTRS

$$\mathcal{R} = \left\{ \begin{array}{lll} 0 \le y \to tt & s(x) \le 0 \to f\!f & s(x) \le s(y) \to x \le y \\ & f(x, f(y, ys)) \to f(y, f(x, ys)) \Leftarrow x \le y \to^* f\!f & \end{array} \right\}$$

The (linear) lhs of the conditional rule $l_0 = f(x, f(y, ys))$. The conditional rule overlaps into itself at position $q = 2$ such that $P_1 = \{2\}$ and $l_1 = f(x, z)$. Now only the conditional rule itself overlaps into l_1 at root position, therefore $\bar{l} = l_1 = f(x, z)$. Since $\bar{l}$ overlaps into the lhs of the conditional rule at some non-root position $P_\rho = \{2\} \cup \{\epsilon\}$. For both positions the root symbol is f and the arity of f is increased by 2. The transformed TRS then is

$$\mathcal{R}' = \left\{ \begin{array}{l} 0 \le y \to tt \qquad s(x) \le 0 \to f\!f \qquad s(x) \le s(y) \to x \le y \\ f(x, f(y, ys, z_1, z_2), z_3, \perp) \to f(x, f(y, ys, \langle x \le y \rangle, z_2), z_3, \langle x \le y \rangle) \\ f(x, f(y, ys, \perp, z_2), z_3, z_4) \to f(x, f(y, ys, \langle x \le y \rangle, z_2), z_3, \langle x \le y \rangle) \\ f(x, f(y, ys, \langle f\!f \rangle, z_2), z_3, \langle f\!f \rangle) \to f(y, f(x, ys, \perp, \perp), \perp, \perp) \end{array} \right\}$$

4.3 Properties of the Transformation

In the following we assume that $\mathcal{R}$ always denotes a normal 1-CTRS, $\mathcal{R}'$ its transformed TRS using our transformation T and ρ a conditional rule with lhs l that leads to k conditional positions $p_1.i_1, \ldots, p_k.i_k$.

The following result contains a selection of syntactical preservation properties of our transformation. Properties (2) - (5) are not satisfied by the transformation of [6].

Lemma 1 (Syntactic properties)

(1) The transformation is sound and complete for being left-linear.
(2) The transformation is sound and complete for being non-collapsing.

(3) If $\mathcal{R}$ is non-overlapping, then $\mathcal{R}'$ is weakly non-overlapping.
(4) The transformation is sound and complete for being an overlay system.
(5) If $\mathcal{R}$ is orthogonal, $\mathcal{R}'$ is weakly orthogonal.

In order to show that our transformation has nice preservation properties for certain CTRSs, we have to guarantee that inconsistencies of conditional arguments and original arguments do not occur or are not "critical".

In a derivation D starting from some initialized term every conditional argument originates from an introduction step. A conditional argument originating from a certain introduction step may be viewed as being inconsistent if the redex of the introduction step is modified at some original argument that is not inside the matcher of the introduction step (rewrite steps in the matcher can be reconstructed in the conditional arguments, at least directly after their introduction). The redex of the "modifying" rewrite step overlaps with the redex of the introduction step so that, according to our transformation, there is (at least) one conditional argument inside the matcher of the "modifying" redex (unless the rewrite step is not potentially dangerous, i.e., it does not modify any variable that occurs in the conditions or it is an introduction step). We will only consider those conditional arguments as being inconsistent, that are inside such a redex, and refer to the overlapping rewrite step as the rewrite step in which the conditional arguments *become* inconsistent.

Definition 6 (Inconsistent conditional arguments). *Let D be a derivation $\phi(s) \rightarrow^* u_0 \rightarrow_{q_0,\rho_0'} u_1 \rightarrow_{q_1,\rho_1'} \cdots (s \in \mathcal{T})$ in $\mathcal{R}'$. Let ρ_0' be an introduction rule of ρ and $u_n|_{q.i_j}$ be a conditional argument that is a descendant of $u_1|_{q_0.p_j.i_j}$. The conditional argument in $u_n|_{q.i_j}$ is inconsistent w.r.t. D, if there is an elimination step $u_m = u_m[l_m'\sigma]_{q_m} \rightarrow_{q_m,\rho_m'} u_m[r_m'\sigma]_{q_m} = u_{m+1}$ in D such that $u_m|_{q_m'.p_j.i_j}$ is a descendant of $u_1|_{q_0.p_j.i_j}$ and an ancestor of $u_n|_{q.i_j}$, and the elimination step "overlaps with" the descendant $u_m|_{q_m'}$ of $u_0|_{q_0}$ above p_j, i.e., $q_m < q_m'.p_j$ and there is no $q' \in \mathcal{O}_{\mathcal{X}}(l_m')$ such that $q_m.q' \leq q_m'$. A conditional argument that is not inconsistent (w.r.t. D) is consistent (w.r.t. D).*

Lemma 2 (Iterative abstraction, inconsistent conditional arguments)
Let $D\colon \phi(s) \rightarrow^*_{\mathcal{R}'} s' \rightarrow_{\mathcal{R}'} t' (s \in \mathcal{T})$ be a derivation in $\mathcal{R}'$ such that $\mathrm{root}(s'|_q) = \mathrm{root}(\phi(l|_{p_j}))$ and $s'|_{q.i_j} \neq \bot$ for some $j \in \{1,\dots,k\}$. If the conditional argument $s'|_{q.i_j}$ becomes inconsistent w.r.t. D in the last rewrite step $s' \rightarrow t'$ of D, then there is a $q' \in \mathcal{O}(s')$ such that $q'.p_j = q$ and in the iteration of Definition 5 for ρ we obtain some l_i that is unifiable with $\psi(s'|_{q'})$, and there is some conditional position $p.i \in \{p_1.i_1, \dots, p_k.i_k\}$ of ρ such that $p < p_j$.

If inconsistent conditional arguments "block" or "are used in" elimination steps, we may obtain incompleteness for confluence or unsoundness. We will refer to those CTRSs where this cannot happen as *consistently transformable* ones.

Definition 7 (Consistently transformable). $\mathcal{R}$ *is* consistently transformable, *if for every derivation $D\colon \phi(s) \rightarrow^*_{\mathcal{R}'} s' (s \in \mathcal{T})$ such that $\psi(s'|_q) = l\sigma$, either $s'|_{q.p_j.i_j} = \bot$ for some j or $s'|_{q.p_j.i_j}$ is consistent w.r.t. D for all $j \in \{1,\dots,k\}$.*

Unfortunately, not all CTRSs are consistently transformable:

Example 8 (Inconsistent conditional arguments in collapsing systems). Consider the CTRS

$$\mathcal{R} = \left\{ \begin{array}{cc} i(a, a) \to a & g(f(x, b)) \to x \\ f(g(x), y) \to h(x) \Leftarrow i(x, y) \to^* a \end{array} \right\}$$

We obtain $P_2 = \{\epsilon, 1\}$ and hence the transformed TRS is

$$\mathcal{R}' = \left\{ \begin{array}{c} i(a, a) \to a \qquad g'(f'(x, b, z_1), z_2) \to x \\ f'(g'(x, \perp), y, z_2) \to f'(g'(x, \langle i(x, y) \rangle), y, \langle i(x, y) \rangle) \\ f'(g'(x, z_1), y, \perp) \to f'(g'(x, \langle i(x, y) \rangle), y, \langle i(x, y) \rangle) \\ f'(g'(x, \langle a \rangle), y, \langle a \rangle) \to h(x) \end{array} \right\}$$

$\mathcal{R}$ has two infeasible conditional critical pairs and is therefore confluent (it is also decreasing). Yet, in $\mathcal{R}'$ we obtain $D: f'(g'(f'(g'(a, \perp), b, \perp), \perp), a, \perp) \to^*$ $f'(g'(f'(g'(a, t_1), b, t_1), t_2), a, t_2) \to f'(g'(a, t_1), t_2)$ where $t_1 = \langle i(a, b) \rangle$ and $t_2 = \langle i(f'(g'(a, \perp), b, \perp), a) \rangle$. The inner conditional argument is inconsistent w.r.t. D while the outer conditional argument is consistent. Usually, we would expect that the inner conditional argument is set to $\perp$, but since the g-rule is collapsing, no conditional argument is set to $\perp$ in its rhs. Hence, the inconsistent (w.r.t. D) conditional argument blocks further reductions.

Lemma 3 (Inconsistent conditional arguments and collapsing rules)
Let $D: \phi(s) \to^*_{\mathcal{R}'} s'$ be a derivation in $\mathcal{R}'$ such that $\psi(s'|_q) = l\sigma$ and $s'|_{q.p_j.i_j} \neq \perp$ for all $j \in \{1, \ldots, k\}$. If some conditional argument $s'|_{q.p_j.i_j}$ is inconsistent w.r.t. D, then $\mathcal{R}$ is collapsing.

Theorem 1 (Sufficient syntactic conditions for consistent transformability (a)). *$\mathcal{R}$ is consistently transformable, if $\mathcal{R}$ is non-collapsing or all $\rho \in \mathcal{R}$ only lead to pairwise parallel subterms encoding conditional arguments.*

Theorem 2 (Sufficient syntactic conditions for consistent transformability (b)). *$\mathcal{R}$ is consistently transformable, if it is non-collapsing, a constructor system, a system where all left-hand sides of conditional rules are constructor terms, or a left-linear overlay system.*

Theorem 3 (Soundness w.r.t. reachable terms). *If $\mathcal{R}$ is consistently transformable and confluent, then for all reachable $s', t' \in \mathcal{T}_r'$ $s' \to^*_{\mathcal{R}'} t' \Rightarrow \psi(s') \to^*_{\mathcal{R}} \psi(t')$.*

Theorem 4 (Completeness w.r.t. reachable terms). *If $\mathcal{R}$ is consistently transformable, left-linear and confluent, then for all $s' \in \mathcal{T}_r', t \in \mathcal{T}$ such that $\psi(s') \to_{\mathcal{R}} t$ there is a $t' \in \mathcal{T}_r'$ with $s' \to^+_{\mathcal{R}'} t'$ and $\psi(t') = t$.*

Theorem 5 (Soundness for preserving normal forms w.r.t. reachable terms). *If $\mathcal{R}$ is consistently transformable, left-linear and confluent, then $\mathcal{R}'$ is sound for preserving normal forms w.r.t. reachable terms.*

Theorem 6 (Preservation of termination). *If $\mathcal{R}$ is consistently transformable, decreasing and confluent, then $\mathcal{R}'$ is terminating on reachable terms.*

Observe that in Theorem 4 and, consequently, also in Theorem 5 we have left-linearity as additional assumption! The reason is that non-left-linear rules may lead to an incomplete (w.r.t. reachable terms) behaviour of our transformation.

Example 9 (Non-left-linearity may lead to incompleteness w.r.t. reachable terms)
Consider the CTRS $\mathcal{R} = \{g(0) \to 0, \; f(x,x) \to a, \; f(g(x),y) \to a \; \Leftarrow \; x \to^* 0, y \to^* 0\}$ that is transformed into $\mathcal{R}'$ consisting of

$$g'(0,z) \to 0 \quad f'(x,x,z) \to a \quad f'(g'(x,\bot),y,z_2) \to f'(g'(x,\langle x,y\rangle),y,\langle x,y\rangle)$$
$$f'(g'(x,z_1),y,\bot) \to f'(g'(x,\langle x,y\rangle),y,\langle x,y\rangle) \quad f'(g'(x,\langle 0,0\rangle),y,\langle 0,0\rangle) \to a$$

For $s = f(g(a),g(a))$ we have in $\mathcal{R}$ just one normalizing step $f(g(a),g(a)) \to_{\mathcal{R}} a$. In $\mathcal{R}'$ we get the corresponding reduction $\phi(s) = f'(g'(a,\bot),g'(a,\bot),\bot) \to_{\mathcal{R}'} a$, but also $\phi(s) \to_{\mathcal{R}'} f'(g'(a,\langle a,g'(a,\bot)\rangle),g'(a,\bot),\langle a,g'(a,\bot)\rangle)$ where the latter term does not rewrite to a and is even irreducible w.r.t. $\mathcal{R}'$. Hence, in this example the transformation is neither complete w.r.t. reachable terms nor sound for preserving normal forms w.r.t. reachable terms.

Our transformation satisfies many properties only for consistently transformable CTRSs. Although it is undecidable, whether a CTRS is consistently transformable, we can show for large sub-classes of CTRSs that they are consistently transformable, cf. e.g. Theorem 2. According to Theorem 1 every system, that is not consistently transformable, must in particular be collapsing. For such CTRSs we will now show how to handle them via an additional preprocessing transformation that makes all rules non-collapsing, but retains a one-to-one-correspondence between rewrite steps.

4.4   Transformation for Non-consistently Transformable CTRSs

For the fully general case, consider a collapsing rule $l \to x \; \Leftarrow \; c$ of a CTRS $\mathcal{R}$. An intuitive method to transform $\mathcal{R}$ into a non-collapsing CTRS is to wrap x into a new symbol $C: l \to C(x) \; \Leftarrow \; c$. In order to retain a one-to-one correspondence of rewrite steps, it is necessary to wrap all non-variable terms in C consistently.

Definition 8 (Transformation into non-collapsing CTRSs T_C) *Let $\mathcal{R}$ be some CTRS with signature $\mathcal{F}$, $C \notin \mathcal{F}$ be some new unary function symbol, ρ be a rule $l \to r \; \Leftarrow \; s_1 \to^* t_1, \ldots, s_n \to^* t_n \in \mathcal{R}$ and let ϕ'_ρ be the auxiliary mapping*

$$\phi'_\rho(t) = \begin{cases} C(f(\phi'_\rho(t_1), \ldots, \phi'_\rho(t_n))) & \text{if } t = f(t_1, \ldots, t_n) \\ C(r) & \text{if } t = r \text{ and } r \text{ is a variable} \\ x & \text{if } t = x \text{ and } x \neq r \end{cases}$$

Then the transformed rule ρ_C of ρ is

$$\phi'_\rho(l) \to \phi'_\rho(r) \; \Leftarrow \; \phi'_\rho(s_1) \to^* \phi'_\rho(t_1), \ldots, \phi'_\rho(s_n) \to^* \phi'_\rho(t_n)$$

The transformed CTRS $\mathcal{R}_C$ then is $\mathcal{R}_C = \{\rho_C \mid \rho \in \mathcal{R}\}$ with signature $\mathcal{F} \cup \{C\}$ and the initialization mapping ϕ_C and backtranslation ψ_C are

$$\phi_C(t) = \begin{cases} C(f(\phi_C(t_1),\ldots,\phi_C(t_n))) & \text{if } t = f(t_1,\ldots,t_n) \\ C(t) & \text{if } t \text{ is a variable} \end{cases}$$

$$\psi_C(t) = \begin{cases} f(\psi_C(t_1),\ldots,\psi_C(t_n)) & \text{if } t = f(t_1,\ldots,t_n) \\ \psi_C(t') & \text{if } t = C(t') \\ t & \text{if } t \text{ is a variable} \end{cases}$$

Using this transformation collapsing rules are replaced by non-collapsing rules into which all other rules overlap. For reachable terms of T_C (these are all $\phi_C(s)$), these overlaps are joinable within one step (from both sides).

Definition 9 (Combined transformation $T \circ T_C$). *Let $\mathcal{R}$ be a CTRS such that $T_C(\mathcal{R}) = (\mathcal{R}_C, \phi_C, \psi_C)$ and $T(\mathcal{R}_C) = (\mathcal{R}'_C, \phi', \psi')$. Then the combined transformation $T \circ T_C$ is $(\mathcal{R}'_C, \phi' \circ \phi_C, \psi_C \circ \psi')$.*

Using this combined transformation, it is easily verified that the "blocking problem" in Example 8 disappears.[12]

Theorem 7 (Properties of the combined transformation). *Let $\mathcal{R}$ be a normal 1-CTRS such that $T \circ T_C(\mathcal{R}) = (\mathcal{R}'_C, \phi, \psi)$.*

(1) If $\mathcal{R}$ is confluent, then $\mathcal{R}'_C$ is sound w.r.t. reachable terms.

(2) If $\mathcal{R}$ is left-linear and confluent, then $\mathcal{R}'_C$ is complete w.r.t. reachable terms.

(3) If $\mathcal{R}$ is left-linear and confluent, then $\mathcal{R}'_C$ is sound for preserving normal forms w.r.t. reachable terms.

(4) If $\mathcal{R}$ is decreasing and confluent, then $\mathcal{R}'_C$ is terminating on reachable terms.

5   Experiments, Related Work and Discussion

In our experiments we compared our transformation (T) with other transformations, especially the one of [6] (T_{SR}). The "sorting list" of Example 7 is representative for most of our results. Sorting the descending list $f(s^{n-1}(0), f(s^{n-2}(0), \cdots, f(0, nil) \cdots))$ needs the following number of rewrite steps to obtain the sorted, irreducible list:

---

[12] As remarked by one of the referees, our construction of introducing separating C-layers in terms combined with the previous transformation appears to have some similarity with another complex transformation in [5, 3.4] (based on a different approach), an early forerunner of [2] and [6]. However, in [5, 3.4] these layers are used to propagate reset information to outer positions similar to [6].

	n	no sharing				maximal sharing			
		innermost		outermost		innermost		outermost	
		left	right	left	right	left	right	left	right
$T_{SR}(\mathcal{R})$	34	15893	15893	39269	125509	15893	15893	33285	125509
$T(\mathcal{R})$		27863	27863	34967	46903	14773	14773	9349	19043
$T_{SR}(\mathcal{R})$	55	62863	62863	166319	820763	62863	62863	140084	820763
$T(\mathcal{R})$		115335	115335	144538	196955	59895	59895	35144	76537

When using outermost rewriting, T_{SR} requires more rewrite steps, because it resets conditional arguments "too often": In $T(\mathcal{R})$ every condition is evaluated and if it is not satisfied, the conditional arguments is "cached" in terms like $f(0, f(s(0), f(s(s(0)), \ldots), \langle tt \rangle, \langle tt \rangle), \bot, \langle tt \rangle)$. In $T_{SR}(\mathcal{R})$ this term corresponds to $\{f(0, f(s(0), f(s(s(0)), \ldots), \langle tt \rangle), \langle tt \rangle)\}$. If we exchange two elements at inner positions, the reset-operator is propagated to outer positions in $T_{SR}(\mathcal{R})$, so that all conditional arguments are reset and must be reintroduced and reevaluated: $f(0, f(s(0), \{\ldots\}, \langle tt \rangle), \langle tt \rangle) \to^* \{f(0, f(s(0), f(s(s(0)), \ldots), \bot), \bot)\}$.

Our transformation is rather complex, because we have to iteratively check terms for unifiability in order to determine the subterms in which we have to encode conditional arguments. We can approximate these subterms via defined symbols: For $\rho : l \to r \;\Leftarrow\; s_1 \to^* t_1, \ldots, s_n \to^* t_n$ just take $P_\rho = \{p \in \overline{\mathcal{O}}(l_\rho) \mid root(l_\rho|_p) \in \mathcal{D}, \mathcal{V}ars(l_\rho|_p) \cap \mathcal{V}ars(s_{\rho,1}, \ldots, s_{\rho,n_\rho}) \neq \emptyset\}$ for encoding. This approximation clearly yields an "approximation from above", cf. Definition 5.

In order to transform *deterministic CTRSs* (DCTRSs) we can use a strategy that is easily adaptable to other transformations like e.g. [6] or [15]. A deterministic rule is a rule $l \to r \;\Leftarrow\; s_1 \to^* t_1, \ldots, s_n \to^* t_n$ that may contain extra variables, yet all extra variables "depend" directly or indirectly on variables in l: $\mathcal{V}ars(s_i) \subseteq \mathcal{V}ars(l, t_1, \ldots, t_{i-1})$ and $\mathcal{V}ars(r) \subseteq \mathcal{V}ars(l, t_1, \ldots, t_n)$. Let w.l.o.g. the conditions $s_1 \to^* t_1, \ldots, s_m \to^* t_m$ be those satisfying $\mathcal{V}ars(s_1, \ldots, s_m) \subseteq \mathcal{V}ars(l)$. By transforming the rule $\rho \colon l \to r \;\Leftarrow\; s_1 \to^* t_1, \ldots, s_m \to^* t_m$ we obtain introduction rules $\rho'_1, \ldots, \rho'_k$ and an elimination rule $\rho'_{k+1} \colon l' \to r'$ without extra variables. By adding the remaining conditions to the elimination rule we obtain the deterministic conditional rule $l' \to r' \;\Leftarrow\; \phi(s_{m+1}) \to^* \phi^{\mathcal{X}}(t_{m+1}), \ldots, \phi(s_n) \to^* \phi^{\mathcal{X}}(t_n)$ with strictly less conditions than the original rule. By repeatedly applying the above strategy, we finally obtain an unconditional TRS. In [13] a similar iterative approach for unravelings is presented.

Regarding future work, we intend to investigate various further aspects of our transformation, especially whether soundness for left-linear (consistently transformable) normal 1-CTRSs holds, and whether we can somehow get rid of the left-linearity requirement in Theorems 4, 5 and 7(2)-(3). Moreover we want to explore the optimizations and refinements sketched above. We also plan to extend our preliminary practical evaluations and comparison with related approaches. Another perspective is to analyze possible applications of our approach like conditional narrowing or inversion of rewrite systems ([14]).

6 Conclusion

We have presented a general framework for analyzing transformations of CTRSs into TRSs as well as a new approach whose characteristic feature is backtracking-freeness. It works for left-linear confluent normal 1-CTRSs and extends the approach of [2] to non-constructor systems. Compared to [6] (which also works for confluent, but not necessarily left-linear systems) with a "reset"-operator and an explicit (rule-based) propagation of "reset"-information, our approach directly incorporates necessary reset information in the transformation.

Acknowledgements. We are grateful to the anonymous referees for their detailed feedback, hints and criticisms. In particular, one of them exhibited examples (similar to Ex. 8 and 9) that triggered a partial revision of our original analysis in Section 4.

References

[1] Aida, H., Goguen, J., Meseguer, J.: Compiling concurrent rewriting onto the rewrite rule machine. In: Okada, M., Kaplan, S. (eds.) CTRS 1990. LNCS, vol. 516, pp. 320–332. Springer, Heidelberg (1991)

[2] Antoy, S., Brassel, B., Hanus, M.: Conditional narrowing without conditions. In: Proc. PPDP (2003), August 27-29, pp. 20–31. ACM Press, New York (2003)

[3] Baader, F., Nipkow, T.: Term rewriting and All That. Cambridge University Press, Cambridge (1998)

[4] Bergstra, J., Klop, J.: Conditional rewrite rules: Confluence and termination. Journal of Computer and System Sciences 32(3), 323–362 (1986)

[5] Braßel, B.: Bedingte Narrowing-Verfahren mit verzögerter Auswertung. Master's thesis, RWTH Aachen (1999)

[6] Şerbănuţă, T.-F., Roşu, G.: Computationally equivalent elimination of conditions. In: Pfenning, F. (ed.) RTA 2006. LNCS, vol. 4098, pp. 19–34. Springer, Heidelberg (2006)

[7] Dershowitz, N., Okada, M., Sivakumar, G.: Confluence of conditional rewrite systems. In: Kaplan, S., Jouannaud, J.-P. (eds.) CTRS 1987. LNCS, vol. 308, pp. 31–44. Springer, Heidelberg (1988)

[8] Dershowitz, N., Plaisted, D.: Equational programming. In: Hayes, J.E., Michie, D., Richards, J. (eds.) Machine Intelligence 11: The logic and acquisition of knowledge ch. 2, pp. 21–56 (1988)

[9] Giovanetti, E., Moiso, C.: Notes on the elimination of conditions. In: Kaplan, S., Jouannaud, J.-P. (eds.) CTRS 1987. LNCS, vol. 308, pp. 91–97. Springer, Heidelberg (1988)

[10] Lucas, S., Meseguer, J., Marché, C., Urbain, X.: Proving operational termination of membership equational programs. Higher-Order and Symbolic Computation 21(10), 59–88 (2008)

[11] Marchiori, M.: Unravelings and ultra-properties. Technical Report 8, University of Padova, Italy (1995)

[12] Marchiori, M.: Unravelings and ultra-properties. In: Hanus, M., Rodríguez-Artalejo, M. (eds.) ALP 1996. LNCS, vol. 1139, pp. 107–121. Springer, Heidelberg (1996)

[13] Nishida, N., Mizutani, T., Sakai, M.: Transformation for refining unraveled conditional term rewriting systems. ENTCS, vol. 174(10), pp. 75–95 (2007)

[14] Nishida, N., Sakai, M., Sakabe, T.: Partial inversion of constructor term rewriting systems. In: Giesl, J. (ed.) RTA 2005. LNCS, vol. 3467, pp. 264–278. Springer, Heidelberg (2005)

[15] Ohlebusch, E.: Advanced Topics in Term Rewriting. Springer, Heidelberg (2002)

[16] Rosu, G.: From conditional to unconditional rewriting. In: Fiadeiro, J.L., Mosses, P.D., Orejas, F. (eds.) WADT 2004. LNCS, vol. 3423, pp. 218–233. Springer, Heidelberg (2005)

[17] Schernhammer, F., Gramlich, B.: On proving and characterizing operational termination of deterministic conditional rewrite systems. In: Hofbauer, D., Serebrenik, A. (eds.) Proc. WST (2007), pp. 82–85 (2007)

[18] Viry, P.: Elimination of conditions. J. Symb. Comput. 28(3), 381–401 (1999)

Towards a Module System for K*

Mark Hills and Grigore Roşu

Department of Computer Science
University of Illinois at Urbana-Champaign, USA
201 N Goodwin Ave, Urbana, IL 61801
{mhills,grosu}@cs.uiuc.edu
http://fsl.cs.uiuc.edu

Abstract. Research on the semantics of programming languages has yielded a wide array of notations and methodologies for defining languages and language features. An important feature many of these notations and methodologies lack is *modularity*: the ability to define a language feature once, insulating it from unrelated changes in other parts of the language, and allowing it to be reused in other language definitions. This paper introduces ongoing work on modularity features in K, an algebraic, rewriting logic based formalism for defining language semantics.

Keywords: language semantics, rewriting logic, modularity, K.

1 Introduction

One important aspect of formalisms for defining the semantics of programming languages is modularity. Modularity is generally expressed as the ability to add new language features, or modify existing features, without having to modify unrelated semantic rules. For instance, when designing a simple expression language, one may want to use structural operational semantics (SOS) [29] to define the semantics of addition:

$$\frac{e_1 \rightarrow e_1'}{e_1 + e_2 \rightarrow e_1' + e_2} \qquad \text{(EXP-PLUS-L)}$$

$$\frac{e_2 \rightarrow e_2'}{n_1 + e_2 \rightarrow n_1 + e_2'} \qquad \text{(EXP-PLUS-R)}$$

$$n_1 + n_2 \rightarrow n, \text{where } n = n_1 + n_2 \qquad \text{(EXP-PLUS)}$$

Further extending the language, one may want to add variables. The standard way to do this is to define a store, mapping names to values, with rules for binding values to names (not shown here) and to retrieve the current value of a binding:

---

* Supported in part by NSF grants CCF-0448501, CNS-0509321 and CNS-0720512, by NASA contract NNL08AA23C, by the Microsoft/Intel funded Universal Parallel Computing Research Center at UIUC, and by several Microsoft gifts.

A. Corradini and U. Montanari (Eds.): WADT 2008, LNCS 5486, pp. 187–205, 2009.

$$\langle x, \sigma \rangle \rightarrow \langle n, \sigma \rangle, \text{where } n = \sigma(x) \qquad \text{(VAR-LOOKUP)}$$

With this change to the language, even though the rules for plus do not actually reference the store they must still be modified to include it as part of the configuration. As an example, rule `EXP-PLUS-L` becomes[1]:

$$\frac{\langle e_1, \sigma \rangle \rightarrow \langle e_1', \sigma \rangle}{\langle e_1 + e_2, \sigma \rangle \rightarrow \langle e_1' + e_2, \sigma \rangle} \qquad \text{(EXP-PLUS-L)}$$

Similar types of changes to existing rules need to be made to accommodate other unrelated language features, such as exceptions or function returns. Alternatively, similar changes may need to be made to add addition expressions to a different language with a different configuration, even if the different elements of the configuration are not used in the rules for addition. All these changes are required because SOS is not modular. Improved support for modularity eliminates the need to make these changes, offering several advantages:

– Modular definitions of language features allow other parts of a language to change more easily by allowing existing feature definitions to remain unchanged in the face of unrelated modifications or additions;
– A modular definition of a language feature can be more easily reused in the definition of a different language which may be structured much differently;
– Modular definitions are easier to understand, since the rules given for a language construct only need to include the information needed by the rule, instead of including extraneous information used in other parts of the language (such as the store in the rules for plus).

For these reasons, improving modularity of language definitions has been a focus of research across multiple semantic formalisms. One example is modular structural operational semantics (MSOS) [25, 26], which solves the problem shown above by leveraging the labels on rule transitions, not normally used in SOS definitions of programming languages, to encode configuration elements, with the ability to elide unused parts of the configuration. This is discussed further with other related work in Section 4.

With a tool supported semantics, modularity can also be expressed as the ability to package language features into discreet reusable units, which can then be assembled when defining a language. This form of modularity depends on the first: it should be possible to plug the same feature into multiple definitions, even in cases where (unused) parts of the configuration are different. Additionally, it should be possible to provide clean interfaces to language features and to different parts of the configuration, something not required in monolithic definitions, or even in modular definitions written on paper.

---

[1] A more general version of this rule would use σ on the left and σ' on the right; here, by using σ on both left and right, we state that expressions do not alter the store, i.e. they do not have side effects.

This paper provides a high-level overview of ongoing work on adding modularity features to K [30], an algebraic, rewriting logic based formalism for programming language semantics. This work is focused on both aspects of modularity mentioned above, allowing the packaging of language features for reuse while insulating existing features from unrelated changes to the language definition. Novel aspects of the module system include the use of context transformers, described along with K in Section 2; the incorporation of some features, such as the explicit hiding or requiring of operations and sorts, which are common in the module systems of programming languages but do not appear to be common in systems for modular language definition; and the methods used to assemble the final language and define the shape of the configuration, chosen to support having multiple distinct semantics for a language and with an eye towards future improved tool support, including visualization.

The remainder of this paper is organized as follows. First, we provide a brief overview of term rewriting, equational logic, rewriting logic, and especially K in Section 2. Next, Section 3 introduces the module system through fragments of a simple imperative language and illustrates the ability to reuse modules in language extensions. Section 4 then reviews related work, while in Section 5 we conclude and discuss future work.

2 Rewriting Logic

This section provides a brief introduction to term rewriting, rewriting logic, rewriting logic semantics, and K. Term rewriting is a standard computational model supported by many systems; rewriting logic [18, 19] organizes term rewriting modulo equations as a complete logic and serves as a foundation for programming language semantics using rewriting logic semantics [21, 22]. K [30] is a rewrite-based method for formally defining computation, here used to provide formal definitions for programming languages.

2.1 Term Rewriting

Term rewriting is a method of computation that works by progressively changing (rewriting) a term. This rewriting process is defined by a number of rules – potentially containing variables – which are each of the form: $l \to r$. A rule can apply to the entire term being rewritten or to a subterm of the term. First, a match within the current term is found. This is done by finding a substitution, θ, from variables to terms such that the left-hand side of the rule, l, matches part or all of the current term when the variables in l are replaced according to the substitution. The matched subterm is then replaced by the result of applying the substitution to the right-hand side of the rule, r. Thus, the part of the current term matching $\theta(l)$ is replaced by $\theta(r)$. The rewriting process continues as long as it is possible to find a subterm, rule, and substitution such that $\theta(l)$ matches the subterm. When no matching subterms are found, the rewriting process terminates, with the final term being the result of the computation.

Rewriting, like other methods of computation, can continue forever. There exist a plethora of term rewriting engines, including ASF+SDF [34], Elan [3], Maude [6], and OBJ [11]. Rewriting is also a fundamental part of existing languages, including Tom [2],which integrates rewriting with Java.

2.2   Rewriting Logic

Rewriting logic [18, 19] is a computational logic built upon equational logic which provides support for concurrency. In equational logic, a number of *sorts* (types) and *equations* are defined. The equations specify which terms are considered to be equal. All equal terms can then be seen as members of the same equivalence class of terms, a concept similar to that from the λ calculus with equivalence classes based on α and β equivalence. Rewriting logic provides *rules* in addition to equations, used to transition between equivalence classes of terms. This allows for concurrency, where different orders of evaluation could lead to non-equivalent results, such as in the case of data races. The distinction between rules and equations is crucial for analysis, since terms which are equal according to equational deduction can all be collapsed into the same analysis state. Rewriting logic is connected to term rewriting in that the equations and rules of rewriting logic, of the form $l = r$ and $l \Rightarrow r$, respectively, can be transformed into term rewriting rules by orienting them properly (necessary because equations can be used for deduction in either direction), transforming both into $l \rightarrow r$. This provides a means of taking a definition in rewriting logic and a term and "executing" it.

2.3   Rewriting Logic Semantics

Rewriting logic semantics (RLS) [21, 22] builds upon the observation that programming languages can be defined as rewriting logic theories. By doing so, one gets essentially "for free" not only an interpreter and an initial model semantics for the defined language, but also a series of formal analysis tools obtained as instances of existing tools for rewriting logic. The work discussed in this paper has grown out of a style of RLS called *Continuation-Based Semantics* [21, 22] which allows the natural modeling of complex control flow constructs, like exceptions and continuations, by treating computations as first-class semantic entities.

2.4   K

K [30], a general notation and technique for defining computation, is based on insights developed in the rewriting logic semantics project [21, 22], with some concepts inspired by abstract state machines (ASMs) [12], the chemical abstract machine (CHAM) [10], and continuations [32].[2] K provides some domain-specific abstractions and assumptions, exploited in this paper, to ease the definition of programming languages.

---

[2] The name K comes from the traditional name of the operator or cell containing the current control context, k.

The idea underlying language semantics in K is to represent the program configuration as a *computational structure*. This structure contains the context needed for the computation, with elements of the context represented as multisets or lists each stored inside a K *cell*. Contexts can also be hierarchical, with one cell containing others. The context generally includes standard items found in configurations in other formalisms, such as environments, stores, etc, as well as items specific to the given semantics, including such items as analysis results for a semantics focused on program analysis. One regularly used cell, referred to as k, represents the current computation as a $\curvearrowright$-separated list of computational tasks, such as $t_1 \curvearrowright t_2 \curvearrowright ... \curvearrowright t_n$. Another, $\top$, represents the entire computational structure. In the rest of the paper, the computational structure will be referred to as just a computation.

A K definition consists of two types of sentences: structural equations and rewrite rules. Structural equations carry no computational meaning; instead, borrowing a concept from CHAMs, structural equations can *heat* and *cool* computations. When a computation is heated, it breaks into smaller pieces, exposing subexpressions of more complex expressions for evaluation. Cooling reverses this process, reassembling the (potentially modified) pieces into a computation with the same "shape". The following are examples of structural equations, with heating represented as going from left to right and cooling from right to left:

$$a_1 + a_2 \rightleftharpoons a_1 \curvearrowright \square + a_2$$
$$\text{if } b \text{ then } s_1 \text{ else } s_2 \rightleftharpoons b \curvearrowright \text{if } \square \text{ then } s_1 \text{ else } s_2$$

Note that, unlike in evaluation contexts, $\square$ does not represent a context, but is instead part of the operator definition, providing visual intuition about what is being evaluated and where the result will go upon cooling; a different scheme could be used instead. The operators involving $\square$ above are $\square + \_$ (in the first equation) and if $\square$ then_ else_ (in the second).

Many structural equations can be automatically generated by annotating constructs in the language syntax with *strict* attributes: a *strict* construct generates the appropriate equations for each strict argument. If an operator is intended to be strict in only some of its arguments, then the positions of the strict arguments are listed as arguments of the *strict* attribute; for example, the two equations directly above correspond to the attributes *strict* for $\_ + \_$ (i.e., strict in all arguments, with the heating/cooling equations for the second operand not shown) and *strict*(1) for if_then_else_.

Rewrite rules represent actual steps of computation:

$$i_1 + i_2 \rightarrow i, \text{ where } i \text{ is the sum of } i_1 \text{ and } i_2$$
$$\text{if true then } s_1 \text{ else } s_2 \rightarrow s_1$$
$$\text{if false then } s_1 \text{ else } s_2 \rightarrow s_2$$
$$(\!| X = V |\!)_k \; (\!|(X, L)|\!)_{env} \; (\!|(L, \_)|\!)_{mem} \rightarrow (\!| \cdot |\!)_k \; (\!|(X, L)|\!)_{env} \; (\!|(L, V)|\!)_{mem}$$

Quite often structural equations would be used between applications of rewrite rules. For example, given an expression $a_1 + a_2$, the first equation for $\_ + \_$ can be applied left-to-right to "schedule" a_1 for processing (which may involve the use of one or more rewrite rules); once evaluated to i_1, the equation is applied in reverse to "plug" the result back in context, leaving $i_1 + a_2$. a_2 would be handled similarly to a_1, yielding expression $i_1 + i_2$, which can then be processed using the first rewrite rule shown above. Note that this step, going from $i_1 + i_2$ to their sum, is not reversible, in contrast to the structural equations for $+$ shown above. The last rule shows an example with multiple cells: here, the value V is being assigned to name X. The round bracket at the left, $\langle$, represents the head of the list, forcing this rule to apply only when it will be the next step of this computation. The "pointed" bracket at the right, $\rangle$, represents the rest of the list, i.e. the remainder of the computation (intuitively, it is pointed as a reminder that the list keeps going in that direction). Multisets are bracketed with $\langle$ and $\rangle$, indicating they conceptually "continue" in either direction. This is used here for both *env* and *mem*: *env* is a multiset of Name $\times$ Location pairs, while *mem* is a multiset of Location $\times$ Value pairs. Other notation includes $\_$, which represents an unnamed value (like in many functional languages), and $\cdot$, representing the identity (here, the list identity). Given that, this rule states: when X := V is the next computational step in this computation, if X is at location L in the environment, change the value at location L in the store to V (while ignoring the current value), and then "dissolve" the current computation, leaving the next item in k (not shown, but to the "right" of $\rangle$) as the next computational step.

Context Transformers: To ensure that rules are modular, it should be possible to continue using a rule, unchanged, when parts of the context not mentioned in the rule are modified or replaced. Given a specific rule, the easiest case to deal with is when the subterm matched by a rule remains the same but the surrounding context changes (for instance, by adding a new top-level cell). This case is handled naturally by term rewrite systems, since it is possible to match a subterm of the term, leaving the rest unnamed in the rule. This handles many common cases, including the motivating example given in Section 1. However, this does not handle changes to the hierarchical organization of the context. For instance, adding threads to a language requires having multiple k cells, representing the computation occurring in each thread, but only one store, leading to a revised rule for assignment like the following:

$$\langle\!\langle X \;=\; V \rangle_k \; \langle\!\langle (X, L) \rangle\!\rangle_{env} \rangle_t \; \langle\!\langle (L, \_) \rangle\!\rangle_{mem} \rightarrow \langle\!\langle \cdot \rangle_k \; \langle\!\langle (X, L) \rangle\!\rangle_{env} \rangle_t \; \langle\!\langle (L, V) \rangle\!\rangle_{mem}$$

Beyond this, the configurations used in different languages will generally be quite different, and may be very complex. To allow rules to be reused, both as a language evolves and in other languages, K uses *context transformers*. Using context transformers, only those portions of the configuration actually used in a rule need to be mentioned. For instance, the assignment rule shown originally can remain as is, without the need to explicitly add the thread cell, used only to provide context for the match. To do this, the transformer uses the declared language configuration (shown in Section 3 in a **Language** module) to determine which cells, used only for context, have been elided; these cells can then be added automatically, using either variables or K brackets to represent unmentioned parts of the added cells. Several sanity conditions are used to ensure that a unique transform is possible, including rules about valid paths between cells and the reuse of cell names. Planned K tool support will provide appropriate warnings in cases where ambiguity prevents the context transformers from deriving a unique transformed version of a rule.

3   The K Module System

While context transformers focus on making individual K rules modular, they do nothing to address the practical challenge of packaging up rules into reusable units. This is the purpose of the K module system.

The module system in K is being designed to support a general module syntax incorporating the entire range of functionality needed when defining the semantics of a language, including the definition of abstract syntax, configuration items, the semantics of language features, and the final collection of features that make up a specific language. In theory, this would allow a single, monolithic module to include definitions of all aspects of the semantics. However, to provide for a better separation of these constructs into more granular modules, and to allow for construct-specific defaults and syntax, specialized module formats for various constructs are being defined, with a translation into the more general syntax. These module formats are illustrated with fragments of the definition of a simple imperative language, IMP. A complete definition of IMP, without the extensions presented at the end of this section, is given in Appendices A and B; note that the definition given in this section differs slightly to better illustrate features of the module system.

3.1   Semantic Entities

Semantic entities in K definitions include configuration items, such as environments and stores, and sorts or operations used during computations, such as computation items and values. A simple example is shown in Figure 1, which uses subsorting to allow K integers (modules starting with K/ are built-ins) to be treated

```
module Int
  imports K/Value, K/Int .
  subsort Int < Value .
end module
```

Fig. 1. Integer Values

as K values. By default, this declaration is available in any other module that imports Int.

Another example is shown in Figure 2. This shows the definition of an environment, which provides a mapping from names to locations (a store then maps locations to values; the separation easily allows features like nested scopes and reference parameters for functions). Like in Figure 1, an existing K definition, in this case for sort Name, is imported. Instead of similarly

```
module Env
  imports K/Name .
  requires sort Loc .
  sortalias Env = Map(Name,Loc) .
  var Env[0-9']* : Env .
end module
```

Fig. 2. Environments

importing a specific definition of locations (sort Loc), module Env uses **requires**, meaning that, when the language is finally assembled, one module must provide sort Loc. This allows the module to state a requirement without stating the module that satisfies that requirement, allowing different modules to be used in different languages. Since K provides lists, multisets, and maps by default, we can immediately refer to maps from sort Name to sort Loc; sortalias lets us give this sort a name, Env, which can then be used in the remainder of the definition. The var declaration allows the definition of a variable pattern: Env, followed by 0 or more numbers or primes, will be used to represent entities of sort Env (e.g., Env, Env8, Env', etc.). Variables used in modules that import Env and that match this pattern will then be identified as being of sort Env.

3.2   Abstract Syntax

Before defining the semantics of language constructs, the abstract syntax of those constructs needs to be defined. This is done using
abstract syntax modules, which are defined using a tag of [Syntax] after the module name.
A first example of an abstract syntax module is the syntax for arithmetic expressions,
shown in Figure 3. One way to define the sort

```
module Exp/AExp[Syntax]
  imports Exp[Syntax] with
      { sort Exp renamed AExp } .
  # sort AExp . subsort AExp < Exp .
  var AE[0-9'a-zA-Z]* : AExp .
end module
```

Fig. 3. Arithmetic Expressions

of arithmetic expressions would be to define a new sort which could be made a subsort of Exp, illustrated in a comment in the module (comments start with #); here, instead, the sort Exp, imported from module Exp, is renamed to AExp using a sort renaming directive on the import of module Exp. A var pattern to refer to arithmetic expressions is then defined.

A second abstract syntax module, defining the
addition construct, is shown in Figure 4. Syntax
is defined using mixfix notation with an algebraic
notation similar to that used in Maude or SDF
(although note that op is not required on syntax
definitions). To increase modularity, it is recom-

```
module Exp/AExp/Plus[Syntax]
  imports Exp/AExp[Syntax] .
  _+_ : AExp AExp -> AExp .
end module
```

Fig. 4. Plus Expressions

mended that each module define only one language construct, although it is possible to define multiple constructs in the same module.

3.3   Semantic Rules

Once the syntax has been defined, the semantics of each construct need to be defined as well.
One explicit goal of the module
system is to allow different semantics to be easily defined for
each language construct. For in-

```
module Exp/AExp/Plus[Dynamic]
  imports Exp/AExp/Plus[Syntax]
    with { op _+_ now strict,
                extends + [Int * Int -> Int] } .
end module
```

Fig. 5. Dynamic Semantics: Plus

stance, it should be possible to define a standard dynamic/execution semantics, a static/typing semantics, and potentially other semantics manipulating different notions of value (for instance, various notions of abstract value used during analysis).

Figure 5 shows an example of a module defining the dynamic semantics of a language feature, here integer addition. Normally a semantics module will implicitly import the related syntax module. Here, since we are modifying the attributes on an imported operator, we need to explicitly import the syntax module. Two attributes are modified. First, we note that the operator is now strict in all arguments, which will automatically generate the structural heating and cooling equations. Second, we use extends to automatically "hook" the semantics of the feature to the builtin definition of integer addition. This completely defines integer addition in the language, so no rules are needed.

Figure 6 shows semantics for the same feature, but this time the static semantics (for type checking) are defined. Like in Figure 5, the operator for plus is changed to be strict. In

```
module Exp/AExp/Plus[Static] is
  imports Exp/AExp/Plus[Syntax] with { op _+_ now strict } .
  imports Types .
  rl int + int => int .
  rl T + T' => fail [where T =/= int or T' =/= int] .
end module
```

Fig. 6. Static Semantics: Plus

this case, though, the values being manipulated are types, not integers, so we also need to import the types and use them in the two rules shown. Here, the first rule is for when an expression is type correct: the two operands are both integers, so the result of adding them is also an integer. If one of the operands is not an integer (checked in the side-condition), the rule will cause a type called fail, representing a type error, to propagate.[3]

Finally, Figure 7 shows the dynamic semantics of blocks. Here, no changes are made to the imported syntax, so there is no need to import the Stmt/Block[Syntax] module explicitly. In this language, blocks provide

```
module Stmt/Block[Dynamic] is
  imports Stmt[Syntax], K/K, Env .
  rl k(| begin S end        |> env(| Env |)
    => k(| S -> restoreEnv(Env) |> env(| Env |) .
end module
```

Fig. 7. Dynamic Semantics: Blocks

for nested scoping, so we want to ensure that the current environment is restored after the code inside the block executes. This is done by capturing the current environment, Env, and placing it on the computation in a restoreEnv computation item. The rule for restoreEnv, not shown here, will replace the current environment with its saved environment when it becomes the first item in the computation.

3.4 Language Definitions

Once the semantic entities, abstract syntax, and language semantics have been defined, they can be assembled into a language module, tagged Language. An example is shown in Figure 8. The line config = defines the language configuration

---

[3] An alternative would be to issue an error message and return the expected type in the hope of finding additional errors.

as a multiset, with each K cell given a name (such as store or env) and the sort of information in the cell (such as Store or Env). Cells can be nested, to represent the hierarchies of information that can be formed. Next, the [[_]] operator initializes this configuration, given an initial computation (K) representing the program to run. Finally, all the modules used in the semantics are im-

```
module Imp[Language]
  imports K/Configuration, K/K, K/Location,
          K/Value, Env, Store, Int, Bool .
  config = top(store(Store) env(Env)
              k(K) nextLoc(Loc)) .

  op [[_]] : K -> Configuration .
  eq [[ K ]] = top(store(empty) env(empty)
              k(K) nextLoc(initLoc)) .

  imports type=Syntax Exp/AExp/Num, Exp/BExp/Bool .
  imports type=Dynamic Exp/AExp/Name, Exp/AExp/Plus,
    Exp/BExp/LessThanEq, Exp/BExp/Not, Exp/BExp/And,
    Stmt/Sequence, Stmt/Assign, Stmt/IfThenElse,
    Stmt/While, Stmt/Halt, Pgm .
end module
```

Fig. 8. Language Definition: IMP

ported. type=Dynamic is a directive that states that all modules in this imports are tagged with the Dynamic tag, and is equivalent to imports Exp/AExp/Name[Dynamic], Exp/AExp/Plus[Dynamic], etc.

3.5   Taking Advantage of Modularity

The goal of ensuring that K is modular is to allow defined language features to be reused in new languages and in extensions to existing languages. To illustrate this, two extensions to IMP, one for exceptions and one for procedures, are defined, generating two new versions of IMP. These extensions are then combined to create a third version of IMP, showing that the existing definitions can be directly leveraged. To save space, the following are not shown: abstract syntax modules for the new features; the Language modules for IMP with just exceptions or with just procedures; and imports clauses in module definitions.

Exceptions: The exceptions extension assumes an abstract syntax similar to that for Java: a try/catch statement is used to specify an exception handler, while throw is used to manually throw an exception. Figure 9 shows the semantics needed for exceptions. In the case of a try/catch,

```
module Stmt/TryCatch[Dynamic]
  requires op addHandler : Name Stmt -> Computation .
  requires op removeHandler : -> Computation .
  rl k(| try S catch X in S' |>
  => k(| addHandler(X,S') -> S -> removeHandler |> .
end module
module Stmt/Throw[Dynamic]
  requires op handleException : -> Computation .
  rl k(| throw V |> => k(| V -> handleException |> .
end module
```

Fig. 9. Exception Semantics

the information needed for the handler is saved with addHandler before the try statement is executed. If the try body (S) is executed without throwing an exception, removeHandler removes the handler information. If an exception is thrown, either implicitly or with throw, the handler will be triggered, in the case of throw through using handleException.

Figure 10 then shows the state operations used in the exception semantics in Figure 9. An exception handler is defined as a triple, with a computation, an environment, and arbitrary other state (K cells). These are stored in an exception handler stack, defined as a list of exception handlers. Rules then provide semantics for the operations: removeHandler just pops the

```
module Exceptions/State
  sortalias ExHandler = Tuple(K,Env,State) .
  sortalias ExStack = List(ExHandler) .
  rl k(| removeHandler |> es(| EH |>
  => k(| . |> es(| . |> .
  rl k(| addHandler(X,S) |> env(| Env |)
       cn(| es(| ES |) CN |)
  => k(| . |> env(| Env |)
       cn(| es(| [assignTo(X) -> S -> restoreEnv(Env),
                  Env, CN] ES |) CN |) .
  rl k(| V -> handleException |>
       env(| _ |) cn(| es(| [K,Env,CN] |> |>
  => k(| V -> K |) env(| Env |) cn(| es(| . |> CN |) .
end module
```

Fig. 10. Exception State Manipulation

stack, while addHandler creates the exception handler (assignTo will take the thrown value, which will be at the head of the computation, and assign it to the identifier given in the catch statement, while restoreEnv will then restore the environment back to the given environment, removing this name mapping). Note the use of two new K cells in these two rules: es, for the exception stack, and cn, for control context information, like that used in exception handlers (this extra level of grouping will prove useful later). Finally, handleException will use the handler to handle the exception, restoring the saved computation, environment, and control context in the process.

Procedures: The semantics for procedures are similar to those for exceptions. When a procedure is called with call, it will use invoke to invoke the procedure and save the current state. The semantics for **return** are similar to **throw**, using popCallStack to remove the current procedure context and restore state saved at the time of the call. Figure 11 shows the semantics for both call and return.

```
module Stmt/Call[Dynamic]
  requires op invoke : Name ValueList -> K .
  rl k(| call X(VL) |>
  => k(| invoke(X,VL) |> .
end module
module Stmt/Return[Dynamic]
  requires op popCallStack : -> K .
  rl k(| return |> => k(| popCallStack |> .
end module
```

Fig. 11. Procedure Semantics

Figure 12 then shows the state manipulation rules, similar in many ways to those for exceptions. Note that there are two new K cells here as well: cs, for the call stack, and pm, for the procedure map (from procedure names to procedure definitions). cn, for control context, is also used. invoke saves the current computation, environment, and other control context in the call stack while as-

```
module Procedures/State
  sortalias CallState = Tuple(K,Env,State) .
  sortalias CallStack = List(CallState) .
  rl k(| invoke(X,VL) -> K |) env(| Env |)
       cn(| cs(| CS |) CN |) pm(PM)
  => k(| VL -> assignTo(XL) -> S |) env(| . |)
       cn(| cs(| [K,Env,CN] CS |) CN |) pm(PM)
       [where proc(XL,S) := lookup(PM,X)] .
  rl k(| popCallStack |> env(| _ |)
       cn(| cs(| [K,Env,CN] |> |>
  => k(| K |) env(| Env |) cn(| cs(| . |> CN |) .
end module
```

Fig. 12. Procedure State Manipulation

signing argument values (VL) to parameter names (XL) and then running the procedure body (S), all in the context of an empty environment (i.e., there are no global variables). The procedure definition is looked up as part of a side condition (with **where** in the brackets following the rest of the rule). popCallStack restores the saved control context, environment, and computation, representing the return from the procedure.

```
module ImpWCallWExceptions[Language]
   imports K/Configuration, K/K, K/Location, K/Value,
           Env, Store, Int, Bool, Procedures/State,
           Exceptions/State .

   config = top(store(Store) env(Env) k(K) nextLoc(Loc) cn(cs(CallStack)
              es(ExStack)) pm(ProcMap)) .

   op [[_,_]] : ProcList K -> Configuration .
   eq [[PL,K]] = top(store(empty) env(empty) k(processEach(PL) -> K)
                   nextLoc(initLoc) cn(cs(empty) es(empty)) pm(empty)) .

   imports type=Syntax Exp/AExp/Num, Exp/BExp/Bool .
   imports type=Dynamic Exp/AExp/Name, Exp/AExp/Plus,
     Exp/BExp/LessThanEq, Exp/BExp/Not, Exp/BExp/And,
     Stmt/Sequence, Stmt/Assign, Stmt/IfThenElse,
     Stmt/While, Stmt/Halt, Stmt/Call, Stmt/Return, Stmt/TryCatch,
     Stmt/Throw, Procedure, Pgm.
end module
```

Fig. 13. IMP with Procedures and Exceptions

Combining Exceptions and Procedures: It should be possible to reuse the definitions of exceptions and procedures without needing to revisit each. This can in fact be done, without requiring any changes to the defined language features. Figure 13 shows the language module, including K cells, that extends IMP with both procedures and exceptions. The use of cell **cn** to group context information, along with the use of context transformers to transform rules that mention both **k** and (for instance) **cs** into ones that also use **cn** and the other context held therein allows the language to be assembled without modifying any existing module. Note that this is not always possible, as different language features may need to be aware of one another. For instance, a definition of a loop **break** feature may need to be aware of functions, since it is generally not possible to use **break** inside a called function to break out of a loop inside the callee.

4   Related Work

Modularity has long been a topic of interest in the language semantics community. Listed below are some of the more significant efforts, including comparisons with the work described in this paper where appropriate.

Action Semantics: One focus of Action Semantics [24] has been on creating modular definitions. The notation for writing action semantics definitions uses a module structure, while language features use *facets* to separate different language construct "concerns", such as updating the store or communicating between processes. A number of tools have been created for working with modular Action Semantics definitions, such as ASD [35], the Action Environment [33], the Maude Action Tool [7], an implementation using Montages [1], and Modular Monadic Action Semantics [37]. Other work has focused specifically on ensuring modules can be easily reused without change, both by using small, focused modules [8] (the approach taken in the K module system) and by creating a number of simpler reusable constructs generic to a large number of languages [15, 27].

ASMs: Montages [16] provides a modular way to define language constructs using Abstract State Machines (ASMs) [14, 31]. Each Montage (i.e., module) combines a graphical depiction of a language construct with information on the static and dynamic semantics of the feature. This has the advantage of keeping all information on a feature in one place, but limits extensibility, since it is not possible (as it is in K) to provide multiple types of dynamic or static semantics to the same feature without creating a new Montage.

Denotational Semantics: One effort to improve modularity in denotational semantics definitions has been the use of Monads [23]. This has been most evident in work on modular, semantics-based definitions of language interpreters and compilers, especially in the context of languages such as Haskell [17, 36] and Scheme [9]. Monads have also been used to improve the modularity of other semantic formalisms, such as Modular Monadic Action Semantics [37], which provided a monadic semantics in place of the original, non-modular SOS semantics underlying prior versions of Action Semantics [24].

MSOS: The focus of MSOS [25, 26] has been on keeping the benefits of SOS definitions while defining rules in a modular fashion. This is done by moving information stored in SOS configurations, such as stores, into the labels on transition rules, which traditionally have not been used in SOS definitions of languages. This, along with techniques that allow parts of the label to be elided if not used by a rule, allow the same rule to be used both when unrelated parts of the configuration change and when the rule is introduced into a language with a different configuration. A recent innovation, Implicitly-Modular SOS (I-MSOS) [28], allows more familiar SOS notation while still providing the benefits of MSOS.

Rewriting Logic Semantics: Beyond the work done on K, Maude has also been used as a platform to experiment with other styles of semantics, enabling the creation of modular language definitions. This includes work on action semantics, with the Maude Action Tool cited above, and MSOS, using the Maude MSOS Tool [5]. Work on defining Eden [13], a parallel variant of Haskell, has focused on modularity to allow for experimentation with the degree of parallelism and the scheduling algorithm used to select processes for execution. General work on modularity of rewriting logic semantics definitions [4, 20] has focused on defining modular features that need not change as a language is extended.

5   Conclusion and Future Work

In this paper we have presented ongoing work on modularity in K language definitions. This includes work on the modularity of individual K rules, ensuring they can be defined once and reused in a variety of contexts, and work on an actual module system for K, providing a technique to easily package and reuse individual language features while building a language.

One major, ongoing component of this work is developing tool support for the module system. Although small modules improve reuse, the large number of modules this leads to can make it challenging to work with language definitions, something noted in similar work on tool support for Action Semantics [33]. For K, work on tool support includes the ongoing development of an Eclipse plugin to provide a graphical environment for the creation and manipulation of K modules. This will initially include editor support, a graph view of module dependencies, and the ability to view both the language features used to define a language and the various semantics defined for a specific language feature. Longer-term goals include the graphical assembly of language configurations and links to an online database of reusable modules. Another part of this work is moving over existing language definitions to the new, modular format. This has already started for a number of pedagogical languages defined in K that are used in the classroom; work on larger languages is waiting on improved tool support.

References

1. Anlauff, M., Chakraborty, S., Kutter, P.W., Pierantonio, A., Thiele, L.: Generating an action notation environment from Montages descriptions. International Journal on Software Tools for Technology Transfer 3(4), 431–455 (2001)
2. Balland, E., Brauner, P., Kopetz, R., Moreau, P.-E., Reilles, A.: Tom: Piggybacking Rewriting on Java. In: Baader, F. (ed.) RTA 2007. LNCS, vol. 4533, pp. 36–47. Springer, Heidelberg (2007)
3. Borovanský, P., Kirchner, C., Kirchner, H., Moreau, P.-E., Ringeissen, C.: An overview of ELAN. In: Proceedings of WRLA 1998. ENTCS, vol. 15 (1998)
4. Braga, C., Meseguer, J.: Modular Rewriting Semantics in Practice. In: Proceedings of WRLA 2004. ENTCS, vol. 117, pp. 393–416. Elsevier, Amsterdam (2005)
5. Chalub, F., Braga, C.: Maude MSOS Tool. In: Proceedings of WRLA 2006. ENTCS, vol. 176, pp. 133–146. Elsevier, Amsterdam (2007)
6. Clavel, M., Durán, F., Eker, S., Lincoln, P., Martí-Oliet, N., Meseguer, J., Talcott, C.L. (eds.): All About Maude - A High-Performance Logical Framework, How to Specify, Program and Verify Systems in Rewriting Logic. LNCS, vol. 4350. Springer, Heidelberg (2007)
7. de Braga, C.O., Haeusler, E.H., Meseguer, J., Mosses, P.D.: Maude Action Tool: Using Reflection to Map Action Semantics to Rewriting Logic. In: Rus, T. (ed.) AMAST 2000. LNCS, vol. 1816, pp. 407–421. Springer, Heidelberg (2000)
8. Doh, K.-G., Mosses, P.D.: Composing programming languages by combining action-semantics modules. Science of Computer Programming 47(1), 3–36 (2003)
9. Espinosa, D.A.: Semantic Lego. PhD thesis (1995)
10. Berry, G., Boudol, G.: The Chemical Abstract Machine. In: Proceedings of POPL 1990, pp. 81–94. ACM Press, New York (1990)

11. Goguen, J., Winkler, T., Meseguer, J., Futatsugi, K., Jouannaud, J.-P.: Introducing OBJ. In: Software Engineering with OBJ: algebraic specification in action. Kluwer, Dordrecht (2000)
12. Gurevich, Y.: Evolving Algebras 1993: Lipari Guide. In: Specification and Validation Methods, pp. 9–36. Oxford University Press, Oxford (1995)
13. Hidalgo-Herrero, M., Verdejo, A., Ortega-Mallén, Y.: Using Maude and Its Strategies for Defining a Framework for Analyzing Eden Semantics. In: Proceedings of WRS 2006. ENTCS, vol. 174, pp. 119–137. Elsevier, Amsterdam (2007)
14. Huggins, J.: ASM-Based Programming Language Definitions, http://www.eecs.umich.edu/gasm/proglang.html
15. Iversen, J., Mosses, P.D.: Constructive Action Semantics for Core ML. IEE Proceedings - Software 152(2), 79–98 (2005)
16. Kutter, P.W., Pierantonio, A.: Montages Specifications of Realistic Programming Languages. Journal of Universal Computer Science 3(5), 416–442 (1997)
17. Liang, S., Hudak, P., Jones, M.P.: Monad Transformers and Modular Interpreters. In: Proceedings of POPL 1995, pp. 333–343. ACM Press, New York (1995)
18. Martí-Oliet, N., Meseguer, J.: Rewriting logic: roadmap and bibliography. Theoretical Computer Science 285, 121–154 (2002)
19. Meseguer, J.: Conditional rewriting logic as a unified model of concurrency. Theoretical Computer Science 96(1), 73–155 (1992)
20. Meseguer, J., de Braga, C.O.: Modular Rewriting Semantics of Programming Languages. In: Rattray, C., Maharaj, S., Shankland, C. (eds.) AMAST 2004. LNCS, vol. 3116, pp. 364–378. Springer, Heidelberg (2004)
21. Meseguer, J., Roşu, G.: Rewriting Logic Semantics: From Language Specifications to Formal Analysis Tools. In: Basin, D., Rusinowitch, M. (eds.) IJCAR 2004. LNCS (LNAI), vol. 3097, pp. 1–44. Springer, Heidelberg (2004)
22. Meseguer, J., Roşu, G.: The rewriting logic semantics project. Theoretical Computer Science 373(3), 213–237 (2007); Also appeared in SOS 2005, ENTCS. vol 156(1), pp. 27–56 (2006)
23. Moggi, E.: An abstract view of programming languages. Technical Report ECS-LFCS-90-113, Edinburgh University, Department of Computer Science (June 1989)
24. Mosses, P.D.: Action Semantics. Cambridge Tracts in Theoretical Computer Science, vol. 26. Cambridge University Press, Cambridge (1992)
25. Mosses, P.D.: Foundations of Modular SOS. In: Kutyłowski, M., Wierzbicki, T., Pacholski, L. (eds.) MFCS 1999. LNCS, vol. 1672, pp. 70–80. Springer, Heidelberg (1999)
26. Mosses, P.D.: Pragmatics of Modular SOS. In: Kirchner, H., Ringeissen, C. (eds.) AMAST 2002. LNCS, vol. 2422, pp. 21–40. Springer, Heidelberg (2002)
27. Mosses, P.D.: A Constructive Approach to Language Definition. Journal of Universal Computer Science 11(7), 1117–1134 (2005)
28. Mosses, P.D., New, M.J.: Implicit Propagation in Structural Operational Semantics. In: SOS 2008. Final version to appear in ENTCS (2008)
29. Plotkin, G.D.: A Structural Approach to Operational Semantics. Journal of Logic and Algebraic Programming 60-61, 17–139 (2004)
30. Roşu, G.: K: A Rewriting-Based Framework for Computations – Preliminary version. Technical Report Department of Computer Science UIUCDCS-R-2007-2926 and College of Engineering UILU-ENG-2007-1827, University of Illinois at Urbana-Champaign (2007)
31. Stärk, R., Schmid, J., Börger, E.: Java^{TM} and the Java^{TM} Virtual Machine: Definition, Verification, Validation. Springer, Heidelberg (2001)

32. Strachey, C., Wadsworth, C.P.: Continuations: A Mathematical Semantics for Handling Full Jumps. Higher-Order and Symbolic Computation 13(1/2), 135–152 (2000)
33. van den Brand, M., Iversen, J., Mosses, P.D.: An Action Environment. Science of Computer Programming 61(3), 245–264 (2006)
34. van den Brand, M.G.J., Heering, J., Klint, P., Olivier, P.A.: Compiling language definitions: the ASF+SDF compiler. ACM TOPLAS 24(4), 334–368 (2002)
35. van Deursen, A., Mosses, P.D.: ASD: The Action Semantic Description Tools. In: Nivat, M., Wirsing, M. (eds.) AMAST 1996. LNCS, vol. 1101, pp. 579–582. Springer, Heidelberg (1996)
36. Wadler, P.: The Essence of Functional Programming. In: Proceedings of POPL 1992, pp. 1–14. ACM Press, New York (1992)
37. Wansbrough, K., Hamer, J.: A Modular Monadic Action Semantics. In: Proceedings of DSL 1997. USENIX (1997)

A   The K Definition of IMP

Figure 14 shows the K definition of the IMP language, a version of which has been used as the running example for the presentation of the module system in this paper. This definition is discussed more fully in a technical report on K [30].

K-Annotated Syntax of IMP

$Int ::= \ldots$ all integer numbers
$Bool ::=$ true | false
$Name ::=$ all identifiers; to be used as names of variables
$Val ::= Int$
$AExp ::= Val \mid Name$
$\quad\quad\quad\mid AExp + AExp$ $[strict,\ extends\ +_{Int\times Int\to Int}]$
$BExp ::= Bool$
$\quad\quad\quad\mid AExp \leq AExp$ $[seqstrict,\ extends\ \leq_{Int\times Int\to Bool}]$
$\quad\quad\quad\mid$ not $BExp$ $[strict,\ extends\ \neg_{Bool\to Bool}]$
$\quad\quad\quad\mid BExp$ and $BExp$ $[strict(1)]$
$Stmt ::= Stmt; Stmt$ $[s_1; s_2 = s_1 \curvearrowright s_2]$
$\quad\quad\quad\mid Name := AExp$ $[strict(2)]$
$\quad\quad\quad\mid$ if $BExp$ then $Stmt$ else $Stmt$ $[strict(1)]$
$\quad\quad\quad\mid$ while $BExp$ do $Stmt$
$\quad\quad\quad\mid$ halt $AExp$ $[strict]$
$Pgm ::= Stmt; AExp$

K Configuration and Semantics of IMP

$$\frac{(\!|\ x\ |\!)_k\ (\!|\sigma|\!)_{state}}{\sigma[x]}$$

$KResult ::= Val$
$K ::= KResult \mid List_{\curvearrowright}[K]$
$Config ::= (\!|K|\!)_k \mid (\!|State|\!)_{state}$
$\quad\quad\quad\mid Val \mid [\![K]\!] \mid (\!|Set[Config]|\!)_\top$

true and $b \to b$
false and $b \to$ false
$$\frac{(\!|x := v|\!)_k\ (\!|\ \ \sigma\ \ |\!)_{state}}{\cdot\qquad\sigma[v/x]}$$
if true then s_1 else $s_2 \to s_1$
if false then s_1 else $s_2 \to s_2$

$[\![p]\!] = (\!|(\!|p|\!)_k\ (\!|\emptyset|\!)_{state}|\!)_\top$
$(\!|(\!|v|\!)_k|\!)_\top = v$

(while b do s)$_k$ = (if b then $(s;$ while b do $s)$ else $\cdot$)$_k$
(halt i)$_k$ = (i)$_k$

Fig. 14. K definition of IMP

B   The Modular K Definition of IMP

A number of modules make up the definition of the IMP language. The first modules shown below make up semantic entities used in the definition.

```
module Env                              module Int
  requires sort Name, sort Loc .          imports K/Int .
  sortalias Env = Map(Name,Loc) .         requires sort Value .
  var Env[0-9']* : Env .                  subsort Int < Value .
end module                              end module

module Store
  requires sort Loc, sort Value .       module Bool
  sortalias Store = Map(Loc,Value) .      sort Bool .
  var Store[0-9']* : Store .              ops true false : -> Bool .
end module                              end module
```

The next modules define the abstract syntax for IMP. This includes sorts for arithmetic expressions, boolean expressions, statements, and programs, as well as a number of syntactic entities (i.e., productions). Note that modules can import other modules of the same "type" (Syntax, Dynamic, etc) without needing to specify the type. If this would lead to an ambiguous import a warning message will be generated.

```
module Exp/AExp[Syntax]                 module Exp/BExp/And[Syntax]
  sort AExp .                             imports Exp/BExp .
  var AE[0-9'a-zA-Z]* : AExp .            _and_ : BExp BExp -> BExp .
end module                              end module

module Exp/AExp/Num[Syntax]             module Stmt[Syntax]
  imports Int, Exp/AExp .                 sort Stmt .
  subsort Int < AExp .                  end module
end module

                                        module Stmt/Sequence[Syntax]
module Exp/AExp/Name[Syntax]              imports Stmt .
  imports Exp/AExp .                       _;_ : Stmt Stmt -> Stmt .
  requires sort Name .                   end module
  subsort Name < AExp .
end module                              module Stmt/Assign[Syntax]
                                          imports Stmt, Exp/AExp .
module Exp/AExp/Plus[Syntax]              requires sort Name .
  imports Exp/AExp .                       _:=_ : Name AExp -> Stmt .
  _+_ : AExp AExp -> AExp .              end module
end module
                                        module Stmt/IfThenElse[Syntax]
module Exp/BExp[Syntax]                    imports Stmt, Exp/BExp .
  sort BExp .                             if_then_else_ : BExp Stmt Stmt -> Stmt .
  var BE[0-9'a-zA-Z]* : BExp .          end module
end module

module Exp/BExp/Bool[Syntax]            module Stmt/While[Syntax]
  imports Bool, Exp/BExp .                imports Stmt, Exp/BExp .
  subsort Bool < BExp .                   while_do_ : BExp Stmt -> Stmt .
end module                              end module

module Exp/BExp/LessThanEq[Syntax]      module Stmt/Halt[Syntax]
  imports Exp/AExp, Exp/BExp .            imports Stmt, Exp/AExp .
  _<=_ : AExp AExp -> BExp .              halt_ : AExp -> Stmt .
end module                              end module

module Exp/BExp/Not[Syntax]             module Pgm[Syntax]
  imports Exp/BExp .                      imports Stmt, Exp/AExp .
  not_ : BExp -> BExp .                   sort Pgm .
end module                                _;_ : Stmt AExp -> Pgm .
                                        end module
```

Using the abstract syntax, a number of modules are used to define the evaluation semantics, with one semantics module per language feature. As a reminder, a semantics module will automatically import the syntax module of the same name; explicit imports of syntax modules are used in cases where attributes, such as strictness, need to be changed. Added strictness information will cause heating and cooling rules to be automatically generated. The `seqstrict` attribute, used on less than, is identical to `strict`, except it enforces a left to right evaluation order on the arguments.

```
module Exp/AExp/Name[Dynamic]
  requires sort Name, sort Loc, sort Value .
  imports Env, Store, K/K .
  rl k(| X |> env<| (X,L) |> store<| (L,V) |>
  => k(| V |> env<| (X,L) |> store<| (L,V) |> .
end module

module Exp/AExp/Plus[Dynamic]
  imports Exp/AExp/Plus[Syntax]
     with { op _+_ now strict,
             extends + [Int * Int -> Int] } .
end module

module Exp/BExp/LessThanEq[Dynamic]
  imports Exp/BExp/LessThanEq[Syntax]
     with { op _<=_ now seqstrict,
             extends <= [Int * Int -> Bool] } .
end module.

module Exp/BExp/Not[Dynamic]
  imports Exp/BExp/Not[Syntax]
     with { op not_ now strict,
             extends not [Bool -> Bool] } .
end module

module Exp/BExp/And[Dynamic]
  imports Exp/BExp/And[Syntax]
     with { op _and_ now strict(1) } .
  rl true and BE => BE .
  rl false and BE => false .
end module

module Stmt/Sequence[Dynamic]
  eq S ; S' = S -> S' .
end module
```

```
module Stmt/Assign[Dynamic]
  imports Stmt/Assign[Syntax]
     with { op _:=_ now strict(2) } .
  requires sort Name, sort Value, sort Loc .
  imports K/K, Env, Store .
  rl k(| X := V |> env<| (X,L) |>
     store<| (L,_) |>
  => k(| .       |> env<| (X,L) |>
     store<| (L,V) |> .
end module

module Stmt/IfThenElse[Dynamic]
  imports Stmt/IfThenElse[Syntax]
     with { op if_then_else_ now strict(1) } .
  imports Bool .
  rl if true then S else S' => S .
  rl if false then S else S' => S' .
end module

module Stmt/While[Dynamic]
  imports Stmt/IfThenElse, Exp/BExp[Syntax] .
  eq k(| while BE do S |>
   = k(| if BE then (S ;
             while BE do S) else . |> .
end module

module Stmt/Halt[Dynamic]
  imports Stmt/Halt[Syntax]
     with {op halt_ now strict } .
  imports Int .
  eq k(| halt i |> = k(| i |) .
end module

module Pgm[Dynamic]
  requires sort Value .
  eq top (| k(| V |) |> = V .
end module
```

Finally, the various modules are assembled together into a language module, representing the entire programming language.

```
module Imp[Language]
  imports K/Configuration, K/K, K/Location, K/Value,
        Env, Store, Int, Bool .
  config = top(store(Store) env(Env) k(K) nextLoc(Loc)) .

  op [[_]] : K -> Configuration .
  eq [[ K ]] = top(store(empty) env(empty) k(K) nextLoc(initLoc)) .

  imports type=Syntax Exp/AExp/Num, Exp/BExp/Bool .
  imports type=Dynamic Exp/AExp/Name, Exp/AExp/Plus, Exp/BExp/LessThanEq, Exp/BExp/Not,
    Exp/BExp/And, Stmt/Sequence, Stmt/Assign, Stmt/IfThenElse, Stmt/While, Stmt/Halt, Pgm .
end module
```

Property Preserving Refinement for CSP-CASL

Temesghen Kahsai and Markus Roggenbach[*]

Department of Computer Science, Swansea University, UK
{csteme, csmarkus}@swan.ac.uk

Abstract. In this paper we present various notions of the combined refinement for data and processes within the specification language CSP-CASL. We develop proof support for our refinement notions and demonstrate how to employ them for system development and for system analysis. Finally, we apply our technique to an industrial standard for an electronic payment system.

1 Introduction

System development in a step-by-step fashion has been central to software engineering at least since Wirth's seminal paper on program development [22] in 1971. Such a development starts with an abstract specification, which defines the general setting, e.g. it might define the components and interfaces involved in the system. In several design steps this abstract specification is then further developed towards a design specification which can be implemented directly. In each of these steps some design decisions are taken and implementation issues are resolved. A design step can for instance refine the type system, or it might set up a basic dialogue structure. It is essential, however, that these design steps preserve properties. This idea is captured by the notion of *refinement*.

In industrial practice, stepwise development usually is carried out informally. In this paper, we capture such informal developments with formal notions of refinement within the specification language CSP-CASL [19]. CSP-CASL allows one to specify data and processes in an integrated way, where CASL [17] is used to describe data and CSP [7,20] is used to specify the reactive side.

Our notions of refinement for CSP-CASL are based on refinements developed in the context of the single languages CSP and CASL. In the context of algebraic specification, e.g., [4] provides an excellent survey on different approaches. For CSP, each of its semantical models comes with a refinement notion of its own. There are for instance traces refinement, failure/divergences refinement, and stable failures refinement [20]. For *system development* one often is interested in liberal notions of refinements, which allow substantial changes in the design. For *system verification*, however, it is important that refinement steps preserve properties. The latter concept allows one to verify properties already on abstract specifications – which in general are less complex than the more concrete ones. The properties, however, are preserved over the design steps. These two purposes motivate our study of various refinement notions.

In this paper, we develop proof methods for CSP-CASL. To this end, we decompose a CSP-CASL refinement into a refinement over CSP and a refinement over CASL alone.

---

[*] This work was supported by EPSRC under the grant EP/D037212/1.

A. Corradini and U. Montanari (Eds.): WADT 2008, LNCS 5486, pp. 206–220, 2009.

We show how to use existing tools to discharge the arising proof obligations. Reactive systems often exhibit the undesirable behaviour of deadlock or divergence (livelock), which both result in lack of progress in the system. Here, we develop proof techniques based on refinement for proving deadlock freeness and divergence freeness.

The language CSP dates back at least to 1985 [7]; an excellent reference is the book [20] (updated 2005). CASL was designed by the CoFI initiative [17]. Tools for CASL are developed, e.g., in [11,12,13]. The combination CSP-CASL was introduced in [19] and used for specifying an electronic payment system in [6]. A tool for CSP refinement was developed in [9]. [5] implements a parser for CSP-CASL, [18] describes a theorem prover for CSP-CASL. In [16], Mossakowski et al. define a refinement language for CASL architectural specifications. In [1] Woodcock et al. discuss a proof-by-refinement technique in the area of Z specification. Deadlock analysis in CSP has been studied in [20], and an industrial application has been described in [3]. Tools for deadlock analysis are developed in [8,10]. Livelock analysis in CSP has been applied to an industrial application in [21].

In the next section we give an overview of the specification language CSP-CASL and refinement based on model class inclusion. Section 3 we develop proof support for CSP-CASL refinement and describe how refinement can be employed for deadlock and divergence analysis In CSP-CASL. In Section 4 we demonstrate that the presented theoretical results are applicable in an industrial setting.

2   CSP-CASL

CSP-CASL [19] is a specification language which combines *processes* written in CSP [7,20] with the specification of *data types* in CASL [17]. The general idea is to describe reactive systems in the form of processes based on CSP operators, where the communications of these processes are the values of data types, which are loosely specified in CASL. All standard CSP operators are available, such as multiple prefix, the various parallel operators, operators for non-deterministic choice, communication over channels. Concerning CASL features, the full language is available to specify data types, namely many-sorted first order logic with sort-generation constraints, partiality, and sub-sorting. Furthermore, the various CASL structuring constructs are included, where the structured **free** construct adds the possibility to specify data types with initial semantics.

Syntactically, a CSP-CASL specification with a name Sp consists of a data part D, which is a structured CASL specification, an (optional) channel part Ch to declare channels, which are typed according to the data part, and a process part P written in CSP, within which CASL terms are used as communications, CASL sorts denote sets of communications, relational renaming is described by a binary CASL predicate, and the CSP conditional construct uses CASL formulae as conditions:

ccspec $Sp =$ **data** D **channel** Ch **process** P **end**

For concrete examples of CSP-CASL specifications see Section 4. The CSP-CASL channel part is syntactic sugar over the data part, see [19] for the details of the encoding into CASL. In our practical examples we will make use of channels. For our semantical considerations, however, we will study specifications only without channels. We often write such specifications shortly as $Sp = (D, P)$.

Semantically, a CSP-CASL specification $Sp = (D, P)$ is a family of process denotations for a CSP process, where each model of the data part D gives rise to one process denotation. CSP-CASL has a 2-steps semantics. In the first step we construct for each CASL model $M \in \mathbf{Mod}(D)$ a CSP process $[\![P]\!]_M$, which communicates in an alphabet $\mathcal{A}(M)$ constructed out of the CASL model M. In the second step we point-wise apply a denotational CSP semantics. This translates a process $[\![P]\!]_M$ into its denotation d_M in the semantic domain of the chosen CSP model. The overall semantical construction is written $([\![[\![P]\!]_M]\!]_{\mathrm{CSP}})_{M \in \mathbf{Mod}(D)}$. For a denotational CSP model with domain $\mathcal{D}$, the semantic domain of CSP-CASL consists of families of process denotations $d_M \in \mathcal{D}$ over some index set I, $(d_M)_{M \in I}$, where I is a class of CASL models over the same signature.

CSP-CASL refinement is based on refinements developed in the context of the single languages CSP and CASL. Intuitively, a refinement step, which we write here as '$\rightsquigarrow$', reduces the number of possible implementations. Concerning data, this means a reduced model class, concerning processes this mean less non-deterministic choice:

Definition 1 (Model class inclusion). *For families $(d_M)_{M \in I}$ and $(d'_{M'})_{M' \in I'}$ of process denotations we write $(d_M)_{M \in I} \rightsquigarrow_{\mathcal{D}} (d'_{M'})_{M' \in I'}$ iff $I' \subseteq I \wedge \forall M' \in I' : d_{M'} \sqsubseteq_{\mathcal{D}} d'_{M'}$.*

Here, $I' \subseteq I$ denotes inclusion of model classes over the same signature, and $\sqsubseteq_{\mathcal{D}}$ is the refinement notion in the chosen CSP model $\mathcal{D}$. In the traces model $\mathcal{T}$ we have for instance $P \sqsubseteq_{\mathcal{T}} P' \Leftrightarrow traces(P') \subseteq traces(P)$, where $traces(P)$ and $traces(P')$ are prefixed closed sets of traces. Here we follow the CSP convention, where P' *refines* P is written as $P \sqsubseteq_{\mathcal{D}} P'$, i.e. the more specific process is on the right-hand side of the symbol. The definitions of CSP refinements for $\mathcal{D} \in \{\mathcal{T}, \mathcal{N}, \mathcal{F}, \mathcal{I}, \mathcal{U}\}$, c.f. [20], which are all based on set inclusion, yield that CSP-CASL refinement is a preorder.

Given CSP-CASL specifications $Sp = (D, P)$ and $Sp' = (D', P')$, by abuse of notation we also write $(D, P) \rightsquigarrow_{\mathcal{D}} (D', P')$ if the above refinement notion holds for the denotations of Sp and Sp', respectively.

On the syntactic level of specification text, we additionally define the notions of data refinement and process refinement in order to characterize situations, where one specification part remains constant. In a *data refinement*, only the data part changes:

$$\left.\begin{array}{c} \textbf{data } D \textbf{ process } P \textbf{ end} \\ \overset{\text{data}}{\rightsquigarrow} \\ \textbf{data } D' \textbf{ process } P \textbf{ end} \end{array}\right\} \quad \text{if} \quad \left\{\begin{array}{l} 1.\ \Sigma(D) = \Sigma(D'), \\ 2.\ \mathbf{Mod}(D') \subseteq \mathbf{Mod}(D) \end{array}\right.$$

Here, $\Sigma(D)$ denotes the CASL signature of D. As in a data refinement the process part remains the same, there is no need to annotate data refinement with a specific process model: all CSP refinements notions are reflexive. In a *process refinement*, the data part is constant:

$$\left.\begin{array}{c} \textbf{data } D \textbf{ process } P \textbf{ end} \\ \overset{\text{proc}}{\rightsquigarrow_{\mathcal{D}}} \\ \textbf{data } D \textbf{ process } P' \textbf{ end} \end{array}\right\} \quad \text{if} \quad \left\{\begin{array}{l} \text{for all } M \in \mathbf{Mod}(D) : \\ [\![[\![P]\!]_M]\!]_{\mathrm{CSP}} \sqsubseteq_{\mathcal{D}} [\![[\![P']\!]_M]\!]_{\mathrm{CSP}} \end{array}\right.$$

Clearly, both these refinements are special forms of CSP-CASL refinement in general.

3 A Basic Theory of CSP-CASL Refinement

In this section we develop proof support for CSP-CASL refinement, and theories of how to analyse CSP-CASL specifications.

3.1 Proof Support

Proof support for the CSP-CASL refinement is based on a decomposition theorem. This decomposition theorem gives rise to a proof method for CSP-CASL, namely, we study CSP-CASL refinement in terms of CASL refinement and CSP refinement separately. With regard to CSP-CASL refinement, data turns out to dominate the processes: While any CSP-CASL refinement can be decomposed into first a data refinement followed by a process refinement, there is no such decomposition result possible for the reverse order. This insight is in accordance with the 2-step semantics of CSP-CASL, where in the first step we evaluate the data part and only in the second step apply the process semantics.

First, we present our positive result concerning decomposition:

Theorem 1. *Let $Sp = (D, P)$ and $Sp' = (D', P')$ be CSP-CASL specifications, where D and D' are data specifications over the same signature. Let (D', P) be a new CSP-CASL specification. For these three specifications holds: $(D, P) \leadsto_{\mathcal{D}} (D', P')$ iff $(D, P) \overset{data}{\leadsto} (D', P)$ and $(D', P) \overset{proc}{\leadsto}_{\mathcal{D}} (D', P')$.*

This result forms the basis for the CSP-CASL tool support developed in [18]. In order to prove that a CSP-CASL refinement $(D, P) \leadsto_{\mathcal{D}} (D', P')$ holds, first one uses proof support for CASL [13] alone in order to establish $\mathbf{Mod}(D') \subseteq \mathbf{Mod}(D)$. Independently of this, one has then to check the process refinement $P \sqsubseteq_{\mathcal{D}} P'$. In principle, the latter step can be carried out using CSP-Prover, see e.g. [9]. The use of CSP-Prover, however, requires the CASL specification D' to be translated into an alphabet of communications. The tool CSP-CASL-Prover [18] implements this translation and also generates proof support for theorem proving on CSP-CASL.

Changing the order in the above decomposition theorem, i.e., to first perform first a process refinement followed by a data refinement, however, is not possible in general. Often, process properties depend on data, as the following counter example illustrates, in which we have: $(D, P) \leadsto_{\mathcal{T}} (D', P')$ but $(D, P) \not\leadsto_{\mathcal{T}} (D, P')$. Consider the three CSP-CASL specifications ABS, MID and CONC, where MID consists of the data part of ABS and the process part of CONC:

ccspec ABS =	**ccspec** MID =	**ccspec** CONC =
data	**data**	**data**
sorts S	**sort** S	**sort** S
ops $a, b : S$;	**ops** $a, b : S$;	**ops** $a, b : S$;
process	**process**	**axiom** $a = b$
P $= a \rightarrow Stop$	Q $= a \rightarrow Stop \,\|[\, a\,]\|$	**process**
end	$b \rightarrow Stop$	R $= a \rightarrow Stop \,\|[\, a\,]\|$
	end	$b \rightarrow Stop$
		end

Let N be a CASL model of the data part D_{ABS} of ABS with $N(S) = \{\#, *\}, N(a) = \#, N(b) = *$. Concerning the process denotations in the traces model $\mathcal{T}$ relatively to N, for ABS we obtain the denotation[1] $d_{\text{ABS}} = \{\langle\rangle, \langle\#\rangle\}$. In MID, the alphabetized parallel operator requires synchronization only w.r.t. the event a. As $N \models \neg a = b$, the rhs of the parallel operator, which is prepared to engage in b, can proceed with b, which yields the trace $\langle*\rangle$ in the denotation. The lhs, however, which is prepared to engage in a, does not find a partner for synchronization and therefore is blocked. This results in the denotation $d_{\text{MID}} = \{\langle\rangle, \langle*\rangle\}$. As $d_{\text{MID}} \not\subseteq d_{\text{ABS}}$, we have ABS $\not\leadsto_{\mathcal{T}}$ MID.

In CONC, the axiom $a = b$ prevents N to be a model of the data part. This makes it possible to establish ABS $\leadsto_{\mathcal{T}}$ CONC over the traces model $\mathcal{T}$. Using Theorem 1, we first prove the data refinement: CONC adds an axiom to ABS – therefore, D_{ABS} refines to D_{CONC} with respect to CASL; concerning the process refinement, using the equation $a = b$ and the step law for generalized parallel, we obtain $a \rightarrow Stop \,\|[\,a\,]\|\, b \rightarrow Stop =_{\mathcal{T}} a \rightarrow Stop \,\|[\,a\,]\|\, a \rightarrow Stop =_{\mathcal{T}} a \rightarrow (Stop \,\|[\,a\,]\|\, Stop) =_{\mathcal{T}} a \rightarrow Stop$ – thus, over D_{CONC} the process parts of ABS CONC are semantically equivalent and therefore in refinement relation over the traces model $\mathcal{T}$.

3.2 Analysis for Deadlock Freeness

In this section we show how to analyse deadlock freeness in the context of CSP-CASL. To this end, first we recall how deadlock is characterized in CSP. Then we define what it means for a CSP-CASL specification to be deadlock free. Finally, we establish a proof technique for deadlock freeness based on CSP-CASL refinement, which turns out to be complete.

In the CSP context, the stable failures model $\mathcal{F}$ is best suited for deadlock analysis. The stable failures model $\mathcal{F}$ records two observations on processes: the first observation is the set of traces a process can perform, this observation is given by the semantic function *traces*; the second observation are the so-called stable failures, given by semantic function *failures*. A failure is a pair (s, X), where s represents a trace that the process can perform, after which the process can refuse to engage in all events of the set X. We often write (T, F) for such a pair of observations, T denoting the set of traces and F denoting the set of stable failures. Deadlock is represented by the process *STOP*. Let A be the alphabet. Then the process *STOP* has

$$(\{\langle\rangle\}, \{(\langle\rangle, X) \mid X \subseteq A^{\checkmark}\}) \in \mathcal{P}(A^{*\checkmark}) \times \mathcal{P}(A^{*\checkmark} \times \mathcal{P}(A^{\checkmark}))$$

as its denotation in $\mathcal{F}$, i.e., the process *STOP* can perform only the empty trace, and after the empty trace the process *STOP* can refuse to engage in all events. Here, $\checkmark \notin A$ is a special event denoting successful termination, $A^{\checkmark} = A \cup \{\checkmark\}$, and $A^{*\checkmark} = A^* \cup A^* ^\frown \langle\checkmark\rangle$ is the set of all traces over A possibly ending with $\checkmark$. In CSP, a process P is considered to be deadlock free, if the process P after performing a trace s never becomes equivalent to the process *STOP*. More formally: A process P is **deadlock-free** in CSP iff

$$\forall s \in A^*.(s, A^{\checkmark}) \notin \textit{failures}(P).$$

---

[1] For the sake of readability, we write the element of the carrier sets rather than their corresponding events in the alphabet of communications.

This definition is justified, as in the model $\mathcal{F}$ the set of stable failures is required to be closed under subset-relation: $(s, X) \in failures(P) \wedge Y \subseteq X \Rightarrow (s, Y) \in failures(P)$. In other words: Before termination, the process P can never refuse all events; there is always some event that P can perform.

A CSP-CASL specification has a family of process denotations as its semantics. Each of these denotations represents a possible implementation. We consider a CSP-CASL specification to be deadlock free, if it enforces all its possible implementations to have this property. On the semantic level, we capture this idea as follows:

Definition 2. *Let* $(d_M)_{M \in I}$ *be a family of process denotations over the stable failures model, i.e.,* $d_M = (T_M, F_M) \in \mathcal{F}(\mathcal{A}(M))$ *for all* $M \in I$.

- d_M *is deadlock free if* $(s, X) \in F_M$ *and* $s \in \mathcal{A}(M)^*$ *implies that* $X \neq \mathcal{A}(M)^{\checkmark}$.
- $(d_M)_{M \in I}$ *is deadlock free if for all* $M \in I$ *it holds that* d_M *is deadlock-free.*

Deadlock can be analyzed trough refinement checking; that is an implementation is deadlock-free if it is the refinement of a deadlock free specification:

Theorem 2. *Let* $(d_M)_{M \in I} \rightsquigarrow_{\mathcal{F}} (d'_{M'})_{M' \in I'}$ *be a refinement over* $\mathcal{F}$ *between two families of process denotations. If* $(d_M)_{M \in I}$ *is deadlock-free, then so is* $(d'_{M'})_{M' \in I'}$.

Following an idea from the CSP context, we formulate the most abstract deadlock free CSP-CASL specification over a subsorted CASL signature $\Sigma = (S, TF, PF, P, \leq)$ – see [17] for the details – with a set of sort symbols $S = \{s_1, \ldots, s_n\}, n \geq 1$:

ccspec $\mathrm{DF}_\Sigma =$
    data $\ldots$ declaration of Σ $\ldots$
    process $DF_S = \bigsqcap_{s:S} (!x : s \rightarrow DF_S) \sqcap Skip$
end

Here, the process $!x : s \rightarrow DF_S$ internally chooses an element x from the sort s, engages in it, and then behaves like DF_S. We observe:

Lemma 1. DF_Σ *is deadlock free.*

Proof. Let $(d_M)_{M \in I}$ be the denotation of DF_Σ over the stable-failures model, where $d_M = (T_M, F_M)$. For all $M \in I$ holds: $T_M = \mathcal{A}(M)^{*\checkmark}$ and $F_M = \{(t, X) \mid t \in \mathcal{A}(M)^*, X \subseteq \mathcal{A}(M) \vee \exists a \in \mathcal{A}(M). X \subseteq \mathcal{A}(M)^{\checkmark} - \{a\}\} \cup \{(t \frown \langle \checkmark \rangle, Y) \mid t \in \mathcal{A}(M)^*, Y \subseteq \mathcal{A}(M)^{\checkmark}\}$.

This result on DF_Σ extends to a complete proof method for deadlock freeness in CSP-CASL:

Theorem 3. *A* CSP-CASL *specification* (D, P) *is deadlock free iff* $\mathrm{DF}_\Sigma \rightsquigarrow_{\mathcal{F}} (D, P)$. *Here,* Σ *is the signature of D.*

Proof. If $\mathrm{DF}_\Sigma \rightsquigarrow_{\mathcal{F}} (D, P)$, Lemma 1 and Theorem 2 imply that (D, P) is deadlock free. Now let (D, P) be deadlock free. We apply Theorem 1 to our proof goal $\mathrm{DF}_\Sigma \rightsquigarrow_{\mathcal{F}} (D, P)$ and decompose it into a data refinement and a process refinement. The data refinement holds, as the model class of DF_Σ consists of all CASL models over Σ. The process refinement holds thanks to the semantics of DF_Σ as given in the proof of Lemma 1.

3.3   Analysis for Divergence Freeness

For concurrent systems, divergence (or livelock) is regarded as an individual starvation, i.e., a particular process is prevented from engaging in any actions. For CSP, the failures/divergences model $\mathcal{N}$ is considered best to study systems with regard to divergence. The CSP process **div** represents this phenomenon: immediately, it can refuse every event, and it diverges after any trace. **div** is the least refined process in the $\sqsubseteq_{\mathcal{N}}$ model. The main sources for divergence in CSP are *hiding* and ill-formed recursive processes.

In the failures/divergences model $\mathcal{N}$, a process is modeled as a pair (F, D). Here, F represents the *failures*, while D collects all *divergences*. Let A be the alphabet. The process **div** has

$$(A^{*\checkmark} \times \mathcal{P}(A^{\checkmark}), A^{*\checkmark}) \in \mathcal{P}(A^{*\checkmark} \times \mathcal{P}(A^{\checkmark})) \times \mathcal{P}(A^{*\checkmark})$$

as its semantics over $\mathcal{N}$.

Following these ideas, we define what it means for a CSP-CASL specification to be divergence free: Essentially, after carrying out a sequence of events, the denotation shall be different from **div**.

Definition 3. *Let* $(d_M)_{M \in I}$ *be a family of process denotations over the failure divergence model, i.e,* $d_M = (F_M, D_M) \in \mathcal{N}(\mathcal{A}(M))$ *for all* $M \in I$.

- *A denotation* d_M *is* **divergence free** *iff one of the following conditions holds:*
  C1. $\forall s \in \mathcal{A}(M)^*.\{(t, X) \mid (s ^\frown t, X) \in F_M\} \neq \mathcal{A}(M)^{*\checkmark} \times \mathcal{P}(\mathcal{A}(M)^{\checkmark})$
  C2. $\forall s \in \mathcal{A}(M)^*.\{t \mid (s ^\frown t) \in D\} \neq \mathcal{A}(M)^{*\checkmark}$.
- $(d_M)_{M \in I}$ *is* **divergence free** *if for all* $M \in I$ *it holds that* d_M *is divergence free.*

Like in the case of analysis for deadlock freeness, also the analysis for divergence freeness can be checked trough refinement, this time over the model $\mathcal{N}$.

Theorem 4. *Let* $(d_M)_{M \in I} \rightsquigarrow_{\mathcal{N}} (d'_{M'})_{M' \in I'}$ *be a refinement over* $\mathcal{N}$ *between two families of process denotations. Let* $(d_M)_{M \in I}$ *be divergence free. Then* $(d'_{M'})_{M' \in I'}$ *is divergence free.*

As for the analysis of deadlock freeness we formulate the least refined divergence free CSP-CASL specification over a CASL signature Σ with a set of sort of symbols $S = \{s_1, \ldots, s_n\}, n \geq 1$.

ccspec $\mathrm{DIVF}_\Sigma =$
    data ... declaration of Σ ...
    process $DivF = (Stop \sqcap Skip) \sqcap (\Box_{s:S} ?x : s \rightarrow DivF)$
end

$DivF$ may deadlock at any time, it may terminate successfully at any time, or it may perform any event at any time, however, it will not diverge. Figure 1 shows DIVF_Σ as a labelled transition system. One can easily see that this transition system does not have a path of infinite silent actions τ. Here, $[s_i]$ denotes the collection of events constructed over the sort s_i. This observation is reflected in the following lemma:

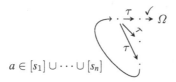

$$a \in [s_1] \cup \cdots \cup [s_n]$$

Fig. 1. An LTS version of DIVF_Σ

Lemma 2. DIVF_Σ *is divergence free.*

Proof. Let $(d_M)_{M \in I}$ be the semantics of DIVF_S over the failures/divergences model $\mathcal{N}$ where $d_M = (F_M, D_M) \in \mathcal{N}(\mathcal{A}(M))$. For all models M holds: $F_M = \mathcal{A}(M)^{*\checkmark} \times \mathcal{P}(\mathcal{A}(M)^\checkmark)$ and $D_M = \emptyset$. Thus, DIVF_Σ is divergence free thanks to conditions **C.2**.

Putting things together, we obtain a complete proof method for divergence freedom of CSP-CASL specifications:

Theorem 5. *A CSP-CASL specification (D, P) is divergence free iff $\text{DIVF}_\Sigma \leadsto_{\mathcal{F}} (D, P)$. Here Σ is the signature of D.*

Proof. If $\text{DIVF}_\Sigma \leadsto_{\mathcal{N}} (D, P)$, Lemma 2 and Theorem 4 imply that (D, P) is divergence free. Now let (D, P) be divergence free. Assume that $\neg(\text{DIVF}_\Sigma \leadsto_{\mathcal{N}} (D, P))$. As the data part of DIVF refines to D, with our decomposition theorem 1 we can conclude that $\neg((D, DivF) \overset{\text{proc}}{\leadsto}_{\mathcal{N}} (D, P))$. Let $(d_M)_{M \in \mathbf{Mod}(D)}$ be the semantics of $(D, DivF)$, where $d_M = (F_M, D_M)$. Let $(d'_M)_{M \in \mathbf{Mod}(D)}$ be the semantics of (D, P), where $d'_M = (F'_M, D'_M)$. By definition of process refinement there exists a model $M \in \mathbf{Mod}(D)$ such that $F'_M \not\subseteq F_M$ or $D'_M \not\subseteq D_M$. As $F'_M = \mathcal{A}(M)^{*\checkmark} \times \mathcal{P}(\mathcal{A}(M)^\checkmark)$, see the proof of Lemma 2, we know that $F'_M \subseteq F_M$ holds. Therefore, we know that $D'_M \not\subseteq D_M$. As $D_M = \emptyset$, there exists a trace $t \in D'_M$ not ending with $\checkmark$, as the healthiness condition $D3$ of the failures/divergences model asserts that for any trace $u' = u \frown \langle \checkmark \rangle \in D_M$ also $u \in D_M$. Applying healthiness condition $D1$ we obtain $t \frown t' \in D'_M$ for all $t' \in \mathcal{A}(M)^{*\checkmark}$. Hence, d'_m is not divergence free, as D'_M violates **C.2** – contradiction to (D, P) divergence free.

3.4 Refinement with Change of Signature

Until now we have analyzed properties of specifications based on a notion of refinement over the same signature. Often, in a refinement step, it is the case that the signature changes. In this section we sketch a first idea of how a theory of refinement with change of signature might look like. In a CSP-CASL institution as discussed in [15], the *semantics* of refinement under change of signature is merely a consequence of our above Definition 1. The aim of this section, however, is to provide a *proof rule* for such a setting. For simplicity, we consider embeddings only and restrict ourselves to the traces model $\mathcal{T}$.

A subsorted CASL signature $\Sigma = (S, TF, PF, P, \leq)$ consists of a set of sort symbols S, a set of total functions symbols TF, a set of partial function symbols PF, a set of predicate symbols P, and a reflexive and transitive subsort relation $\leq \subseteq S \times S$ – see [17] for details.

Definition 4. *We say that a signature* $\Sigma = (S, TF, PF, P, \leq)$ *is **embedded into** a signature* $\Sigma' = (S', TF', PF', P', \leq')$ *if* $S \subseteq S'$, $TF \subseteq TF'$, $PF \subseteq PF'$, $P \subseteq P'$, *and additionally the following conditions regarding subsorting hold:*

preservation and reflection $\leq = \leq' \cap (S \times S)$.
weak non-extension *For all sorts* $s_1, s_2 \in S$ *and* $u' \in S'$:
  if $s_1 \neq s_2$ *and* $s_1, s_2 \leq' u'$ *then there exists* $t \in S$ *with* $s_1, s_2 \leq t$ *and* $t \leq' u$.
sort condition $S' \subseteq \{s' \in S' \mid \exists s \in S : s' \leq' s\}$.

We write $\sigma : \Sigma \to \Sigma'$ for the induced map from Σ to Σ', where $\sigma(s) = s, \sigma(f) = f, \sigma(p) = p$ for all sort symbols $s \in S$, function symbols $f \in TF \cup PF$ and predicate symbols $p \in P$.

The conditions 'preservation and reflection' and 'weak non-extension' are inherited from the CSP-CASL design, see [19]. The 'sort condition' ensures that reducts are defined, see Lemma 3. In a development process, these conditions allow one to refine the type system by the introduction of new subsorts. Operation symbols and predicate symbols can be added without restriction.

Let $\Sigma = (S, TF, PF, P, \leq)$ be embedded into $\Sigma' = (S', TF', PF', P', \leq')$, then every Σ'-model M' defines a the *reduct* Σ-model $M' \mid_\sigma$, such that $(M' \mid_\sigma)_s = M'(\sigma(s)) = M'_s$, $(M' \mid_\sigma)_f = M'(\sigma(s)) = M'_f$, and $(M' \mid_\sigma)_p = M'(\sigma(s)) = M'_p$ for all sort symbols $s \in S$, function symbols $f \in TF \cup PF$ and predicate symbols $p \in P$. On the alphabet level, the map σ induces an injective map $\sigma^{\mathcal{A}}_{M'} : \mathcal{A}(M' \mid_\sigma) \to \mathcal{A}(M')$, where $\sigma^{\mathcal{A}}_{M'}([(s, x)]_{\sim_{M' \mid_\sigma}} = [(\sigma^S(s), x)]_{\sim_{M'}}$ – see [19] for the definition of $\sim$. In the following, we will make use of the partial inverse $\hat{\sigma}^{\mathcal{A}}_{M'} : \mathcal{A}(M') \to^? \mathcal{A}(M' \mid_\sigma)$ of this map. In [13] it is shown that, given an injective map, the canonical extension of its partial inverse to trace sets preserves the healthiness conditions in the traces model $\mathcal{T}$. Applying this result to our setting, we obtain:

$$T' \in \mathcal{T}(\mathcal{A}(M')) \Rightarrow \hat{\sigma}^{\mathcal{A}}_{M'}(T') \in \mathcal{T}(\mathcal{A}(M' \mid_\sigma)).$$

This allows us to lift the maps $\hat{\sigma}^{\mathcal{A}}_{M'}$ to the domain of the traces model and we can define:

Definition 5 (Refinement with change of signature). *Let* Σ *and* Σ' *be signatures such that* Σ *is embedded into* Σ'. *Let* $\sigma : \Sigma \to \Sigma'$ *be the induced signature morphism from* Σ *to* Σ'. *Let* $(d_M)_{M \in I}$ *and* $(d'_{M'})_{M' \in I'}$ *be a families of process denotations over* Σ *and* Σ', *respectively.*

$$d_M \leadsto^\sigma_{\mathcal{T}} d'_{M'} \Leftrightarrow I'|_\sigma \subseteq I \wedge \forall M' \in I' : d_{M'|_\sigma} \sqsubseteq_{\mathcal{T}} \hat{\sigma}^{\mathcal{A}}_{M'}(d'_{M'}).$$

Here, $I'|_\sigma = \{M'|_\sigma \mid M' \in I'\}$, and $\sqsubseteq_{\mathcal{T}}$ denotes CSP traces refinement. Figure 3 summarize the overall idea of refinement with change of signature. Thanks to the 'sort condition' we have:

Lemma 3. *In the above setting, the elements* $\hat{\sigma}^{\mathcal{A}}_{M'}(d'_{M'})$ *are defined over the model* $\mathcal{T}$.

Let $Sp = (D, P)$ and $Sp' = (D', P)$ be two specifications where the signature of D is embedded into the signature of D' and the process parts are syntactically identical. We say that there is a *data refinement with hiding* from Sp to Sp' , in signs $Sp \overset{\text{data}}{\leadsto}_\sigma Sp'$, if $\mathbf{Mod}(D')|_\sigma \subseteq \mathbf{Mod}(D)$.

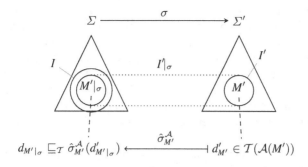

Fig. 2. Refinement with change of signature

Theorem 6. *Let* $Sp = (D, P)$ *and* $Sp' = (D', P')$ *be* CSP-CASL *specifications, such that the signature of* D *is embedded into the signature of* D'. *Then* $(D, P) \overset{data}{\leadsto}_\sigma (D', P)$ *and* $(D', P) \overset{proc}{\leadsto}_T (D', P')$ *imply* $(D, P) \leadsto_T^\sigma (D', P')$.

Proof. (Sketch) Similar to [14], we prove a *reduct property* by structural induction, namely $traces(\llbracket P \rrbracket_{M'|_\sigma}[a_1/x_1, ..., a_k/x_k]) = \hat\sigma_{M'}^{\mathcal{A}}(\llbracket P \rrbracket_{M'}[\sigma_{M'}^{\mathcal{A}}(a_1)/x_1, ..., \sigma_{M'}^{\mathcal{A}}(a_k)/x_k])$. Here, the $x_i : s_i$ are free variables in P, and the a_i ranges over $[s_i]_{\sim_{M'|_\sigma}}$, i.e. the set of values in the alphabet generated by the sort symbol s_i. We then prove that this reduct property also applies to least fixed points (in the case of cpo semantics) and unique fixed points (in the case of cms semantics). Let (D, P) have denotations $(d_M)_{M \in \mathbf{Mod}(D)}$, (D', P) have denotations $(d'_{M'})_{M' \in \mathbf{Mod}(D')}$, and (D', P') have denotations $(d''_{M'})_{M \in \mathbf{Mod}(D')}$. The data refinement immediately gives the required model class inclusion. The process refinement yields $d'_{M'|_\sigma} \sqsubseteq_T d''_{M'}$. As $\hat\sigma_{M'}^{\mathcal{A}}$ is monotonic w.r.t. traces refinement, and using the reduct property, we obtain $d_{M'|_\sigma} = \hat\sigma_{M'}^{\mathcal{A}}(d'_{M'|_\sigma}) \sqsubseteq_T \hat\sigma_{M'}^{\mathcal{A}}(d''_{M'})$.

4   Electronic Payment System: EP2

In this section we apply the theoretical results presented so far in an industrial setting. The EP2 system is an electronic payment system and it stands for 'EFT/POS 2000', short for 'Electronic Fund Transfer/Point Of Service 2000', is a joint project established by a number of (mainly Swiss) financial institutes and companies in order to define EFT/POS infrastructure for credit, debit, and electronic purse terminals in Switzerland (www.eftpos2000.ch). The system consists of seven autonomous entities: Card-Holder, Point of Service, Attendant, POS Management, Acquirer, Service Center and Card. These components are centered around an EP2 *Terminal*. These entities communicate with the Terminal and, to a certain extent, with one another via XML-messages in a fixed format. These messages contain information about authorisation, financial transactions, as well as initialisation and status data. The state of each component heavily depends on the content of the exchanged data. Each component is a reactive system defined by a number of use cases. Thus, there are both reactive parts and data parts which need to be modeled, and these parts are heavily intertwined. The EP2 specification consists of 12 documents, each of which describe the different components or some aspect common to the components. The way that the specifications are written is typical of a number of similar industrial application. The specification consists of a mixture of plain English and semi-formal notation. The top level EP2 documents provide an overview of the data involved, while the presentation of further details for a specific type is delayed to separate low-level documents. CSP-CASL is able to match such a document structure by a library of specifications, where the informal design steps of the EP2 specification are mirrored in terms of a formal refinement relation defined in the previous sections. A first modeling approach of the different levels of EP2 in CSP-CASL has been described in [6].

In this section we consider two levels of the EP2 specification, namely: the *architectural level* (ARCH) and the *abstract component level* (ACL). We choose a dialogue between the *Terminal* and the *Acquirer*. In this dialogue, the Terminal and the Acquirer are supposed to exchange initialization information. For presentation purposes, we study here only a nucleus of the full dialogue, which, however, exhibits all technicalities present in the full version.

4.1   Formal Refinement in EP2

Our notion of CSP-CASL refinement mimics the informal refinement step present in the EP2 documents: There, the first system design sets up the interface between the components (architectural level), then these components are developed further (abstract component level). Here, we demonstrate in terms of a simple example how we can capture such an informal development in a formal way.

We first specify the data involved using CASL only. The data specification of the architectural level (D_ARCH_GETINIT) consists only of one set of data:

spec D_ARCH_GETINIT =
    sort D_SI_Init
end

In the EP2 system, these values are communicated over channels; data of sort D_SI_Init is interchanged on a channel C_SI_Init linking the Terminal and the Acquirer. On the architectural level, both these processes just 'run', i.e., they are always prepared to communicate an event from D_SI_Init or to terminate. We formalize this in CSP-CASL:

ccspec ARCH_INIT =
data D_ARCH_GETINIT
channel $C\_SI\_Init : D\_SI\_Init$
process
    let $Acquirer = EP2Run$ $Terminal = EP2Run$
    in $Terminal \,||[\, C\_SI\_Init \,]||\, Acquirer$
end

Here, $EP2Run = (C\_SI\_Init \,?\, x : D\_SI\_Init \rightarrow EP2Run) \,\square\, SKIP$. On the abstract component level (D_ACL_GETINIT), data is refined by introducing a type system on messages. In CASL, this is realised by introducing subsorts of D_SI_Init. For our nucleus, we restrict ourselves to four subsorts, the original dialogue involves about twelve of them.

spec D_ACL_GETINIT =
    sorts $SesStart, SesEnd, DataRequest, DataResponse < D\_SI\_Init$
    ops $r : DataRequest;\; e : SesEnd$

axioms $\forall x : DataRequest;\ y : SesEnd.\neg(x = y)$
    $\forall x : DataRequest;\ y : SesStart.\neg(x = y)$
    $\forall x : DataResponse;\ y : SesEnd.\neg(x = y)$
    $\forall x : DataResponse;\ y : SesStart.\neg(x = y)$
end

In the above specification, the axioms prevent confusion between several sorts. Using this data, we can specify the ACL level of the Acquirer-Terminal dialogue in CSP-CASL. In the process part the terminal (*TerInit*) initiates the dialogue by sending a message of type *SesStart*; on the other side the Acquirer (*AcqInit*) receives this message. The process *AcqConf* takes the internal decision either to end the dialogue by sending the message *e* of type *SesEnd* or to send another type of message. The Terminal (*TerConf*), waits for a message from the Acquirer, and depending on the type of this message, the Terminal engages in a data exchange. The system as a whole consists of the parallel composition of Terminal and Acquirer.

ccspec ACL_INIT =
data D_ACL_GETINIT
channels $C\_ACL\_Init : D\_SI\_Init$
process
    let $AcqInit = C\_ACL\_Init\ ?\ session : SesStart \to AcqConf$
    $AcqConf = C\_ACL\_Init\ !\ e \to Skip$
            $\sqcap\ C\_ACL\_Init\ !\ r \to C\_ACL\_Init\ ?\ resp : DataResponse$
                $\to AcqConf$
    $TerInit = C\_ACL\_Init\ !\ session : SesStart \to TerConf$
    $TerConf = C\_ACL\_Init\ ?\ confMess \to$
                (**if** $(confMess : DataRequest)$
                    then $C\_ACL\_Init\ !\ resp : DataResponse \to TerConf$
                    else if $(confMess : SesEnd)$ **then** $Skip$ **else** $Stop)$
    in $TerInit\ |[\ C\_ACL\_Init\ ]|\ AcqInit$
end

Theorem 7. ARCH_INIT $\rightsquigarrow^{\sigma}_{\mathcal{T}}$ ACL_INIT

Proof. Using tool support, we establish this refinement by introducing two intermediate specifications RUN_ARCH and ACL_SEQ:

ccspec RUN_ARCH =
 data D_ARCH_GETINIT
 channel $C\_SI\_Init : D\_SI\_Init$
 process *EP2Run*
end

ccspec ACL_SEQ =
 data D_ACL_GETINIT
 channels $C\_ACL\_Init : D\_SI\_Init$
 process
    let $SeqStart = C\_ACL\_Init\ !\ session : SesStart \to SeqConf$
    $SeqConf = C\_ACL\_Init\ !\ e \to Skip$
            $\sqcap\ C\_ACL\_Init\ !\ r$
                $\to C\_ACL\_Init\ !\ resp : DataResponse \to SeqConf$
    in $SeqStart$
end

With CSP-CASL-Prover we proved: ARCH_INIT $=_T$ RUN_ARCH. Now we want to prove that RUN_ARCH $\leadsto^\sigma_T$ ACL_SEQ. To this end, we apply Theorem 6: Using *HETS* [13], we automatically prove the data refinement D_ARCH_GETINIT $\overset{data}{\leadsto}_\sigma$ D_ACL_GETINIT.

Now, we formed the specification (D_ACL_GETINIT, P_{ACL_SEQ}), where P_{ACL_SEQ} denotes the process part of ACL_SEQ. Next we show in CSP-CASL-Prover that, over the traces model T, this specification refines to ACL_SEQ:

```
theorem Arch_ACL_refinement : "EP2Run <=T SeqStart "
apply(unfold EP2Run_def SeqStart_def)
apply (rule cspT_fp_induct_right[of _ _"Seq_to_Run"])
apply simp_all
apply (induct_tac procName)
...
apply (simp add: cspT_semantics)
apply rule
apply (simp add: in_traces)
apply (auto simp add: D_SI_Init_def SesStart_def  SesEnd_def
                      DataRequest_def DataResponse_def)
...
done
```

Fig. 3. Snippet of proof script for RUN_ARCH $\leadsto^\sigma_T$ ACL_SEQ.

Figure 3 shows a snippet of the proof script for (D_ACL_GETINIT, P_{ACL_SEQ}) $\overset{proc}{\leadsto}_T$ ACL_SEQ. We first unfold the definitions of EP2Run and SeqStart. Next, we apply (metric) fixed point induction on the rhs and make a case distinction over the process names, here encoded as induct_tac procName. After rewriting and decomposing both of the processes we compute the trace semantics and check that there is indeed an inclusion of traces.

[18] proves ACL_INIT $=_F$ SEQ_INIT. As stable failure equivalence implies trace equivalence, we obtain ACL_INIT $=_T$ SEQ_INIT. Figure 4 summarizes this proof structure.

4.2   Deadlock Analysis of EP2

As ACL_INIT involves parallel composition, it is possible for this system to deadlock. Furthermore, the process *TerConf* includes the CSP process *STOP* within one branch of its conditional. Should this branch of *TerConf* be reached, the whole system will be in deadlock.

The dialogue between the Terminal and the Acquirer for the exchange of initialization messages have been proven to be deadlock free in [18]. Specifically, it has been proven that the following refinement holds: ACL_SEQ $\overset{proc}{\leadsto}_F$ ACL_INIT, where ACL_SEQ is a sequential system. Sequential system are regarded to be deadlock-free.

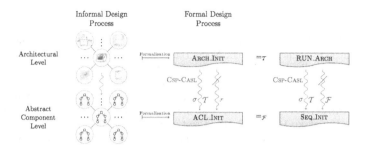

Fig. 4. Refinement in EP2

With our proof method from Section 3.2, we can strengthen this result by actually proving that ACL_SEQ is deadlock free. To this end, we proved with CSP-CASL-Prover that DF $\overset{proc}{\leadsto}_{\mathcal{F}}$ ACL_SEQ where DF is the least refined deadlock-free process described in Section 3.2.

4.3 Analysis of Divergence Freeness of EP2

As described in Section 3.3 divergence freeness is best analysed in the model $\mathcal{N}$. The model $\mathcal{N}$ has not yet been implemented in CSP-CASL-Prover. However, using basic step and distributivity laws we have manually shown that ACL_INIT $=_{\mathcal{N}}$ ACL_SEQ and DIVF $\leadsto_{\mathcal{N}}$ ACL_SEQ, i.e., the EP2 dialogue considered here is divergence free.

5 Conclusions and Future Work

In this paper we have studied various property preserving refinement notions for CSP-CASL. We established proof methods based on decomposition theorems which enable us to reason about refinement using interactive theorem proving. We reduce the analysis of deadlock and divergence freeness to refinement statements.

We showed that our theoretical results apply to a "real world" system. We proved in a systematic way using CSP-CASL-Prover and *HETS* a refinement step from the architectural specification of EP2 to a more detailed one. We proved that at this level of abstraction EP2 is free of deadlock and free of divergence.

Future work will include the extension of our theory on refinement with change of signature to arbitrary signature morphisms as well as the exploration of more "sophisticated" refinement notions for CSP-CASL. In [2], Bidoit et al., present a refinement notion based on observational interpretation of CASL specifications. Following this work we intend to develop observational refinement for CSP-CASL. In the context of EP2 such refinement would be required in order to capture the relations between the more detailed levels.

Acknowledgements. The authors would like to thank Liam P. O'Reilly for good support with CSP-CASL-Prover and Erwin R. Catesbeiana (Jr) for numerous, but often divergent discussions on the topic of refinement.

References

1. Atiya, D.-A., King, S., Woodcock, J.: Simpler reasoning about system properties: a proof-by-refinement technique. Electronic Notes Theoretical Computer Science 137(2) (2005)
2. Bidoit, M., Sannella, D., Tarlecki, A.: Observational interpretation of CASL specifications. Mathematical Structures in Computer Science 18(2) (2008)
3. Buth, B., Kouvaras, M., Shi, H.: Deadlock analysis for a fault-tolerant system. In: Johnson, M. (ed.) AMAST 1997. LNCS, vol. 1349. Springer, Heidelberg (1997)
4. Ehrig, H., Kreowski, H.-J.: Refinement and implementation. In: Astesiano, E., Kreowski, H.-J., Krieg-Brückner, B. (eds.) Algebraic Foundations of Systems Specification. Springer, Heidelberg (1999)
5. Gimblett, A.: Tool support for CSP-CASL. MPhil Thesis, Swansea University (2008)
6. Gimblett, A., Roggenbach, M., Schlingloff, H.: Towards a formal specification of an electronic payment systems in CAP-CASL. In: Fiadeiro, J.L., Mosses, P.D., Orejas, F. (eds.) WADT 2004. LNCS, vol. 3423, pp. 61–78. Springer, Heidelberg (2005)
7. Hoare, C.A.R.: Communicating Sequential Processes. Prentice Hall, Englewood Cliffs (1985)
8. Isobe, Y., Roggenbach, M.: Webpage on CSP-Prover, http://staff.aist.go.jp/y-isobe/CSP-Prover/CSP-Prover.html
9. Isobe, Y., Roggenbach, M.: A generic theorem prover of CSP refinement. In: Halbwachs, N., Zuck, L.D. (eds.) TACAS 2005. LNCS, vol. 3440, pp. 108–123. Springer, Heidelberg (2005)
10. Isobe, Y., Roggenbach, M., Gruner, S.: Extending CSP-Prover by deadlock-analysis: Towards the verification of systolic arrays. In: FOSE 2005, Japanese Lecture Notes Series 31. Kindai-kagaku-sha (2005)
11. Lüth, C., Roggenbach, M., Schröder, L.: CCC —the CASL Consistency Checker. In: Fiadeiro, J.L., Mosses, P.D., Orejas, F. (eds.) WADT 2004. LNCS, vol. 3423, pp. 94–105. Springer, Heidelberg (2005)
12. Lüttich, K., Mossakowski, T.: Reasoning support for CASL with automated theorem proving systems. In: Fiadeiro, J.L., Schobbens, P.-Y. (eds.) WADT 2006. LNCS, vol. 4409, pp. 74–91. Springer, Heidelberg (2007)
13. Mossakowski, T., Maeder, C., Lüttich, K.: The Heterogeneous Tool Set, HETS. In: Grumberg, O., Huth, M. (eds.) TACAS 2007. LNCS, vol. 4424, pp. 519–522. Springer, Heidelberg (2007)
14. Mossakowski, T., Roggenbach, M.: Structured CSP – A Process Algebra as an Institution. In: Fiadeiro, J.L., Schobbens, P.-Y. (eds.) WADT 2006. LNCS, vol. 4409, pp. 92–110. Springer, Heidelberg (2007)
15. Mossakowski, T., Roggenbach, M.: An institution for processes and data. In: WADT 2008 – Preliminary Proceedings, TR-08-15. Università di Pisa (2008)
16. Mossakowski, T., Sannella, D., Tarlecki, A.: A simple refinement language for CASL. In: Fiadeiro, J.L., Mosses, P.D., Orejas, F. (eds.) WADT 2004. LNCS, vol. 3423, pp. 162–185. Springer, Heidelberg (2005)
17. Mosses, P.D. (ed.): CASL Reference Manual. LNCS, vol. 2960. Springer, Heidelberg (2004)
18. O'Reilly, L., Isobe, Y., Roggenbach, M.: CSP-CASL-Prover – a generic tool for process and data refinement. Electronic Notes in Theoretical Computer Science (to appear)
19. Roggenbach, M.: CSP-CASL – A new integration of process algebra and algebraic specification. Theoretical Computer Science 354 (2006)
20. Roscoe, A.: The theory and practice of concurrency. Prentice Hall, Englewood Cliffs (1998)
21. Shi, H., Peleska, J., Kouvaras, M.: Combining methods for the analysis of a fault-tolerant system (1999)
22. Wirth, N.: Program development by stepwise refinement. Communications of the ACM 14(4) (1971)

Reconfiguring Distributed Reo Connectors

Christian Koehler[1],[*],[**], Farhad Arbab[1], and Erik de Vink[2]

[1] CWI, P.O. Box 94079, 1090 GB Amsterdam, The Netherlands
christian.koehler@cwi.nl
[2] Technische Universiteit Eindhoven, Den Dolech 2, Eindhoven, The Netherlands

Abstract. The coordination language Reo defines circuit-like connectors to steer the collaboration of independent components. In this paper, we present a framework for the modeling of distributed, self-reconfigurable connectors based on algebraic graph transformations. Reconfiguring a connector that is composed with others, may involve a change of shared interfaces and may therefore require a reconfiguration of the surrounding connectors as well. We present a method of synchronized local reconfigurations in this setting and discuss a bottom-up strategy for coordinating synchronized reconfigurations in a connector network. We exploit the double-pushout approach for the modeling of reconfigurations, and propose an adaptation of the concept of amalgamation for synchronizing reconfigurations. We use a nondeterministic scheduler as our running example.

1 Introduction

Building software systems using an exogenous coordination language, such as Reo [1], is done by (i) implementing a set of (wrappers for existing) components or services and (ii) composing these entities using a kind of glue code. In the case of Reo, circuit-like *connectors* constitute the glue code. Connectors may consist of other connectors, but are elementarily composed from channels and nodes. Every connector implements a protocol defined by the semantics of its constituents, and the topology of the connector. Reconfiguring a connector means to change its topology and thereby the coordination protocol that it implements. Reconfiguration arises from the need to dynamically adapt the behavior of a system, e.g., in response to a change in its environment, to cope with altered resource availability or to retrofit it for a modified mission. The need for considering *distributed* connectors arises from two concerns. On the one hand, connectors are decomposed into logically separate parts, each of which defines a specific subprotocol. On the other hand, connectors can be deployed on different physical locations in a network. In both cases, the concept of distribution facilitates and promotes the use of black-boxed subconnectors in a larger context.

In this paper, we propose a framework for modeling reconfigurable, distributed Reo connectors. We consider connectors that are distributed over a network and

---

[*] Supported by NWO GLANCE project WoMaLaPaDiA and SYANCO.
[**] Corresponding author.

A. Corradini and U. Montanari (Eds.): WADT 2008, LNCS 5486, pp. 221–235, 2009.
© Springer-Verlag Berlin Heidelberg 2009

are encapsulated, i.e. their internals are hidden from the outside world and communicate only via a published interface. Connectors are linked together via the interfaces that they share. Reconfiguration of a distributed connector is achieved by reconfiguring its subconnectors. Ultimately, reconfiguration is defined and performed locally, that is to say, in the scope of a single connector, but it can be either triggered from the inside or invoked from the outside. Reconfiguring a connector may involve a change in its interfaces and may require connectors in its neighborhood to reconfigure as well. This implies a need for synchronizing such local reconfigurations into a consistent reconfiguration of the connector as a whole. It goes without saying that in a distributed setting, we cannot assume the existence of a (centralized) third party that monitors and coordinates local reconfigurations. Therefore, other mechanisms should be in place to assure the consistency of a reconfigured network.

In short, this paper contributes the following: We propose a model for reconfigurable, distributed connectors. We utilize the well-studied framework of distributed graph transformation [3,4] for this purpose. We show, furthermore, how reconfigurations can be defined and performed locally using a synchronization mechanism based on the notion of amalgamation [5,6,7]. Finally, we propose a distributed strategy to organize the stepwise reconfiguration of large networks.

Related work. Modeling the distribution of systems via embedding interfaces represented as morphisms in a suitable category is also used for open Petri-nets [9]. The explicit modeling of glue code and the exploitation of pushout constructions to deal with composition in this work is similar to ours. However, they do not consider the application of double pushouts and their concept of amalgamation is different. In [9], amalgamation serves the composition of deterministic processes of open Petri-nets, whereas in our approach it is used as a synchronization mechanism for the superposition of local reconfiguration rules.

In [10], Architectural Design Rewriting is proposed as a framework for modeling reconfigurable software architectures. This work deals with hierarchical, non-distributed architectures and uses hyperedge replacement, as opposed to algebraic graph transformations in our work. A general introduction to system modeling and system evolution using graph transformation techniques, including hierarchical and distributed approaches, can be found in [11].

Our approach to coordination achieved by Reo connectors and their dynamic reconfiguration fits in the framework of runtime software adaptation [12,13] for component-based software engineering. Process algebraic treatments include [14,15]. To accommodate dynamic reconfiguration, predicted behavioral changes combined with revision of message translation are captured by so-called contextual mappings. However, the focus in this work is not on distribution, which is a key aspect of our paper. A workflow language extension with the so-called configurable elements, e.g. for YAWL, is proposed in [16], with a semantics based on a variant of Petri-nets, called extended workflow nets (EWF-nets).

Structure of the paper. The rest of this paper is organized as follows. Section 2 contains an overview of the coordination language Reo. We describe basic graph transformation techniques using the double pushout approach and an application to Reo in Section 3. In Section 4, we formally introduce distributed connectors as typed distributed graphs. In Section 5, we introduce reconfigurations for distributed connectors and discuss amalgamation as a synchronization mechanism. Section 6 contains conclusions and future work.

2   Reo Connectors

A Reo connector [1,2] acts as glue code connecting a number of components together. A connector orchestrates the behavior of the components it connects, enforcing a specific interaction pattern. By influencing the timing of I/O operations of the components, it achieves coordination through constraining their behavior. The computational internals of the components are oblivious to the connector. As such, Reo falls in the class of exogenous coordination languages.

Taking various flavors of channels as primitive building blocks, more complex connectors can be composed in Reo from simpler ones. Channels are point-to-point means to communicate that meet at nodes. Channels have two ends, a source and a sink, or two sources or two sinks.

The set of primitive Reo channels is user-defined, but typically includes the channel types in Table 1. The synchronous channel $Sync$, simultaneously takes a data item from its source end and makes it available at its sink end. The synchronous drain $SyncDrain$ has two source ends, but no sink end. If there are data items available at both ends, it consumes and loses both of them simultaneously. The lossy synchronous channel $LossySync$ behaves like the $Sync$ channel, except that it does not block its source when its sink end cannot accept data. Instead it accepts and loses the data item taken from the source. The $FIFO_1$ is an asynchronous and stateful channel, having a buffer of size one. If its buffer is empty and a data item is available at its source end, the I/O operation succeeds and the item is stored in the buffer. The $FIFO_1$ blocks any further write requests until the data item is delivered through its sink end. It then returns back to its empty state. Other channels are allowed as well, e.g. with filtering capability. The only requirement for a channel is that it has exactly two ends.

Reo distinguishes three kinds of nodes: source nodes, sink nodes and mixed nodes. From a connector's point of view, source and sink nodes are also called input and output nodes, respectively. Collectively, they form the boundary nodes of a connector, which interact with its environment. Mixed nodes, on the other hand, are internal and not accessible from the outside. A mixed node has both incoming and outgoing channels, i.e. channels meeting at the node with their sink and source ends. For a mixed node to fire, it needs at least one of its incoming

Table 1. Some primitive channels

$Sync$	$SyncDrain$	$LossySync$	$FIFO_1$
⟶	→ ←	- - - -▶	⊐▭▷

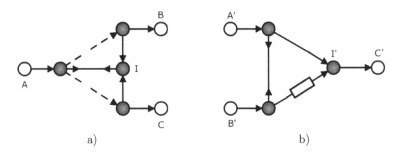

Fig. 1. a) Exclusive router, b) ordering

channels be willing to deliver a data item, while simultaneously all its outgoing
channels are willing to consume it. The node then nondeterministically selects
one of its enabled sink ends and passes its data item to all its source ends. In
all of our examples we will represent mixed nodes as filled circles and boundary
nodes as empty circles.

Example 1. The exclusive router, shown in Figure 1a), routes data from A to
either B or C. The connector can accept data only if there is a write operation at
the source node A, and there is at least one component attached to the sink nodes
B or C, that performs a take operation. If both B and C allow an output, the
choice between B and C is made nondeterministically by the mixed node I. Note
that data flows synchronously through this connector, i.e. there is no buffering
involved. The exclusive router proves to be useful in a number of situations. The
binary exclusive router in Figure 1a) can be generalized straightforwardly to an
n-ary one. The symbol $\otimes$ is used as its shorthand in the sequel.

Example 2. The second connector, shown in Figure 1b), imposes an ordering
on the data flow from the input nodes A' and B' to the output node C'. The
SyncDrain enforces synchronous flows through A' and B'. The empty buffer
together with the *SyncDrain* guarantee that the data item obtained from A' is
delivered to C' whereas the data item obtained from B' is stored in the FIFO
buffer. At this moment, the buffer of the $FIFO_1$ is full and data cannot flow
in, neither through A' nor B', as they are coupled by the synchronous drain.
However, C' can now obtain the data stored in the buffer.

Reo connectors come equipped with a formal data flow semantics based on so-
called connector colorings [17]. The basic idea is to assign to each channel an
admissible communication behavior that is compatible with all nodes. Conceptually, the execution cycle for a connector allows for system reconfiguration after
a consistent coloring has been established and the corresponding data flow has
been accomplished. See [17] for more detail on the coloring semantics, and [18]
and [19] for the semantics of Reo based on constraint automata and on tiles,
respectively. Currently, the software suite for Reo includes a number of development tools, integrated within Eclipse [20], and runtime engines for various

platforms. In particular, a distributed implementation of Reo is available, built on Scala [21], allowing individual nodes to be distributed over the network.[1]

3 Reconfiguration by Graph Transformation

First, we recall the basic definitions for typed graphs and introduce our running example. Next, we show how to model basic reconfigurations of Reo connectors using algebraic graph transformation. We refer to [6,8] for more details on the algebraic approach to graph transformation.

3.1 Typed Graphs

We use a graph model where edges are directed and have identity, i.e. a *graph* is a structure $G = \langle V, E, s, t \rangle$ with V a set of nodes, E a set of edges and $s, t : E \rightarrow V$ source and target functions. A *graph morphism* is a pair of functions $h = \langle h_V, h_E \rangle$ that preserve the source and target functions. Graphs and graph morphisms form the category **Graph**.

When modeling Reo connectors as graphs, nodes in a graph represent Reo nodes and edges represent channels. Intuitively, the set of nodes and channels allowed in a connector is given via a *type graph*. For a fixed type graph T, an *instance graph* over T is a pair $\langle G, type \rangle$ where G is a graph and $type : G \rightarrow T$ a morphism into the type graph. A *typed graph morphism* $h : \langle G_1, type_1 \rangle \rightarrow \langle G_2, type_2 \rangle$ is a graph morphism $h : G_1 \rightarrow G_2$ that preserves the type information, i.e. $type_1 = type_2 \circ h$. Fixing a type graph T, the category of typed graphs over T is denoted with **Graph$_T$**.

For the Reo connectors in our running example, we consider a type graph of four different node types, two types of internal nodes: ordinary Reo nodes ● and exclusive routers ⊗, and two types of interface nodes ○, which are either *start* or *finish* nodes. The type graph further includes an edge for every channel type. Note that, for undirected channels, such as the *SyncDrain*, the source and the target of the edge model both a source or a sink end of the channel. We do not restrict the use of channels, i.e. a channel of any type can be connected to a node of any type.[2] We further include additional edge types for primitive components with two ends. The category of Reo graphs is denoted by **Graph$_{Reo}$**.

Example 3. Figure 2 depicts two connectors, which constitute our running example. The first connector in Figure 2a) is a *nondeterministic scheduler* in its initial state. It consists of four interface nodes: two *start* nodes and two *finish* nodes, and it is capable of scheduling two tasks. In the first step of execution, the scheduler enables –if possible– one of the *start* nodes by moving the token from the $FIFO_1$ in the middle to one of the other two $FIFO_1$s and replicating the token on the selected *start* node. If more than one *start* node can be enabled,

---

[1] See http://homepages.cwi.nl/~proenca/distributedreo/

[2] A more natural way of modeling Reo connectors would use attributed typed graphs with node inheritance (cf. [22]). For simplicity, we restrict to typed graphs here.

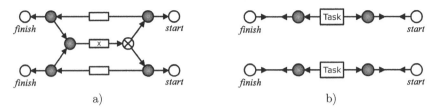

Fig. 2. a) Nondeterministic scheduler for two tasks, b) task repository with two tasks

the choice is made nondeterministically. At this stage, the scheduler waits until it can enable the corresponding *finish* node. When this step is performed, the connector goes back to its original configuration by moving the token back to the middle.

The second connector in our example, depicted in Figure 2b), models a *task repository*. A task is a primitive component that can be either in the idle or the processing state. When it is idle and ready to switch to the processing state, it produces a token on its right-hand end. It signals that it can be stopped again using a token on its left-hand end. The repository wraps every task using two *SyncDrains* and exposes interface nodes for starting and finishing each of them.

In our application scenario, we will connect these two connectors along their interface nodes. However, instead of gluing the nodes and thereby hardwiring the connectors, we will consider them as distributed in a network having an exposed interface to be shared. Before introducing distribution, we first illustrate how to reconfigure connectors using graph rewriting techniques.

3.2 Algebraic Graph Transformation

We follow the double-pushout (DPO) approach to graph transformation. Graphs are transformed by applying graph productions, which we will also refer to as *reconfiguration rules* in our application. In categorical terms, a *production* is a span of injective morphisms $p = L \xleftarrow{\ell} K \xrightarrow{r} R$ in the category **Graph**. The left-hand side L defines the pattern that must be matched to apply the production. K contains all elements that are not removed by the rule and R additionally has those elements of the graph that are created by the rule. Given a production p, a *match* is a morphism $m : L \to M$, where M is the graph to be transformed. A *derivation* $M \xRightarrow{p,m} N$ is an application of p with the match m, formally defined as the following diagram where (1) and (2) are pushouts.

$$L \xleftarrow{\ell} K \xrightarrow{r} R$$
$$\downarrow{m} \quad (1) \quad \downarrow \quad (2) \quad \downarrow$$
$$M \longleftarrow C \longrightarrow N$$

Operationally, the graph M is transformed to the graph N by (i) removing the occurrence of $L \backslash \ell(K)$ in M, yielding the intermediate graph C, and (ii) adding

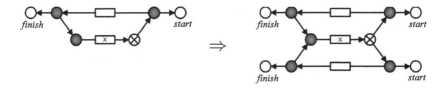

Fig. 3. Reconfiguration rule for extending a nondeterministic scheduler

a copy of $R\backslash r(K)$ to C. Due to the categorical formulation of the transformation concepts, the approach can be directly transferred to typed graphs, which we use in the sequel to model reconfigurations of Reo connectors.

Example 4. To reconfigure the nondeterministic scheduler and the task repository introduced above, we define a number of reconfiguration rules. Figure 3 depicts an example rule that extends the scheduler with a slot for an additional task. The gluing graph K is not drawn here, as in all of our rules. It is defined as the intersection of the left and the right-hand side. The mappings ℓ and r are indicated by the relative positions of the nodes and channels. In the same way, we define a rule for adding a task to a repository. By reversing these rules, we can realize a removal of tasks or slots in the scheduler, respectively.

4   Distributed Connectors

To model reconfigurations of distributed Reo connectors, we use the framework of *Distributed Graph Transformation*, as introduced by Taentzer [3] for graphs and generalized by Ehrig [4] to transformations of distributed objects. In the following, we recall the notions of the framework that are relevant to the present setting and apply them to Reo.

4.1   Distributed Typed Graphs

Distribution of graphs can be described by adding a second level of abstraction, namely by modeling the topology of a system using a so-called *network graph*. The nodes in a network graph consist of *local graphs* and the edges are morphisms of these local graphs. The idea is that a node models a physical or logical location of a local graph, whereas an edge indicates an occurrence of the source graph in the target graph. In particular, multiple outgoing edges from one local graph model the fact that the source graph is shared among the target graphs.

Formally, a distributed graph is a pair (N, D) where N is an ordinary graph, the network graph, and D is a mapping that associates to every node n in N a local graph $D(n)$ and to every edge $n \xrightarrow{e} n'$ in N a graph morphism $D(e)$: $D(n) \to D(n')$. In categorical terms, this mapping corresponds to a functor $D : N \to \mathbf{Graph}$, also called a *diagram*, where the graph N is interpreted as a category. Following [4], this functor is required to be commutative, i.e., for any

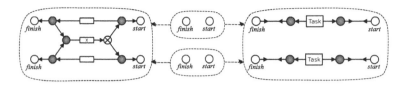

Fig. 4. Nondeterministic scheduler connected to a task repository

two paths $p_1, p_2 : n \xrightarrow{*} n'$ in N, it must hold that $D(p_1) = D(p_2)$. This arises from the assumption that the morphisms associated with edges represent the sharing of the local graphs. Note that due to the categorical formalization, the concept of distribution can be applied to typed graphs as well by considering functors $D : N \to \mathbf{Graph_T}$. Hence, as a first approximation, distributed Reo connectors can be modeled as distributed typed graphs.

Example 5. An example of a distributed connector is depicted in Figure 4. The network graph is drawn using dashed lines. The nondeterministic scheduler and the task repository appear as nodes in the network graph. The two other nodes of the network graph are interfaces, each containing two interface nodes, viz. a *start* and a *finish*. Further, there are four embeddings of the interfaces into the scheduler and the repository. Type preservation by the embedding guarantees that start nodes map to start nodes and similarly for finish nodes. Obviously, although details are suppressed here, the two embeddings are supposed to have disjoint ranges. However, since the interfaces are embedded into both the scheduler and the repository, these connectors are considered to be connected along their interfaces.

Given two distributed typed graphs (N_1, D_1) and (N_2, D_2), a morphism $f = (f_N, f_D) : (N_1, D_1) \to (N_2, D_2)$ consists of a graph morphism $f_N : N_1 \to N_2$ and a natural transformation $f_D : D_1 \to D_2 \circ f_N$. We will just write f for the network morphism f_N. By definition, the natural transformation f_D assigns to every node n of N_1 a graph morphism $f_n : D_1(n) \to D_2(f(n))$; f_n is called the *local* graph morphism of n. Furthermore, for every edge $n \xrightarrow{e} n'$ in N_1 the following diagram commutes.

$$
\begin{array}{ccc}
D_1(n) & \xrightarrow{\ D_1(e)\ } & D_1(n') \\
{\scriptstyle f_n}\big\downarrow & & \big\downarrow{\scriptstyle f_{n'}} \\
D_2(f(n)) & \xrightarrow{\ D_2(f(e))\ } & D_2(f(n'))
\end{array}
$$

Example 6. An example morphism of distributed connectors is given in Figure 5. The unary scheduler in node n' is mapped into a binary scheduler in node $f(n')$. The target network has an interface node and an embedding into the scheduler that is not in the image of the morphism.

The categories of distributed graphs and of distributed typed graphs are denoted by $\mathbf{Dis(Graph)}$ and $\mathbf{Dis(Graph_T)}$. However, for a proper modeling of

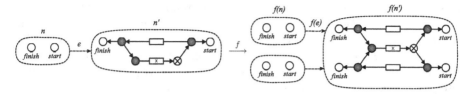

Fig. 5. Example morphism of distributed typed graphs

distributed Reo connectors the typing mechanism of **Dis(Graph$_T$)** is not sufficient, as we will argue in the following.

4.2   Typed Distributed Graphs

When considering Reo connectors as distributed typed graphs, i.e. objects (N, D) with $D : N \rightarrow \mathbf{Graph_T}$, only the local graphs are typed. For our application, we need types at the network level as well, because of the following constraints:

(i) Nodes in a Reo network graph are either *Connectors* or *Interfaces*.
(ii) Edges in a network graph are allowed only from *Interfaces* to *Connectors*.
(iii) An *Interface* is linked to at most two *Connectors*.

A way of dealing with these constraints is to use a typed network graph with multiplicity constraints (cf. [23]). Multiplicity constraints for edges are useful in this context, since they allow to restrict the number of links between interfaces and connectors. Figure 6 depicts a type graph for network graphs that enforces the above constraints. The edge multiplicities make sure that interfaces always connect at most two connectors. Interfaces connecting more than two connectors are avoided, since the merger-replicator semantics of Reo nodes would give the interfaces a non-trivial and potentially unexpected behavior. Hence, using the

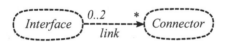

Fig. 6. Network type graph for Reo

network type graph in Figure 6, we can impose the constraints (i)–(iii) above. However, type constraints that relate the local and the network levels cannot be expressed by this formalism. Therefore, we require the following additional constraint for distributed connectors:

(iv) Local graphs assigned to *Interfaces* consist only of *start* and *finish* nodes.

We can model this constraint using *typed distributed graphs*, as opposed to distributed typed graphs. The latter approach, as alluded to above, involves considering objects (N, D) in the category **Dis(Graph$_T$)**, i.e. where D maps the nodes of N to typed graphs and the edges to typed graph morphisms. On the other hand, a *typed distributed graph* is a tuple $(N, D, type)$ where (N, D) is an object in **Dis(Graph)** and $type : (N, D) \rightarrow (N_T, D_T)$ a morphism in **Dis(Graph)** with $T = (N_T, D_T)$ a fixed distributed graph, called the *distributed type graph*. This means, instead of considering the category **Dis(Graph$_T$)**, we use the slice category **Dis(Graph)**$\backslash T = $ **Dis(Graph)$_T$**. This typing mechanism is more expressive, since the type graph and the type morphisms are distributed already. Accordingly, we model distributed connectors as typed distributed graphs in the rest of this paper. The distributed type graph for Reo consists on the network level of the type graph in Figure 6. On the local level, the node type *Interface* is mapped to a graph that consists only of the two interface nodes *start* and *finish*. The node type *Connector* is mapped to the default type graph for Reo, as presented in Section 3.1. The edge type *link* is mapped to a graph morphism, that maps the node *start* in *Interface* to the node *start* in *Connector*, and analogously for *finish*. We denote the category of distributed Reo connectors by **Dis(Graph)$_{Reo}$**.

5   Reconfiguration of Distributed Connectors

In order for the double-pushout approach to apply, we must make sure that **Dis(Graph)$_{Reo}$** has pushouts. Since the category **Dis(Graph)** is cocomplete (cf. [4]), it follows that **Dis(Graph)$_{Reo}$** is cocomplete too, as it is a slice category of **Dis(Graph)**. Consequently, pushouts exist and we can use DPO-rewriting for modeling reconfigurations of distributed Reo connectors. Next, we show that reconfigurations can be defined locally, i.e. in the scope of a single connector, and we discuss how these local reconfigurations can be synchronized.

5.1   Local Reconfigurations

The need for local reconfigurations arises from the distributed setting, where no global knowledge of the system is available. In the following example, we give distributed versions of the reconfiguration rules of our scheduler application.

Example 7. Two distributed reconfiguration rules p_1 and p_2 are depicted in Figure 7. Rule p_1 extends the nondeterministic scheduler by a slot for an additional task and creates a new interface for this slot. Rule p_2 adds another task to the repository and creates an interface for this task. Note that the reconfiguration of the scheduler on the one hand, and the task repository on the other, are modeled by two separate rules, because we assume that connectors are reconfigured locally. In principle, the connectors can see each other as black boxes that publish only their interfaces and reconfigure themselves on demand.

To reconfigure networks using local rules, we need a way to synchronize reconfigurations. As can be seen from the example above, applying the two reconfiguration rules p_1 and p_2 naively to the network of Figure 4, does not give the desired result, since each rule creates a new interface whereas we need just one that is shared by the two connectors.

p_1 :

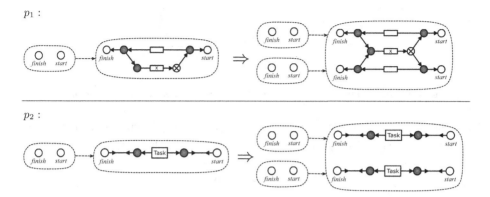

p_2 :

Fig. 7. Rules for extending a distributed scheduler / task repository

5.2   Synchronizing Local Reconfigurations

Following [5], we use the concept of amalgamation for synchronizing local reconfigurations. We first recall the basic definitions for amalgamated graph transformations.

The synchronization of two productions is achieved by identifying a common subproduction and gluing the productions along this subproduction. Let

$$p_i = L_i \xleftarrow{\ell_i} K_i \xrightarrow{r_i} R_i$$

be two productions with $i \in \{0,1\}$. The production p_0, together with graph morphisms $in_L^1 : L_0 \rightarrow L_1$, $in_K^1 : K_0 \rightarrow K_1$, $in_R^1 : R_0 \rightarrow R_1$, are called a *subproduction* of p_1, if in the following diagram (1) and (2) commute. Putting $in^1 = \langle in_L^1, in_K^1, in_R^1 \rangle$, we write $in^1 : p_0 \rightarrow p_1$ for the embedding of p_0 into p_1.

$$
\begin{array}{ccccc}
L_0 & \xleftarrow{\ell_0} & K_0 & \xrightarrow{r_0} & R_0 \\
\downarrow{in_L^1} & (1) & \downarrow{in_K^1} & (2) & \downarrow{in_R^1} \\
L_1 & \xleftarrow{\ell_1} & K_1 & \xrightarrow{r_1} & R_1
\end{array}
$$

The productions p_1 and p_2 are called *synchronized* with respect to p_0, if p_0 is a subproduction of both p_1 and p_2, denoted by $p_1 \xleftarrow{in_1} p_0 \xrightarrow{in_2} p_2$.

Example 8. A non-trivial, i.e. non-empty, subproduction p_0 of the reconfiguration rules p_1 and p_2 is depicted in Figure 8. The rule creates a new interface node in a network graph. As a general property of our reconfiguration approach, the common subproduction of two synchronized rules always describes an interface change of the involved connectors.

By making explicit the change of interface due to an update of connectors, synchronized productions can properly describe reconfigurations in a network.

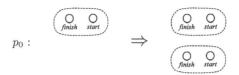

$$p_0 : \qquad\qquad \Rightarrow$$

Fig. 8. Common subproduction of p_1 and p_2 modeling the interface evolution

Execution of synchronized productions can be achieved using amalgamations. Given two synchronized productions $p_1 \xleftarrow{in_1} p_0 \xrightarrow{in_2} p_2$, the *amalgamated production*

$$p_1 \oplus_{p_0} p_2 : L \xleftarrow{\ell} K \xrightarrow{r} R$$

is constructed by gluing p_1 and p_2 along p_0 using the pushouts (1), (2) and (3) in the diagram below, such that all squares commute. The morphisms ℓ and r are induced by the universal property of the pushout (2). Applying $p_1 \oplus_{p_0} p_2$ to a graph G yields an *amalgamated derivation* $G \Rightarrow X$.

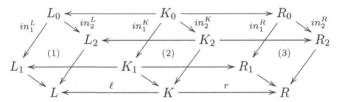

Example 9. Amalgamation of the productions p_1 and p_2 in Figure 7 using the subproduction p_0 in Figure 8, generates the intended rewrite rule for the reconfiguration of the original distributed system in Figure 4. Note that the complete network is reconfigured using local rules in one atomic step.

Even though it provides a proper means of synchronizing local reconfigurations, in general, amalgamation is less-suited for distributed systems. This is because it requires (i) knowledge of all connectors and their reconfiguration rules, and (ii) a centralized entity that is aware of the whole network and that performs the reconfiguration in a non-local fashion. To overcome these problems we show now a different way of applying synchronized rules.

5.3   Local Execution of Synchronized Rules

The reconfiguration mechanism for distributed connectors that we introduce in the following, avoids the problem of amalgamations by performing the reconfiguration asynchronously, where locally only a single connector and its interface are updated at a time. Without repeating the actual constructions, we first recall (the analysis part of) the so-called amalgamation theorem, which provides the means to locally execute synchronized reconfiguration rules. For more details and a proof of the theorem we refer to [5].

Amalgamation Theorem. Let $p_1 \xleftarrow{in_1} p_0 \xrightarrow{in_2} p_2$ be synchronized productions and $G \Rightarrow X$ an amalgamated derivation via $p_1 \oplus_{p_0} p_2$. Then there exist productions

p'_1 and p'_2, called the *remainders* of p_1 and p_2 with respect to p_0, such that the following derivations exist:

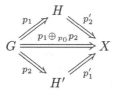

Intuitively, the remainders p'_1 and p'_2 simulate the effect of p_1 and p_2 without performing the action of the subproduction p_0. For our application to distributed Reo connectors, this means that the remainders do the reconfiguration of the connectors without updating the interfaces, or more precisely, assuming that they have been updated already.

Example 10. The remainders of the productions p_1 and p_2 in Figure 7 are the same as the original rules, except that their left-hand sides contain the newly created interface already. The rules merely establish a new connection to the already existing interface, instead of creating it. Hence, the reconfiguration of the distributed connector in Figure 4 can be done by first applying p_1 to update the scheduler including the interfaces and then p'_2 to update the task repository accordingly. Analogously, it can be also the case that first the repository together with the interfaces are updated and then the scheduler.

Using the above approach, a network can be reconfigured by a stepwise updating of its constituent connectors. In particular, the connectors can also be black boxes that reconfigure themselves. On the other hand, these local reconfigurations must be coordinated somehow, since the order of local reconfigurations and the choice of which connector updates the common interface is not clear. For this purpose, we discuss a strategy in the next subsection.

5.4   Coordinating Local Reconfigurations

We informally describe a strategy for organizing local reconfigurations in a network. The central idea is that a reconfiguration is triggered locally at one of the subconnectors and that this creates a cascade of follow-up reconfigurations across the network.

Connectors may define synchronized reconfiguration rules, i.e. rules that describe how the connector itself is changed, and further, how its interfaces are updated. We also assume that a connector reconfigures itself triggered by an external request. For this purposes it may publish the names of its reconfiguration rules. Connectors in the neighborhood can invoke these reconfiguration rules via their shared interface (through a communication channel that is not explicitly modeled here). When a rule is invoked, a connector performs the reconfiguration in three steps:

1. Determine the interface where the request came from and the interfaces of those connectors in the neighborhood that also need to be updated.

2. Send reconfiguration requests to those connectors in the neighborhood that must be updated and block until they are reconfigured.
3. Do the local reconfiguration and reconfigure the interface, if necessary, only where the request came from.

We assume that there is an 'active' party in the network that initiates the reconfiguration by invoking a rule on some connector. Every connector can handle only one reconfiguration request at a time. Hence, the request builds up a reconfiguration dependency tree in the network. The root of the tree is where the reconfiguration was initially invoked. The reconfiguration is then executed bottom-up, starting at the leaves until the root is also reconfigured.

Connectors may also respond to a reconfiguration request with a failure. In that case, the failure is forwarded in the network and all reconfigurations performed so far are rolled-back. This ensures the atomicity of the reconfigurations.

6   Concluding Remarks

We presented a framework, exploiting algebraic graph transformations, for the reconfiguration of distributed Reo connectors. This approach allows a blackboxed view on subconnectors for which reconfigurations can be defined and executed locally. We showed how to synchronize local reconfigurations in the absence of a centralized entity, which is a prime assumption in distributed environments.

Future work includes the formal modeling of the distributed strategy for coordinating local reconfigurations. The typing mechanism used for Reo can be further extended. In particular, constraints that ensure disjointness of multiple interfaces are not modeled at present. Our distribution model can also serve as a basis for describing deployment operations, e.g. transparently moving a connector to another network location. Finally, the dynamic reconfiguration approach presented here, needs to be incorporated in the existing distributed Reo implementation. For this purpose, we have already implemented a reconfiguration engine based on algebraic graph transformation as a part of the Eclipse Coordination Tools [20].

References

1. Arbab, F.: Reo: A Channel-based Coordination Model for Component Composition. Mathematical Structures in Computer Science 14, 329–366 (2004)
2. Arbab, F.: Abstract Behavior Types: A Foundation Model for Components and Their Composition. Science of Computer Programming 55, 3–52 (2005)
3. Taentzer, G.: Distributed Graphs and Graph Transformation. Applied Categorical Structures 7, 431–462 (1999)
4. Ehrig, H., Orejas, F., Prange, U.: Categorical Foundations of Distributed Graph Transformation. In: Corradini, A., Ehrig, H., Montanari, U., Ribeiro, L., Rozenberg, G. (eds.) ICGT 2006. LNCS, vol. 4178, pp. 215–229. Springer, Heidelberg (2006)
5. Boehm, P., Fonio, H.R., Habel, A.: Amalgamation of Graph Transformations: A Synchronization Mechanism. Journal of Computer and System Sciences 34, 377–408 (1987)

6. Corradini, A., Montanari, U., Rossi, F., Ehrig, H., Heckel, R., Löwe, M.: Algebraic Approaches to Graph Transformation I: Basic Concepts and Double Pushout Approach. In: Handbook of Graph Grammars and Computing by Graph Transformation, pp. 163–245. World Scientific, Singapore (1997)
7. Taentzer, G., Beyer, M.: Amalgamated Graph Transformations and Their Use for Specifying AGG. In: Ehrig, H., Schneider, H.-J. (eds.) Dagstuhl Seminar 1993. LNCS, vol. 776, pp. 380–394. Springer, Heidelberg (1994)
8. Ehrig, H., Ehrig, K., Prange, U., Taentzer, G.: Fundamentals of Algebraic Graph Transformation. In: EATCS Monographs in Theoretical Computer Science. Springer, Heidelberg (2006)
9. Baldan, P., Corradini, A., Ehrig, H., Heckel, R.: Compositional Semantics for Open Petri Nets based on Deterministic Processes. Mathematical Structures in Computer Science 15, 1–35 (2005)
10. Bruni, R., Lafuente, A.L., Montanari, U., Tuosto, E.: Style-based Architectural Reconfigurations. Bulletin of the EATCS 94, 180–181 (2008)
11. Engels, G., Heckel, R.: Graph Transformation as a Conceptual and Formal Framework for System Modeling and Model Evolution. In: Welzl, E., Montanari, U., Rolim, J.D.P. (eds.) ICALP 2000. LNCS, vol. 1853, pp. 127–150. Springer, Heidelberg (2000)
12. Yellin, D., Strom, R.: Protocol specifications and component adaptors. ACM Transactions on Programming Languages and Systems 19, 292–333 (1990)
13. Canal, C., Murillo, J., Poizat, P.: Software adaptation. L'Object 12, 9–31 (2006)
14. Brogi, A., Cámera, J., Canal, C., Cubo, J., Pimentel, E.: Dynamic contextual adaptation. Electronics Notes in Theoretical Computer Science 175, 81–95 (2007)
15. Cubo, J., Salaün, G., Cámara, J., Canal, C., Pimentel, E.: Context-based adaptation of component behavioural interfaces. In: Murphy, A.L., Vitek, J. (eds.) COORDINATION 2007. LNCS, vol. 4467, pp. 305–323. Springer, Heidelberg (2007)
16. Gottschalk, F., Aalst, W.v.d., Jansen-Vullers, M., La Rosa, M.: Configurable workflow models. Journal of Cooperative Information Systems 17, 177–221 (2008)
17. Clarke, D., Costa, D., Arbab, F.: Connector Colouring I: Synchronisation and Context Dependency. Science of Computer Programming 66, 205–225 (2007)
18. Baier, C., Sirjani, M., Arbab, F., Rutten, J.: Modeling component connectors in Reo by constraint automata. Science of Computer Programming 61, 75–113 (2006)
19. Arbab, F., Bruni, R., Clarke, D., Lanese, I., Montanari, U.: Tiles for Reo (extended abstract). In: Corradini, A., Gadduci, F. (eds.) WADT 2008, preliminary proceedings, Technical Report TR–08–15, Dipartemento di Informatica, Università di Pisa, pp. 21–24 (2008)
20. Arbab, F., Koehler, C., Maraikar, Z., Moon, Y.J., Proenca, J.: Modeling, Testing and Executing Reo Connectors with the Eclipse Coordination Tools. In: Canal, C., Pasareanu, C. (eds.) Proc. FACS 2008 (to appear)
21. Odersky, M.: The Scala experiment: Can we provide better language support for component systems? In: Morrisett, J., Peyton Jones, S. (eds.) Proc. POPL, pp. 166–167. ACM, New York (2006)
22. de Lara, J., Bardohl, R., Ehrig, H., Ehrig, K., Prange, U., Taentzer, G.: Attributed Graph Transformation with Node Type Inheritance. Theoretical Computer Science 376, 139–163 (2007)
23. Taentzer, G., Rensink, A.: Ensuring structural constraints in graph-based models with type inheritance. In: Cerioli, M. (ed.) FASE 2005. LNCS, vol. 3442, pp. 64–79. Springer, Heidelberg (2005)

A Rewrite Approach for Pattern Containment

Barbara Kordy

LIFO - Université d'Orléans, France
`barbara.kordy@univ-orleans.fr`

Abstract. In this paper we introduce an approach that allows to handle the containment problem for the fragment XP(/,//,[ ],*) of XPath. Using rewriting techniques we define a necessary and sufficient condition for pattern containment. This rewrite view is then adapted to query evaluation on XML documents, and remains valid even if the documents are given in a compressed form, as dags.

1 Introduction

The focus in this paper is on the containment problem ([1,2]) for the fragment XP(/,//,[ ],*) of XPath. XPath ([3]) is the main language for navigating and selecting nodes in XML documents. The segment XP(/,//,[ ],*) defines a class of Core XPath queries expressing descendant relationships between nodes, possibly containing filters, and allowing to use the don't–care (or wildcard) symbol '*'. The queries of this fragment can be modeled by patterns: tree like graphs having two types of edges *child* and *descendant*. Every XML document t is an unranked tree $t = (Nodes_t, Edges_t)$, and can also be seen as a pattern. For any two patterns P and Q, we say that P is contained in Q ($P \subseteq Q$), iff the query represented by Q is more general than the one represented by P. For example, a/b is contained in $a//b$, since `child` (/) is a particular case of `descendant` (//).

The big interest in the query containment problem ([1,2,4,5]) is motivated by its applications. Using the notion of pattern containment we can define queries which are equivalent, i.e., that on any XML document, select the same set of nodes. The query equivalence problem is closely linked to the query minimization problem, which is essential for data base researchers. Since the time required for the evaluation of a given query Q is linear with respect to the size of Q ([6]), the minimization — possibility of replacing Q by an equivalent query of smaller size — is of interest from the point of view of complexity ([7,8,9,10]).

We propose to handle the containment problem using a rewrite approach. We define a set of rewrite rules based on the semantics of XP(/,//,[ ],*)–query containment, and show that for any two patterns P and Q, P is contained in Q *if and only if* we can rewrite P to Q using these rules. This provides a characterization of the containment problem using algebraic techniques, which was missing in the literature. Such a rewrite view gives us a uniform framework to treat also other problems, for instance query evaluation. We extend our approach on compressed documents encoded as straightline regular grammars, and apply

A. Corradini and U. Montanari (Eds.): WADT 2008, LNCS 5486, pp. 236–250, 2009.

our rewrite technique in order to evaluate XP(/,//,[ ],*)–queries on compressed or unfolded (arborescent) XML documents.

This paper is organized as follows: In Section 2 we introduce terminology and notation, and recall some results on the pattern containment problem. Our rewrite method is presented in Section 3. Finally, in Section 4 we show how to adapt this rewrite approach to the query evaluation problem.

2   The Pattern Containment Problem

Let Σ be an alphabet containing the element names of all XML documents considered. In this work we consider the fragment XP(/,//,[ ],*) of XPath, which consists of: node tests (symbols from $\Sigma \cup \{*\}$), child axis (/), descendant axis (//), and qualifiers also called filters ([...]). Any element of XP(/,//,[ ],*) is a query that can be represented as a rooted tree structure graph over $\Sigma \cup \{*\}$, called *unary pattern*, having:

- edges of two types: simple for *child*, and double for *descendant*,
- nodes labeled by the symbols from $\Sigma \cup \{*\}$,
- one distinguished node marked with a special selection symbol 's' representing the output information (located at the end of the main path in the query considered).

For instance, the unary pattern in Figure 1 represents the XP(/,//,[ ],*) query /a//b[./b/c/d]/c[./*//d]. The notion of unary patterns is easily extended to that of n–ary patterns, where we have n distinguished nodes, that model n–ary queries selecting n–tuples of nodes. Miklau and Suciu show in [1] that, for the purpose of the containment problem, it is sufficient to consider only the patterns of arity zero, called *boolean*, where there are no distinguished nodes. Thus, all patterns considered in the sequel will be boolean, and they will be simply called patterns.

For a given pattern P, we denote by $Nodes_P$ the set of all its nodes. For any $u \in Nodes_P$, $name_P(u)$ stands for the element of $\Sigma \cup \{*\}$ labeling the node u. By $Edges_{\downarrow}(P)$ and $Edges_{\Downarrow}(P)$ we mean respectively the set of child and descendant edges of P. We define the *size* of P (denoted by $|P|$) to be the number of all edges in P.

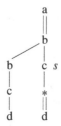

Fig. 1. Unary pattern representing query /a//b[./b/c/d]/c[./*//d]

Definition 1. *An XML tree t is a model of a pattern P iff there exists an embedding from P to t; i.e., a function $\mathbf{e} \colon Nodes_P \to Nodes_t$, satisfying the following conditions:*

1. $\mathbf{e}$ *preserves the root:* $\mathbf{e}(root_P) = root_t$;
2. $\mathbf{e}$ *preserves the names:*
   $\forall u \in Nodes_P, \; name_P(u) = *, \; or \; name_P(u) = name_t(\mathbf{e}(u))$;
3. $\mathbf{e}$ *preserves the relation* child:
   $\forall (u, v) \in Edges_\downarrow(P), \; (\mathbf{e}(u), \mathbf{e}(v)) \in Edges_t$;
4. $\mathbf{e}$ *preserves the relation* descendant:
   $\forall (u, v) \in Edges_\Downarrow(P), \; (\mathbf{e}(u), \mathbf{e}(v)) \in (Edges_t)^+$,

where $(Edges_t)^+$ is the transitive closure of the relation $Edges_t$.

The notion of model is illustrated in Figure 2.

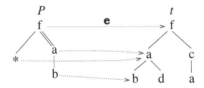

Fig. 2. Pattern P, its model t, and embedding $\mathbf{e}$ from P to t

Definition 2. *Given two patterns P and Q, we say that P is contained in Q ($P \subseteq Q$) iff every model of P is also a model of Q. The patterns P and Q are equivalent ($P \equiv Q$) iff $P \subseteq Q$ and $Q \subseteq P$.*

Figure 3 represents two patterns which are easily seen to be equivalent.

Fig. 3. Equivalent patterns P and Q

Miklau and Suciu prove in [1] that the containment problem for XP(/,//,[ ],*) is CoNP–complete. They also give a sufficient — but not necessary — condition for pattern containment. For that purpose, they extend the notion of embedding to pattern homomorphism:

Definition 3. *Given two patterns P and Q, a homomorphism from Q to P is a function $\varphi \colon Nodes_Q \to Nodes_P$, which is:*

- *root and name preserving;*
- *child preserving:*
   $\forall(u, v) \in Edges_\downarrow(Q), \ (\varphi(u), \varphi(v)) \in Edges_\downarrow(P);$
- *descendant preserving:*
   $\forall(u, v) \in Edges_\Downarrow(Q), \ (\varphi(u), \varphi(v)) \in (Edges_\downarrow(P) \cup Edges_\Downarrow(P))^+.$

The authors of [1] prove that if there exists a homomorphism from Q to P, then P is contained in Q. They give an algorithm which for two given patterns P and Q verifies, in time $\mathcal{O}(|P||Q|)$, whether there exists a homomorphism from Q to P. Figure 4 shows the patterns P and Q, and the homomorphism φ from Q to P proving that $P \subseteq Q$. Nevertheless, the existence of a homomorphism from Q to P is not a necessary condition for $P \subseteq Q$ (as is easily checked for the patterns P and Q given in Figure 3, which are equivalent, but there is no homomorphism neither from Q to P, nor from P to Q). In the following example we show a way to prove the containment $P \subseteq Q$, if there is no homomorphism from Q to P.

Fig. 4. Homomorphism φ from Q to P proving that $P \subseteq Q$

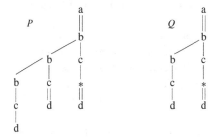

Fig. 5. Patterns s.t. $P \subseteq Q$, but no homomorphism from Q to P

Example 1. In Figure 5 we have presented two patterns P and Q (borrowed from [1]) satisfying $P \subseteq Q$, such that there is no homomorphism from Q to P.

Here, to show the containment $P \subseteq Q$, we have to reason by cases. Let t be a model of P, and consider the middle edge $c//d$ of pattern P. This edge can be realized on t:

- either by the child edge c/d (as in Figure 6),
- or by a path $c/*/\ldots/d$, having length ≥ 2 (as in Figure 7).

Such an analysis shows that any model of P is also a model of Q, thus $P \subseteq Q$. However, it is impossible to define one general homomorphism from Q to P, as the right branch $a//b/c/*//d$ of Q corresponds in each case to a different branch of P.

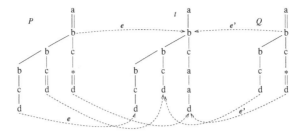

Fig. 6. Model of P (and Q), where $c//d$ is realized by c/d

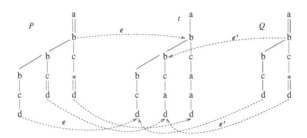

Fig. 7. Model of P (and Q), where $c//d$ is realized by $c/a/d$

3   Pattern Containment via Rewriting

We propose to handle the pattern containment problem using an approach based on rewriting techniques. A key idea is that checking containment requires case analysis in general, and this can be encoded as rewriting (as we illustrate in Example 2 below). We construct a rewrite system $\mathcal{R}$ that permits to define a *necessary* and sufficient condition (see Theorem 1) for pattern containment on the fragment XP(/,//,[ ],*).

We start by giving a formal definition of pattern, alternative to that used in the previous sections.

Definition 4. *We define patterns over an alphabet Σ as the expressions P derived from the grammar of Table 1, where $\downarrow$ and $\Downarrow$ stand respectively for* child *and* descendant*, $\omega \in \Sigma \cup \{*\}$, and '$*$' is the* don't–care *symbol of XPath that can replace any σ of Σ.*

This grammar produces precisely the patterns as defined in [1,7]. For instance, the graph P in Figure 8 corresponds to the expression

$$P = a \Downarrow b\{\downarrow b\{\downarrow b \downarrow c \downarrow d, \downarrow c \Downarrow d\}, \downarrow c \downarrow * \Downarrow d\}$$

derived from the grammar of Table 1.

By a *term* we mean any expression of the type M, S or P derived from the grammar of Table 1, as well as any finite disjunction $P_1 \vee P_2 \vee \cdots \vee P_n$ of patterns. The terms of the type M correspond to the linear paths without branching, they

Table 1. Grammar for patterns

$$
\begin{array}{lll}
M : & \varepsilon \mid\; \downarrow\omega \mid\; \Downarrow\omega \mid\; MM & // \text{ path} \\
S : & \emptyset \mid\; \{MS\} \mid\; S \cup S & // \text{ set of sibling unrooted terms} \\
P : & \omega MS & // \text{ patterns}
\end{array}
$$

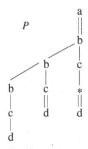

Fig. 8. Pattern

start by a modal symbol $\xi \in \{\downarrow, \Downarrow\}$; those of the type S represent a set of terms having a common parent node; and those of the type P are patterns. The terms in P are *rooted* (they start by a symbol from $\Sigma \cup \{*\}$), those in M and S are *unrooted*. To simplify, we will often identify the singleton $\{MS\}$ with the term MS. Given patterns P and P_i, for $1 \le i \le n$, the terms of the form ε, P, or $P_1 \vee \cdots \vee P_n$, will also be called *d–patterns*. A tree t is a model of a d–pattern $P_1 \vee \cdots \vee P_n$ iff t is a model of at least one pattern P_i, for $1 \le i \le n$. Definition 2 of pattern containment is extended in a natural way to a d–pattern containment. A disjunctive d–pattern will be used in case analysis to represent different models of a given pattern with a unique term, as in the following example.

Example 2. Consider the patterns

$$P = f \downarrow * \Downarrow a \qquad \text{and} \qquad Q = f \Downarrow * \downarrow a$$

given in Figure 3. We know that $P \equiv Q$, thus in particular $P \subseteq Q$, but there is no homomorphism which proves it. Using the rules of our system $\mathcal{R}$ defined below we will be able to rewrite P to Q, and prove the containment $P \subseteq Q$. The idea is that every descendant is either a child or has a depth ≥ 2; thus, the edge $* \Downarrow a$ of P can be realized either by the child edge $* \downarrow a$, or by a path having at least one additional node between '$*$' and 'a', that we can denote by $* \Downarrow * \downarrow a$. We will then rewrite the pattern P to the d–pattern

$$f \downarrow * \downarrow a \;\; \vee \;\; f \downarrow * \Downarrow * \downarrow a$$

depicting the two cases mentioned. The two pattern components of this d–pattern will then be rewritten in parallel. A child, as well as a descendant of depth ≥ 2 are particular cases of descendant. As a consequence, the edge $f \downarrow *$ will be rewritten to $f \Downarrow *$, idem for the path $f \downarrow * \Downarrow *$. This will give us the following term:

$$f \Downarrow * \downarrow a \;\; \vee \;\; f \Downarrow * \downarrow a,$$

which will be finally rewritten to Q, since each pattern composing this d–pattern is exactly the pattern Q.

To formalize the idea employed in the examples above, we introduce a set $\mathcal{R}$ of rules that serve to rewrite rooted and unrooted terms. Let M, S (possibly with primes, subscripts) be as in the grammar of Table 1; $\xi, \xi' \in \{\downarrow, \Downarrow\}$, $\sigma \in \Sigma$, and $\omega, \omega' \in \Sigma \cup \{*\}$:

1. $S \longrightarrow \emptyset$, $M \longrightarrow \varepsilon$ //cut;
2. $M\sigma S \longrightarrow M * S$ //replace any symbol of Σ by the '*' of XPath;
3. $\downarrow \omega S \longrightarrow \Downarrow \omega S$ //every child is also a descendant;
4. $\xi \omega \xi' \omega' S \longrightarrow \Downarrow \omega' S$ //ignore an intermediate node;
5. $M\{S_1, S_2\} \longrightarrow \{MS_1, MS_2\}$ //left distributivity;
6. $S \longrightarrow S \cup S'$, where $S \longrightarrow S'$ //add new siblings;
7. $S \cup S_1 \longrightarrow S' \cup S_1$, if $S \longrightarrow S'$ //rewrite some of the siblings;
8. $\Downarrow \omega S \longrightarrow (\downarrow \omega S) \vee (\downarrow * \Downarrow \omega S)$ //case analysis: descendant is either a child or has depth ≥ 2;
9. $\Downarrow \omega S \longrightarrow (\downarrow \omega S) \vee (\Downarrow * \downarrow \omega S)$ //idem.

By *context–pattern* we mean any pattern having a special additional *hole* symbol $\Diamond$ that replaces one of its unrooted sub–terms. Let us consider a context–pattern $\mathcal{C}$ and an unrooted term X. We define the *fill–in* of $\mathcal{C}$ with X (denoted as $\mathcal{C}\blacklozenge X$) to be the pattern obtained from $\mathcal{C}$ by replacing its hole symbol with the term X; e.g. for the context–pattern $\mathcal{C} = f\{\downarrow a, \Downarrow b\{\Diamond, \downarrow d\}, \Downarrow *\}$, and the unrooted term $X = \Downarrow x\{\downarrow y, \Downarrow z\}$, we get the fill–in:

$$\mathcal{C}\blacklozenge X = f\{\downarrow a, \Downarrow b\{\Downarrow x\{\downarrow y, \Downarrow z\}, \downarrow d\}, \Downarrow *\}.$$

We also suppose that for any context–pattern $\mathcal{C}$ and unrooted terms X and X', the notation $\mathcal{C}\blacklozenge(X \vee X')$ stands for the disjunctive d–pattern $\mathcal{C}\blacklozenge X \vee \mathcal{C}\blacklozenge X'$.

To rewrite patterns with the rules of $\mathcal{R}$ given above, we use *suffix rewriting*:

Definition 5. *Given a pattern P and a pattern or a d–pattern Q, we say that P can be rewritten to Q in one step using suffix rewriting, if there exist a context–pattern $\mathcal{C}$ and two unrooted terms X and X', such that: $P = \mathcal{C}\blacklozenge X$, $Q = \mathcal{C}\blacklozenge X'$, and $X \longrightarrow X'$ is an instance of a rule in $\mathcal{R}$.*

Moreover, disjunctive terms can be rewritten using the following additional two rules, where P is a pattern, and D, D_1, D_2 stand for d–patterns:

10. $D_1 \vee D \longrightarrow D_2 \vee D$, if D_1 can be rewritten to D_2 //case rewriting;
11. $P \vee P \vee D \longrightarrow P \vee D$ //consider any given case only once.

Rules 10 and 11 are used as follows: if a d–pattern L is an instance (modulo commutativity) of the LHS of rule 10 or 11, and a d–pattern R is an instance (modulo commutativity) of the RHS of the same rule, then L can be rewritten to R. We will denote by $A \longrightarrow_{\mathcal{R}} B$ the fact that a pattern or a d–pattern A is rewritten in one step to a pattern or a d–pattern B, by using the rules of $\mathcal{R}$.

The main result of our work is the following:

Theorem 1. *For any two patterns P and Q, P is contained in Q if and only if $P \xrightarrow{*}_{\mathcal{R}} Q$, i.e., P can be rewritten to Q using the rules of $\mathcal{R}$ in zero or finitely many steps.*

Proof. The semantics of the rules in $\mathcal{R}$ guarantee that $P \xrightarrow{*}_{\mathcal{R}} Q$ implies $P \subseteq Q$. Indeed, if $X \longrightarrow X'$ is an instance of one of the rules $1-9$, then for every context–pattern $\mathcal{C}$, we have $\mathcal{C} \blacklozenge X \subseteq \mathcal{C} \blacklozenge X'$; if $L \longrightarrow R$ is an instance of rule 10 or 11, we obviously have $L \subseteq R$.

To show the converse, we start with the following lemma:

Lemma 1. *For any patterns P and Q, if there exists a homomorphism from Q to P, then $P \xrightarrow{*}_{\mathcal{R}} Q$.*

Proof. Given a homomorphism φ from Q to P, we construct a pattern P', such that $P \xrightarrow{*}_{\mathcal{R}} P' \xrightarrow{*}_{\mathcal{R}} Q$, as follows:

(a) for every node u of Q, we construct a corresponding node u' of P', and we set $name_{P'}(u') = name_P(\varphi(u))$;
(b) we construct a child edge $(u',v') \in Edges_{\downarrow}(P')$, if and only if $(u,v) \in Edges_{\downarrow}(Q)$;
(c) we construct a descendant edge $(u',v') \in Edges_{\Downarrow}(P')$, if and only if $(u,v) \in Edges_{\Downarrow}(Q)$.

The cost of such a construction is linear with respect to the size of Q. The pattern P' can be rewritten to the pattern Q using rule 2 of $\mathcal{R}$. Indeed, the structures (nodes, simple and double edges) of P' and Q are the same, but the names of some $u \in Nodes_Q$ and the corresponding node $u' \in Nodes_{P'}$ may be different. Condition (a) implies that: either $name_Q(u) = name_{P'}(u') = name_P(\varphi(u))$, or $name_Q(u) \neq name_{P'}(u') = name_P(\varphi(u))$. In the second case we have (see Definition 3): $name_Q(u) = *$, and $name_{P'}(u') \in \Sigma$, thus to rewrite P' to Q we have to use rule 2.

It remains to be shown that P can be rewritten to P':

- using rules 1 and 7 (with $S' = \emptyset$), we can ignore all sub–branches of P which do not contain the nodes images under φ;
- if some node w of P is an image of m distinct nodes $u_1, \ldots, u_m$ of Q, then we rewrite the unique node w of P to m nodes $u'_1, \ldots, u'_m$ of P', by using rule 6 (with $S' = S$) and/or rule 5;
- case when edge (u',v') is in $Edges_{\downarrow}(P')$: from condition (b) we know that $(u,v) \in Edges_{\downarrow}(Q)$, thus by Definition 3 we have $(\varphi(u), \varphi(v)) \in Edges_{\downarrow}(P)$ (we have nothing to do with the edge $(\varphi(u), \varphi(v))$ when rewriting P to P');
- case when edge (u',v') is in $Edges_{\Downarrow}(P')$: from condition (c) and Definition 3 we can deduce that there exist $k \geq 1$ and $w_0, \ldots w_k \in Nodes_P$, such that: $w_0 = \varphi(u), w_k = \varphi(v)$, and $\forall\, i \in \{0, \ldots, k-1\}$ we have $(w_i, w_{i+1}) \in Edges_{\downarrow}(P) \cup Edges_{\Downarrow}(P)$. If $k = 1$ and $(\varphi(u), \varphi(v)) \in Edges_{\downarrow}(P)$, then we can rewrite P to P' using rule 3. If $k \geq 2$, then we use $(k-1$ times) rule 4 to ignore the nodes $w_1, \ldots w_{k-1}$ while rewriting P to P'.

Finally, we obtain $P \xrightarrow{*}_{\mathcal{R}} P' \xrightarrow{*}_{\mathcal{R}} Q$.   □

Note that if P is a tree, we also have the converse of Lemma 1. Indeed, it is sufficient to remark that if $P \xrightarrow{*}_{\mathcal{R}} Q$ and P is a tree, then one can rewrite P to Q by using only rules $1-7$; if $X \longrightarrow X'$ is an instance of one of those rules, then for every context–pattern $\mathcal{C}$, there exists a homomorphism from $\mathcal{C} \blacklozenge X'$ to $\mathcal{C} \blacklozenge X$. Of course, in the case when P is a tree, a homomorphism from Q to P is an embedding from the pattern Q to the tree P. The above considerations give us the following characterization:

Remark 1. A tree t is a model of a pattern Q iff $t \xrightarrow{*}_{\mathcal{R}} Q$.

By a *homomorphism* from a pattern Q to a d–pattern $D = P_1 \vee \cdots \vee P_n$, we mean a function which is a homomorphism from Q to P_i, for *every* $1 \leq i \leq n$. Thus, using Lemma 1, we obtain the following corollary:

Corollary 1. *For any given pattern Q and a d–pattern D, if there exists a homomorphism from Q to D, then $D \xrightarrow{*}_{\mathcal{R}} Q$.*

Proof. It suffices to remark that rules 10 and 11 imply that a d–pattern $P_1 \vee \cdots \vee P_n$ can be rewritten to a pattern Q if and only if, for every $1 \leq i \leq n$, we have $P_i \xrightarrow{*}_{\mathcal{R}} Q$. □

To finish the proof of Theorem 1, we use the following proposition:

Proposition 1. *For two patterns P and Q, if $P \subseteq Q$, then one can construct a d–pattern D verifying $P \xrightarrow{*}_{\mathcal{R}} D$, such that there exists a homomorphism from Q to D.*

Proof. From the result of Miklau and Suciu ([1]) we know that it is possible to check if there exists a homomorphism from Q to P. If it is the case, the d–pattern D satisfying the proposition is equal to P (see Lemma 1). If not, a disjunctive d–pattern D satisfying the proposition can be constructed by using rules 8 and 9 finitely many times. We know that every model of P is also a model of Q. The idea is to represent all models of P by an equivalent d–pattern $D = P_1 \vee \cdots \vee P_n$ representing case analysis, such that for every $1 \leq i \leq n$, there exists a homomorphism from Q to P_i. □

This terminates the proof of Theorem 1.

The rewrite system $\mathcal{R}$ is non–deterministic; nevertheless if P and Q are given, there exists a well–defined, goal–directed strategy for rewriting P to Q. The idea is to use *only* those rules among $1-11$ that permit to converge to Q. We illustrate this strategy in the following example:

Example 3. Let P and Q be the patterns represented in Figure 5. We show how to rewrite P to Q, and thus prove the containment $P \subseteq Q$. The pattern $P = a \Downarrow b\{\downarrow b\{\downarrow b \downarrow c \downarrow d, \downarrow c \Downarrow d\}, \downarrow c \downarrow * \Downarrow d\}$ can be seen as the fill–in

$$a \Downarrow b\{\downarrow b\{\downarrow b \downarrow c \downarrow d, \downarrow c \Diamond\}, \downarrow c \downarrow * \Downarrow d\} \ \blacklozenge \ \underline{\Downarrow d}.$$

Using rule 8 for the underlined term, we encode the cases depicted in Example 1:

$$a \Downarrow b\{\downarrow b\{\downarrow b \downarrow c \downarrow d, \downarrow c \Diamond\}, \downarrow c \downarrow * \Downarrow d\} \; \blacklozenge \;\; \downarrow d$$
$$\vee \quad a \Downarrow b\{\downarrow b\{\downarrow b \downarrow c \downarrow d, \downarrow c \Diamond\}, \downarrow c \downarrow * \Downarrow d\} \; \blacklozenge \;\; \downarrow * \Downarrow d.$$

We obtain the d–pattern

$$a \Downarrow b\{\downarrow b\{\downarrow b \downarrow c \downarrow d, \downarrow c \downarrow d\}, \downarrow c \downarrow * \Downarrow d\}$$
$$\vee \quad a \Downarrow b\{\downarrow b\{\downarrow b \downarrow c \downarrow d, \downarrow c \downarrow * \Downarrow d\}, \downarrow c \downarrow * \Downarrow d\},$$

which can be seen under the form

$$a \Downarrow b\{\downarrow b\{\Diamond, \downarrow c \downarrow d\}, \downarrow c \downarrow * \Downarrow d\} \; \blacklozenge \; \underline{\downarrow b \downarrow c \downarrow d}$$
$$\vee \quad a \Downarrow b\{\downarrow b\{\downarrow b \downarrow c \downarrow d, \downarrow c \downarrow * \Downarrow d\}, \Diamond\} \; \blacklozenge \; \underline{\downarrow c \downarrow * \Downarrow d}.$$

We rewrite it using rule 10. We cut (rule 1) the underlined parts, and get

$$a \Downarrow b\{\downarrow b\{\downarrow c \downarrow d\}, \downarrow c \downarrow * \Downarrow d\} \quad \vee \quad a \Downarrow b\{\downarrow b\{\downarrow b \downarrow c \downarrow d, \downarrow c \downarrow * \Downarrow d\}\}.$$

The d–pattern that we have obtained is then identified with

$$a \Downarrow b\{\downarrow b \downarrow c \downarrow d, \downarrow c \downarrow * \Downarrow d\} \quad \vee \quad a \Downarrow b \downarrow b\{\downarrow b \downarrow c \downarrow d, \downarrow c \downarrow * \Downarrow d\}.$$

Its first component is equal to the pattern Q. To the second one, seen as the fill–in $a\Diamond \; \blacklozenge \; \Downarrow b \downarrow b\{\downarrow b \downarrow c \downarrow d, \downarrow c \downarrow * \Downarrow d\}$, we apply rule 4, and get the term $a\Diamond \; \blacklozenge \; \Downarrow b\{\downarrow b \downarrow c \downarrow d, \downarrow c \downarrow * \Downarrow d\} = a \Downarrow b\{\downarrow b \downarrow c \downarrow d, \downarrow c \downarrow * \Downarrow d\}$. Thus we obtain the d–pattern

$$a \Downarrow b\{\downarrow b \downarrow c \downarrow d, \downarrow c \downarrow * \Downarrow d\} \quad \vee \quad a \Downarrow b\{\downarrow b \downarrow c \downarrow d, \downarrow c \downarrow * \Downarrow d\} = Q \vee Q,$$

that is finally rewritten to Q using rule 11.

Remark 2. Our approach is no longer valid, if it is not based on suffix rewriting; e.g. for $P = * \Downarrow *$ and $Q = * \downarrow *$, we have $P \subseteq Q$ ($P \xrightarrow{*}_{\mathcal{R}} Q$ using rules 9, 1), but $P\Diamond \; \blacklozenge \; \downarrow a = * \Downarrow * \downarrow a$ is *not* contained in $Q\Diamond \; \blacklozenge \; \downarrow a = * \downarrow * \downarrow a$: for instance, the tree $t = f \downarrow g \downarrow b \downarrow a$ is a model of $* \Downarrow * \downarrow a$, but not of $* \downarrow * \downarrow a$.

4   Applications

The objective of this section is to show that our rewrite approach remains valid even if the models of patterns are given in a compressed form (as dags), and that it can be adapted for query evaluation on XML documents.

4.1   Case of Compressed Documents

To model compressed documents we use rooted dags instead of trees (as in [11,12,13,14]). Figure 9 represents three formats of the same document: tree, fully and partially compressed format (see [11] for formal definitions). In the sequel, by

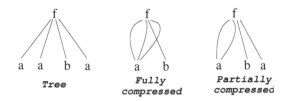

Fig. 9. Tree, fully compressed format, partially compressed format

t we will denote any given representation (tree or dag) of the document considered. To distinguish between different formats of the same document we use regular tree grammars. Given a document t, we call *normalized grammar* for t a regular tree grammar G_t:

- which recognizes only t,
- where every node of t is represented by exactly one non–terminal,
- the indexes of non–terminals for children nodes are greater then the indexes of non–terminals for parent nodes.

Such normalized grammars are *straightline* in the sense defined in [15], i.e., there is no cycle on their dependency graph. For this reason we will refer to them as SLR grammars.

Example 4. The SLR grammars for the three dags from Figure 9 are respectively:

$$X_0 \to f(X_1, X_2, X_3, X_4) \quad Y_0 \to f(Y_1, Y_1, Y_2, Y_1) \quad Z_0 \to f(Z_1, Z_1, Z_2, Z_3)$$
$$X_1 \to a \qquad\qquad\qquad Y_1 \to a \qquad\qquad\quad Z_1 \to a$$
$$X_2 \to a \qquad\qquad\qquad Y_2 \to b \qquad\qquad\quad Z_2 \to b$$
$$X_3 \to b \qquad\qquad\qquad\qquad\qquad\qquad\qquad Z_3 \to a.$$
$$X_4 \to a$$

We extend the notion of SLR grammar to patterns. To define a normalized grammar G_P for pattern P, it is sufficient that every non–terminal X_i appearing on the right hand side of any production of G_P, is preceded by a modal symbol $\downarrow$ or $\Downarrow$, corresponding to the type of edge pointing to the node represented by X_i on P. In order to have a uniform notation that covers patterns as well as documents, we will do the same on the normalized grammar G_t, for any document t: every non–terminal X_i appearing on the right hand side of some production in G_t, will be preceded by $\downarrow$. For instance, the grammars G_P and G_t respectively for the pattern P and the tree t of Figure 11, are given in Figure 10.

To define an embedding **e** from a pattern P to a dag t, we replace the conditions 3 and 4 of Definition 1 respectively by:

3. $\forall u, v \in Nodes_P$, such that $(u, v) \in Edges_{\downarrow}(P)$, there exists an edge going form **e**(u) to **e**(v) on t;
4. $\forall u, v \in Nodes_P$, such that $(u, v) \in Edges_{\Downarrow}(P)$, there exists a path going from **e**(u) to **e**(v) in $(Edges_t)^+$.

$$P_0 \rightarrow *(\downarrow P_1, \Downarrow P_2) \qquad\qquad X_0 \rightarrow f(\downarrow X_2, \downarrow X_1)$$
$$P_1 \rightarrow a \qquad\qquad\qquad\qquad X_1 \rightarrow b(\downarrow X_2)$$
$$P_2 \rightarrow a \qquad\qquad\qquad\qquad X_2 \rightarrow a.$$

Fig. 10. SLR grammars G_P and G_t for P and t from Figure 11

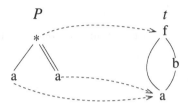

Fig. 11. Pattern P, its compressed model t, and embedding from P to t

The notion of (dag) model of a pattern and the pattern containment problem are defined in the same way as in the case of tree models. Figure 11 shows a pattern P, its compressed model t, and an embedding from P to t.

SLR grammars can be used in our rewrite approach. To prove that a given dag t is a model of a pattern P, it is sufficient (according to Remark 1) to rewrite the grammar G_t representing t to the grammar G_P representing P. We illustrate this idea in the following example.

Example 5. Consider the grammars G_P and G_t given in Figure 10. We show how to rewrite G_t to G_P using rules of $\mathcal{R}$:

$$X_0 \rightarrow f(\downarrow X_2, \downarrow X_1) \xrightarrow{2} \quad X_0 \rightarrow *(\downarrow X_2, \downarrow X_1) \xrightarrow{1} \quad X_0 \rightarrow *(\downarrow X_2) \xrightarrow{6}$$
$$X_1 \rightarrow b(\downarrow X_2) \qquad\qquad X_1 \rightarrow b(\downarrow X_2) \xrightarrow{1}$$
$$X_2 \rightarrow a \qquad\qquad\qquad X_2 \rightarrow a \qquad\qquad\qquad X_2 \rightarrow a$$

The first production of G_t is first rewritten using rule 2; then we cut a branch represented by X_1 (rule 1). At the same time, we can eliminate from G_t the production $X_1 \rightarrow b(\downarrow X_2)$, since it has become unproductive (there is no more production having X_1 on their right hand sides).

$$X_0 \rightarrow *(\downarrow X_2, \downarrow X_2') \xrightarrow{3} \quad X_0 \rightarrow *(\downarrow X_2, \Downarrow X_2') \quad \approx \quad P_0 \rightarrow *(\downarrow P_1, \Downarrow P_2)$$
$$X_2 \rightarrow a \qquad\qquad\qquad X_2 \rightarrow a \qquad\qquad \approx \quad P_1 \rightarrow a$$
$$X_2' \rightarrow a \qquad\qquad\qquad X_2' \rightarrow a \qquad\qquad \approx \quad P_2 \rightarrow a.$$

Then, using rule 6 we double the number of children of X_0; we introduce a new non–terminal X_2', which produces the same sub–pattern as X_2. Finally, by Rule 3, we get a grammar which is equal, up to non–terminal renaming, to G_P.

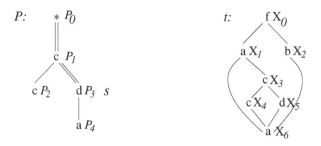

Fig. 12. Unary pattern P and its compressed model t

4.2   Query Evaluation

SLR grammars help us to adapt the rewrite approach of Section 3 to XP(/,//,[ ],*)–query evaluation on (compressed) documents. To represent unary queries, we use unary patterns (see Section 2). Let us consider the unary pattern P representing the query $P = $ /*//c[./c]//d[./a], and the compressed document t, given in Figure 12. The corresponding SLR grammars G_P and G_t are respectively:

$$P_0 \rightarrow *(\Downarrow P_1)$$
$$P_1 \rightarrow c(\downarrow P_2, \Downarrow P_3)$$
$$P_2 \rightarrow c$$
$$P_3(s) \rightarrow d(\downarrow P_4)$$
$$P_4 \rightarrow a$$

$$X_0 \rightarrow f(\downarrow X_1, \downarrow X_2)$$
$$X_2 \rightarrow b(\downarrow X_6)$$
$$X_1 \rightarrow a(\downarrow X_6, \downarrow X_3)$$
$$X_3 \rightarrow c(\downarrow X_4, \downarrow X_5)$$
$$X_4 \rightarrow c(\downarrow X_6)$$
$$X_5 \rightarrow d(\downarrow X_6)$$
$$X_6 \rightarrow a.$$

The non–terminal P_3 of G_P is marked 's', since it represents the output node of P. To find an answer for P on t, we rewrite the grammar G_t to the grammar G_P, using the rules of $\mathcal{R}$. The non–terminal of G_t which will be rewritten to the selecting non–terminal P_3 of G_P, will represent an answer for P on t. We illustrate this reasoning below:

$$X_0 \rightarrow f(\downarrow X_1, \downarrow X_2) \xrightarrow{1} \quad X_0 \rightarrow f(\downarrow X_1) \xrightarrow{2} \quad X_0 \rightarrow *(\downarrow X_1) \xrightarrow{4}$$
$$X_2 \rightarrow b(\downarrow X_6)$$
$$X_1 \rightarrow a(\downarrow X_6, \downarrow X_3) \xrightarrow{1} \quad X_1 \rightarrow a(\downarrow X_3) \quad X_1 \rightarrow a(\downarrow X_3)$$
$$X_3 \rightarrow c(\downarrow X_4, \downarrow X_5) \quad X_3 \rightarrow c(\downarrow X_4, \downarrow X_5) \xrightarrow{3} \quad X_3 \rightarrow c(\downarrow X_4, \Downarrow X_5)$$
$$X_4 \rightarrow c(\downarrow X_6) \xrightarrow{1} \quad X_4 \rightarrow c \quad X_4 \rightarrow c$$
$$X_5 \rightarrow d(\downarrow X_6) \quad X_5 \rightarrow d(\downarrow X_6) \quad X_5 \rightarrow d(\downarrow X_6)$$
$$X_6 \rightarrow a \quad X_6 \rightarrow a \quad X_6 \rightarrow a$$

$$X_0 \to *(\Downarrow X_3) \qquad\qquad \approx \qquad\qquad P_0 \to *(\Downarrow P_1)$$
$$X_3 \to c(\downarrow X_4, \Downarrow X_5) \qquad \approx \qquad\qquad P_1 \to c(\downarrow P_2, \Downarrow P_3)$$
$$X_4 \to c \qquad\qquad\qquad \approx \qquad\qquad P_2 \to c$$
$$X_5 \to d(\downarrow X_6) \qquad\qquad \approx \qquad\qquad P_3(s) \to d(\downarrow P_4)$$
$$X_6 \to a \qquad\qquad\qquad \approx \qquad\qquad P_4 \to a,$$

We have obtained an SLR grammar, which is (up to non–terminal renaming) the SLR grammar G_P for P. The non–terminal X_5 of G_t has been rewritten to the non–terminal P_3, thus the node represented by X_5 is an answer for P on t.

Note that, as any query P of the fragment XP(/,//,[ ],*) is purely descendant, the answer for P on a document t does *not* depend on the form under which t is given (tree or dag); this is no longer valid for queries containing ascendant axes (cf.[11]). Remark also that our rewrite approach can be extended to any n–ary query of XP(/,//,[ ],*); an n–ary query selects a set of n–tuples of nodes ([16]), and is easily represented as an n–ary pattern.

5   Conclusion

We have presented an approach based on rewrite techniques, that allows to handle the problem of query containment for the segment XP(/,//,[ ],*) of XPath. Such a rewrite view is also appropriate for compressed documents modeled as dags, and can be adapted to (unary as well as n–ary) query evaluation on (compressed) documents.

Straightline regular tree grammars can provide an exponential space compression. Nevertheless there exist more efficient compression techniques, like those based on staightline context–free grammars (SLCF, [15]), giving better (up to doubly exponential) compression rates. Currently we are studying the possibility of extending our rewrite approach to such more efficient compressions. We also hope to adapt our results to larger fragments of XPath, containing queries modeled by more general patterns, having both descendant and ascendant edges.

References

1. Miklau, G., Suciu, D.: Containment and Equivalence for a Fragment of XPath. J. ACM 51(1), 2–45 (2004)
2. Neven, F., Schwentick, T.: XPath Containment in the Presence of Disjunction, DTDs, and Variables. In: Calvanese, D., Lenzerini, M., Motwani, R. (eds.) ICDT 2003. LNCS, vol. 2572, pp. 312–326. Springer, Heidelberg (2002)
3. W3C: XML Path Language (1999), http://www.w3.org/TR/xpath
4. Wood, P.T.: On the Equivalence of XML Patterns. In: Palamidessi, C., Moniz Pereira, L., Lloyd, J.W., Dahl, V., Furbach, U., Kerber, M., Lau, K.-K., Sagiv, Y., Stuckey, P.J. (eds.) CL 2000. LNCS, vol. 1861, pp. 1152–1166. Springer, Heidelberg (2000)
5. Schwentick, T.: XPath Query Containment. SIGMOD Rec. 33(1), 101–109 (2004)

6. Gottlob, G., Koch, C., Pichler, R.: Efficient Algorithms for Processing XPath Queries. ACM Trans. Database Syst. 30(2), 444–491 (2005)
7. Flesca, S., Furfaro, F., Masciari, E.: On the Minimization of XPath Queries. J. ACM 55(1) (2008)
8. Amer-Yahia, S., Cho, S., Lakshmanan, L.V.S., Srivastava, D.: Tree Pattern Query Minimization. VLDB J. 11(4), 315–331 (2002)
9. Kimelfeld, B., Sagiv, Y.: Revisiting Redundancy and Minimization in an XPath Fragment. In: EDBT 2008: Proceedings of the 11th International Conference on Extending Database Technology, pp. 61–72. ACM, New York (2008)
10. Ramanan, P.: Efficient Algorithms for Minimizing Tree Pattern Queries. In: SIGMOD 2002: Proceedings of the 2002 ACM SIGMOD International Conference on Management of Data, pp. 299–309. ACM, New York (2002)
11. Fila, B., Anantharaman, S.: Automata for Positive Core XPath Queries on Compressed Documents. In: Hermann, M., Voronkov, A. (eds.) LPAR 2006. LNCS (LNAI), vol. 4246, pp. 467–481. Springer, Heidelberg (2006)
12. Buneman, P., Grohe, M., Koch, C.: Path Queries on Compressed XML. In: VLDB 2003, pp. 141–152. Morgan Kaufmann, San Francisco (2003)
13. Marx, M.: XPath and Modal Logics of Finite DAG's. In: Cialdea Mayer, M., Pirri, F. (eds.) TABLEAUX 2003. LNCS (LNAI), vol. 2796, pp. 150–164. Springer, Heidelberg (2003)
14. Frick, M., Grohe, M., Koch, C.: Query Evaluation on Compressed Trees (Extended Abstract). In: LICS 2003: Proceedings of the 18th Annual IEEE Symposium on Logic in Computer Science, p. 188. IEEE Computer Society, Washington (2003)
15. Busatto, G., Lohrey, M., Maneth, S.: Efficient Memory Representation of XML Document Trees. Inf. Syst. 33(4-5), 456–474 (2008)
16. Niehren, J., Planque, L., Talbot, J.M., Tison, S.: N–ary Queries by Tree Automata. In: Bierman, G., Koch, C. (eds.) DBPL 2005. LNCS, vol. 3774, pp. 217–231. Springer, Heidelberg (2005)

A Coalgebraic Characterization of Behaviours in the Linear Time – Branching Time Spectrum

Luís Monteiro

CITI, Departamento de Informática,
Faculdade de Ciências e Tecnologia, Universidade Nova de Lisboa, Portugal
lm@di.fct.unl.pt

Abstract. The paper outlines an approach for characterizing several kinds of behaviours for transition systems in coalgebraic terms and illustrates the approach with some behaviours in the linear time – branching time spectrum, namely, traces, ready-traces and failures. The approach is based on an abstract notion of "behaviour object" that can be defined in any (concrete) category and enjoys a uniqueness property similar to the uniqueness of morphisms to final objects. That property makes behaviour objects final in a suitable extension of the given category with additional morphisms, which allows to define the behaviours of arbitrary objects by the unique morphisms to the behaviour objects. The main purpose of the paper is to show how trace, ready-trace and failure semantics can be characterized in terms of behaviour objects.

1 Introduction

When coalgebras are viewed as models of general dynamical systems [1], final coalgebras appear as abstract descriptions of the observable behaviours of such systems. For many systems, however, several notions of behaviour have been proposed, depending on the point of view or on the intended application. A case in point is the class of (labeled) transition systems: in [2], twelve notions of behaviour are presented and hierarchically organized in the so-called linear time – branching time spectrum. The problem is then how to characterize in coalgebraic terms different notions of behaviour for a given type of system, since final coalgebras, when they exist, are unique up to isomorphism and so can not capture more than one notion of behaviour. In this paper we introduce an abstract notion of "behaviour object" in any (concrete) category, enjoying a uniqueness property similar to the uniqueness of morphisms to final objects; not surprisingly, final objects are themselves instances of behaviour objects. We then proceed to show how traces, ready-traces and failures give rise to behaviour objects in a category of transition systems. We believe most other kinds of behaviours in the spectrum can be treated in a similar way.

Not much work has been done on general notions of behaviour for coalgebras beyond those based on final coalgebras. It is worth mentioning in this respect some work on a characterization of traces for some categories of coalgebras that appear in [3, 4, 5]. Our approach differs from the one in the cited papers, which is

A. Corradini and U. Montanari (Eds.): WADT 2008, LNCS 5486, pp. 251–265, 2009.

based on monads and distributive laws. We start from a category $\mathbf{C}$ (of transition systems) equipped with a faithful functor $U : \mathbf{C} \to \mathbf{Set}$ where we assume $Uf = f$ for arrows f. For each type of behaviours (traces, ready-traces or failures) we define a full subcategory $\mathbf{D}$ whose behaviours are precisely those we wish to capture. More precisely, the given behaviours can be structured as a final object $\mathbb{Z}$ in $\mathbf{D}$. To associate behaviours of the given kind with an arbitrary transition system $\mathbb{S}$ and not with just those in $\mathbf{D}$, we introduce a functor T from $\mathbf{C}$ to $\mathbf{C}$ whose image is contained in $\mathbf{D}$. There is a unique morphism $\beta : T\mathbb{S} \to \mathbb{Z}$, but our interest lies in $\mathbb{S}$, not $T\mathbb{S}$. So the last step is to define $\eta_{\mathbb{S}} : U\mathbb{S} \to UT\mathbb{S}$ which combined with β gives the "behaviour map" $\mathrm{beh}_{\mathbb{S}}^{\mathbb{Z}} = \beta \circ \eta_{\mathbb{S}}$ from $U\mathbb{S}$ to $U\mathbb{Z}$. Ideally, we would like η to be a natural transformation $\mathrm{Id}_{\mathbf{C}} \to T$, but as it turns out $\eta_{\mathbb{S}}$ is not in general a morphism in $\mathbf{C}$. Thus, η is just a natural transformation $U \to UT$. We may assume, however, that $\eta_{\mathbb{S}}$ is a morphism $\mathbb{S} \to T\mathbb{S}$ if $\mathbb{S}$ is in $\mathbf{D}$; in that case $\mathrm{beh}_{\mathbb{S}}^{\mathbb{Z}}$ is a morphism, and in fact the unique morphism $\mathbb{S} \to \mathbb{Z}$, thus showing that our notion of behaviour is a conservative extension of the one based on final objects.

Thus, the behaviour object $\mathbb{Z}$ is a final object in the subcategory $\mathbf{D}$; but in fact it is possible to introduce a new notion of morphism that depends on T, to be called a "T-morphism", and $\mathbb{Z}$ is also final in the supercategory $\mathbf{C}_T$ of $\mathbf{C}$ that has T-morphisms as morphisms. The behaviour function $\mathrm{beh}_{\mathbb{S}}^{\mathbb{Z}}$ is then the unique T-morphism from $\mathbb{S}$ to $\mathbb{Z}$. The categories of the form $\mathbf{C}_T$ will play a crucial role in comparing behaviour equivalences with respect to distinct behaviour objects.

The paper is organized as follows. In the next section we present an abstract framework where we introduce the notion of behaviour object in the setting of an arbitrary (concrete) category; we also present the general results that will be used in the concrete cases to be considered next. The three sections that follow instantiate the framework by considering an appropriate category of transition systems and studying in turn the cases of traces, ready-traces and failures. A short section then compares the three behaviour equivalences associated with the three types of behaviours. The paper ends with some conclusions.

As prerequisites we assume familiarity with the notions of trace, ready-trace and failure as can be found in [2]; these notions will be defined in the present paper but no motivation for them will be supplied. From category theory we only assume knowledge of the basic notions of category, functor, natural transformation and final object. No knowledge of the theory of coalgebras will be required; when we speak of coalgebras it will always refer to a particular category of transition systems and the reader may safely ignore all mentions of coalgebras. Finally, most results are proved in the paper; some proofs have been omitted by lack of space, but mostly in cases of routine verification of properties. An expanded version of this paper containing all proofs appears in [6].

2   The General Framework

Let $\mathbf{C}$ be a category and $U : \mathbf{C} \to \mathbf{Set}$ a faithful functor; for simplicity, we identify in the sequel arrows f in $\mathbf{C}$ with the functions Uf, that is, we assume that morphisms in **C** are functions on the underlying sets. These data will be fixed throughout this section.

Definition 1. *An object* $\mathbb{Z}$ *of* **C** *will be called a* behaviour object *if there is a full subcategory* **D** *of* **C**, *an endofunctor* T *on* **C** *and a natural transformation* $\eta : U \rightarrow UT$ *such that* $\mathbb{Z}$ *is final in* **D**, *the image of* T *is contained in* **D** *and* $\eta_\mathbb{S} : U\mathbb{S} \rightarrow UT\mathbb{S}$ *is a morphism* $\mathbb{S} \rightarrow T\mathbb{S}$ *whenever* $\mathbb{S}$ *is in* **D**. *The composite* $\mathrm{beh}_\mathbb{S}^\mathbb{Z} = \beta \circ \eta_\mathbb{S} : U\mathbb{S} \rightarrow U\mathbb{Z}$ *is the* behaviour function *determined by* $\mathbb{Z}, \mathbf{D}, T, \eta$; *the kernel of the behaviour function is the* behavioural equivalence $\equiv_\mathbb{S}^\mathbb{Z}$, *that is,* $s \equiv_\mathbb{S}^\mathbb{Z} s'$ *iff* $\mathrm{beh}_\mathbb{S}^\mathbb{Z}(s) = \mathrm{beh}_\mathbb{S}^\mathbb{Z}(s')$ *for all* $s, s' \in U\mathbb{S}$.

The behaviour function is not in general a morphism of **C**, but it will prove useful to view it as a morphism of a supercategory extending **C** with additional morphisms called "T-morphisms." Note that a final object of **C**, if there is one, is a behaviour object, for we may take **D** to be **C**, T to be the identity functor and η to be the identity natural transformation of U; in that case, $\mathrm{beh}_\mathbb{S}^\mathbb{Z}$ is the only morphism from $\mathbb{S}$ to $\mathbb{Z}$.

Definition 2. *Let* $\mathbf{D}, T, \eta$ *be as in the previous definition. If* $\mathbb{S}_1$ *and* $\mathbb{S}_2$ *are objects in* **C**, *by a* T-morphism $f : \mathbb{S}_1 \rightarrow \mathbb{S}_2$ *we mean a function* $f : U\mathbb{S}_1 \rightarrow U\mathbb{S}_2$ *for which there exists a morphism* $f' : T\mathbb{S}_1 \rightarrow T\mathbb{S}_2$ *such that the following diagram commutes:*

$$
\begin{array}{ccc}
U\mathbb{S}_1 & \xrightarrow{\;f\;} & U\mathbb{S}_2 \\
\downarrow{\scriptstyle \eta_{\mathbb{S}_1}} & & \downarrow{\scriptstyle \eta_{\mathbb{S}_2}} \\
UT\mathbb{S}_1 & \xrightarrow{\;f'\;} & UT\mathbb{S}_2 .
\end{array}
$$

Note that f' needs not be unique. If f is a morphism, then f is also a T-morphism; indeed, take $f' = Tf$ in the previous diagram and observe that the diagram commutes by the naturality of η. Also, every $\eta_\mathbb{S}$ is a T-morphism, because we may take $\eta_\mathbb{S}' = \eta_{T\mathbb{S}}$, which is a morphism since $T\mathbb{S}$ is in **D**. The composition of two T-morphisms is a T-morphism and the identity $1_\mathbb{S}$, being a morphism, is also a T-morphism. This gives a category $\mathbf{C}_T$ which extends **C**, with the same objects as **C** and with T-morphisms as morphisms.

Theorem 1. *Let* **D** *be a full subcategory of* **C** *containing the image of an endofunctor* T *on* **C** *and* $\eta : U \rightarrow UT$ *a natural transformation such that* $\eta_\mathbb{S} : U\mathbb{S} \rightarrow UT\mathbb{S}$ *is a morphism for every object* $\mathbb{S}$ *in* **D**. *An object* $\mathbb{Z}$ *of* **D** *is final in* **D** *iff* $\mathbb{Z}$ *is final in* $\mathbf{C}_T$ *and the unique* T-morphism $T\mathbb{Z} \rightarrow \mathbb{Z}$ *is a morphism.*

Proof. Assume $\mathbb{Z}$ is final in **D**. Since $T\mathbb{Z}$ is in **D**, there is a unique morphism $\zeta : T\mathbb{Z} \rightarrow \mathbb{Z}$. Since $\eta_\mathbb{Z}$ is a morphism, because $\mathbb{Z}$ is in **D**, we have $\zeta \circ \eta_\mathbb{Z} = 1_{U\mathbb{Z}}$, by the finality of $\mathbb{Z}$. Given any object $\mathbb{S}$ in **C**, we must show that there is a unique T-morphism $\mathbb{S} \rightarrow \mathbb{Z}$. The unique morphism $\beta : T\mathbb{S} \rightarrow \mathbb{Z}$ gives a T-morphism

$\beta \circ \eta_{\mathbb{S}} : \mathbb{S} \to \mathbb{Z}$. If $\alpha : \mathbb{S} \to \mathbb{Z}$ is another T-morphism, we show that $\alpha = \beta \circ \eta_{\mathbb{S}}$. By definition of T-morphism, there is a morphism $\alpha' : T\mathbb{S} \to T\mathbb{Z}$ such that $\eta_{\mathbb{Z}} \circ \alpha = \alpha' \circ \eta_{\mathbb{S}}$, as depicted in the next diagram:

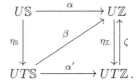

Using the identities $\zeta \circ \eta_{\mathbb{Z}} = 1_{U\mathbb{Z}}$, as noted above, and $\zeta \circ \alpha' = \beta$, by the uniqueness of β, we calculate: $\beta \circ \eta_{\mathbb{S}} = \zeta \circ \alpha' \circ \eta_{\mathbb{S}} = \zeta \circ \eta_{\mathbb{Z}} \circ \alpha = \alpha$.

Conversely, suppose $\mathbb{Z}$ is final in $\mathbf{C}_T$ and ζ is a morphism. Given any object $\mathbb{S}$ in $\mathbf{D}$, we must show that there is a (necessarily unique) morphism $\mathbb{S} \to \mathbb{Z}$. By hypothesis, there is a (unique) T-morphism $\alpha : \mathbb{S} \to \mathbb{Z}$. By definition of T-morphism, $\eta_{\mathbb{Z}} \circ \alpha = \alpha' \circ \eta_{\mathbb{S}}$ for some morphism $\alpha' : T\mathbb{S} \to T\mathbb{Z}$. Since ζ is a morphism, $\zeta \circ \alpha' : T\mathbb{S} \to \mathbb{Z}$ is a morphism. Composing with $\eta_{\mathbb{S}} : U\mathbb{S} \to UT\mathbb{S}$, which is a morphism by hypothesis, gives a morphism $\zeta \circ \alpha' \circ \eta_{\mathbb{S}}$ from $\mathbb{S}$ to $\mathbb{Z}$.

In the conditions of the theorem, $\mathbb{Z}$ is a behaviour object and $\mathrm{beh}_{\mathbb{S}}^{\mathbb{Z}} : \mathbb{S} \to \mathbb{Z}$ is the *only T-morphism* from $\mathbb{S}$ to $\mathbb{Z}$, which is a morphism if $\mathbb{S}$ is in $\mathbf{D}$. By uniqueness, $\mathrm{beh}_{\mathbb{S}}^{\mathbb{Z}} = \mathrm{beh}_{T\mathbb{S}}^{\mathbb{Z}} \circ \eta_{\mathbb{S}}$; these functions showed up as $\alpha = \mathrm{beh}_{\mathbb{S}}^{\mathbb{Z}}$ and $\beta = \mathrm{beh}_{T\mathbb{S}}^{\mathbb{Z}}$ in the proof of the previous theorem.

What is the relationship between $\mathbb{Z}$ and any final object that $\mathbf{C}$ might have? And between the corresponding behaviour equivalences? The next proposition shows that general behaviour objects give rise to behaviour equivalences coarser than the behaviour equivalence determined by a final object. In categories of coalgebras, bisimilarity is finer than the behaviour equivalence determined by a final object, so the technique of behaviour objects never gives behaviour equivalences finer than bisimulation.

Proposition 1. *Suppose $\mathbf{C}$ has a final object $\mathbb{W}$ and $\mathbf{C}_T$ has a final object $\mathbb{Z}$. Then $\mathbb{Z}$ is a subobject of $\mathbb{W}$ in $\mathbf{C}$ and a retract of $\mathbb{W}$ in $\mathbf{C}_T$. Furthermore, the equivalence $\overset{\mathbb{W}}{=}_{\mathbb{S}}$ is finer than $\overset{\mathbb{Z}}{=}_{\mathbb{S}}$, that is, $\overset{\mathbb{W}}{=}_{\mathbb{S}} \subseteq \overset{\mathbb{Z}}{=}_{\mathbb{S}}$.*

Proof. There is a unique morphism $\alpha : \mathbb{Z} \to \mathbb{W}$ and a unique T-morphism $\beta : \mathbb{W} \to \mathbb{Z}$. The composition $\beta \circ \alpha : \mathbb{Z} \to \mathbb{Z}$ is a T-morphism, hence must be $1_{\mathbb{Z}}$. In particular, α is injective, so is monic, and the first conclusion follows. For the second statement, just note that $\mathrm{beh}_{\mathbb{S}}^{\mathbb{Z}} = \mathrm{beh}_{\mathbb{W}}^{\mathbb{Z}} \circ \mathrm{beh}_{\mathbb{S}}^{\mathbb{W}}$, by uniqueness.

In the cases considered in this paper $\mathbf{C}$ has not a final object, but we may wish to compare the behaviour equivalences induced by distinct behaviour objects. To perform the comparison we must be able to relate the behaviour objects somehow. In the proposition below we assume that the natural transformation associated with one of the behaviour objects preserves the behaviours with respect to the other.

Proposition 2. *Let $\mathbb{W}$ be a behaviour object of $\mathbf{C}$ with respect to a full subcategory $\mathbf{E}$, an endofunctor F and a natural transformation ζ, and $\mathbb{Z}$ a behaviour* object with respect to $\mathbf{D}, T, \eta$. Assume ζ preserves the behaviours over $\mathbb{Z}$, in the sense that $\mathrm{beh}^{\mathbb{Z}}_{\mathbb{S}} = \mathrm{beh}^{\mathbb{Z}}_{F\mathbb{S}} \circ \zeta_{\mathbb{S}}$. In these conditions, $\mathrm{beh}^{\mathbb{Z}}_{\mathbb{S}} = \mathrm{beh}^{\mathbb{Z}}_{\mathbb{W}} \circ \mathrm{beh}^{\mathbb{W}}_{\mathbb{S}}$, and consequently $\stackrel{\mathbb{W}}{=}_{\mathbb{S}}$ is finer than $\stackrel{\mathbb{Z}}{=}_{\mathbb{S}}$.

Proof. We can write $\mathrm{beh}^{\mathbb{Z}}_{F\mathbb{S}} = \mathrm{beh}^{\mathbb{Z}}_{\mathbb{W}} \circ \mathrm{beh}^{\mathbb{W}}_{F\mathbb{S}}$, by uniqueness, since $\mathrm{beh}^{\mathbb{W}}_{F\mathbb{S}}$ is a morphism so that both sides are T-morphisms. On the other hand, by definition of $\mathrm{beh}^{\mathbb{W}}_{\mathbb{S}}$, we have $\mathrm{beh}^{\mathbb{W}}_{\mathbb{S}} = \mathrm{beh}^{\mathbb{W}}_{F\mathbb{S}} \circ \zeta_{\mathbb{S}}$. From these equalities and the one in the hypothesis of the proposition we derive $\mathrm{beh}^{\mathbb{Z}}_{\mathbb{S}} = \mathrm{beh}^{\mathbb{Z}}_{\mathbb{W}} \circ \mathrm{beh}^{\mathbb{W}}_{\mathbb{S}}$, as required.

3   Traces of Transition Systems

For the rest of this paper we fix a set $A \neq \emptyset$ of *actions*. Recall that a (labelled) transition system with labels in A is a pair $\langle S, \rightarrow \rangle$, where S is a set of *states* and $\rightarrow$ is a ternary relation $\rightarrow \subseteq S \times A \times S$; as usual, we write $s \stackrel{a}{\rightarrow} t$ instead of $(s, a, t) \in \rightarrow$. There are several ways to view a transition system as a coalgebra $\mathbb{S} = \langle S, \psi \rangle$; in this paper we assume that ψ maps S to $\mathcal{P}(S)^A$, where $\psi(s)(a) = \{t \mid s \stackrel{a}{\rightarrow} t\}$ for all $s \in S$ and $a \in A$; the function ψ will sometimes be called the "dynamics" of the transition system. Thus, transition systems are basically coalgebras for the functor $\mathcal{P}(-)^A$; in the sequel we shall still often write $s \stackrel{a}{\rightarrow} t$ as an abbreviation of $t \in \psi(s)(a)$, for clarity; sometimes we write $s \stackrel{a}{\rightarrow}$ to mean that $s \stackrel{a}{\rightarrow} t$ for some t. If $\mathbb{S}' = \langle S', \psi' \rangle$ is another transition system, a *morphism* $f : \mathbb{S} \rightarrow \mathbb{S}'$ is a function $f : S \rightarrow S'$ such that $\psi' \circ f = \mathcal{P}(f)^A \circ \psi$, as in the following commutative diagram:

$$
\begin{array}{ccc}
S & \stackrel{f}{\longrightarrow} & S' \\
\psi \downarrow & & \downarrow \psi' \\
\mathcal{P}(S)^A & \xrightarrow{\mathcal{P}(f)^A} & \mathcal{P}(S')^A .
\end{array}
$$

It is easy to see that this notion is equivalent to the following two conditions taken together:

- whenever $s \stackrel{a}{\rightarrow} t$ in $\mathbb{S}$, then $f(s) \stackrel{a}{\rightarrow} f(t)$ in $\mathbb{S}'$;
- if $f(s) \stackrel{a}{\rightarrow} t'$ in $\mathbb{S}'$, there is $t \in S$ such that $s \stackrel{a}{\rightarrow} t$ in $\mathbb{S}$ and $f(t) = t'$.

The category of transition systems will be the main category of interest in this paper, and so it will be denoted $\mathbf{C}$, to agree with the notation introduced in the previous section. The forgetful functor $U : \mathbf{C} \rightarrow \mathbf{Set}$ maps $\mathbb{S} = \langle S, \psi \rangle$ to S and any morphism to itself as a function.

Let $\mathbb{S} = \langle S, \psi \rangle$ be a transition system. We extend the transition relation to strings $x \in A^*$ by writing $s \stackrel{x}{\rightarrow} t$ if $s \stackrel{a_1}{\rightarrow} \cdots \stackrel{a_n}{\rightarrow} t$ with $x = a_1 \cdots a_n$; of course, if $n = 0$, then x is the null string ε and $s \stackrel{\varepsilon}{\rightarrow} s$. The strings x such that $s \stackrel{x}{\rightarrow} t$ are the *traces* of s; we put $tr_{\mathbb{S}}(s) = \{x \mid \exists t, s \stackrel{x}{\rightarrow} t\}$. The set $tr_{\mathbb{S}}(s)$ is nonempty and prefix-closed; this amounts to say that $\varepsilon \in tr_{\mathbb{S}}(s)$ and whenever $xy \in tr_{\mathbb{S}}(s)$, then $x \in tr_{\mathbb{S}}(s)$.

A *trace language* is a nonempty and prefix-closed subset $L \subseteq A^*$. If Tr is the set of all trace languages, $tr_{\mathbb{S}}$ is then a function from S to Tr. We turn Tr into a transition system $\mathbb{T} = \langle Tr, \zeta_{Tr} \rangle$ by defining $\zeta_{Tr} : Tr \to \mathcal{P}(Tr)^A$ by the transitions

$$L \xrightarrow{a} \{x \mid ax \in L\}$$

for $a \in L$ (to guarantee that $\{x \mid ax \in L\} \neq \emptyset$). This system is "deterministic" in the sense that each $\zeta_{Tr}(L)(a)$ is either empty, in case $a \notin L$, or the singleton $\{\{x \mid ax \in L\}\}$ when $a \in L$. Note that $tr_{\mathbb{S}} : S \to Tr$ is not in general a morphism from $\mathbb{S}$ to $\mathbb{T}$; as we shall see, $tr_{\mathbb{S}}$ is a T_{Tr}-morphism for an appropriate endofunctor T_{Tr} of $\mathbf{C}$. The following results are easily proved.

Proposition 3. *If $f : \mathbb{S}_1 \to \mathbb{S}_2$ is a morphism of transition systems and s is a state of $\mathbb{S}_1$, then s and $f(s)$ have the same traces.*

Proposition 4. *If L is in Tr, then $tr_{\mathbb{T}}(L) = L$.*

Let $\mathbf{D}_{Tr}$ be the full subcategory of $\mathbf{C}$ of deterministics transition systems $\mathbb{S} = \langle S, \psi \rangle$, that is, such that $\psi(s)(a)$ has cardinality at most one for all $s \in S$ and $a \in A$; equivalently, if $s \xrightarrow{a} t_1$ and $s \xrightarrow{a} t_2$ imply $t_1 = t_2$, for all $s, t_1, t_2 \in S$ and $a \in A$.

Theorem 2. *The transition system $\mathbb{T} = \langle Tr, \zeta_{Tr} \rangle$ is final in $\mathbf{D}_{Tr}$. Given $\mathbb{S} = \langle S, \psi \rangle$ in $\mathbf{D}_{Tr}$, the only morphism from $\mathbb{S}$ to $\mathbb{T}$ is $tr_{\mathbb{S}} : S \to Tr$.*

Proof. To see that $tr_{\mathbb{S}}$ is a morphism we only have to check that $s \xrightarrow{a} t$ in $\mathbb{S}$ implies $tr_{\mathbb{S}}(s) \xrightarrow{a} tr_{\mathbb{S}}(t)$, because $\mathbb{S}$ is deterministic. Since $s \xrightarrow{a} t$ is the only transition from s labeled with a, the traces of s that start with a all go through t, that is, $ax \in tr_{\mathbb{S}}(s)$ iff $x \in tr_{\mathbb{S}}(t)$; this means that $tr_{\mathbb{S}}(t) = \{x \mid ax \in tr_{\mathbb{S}}(s)\}$, which is the same as saying that $tr_{\mathbb{S}}(s) \xrightarrow{a} tr_{\mathbb{S}}(t)$. For the uniqueness, if $f : S \to Tr$ is another morphism from $\mathbb{S}$ to $\mathbb{T}$ and $s \in S$, then $tr_{\mathbb{S}}(s) = tr_{\mathbb{T}}(f(s)) = f(s)$, by the previous propositions, so $f = tr_{\mathbb{S}}$.

The endofunctor T_{Tr} on $\mathbf{C}$ is basically the powerset construction that turns an arbitrary system into a deterministic one; in the definition that follows, $\mathcal{P}_{ne}$ is the nonempty powerset functor and $\subseteq_{ne}$ the nonempty set inclusion. On objects $\mathbb{S} = \langle S, \psi \rangle$ let $T_{Tr}\mathbb{S} = \bar{\mathbb{S}} = \langle \bar{S}, \bar{\psi} \rangle$, where $\bar{S} = \mathcal{P}_{ne}(S)$ and $\bar{\psi} : \bar{S} \to \mathcal{P}(\bar{S})^A$ is given by the transitions $M \xrightarrow{a} \{t \mid \exists s \in M, s \xrightarrow{a} t\}$ for all $M \subseteq_{ne} S$ and $a \in A$, provided that the set on the right is not empty; on arrows $f : \mathbb{S}_1 \to \mathbb{S}_2$ put $T_{Tr}(f) = \bar{f} = \mathcal{P}_{ne}(f)$.

Proposition 5. *T_{Tr} is well-defined, is a functor and its image is contained in $\mathbf{D}_{Tr}$.*

The natural transformation $\eta_{Tr} : U \to U T_{Tr}$ is defined by $\eta_{Tr\mathbb{S}}(s) = \{s\}$ for all transition systems $\mathbb{S}$ and all states s of $\mathbb{S}$.

Proposition 6. *$\eta = \eta_{Tr}$ is a natural transformation and $\eta_{\mathbb{S}} : S \to \bar{S}$ is a morphism whenever $\mathbb{S}$ is in $\mathbf{D}_{Tr}$.*

Proof. If $f : \mathbb{S}_1 \to \mathbb{S}_2$ is a morphism, then $\eta_{\mathbb{S}_2} \circ f = \bar{f} \circ \eta_{\mathbb{S}_1}$, since both maps apply any $s \in S_1$ to $\{f(s)\}$; thus, η is a natural transformation. If $\mathbb{S}$ is deterministic, $s \xrightarrow{a} t$ in $\mathbb{S}$ implies $\{s\} \xrightarrow{a} \{t\}$ in $\bar{\mathbb{S}}$, so $\eta_{\mathbb{S}}$ is an isomorphism between $\mathbb{S}$ and the subsystem of $\bar{\mathbb{S}}$ generated by the singletons $\{s\} = \eta_{\mathbb{S}}(s)$.

We noticed before that $tr_{\mathbb{S}} : S \to Tr$ is not a morphism in general; we are now ready to prove that $tr_{\mathbb{S}}$ is not only a T_{Tr}-morphism, but is the unique T_{Tr}-morphism from $\mathbb{S}$ to $\mathbb{T}$.

Theorem 3. *For any transition system* $\mathbb{S}$, $tr_{\mathbb{S}} = \mathrm{beh}_{\mathbb{S}}^{\mathsf{T}}$, *the unique* T_{Tr}-*morphism from* $\mathbb{S}$ *to* $\mathbb{T}$.

Proof. We know by Theorem 1 that $\mathrm{beh}_{\mathbb{S}}^{\mathsf{T}} = \mathrm{beh}_{\bar{\mathbb{S}}}^{\mathsf{T}} \circ \eta_{\mathbb{S}}$, where η abbreviates η_{Tr}; see the remark immediately after the proof of that theorem. On the other hand, $\mathrm{beh}_{\bar{\mathbb{S}}}^{\mathsf{T}} = tr_{\bar{\mathbb{S}}}$, by Theorem 2, since $\bar{\mathbb{S}} = T_{Tr}\mathbb{S}$ is in $\mathbf{D}_{Tr}$. So to prove that $tr_{\mathbb{S}} = \mathrm{beh}_{\mathbb{S}}^{\mathsf{T}}$ we need to show that $tr_{\mathbb{S}}(s) = tr_{\bar{\mathbb{S}}}(\{s\})$ for all $s \in S$, that is, s and $\{s\}$ have the same traces. But this is easy: if $s \xrightarrow{a_1} s_1 \xrightarrow{a_2} \cdots \xrightarrow{a_n} s_n$, there is a sequence of transitions $\{s\} \xrightarrow{a_1} M_1 \xrightarrow{a_2} \cdots \xrightarrow{a_n} M_n$ since each s_i is in M_i, $i = 1, \ldots, n$; conversely, if $\{s\} \xrightarrow{a_1} M_1 \xrightarrow{a_2} \cdots \xrightarrow{a_n} M_n$, select an arbitrary $s_n \in M_n$, then $s_{n-1} \in M_{n-1}$ such that $s_{n-1} \xrightarrow{a_n} s_n$, and so on, so that $s \xrightarrow{a_1} s_1 \xrightarrow{a_2} \cdots \xrightarrow{a_n} s_n$.

4   Ready-Traces of Transition Systems

A *ready-trace* on A is a string $(a_1, X_1) \cdots (a_n, X_n)$ $(n \geq 0)$ on $A \times \mathcal{P}(A)$ such that $a_{i+1} \in X_i$ for all i, $1 \leq i < n$; for brevity, the ready-trace will be written simply $a_1 X_1 \cdots a_n X_n$; if $n = 0$, the *empty* ready-trace is just the empty trace ε. Intuitively, the condition on ready-traces is that any action but the first must have been declared "ready" in the previous step.[1]

Given a transition system $\mathbb{S} = \langle S, \psi \rangle$ and $s \in S$, let $I(s) = \{a \in A \mid s \xrightarrow{a}\}$ be the set of the *initials* of s. In order to characterize the ready-traces of $\mathbb{S}$, it is convenient to define first an auxiliary transition system $\mathbb{S}_I = \langle S, \psi_I \rangle$. This system has the same states as $\mathbb{S}$ and basically the same transitions, except that we add to the label of each $s \xrightarrow{a} t$ the initials of t. Thus, $\mathbb{S}_I$ is a transition system over $A \times \mathcal{P}(A)$, where $\psi_I : S \to \mathcal{P}(S)^{A \times \mathcal{P}(A)}$ is defined by the transitions $s \xrightarrow{a, X} t$ such that $s \xrightarrow{a} t$ and $X = I(t)$ in $\mathbb{S}$. If $f : \mathbb{S} \to \mathbb{S}'$ is a morphism of transition systems over A, then f is also a morphism from $\mathbb{S}_I$ to $\mathbb{S}_I'$ since morphisms clearly preserve the initials of states; conversely, a morphism $\mathbb{S}_I \to \mathbb{S}_I'$ is a also a morphism $\mathbb{S} \to \mathbb{S}'$. Thus, the assignment $\mathbb{S} \mapsto \mathbb{S}'$ on objects and $f \mapsto f$ on arrows is a functor from $\mathbf{C}$ to the category $\mathbf{C}(A \times \mathcal{P}(A))$ of transition systems over $A \times \mathcal{P}(A)$; this functor is injective on objects, full and faithful, so in an embedding, a fact that will be useful later. We shall have occasion to use the final system of trace languages over $A \times \mathcal{P}(A)$, which will be denoted $\mathbb{T}(A \times \mathcal{P}(A))$.

---

[1] Sometimes a ready-trace is required to start with a set X_0 of actions such that $a_1 \in X_0$, but this is unnecessary and inconvenient for our purposes.

The *ready-traces* of $\mathbb{S}$ are by definition the traces of $\mathbb{S}_I$; the set of ready-traces of $s \in S$ is written $rt_{\mathbb{S}}(s)$. The set $rt_{\mathbb{S}}(s)$, besides being nonempty and prefix-closed, is *extension-closed*, in the sense that if $raX \in rt_{\mathbb{S}}(s)$ and $b \in X$, then $raXbY \in rt_{\mathbb{S}}(s)$ for some $Y \subseteq A$. Intuitively, extension-closure states that any action declared explicitly as "ready" may actually occur.

A *ready-trace language* $L \subseteq (A \times \mathcal{P}(A))^*$ is a nonempty, prefix-closed and extension-closed set of ready-traces; the set of all ready-trace languages will be written Rt, so that $rt_{\mathbb{S}}$ is a function from S to Rt. Ready-traces being traces over $A \times \mathcal{P}(A)$, the ready-trace languages inherit transitions $L \xrightarrow{a,X} L'$ from the trace languages; we must be careful though and check that the language L' is in fact a ready-trace language.

Lemma 1. *If L is a ready-trace language, any transition $L \xrightarrow{a,X} L'$ has L' a ready-trace language.*

Thus, the set Rt of ready-trace languages can be seen as a transition system over $A \times \mathcal{P}(A)$; it is a subsystem of the transition system of all trace languages on $A \times \mathcal{P}(A)$, a fact that will be used later. We are more interested, however, in seeing Rt as a transition system $\mathbb{R} = \langle Rt, \zeta_{Rt} \rangle$ over A, defining ζ_{Rt} by the transitions $L \xrightarrow{a} L'$ such that $L \xrightarrow{a,X} L'$ for some $X \subseteq A$.

Proposition 7. *Let L be a ready-trace language.*

1. *The set of initials $I_{\mathbb{R}}(L)$ of L in $\mathbb{R}$ is the set of all elements in A that start ready-traces in L.*
2. *If $L \xrightarrow{a,X} L'$, then $I_{\mathbb{R}}(L') = X$.*
3. *$L \xrightarrow{a,X} L'$ iff $L \xrightarrow{a} L'$ and $I_{\mathbb{R}}(L') = X$.*
4. *Finally, $rt_{\mathbb{R}}(L) = L$.*

Proposition 8. *If $f : \mathbb{S}_1 \to \mathbb{S}_2$ is a morphism of transition systems and s is a state of $\mathbb{S}_1$, then s and $f(s)$ have the same ready-traces.*

Proof. We know that f is also a morphism $\mathbb{S}_{1I} \to \mathbb{S}_{2I}$, so s and $f(s)$ have the same traces over $A \times \mathcal{P}(A)$ by Proposition 3, that is, have the same ready-traces.

We say a transition system $\mathbb{S} = \langle S, \psi \rangle$ is *deterministic with respect to the initials*, or *I-deterministic*, if $\psi_I(s)(a, X)$ has cardinality at most one for all $s \in S$, $a \in A$ and $X \subseteq A$; equivalently, if $s \xrightarrow{a,X} t_1$ and $s \xrightarrow{a,X} t_2$ imply $t_1 = t_2$, or if $s \xrightarrow{a} t_1$, $s \xrightarrow{a} t_2$ and $I(t_1) = I(t_2)$ imply $t_1 = t_2$. Thus, $\mathbb{S}$ in $\mathbf{C}$ is I-deterministic iff $\mathbb{S}_I$ in $\mathbf{C}(A \times \mathcal{P}(A))$ is deterministic. Let $\mathbf{D}_{Rt}$ be the full subcategory of $\mathbf{C}$ of I-deterministic transition systems.

Theorem 4. *The transition system $\mathbb{R} = \langle Rt, \zeta_{Rt} \rangle$ is final in $\mathbf{D}_{Rt}$. Given $\mathbb{S} = \langle S, \psi \rangle$ in $\mathbf{D}_{Rt}$, the only morphism from $\mathbb{S}$ to $\mathbb{R}$ is $rt_{\mathbb{S}} : S \to Rt$.*

Proof. Clearly, $\mathbb{R}$ is in $\mathbf{D}_{Rt}$. To see that $rt_{\mathbb{S}} : S \to Rt$ is the only morphism from $\mathbb{S}$ to $\mathbb{R}$, it is enough to show that $rt_{\mathbb{S}}$ is the only morphism from $\mathbb{S}_I$ to $\mathbb{R}_I$. As $\mathbb{S}_I$ is deterministic by hypothesis, we know by Theorem 2 that $tr_{\mathbb{S}_I} : \mathbb{S}_I \to \mathbb{T}(A \times \mathcal{P}(A))$ is the only morphism from $\mathbb{S}_I$ to $\mathbb{T}(A \times \mathcal{P}(A))$. But $rt_{\mathbb{S}}$ is by definition the restriction of $tr_{\mathbb{S}_I}$ to $\mathbb{S}_I \to \mathbb{R}_I$, which is therefore the unique such morphism, as required.

The endofunctor T_{Rt} on $\mathbf{C}$ applies objects $\mathbb{S} = \langle S, \psi \rangle$ to $\check{\mathbb{S}} = \langle \check{S}, \check{\psi} \rangle$, where $\check{S} = \mathcal{P}_{ne}(S)$ and $\check{\psi} : \check{S} \to \mathcal{P}(\check{S})^A$ has transitions $M \xrightarrow{a} \{t \mid \exists s \in M, s \xrightarrow{a} t$ and $I(t) = X\}$ for all $M \subseteq_{ne} S$, $a \in A$ and $X \subseteq A$ such that the set on the right is not empty; on arrows, T_{Rt} maps $f : \mathbb{S} \to \mathbb{S}'$ to $\check{f} = \mathcal{P}_{ne}(f)$.

Proposition 9. T_{Rt} *is well-defined, is a functor and its image is contained in* $\mathbf{D}_{Rt}$.

The natural transformation $\eta_{Rt} : U \to U T_{Rt}$ is defined as for traces by $\eta_{Rt\mathbb{S}}(s) = \{s\}$, for all transition systems $\mathbb{S}$ and all states s of $\mathbb{S}$.

Proposition 10. $\eta = \eta_{Rt}$ *is a natural transformation and* $\eta_{\mathbb{S}} : S \to \check{S}$ *is a morphism whenever $\mathbb{S}$ is in* $\mathbf{D}_{Rt}$.

Proof. The proof is similar to the proof of Proposition 6.

Finally, we arrive at our main result concerning ready-traces.

Theorem 5. *For any transition system* $\mathbb{S}$, $rt_{\mathbb{S}} = \mathrm{beh}_{\mathbb{S}}^{\mathbb{R}}$, *the unique* T_{Rt}-*morphism from* $\mathbb{S}$ *to* $\mathbb{R}$.

Proof. Like in the proof of Theorem 3, we conclude that we need to show that $rt_{\mathbb{S}}(s) = rt_{\check{\mathbb{S}}}(\{s\})$ for all $s \in S$. We calculate: $rt_{\check{\mathbb{S}}}(\{s\}) = tr_{\check{\mathbb{S}}_I}(\{s\}) = tr_{\overline{\mathbb{S}_I}}(\{s\}) = tr_{\mathbb{S}_I}(s) = rt_{\mathbb{S}}(s)$, where the first and the last equalities are by definition of ready-trace, the second equality because $\check{\mathbb{S}}_I = \overline{\mathbb{S}_I}$ as noted in the proof of Proposition 9, and the third equality by Theorem 3.

5   Failures of Transition Systems

5.1   Failures and Failure-Sets

In this section a pair $(x, X) \in A^* \times \mathcal{P}(A)$ will be called a *failure*; the set A will be fixed throughout, as before. Given a transition system $\mathbb{S} = \langle S, \psi \rangle$ and $s \in S$, a pair $(x, X) \in A^* \times \mathcal{P}(A)$ is a *failure of* s if there exists t such that $s \xrightarrow{x} t$ and $I(t) \cap X = \emptyset$; we shall call x the *trace* and X the *refusal* of the failure. Clearly, $(x, \emptyset)$ is a failure of s iff $s \xrightarrow{x} t$ for some t, so the traces are in bijective correspondence with the failures with empty refusal. The set of failures of s will be written $fl_{\mathbb{S}}(s)$.

Proposition 11. *If $f : \mathbb{S}_1 \to \mathbb{S}_2$ is a morphism of transition systems and s is a state of $\mathbb{S}_1$, then s and $f(s)$ have the same failures.*

A *failure-set* over A is any set $F \subseteq A^* \times \mathcal{P}(A)$ such that the following conditions hold:

F1 $(\varepsilon, \emptyset) \in F$.
F2 $(\varepsilon, X) \in F \Rightarrow \forall a \in X, (a, \emptyset) \notin F$.
F3 $(xy, X) \in F \Rightarrow (x, \emptyset) \in F$.
F4 $(x, X) \in F \wedge Y \subseteq X \Rightarrow (x, Y) \in F$.
F5 $(x, X) \in F \wedge \forall a \in Y, (xa, \emptyset) \notin F \Rightarrow (x, X \cup Y) \in F$.

Proposition 12. *For any state s of a transition system $\mathbb{S}$, $fl_{\mathbb{S}}(s)$ is a failure-set.*

Let Fl be the set of all failure-sets. Our immediate goal is to define a transition relation in Fl as we did for traces and ready-traces; we start by defining transitions labeled by $A \times \mathcal{P}(A)$ and then drop the second component of the label to get transitions labeled by A. Given $F \in Fl$ and $x \in A^*$,

$$C_F(x) = \{a \in A \mid (xa, \emptyset) \in F\}$$

is the set of *continuations* of x in F; if F has the form $fl_{\mathbb{S}}(s)$, we abbreviate $C_{fl_{\mathbb{S}}(s)}(x)$ to $C_s(x)$; clearly, $C_s(x) = \{a \in A \mid s \xrightarrow{xa}\}$, since the conditions $(xa, \emptyset) \in fl_{\mathbb{S}}(s)$ and $s \xrightarrow{xa}$ are equivalent; note that $C_s(\varepsilon) = I(s)$. Conditions **F2** and **F5** of the definition of failure-set can be rewritten in a form that is sometimes useful by using sets of continuations $C_F(x)$:

F2' $(\varepsilon, X) \in F \Rightarrow X \cap C_F(\varepsilon) = \emptyset$.
F5' $(x, X) \in F \wedge Y \cap C_F(x) = \emptyset \Rightarrow (x, X \cup Y) \in F$.

The *primary failures* of F are the failures $(x, X) \in F$ for which $X \subseteq C_F(x)$.

Proposition 13. *Let $F \in Fl$.*

1. *Every $(x, \emptyset)$ in F is a primary failure.*
2. *If (ε, X) is a primary failure, then $X = \emptyset$.*
3. *If (x, X) is a primary failure and $Y \subseteq X$, then (x, Y) is a primary failure.*
4. *$(x, X) \in F$ iff $(x, X \cap C_F(x)) \in F$.*
5. *If F and F' have the same primary failures, then $F = F'$.*

Given $F \in Fl$ and a primary failure $(a, X) \in F$, put

$$F \xrightarrow{a, X} F'$$

where

$$F' = \{(\varepsilon, Y) \mid Y \cap (C_F(a) - X) = \emptyset\}$$
$$\cup$$
$$\{(bx, Y) \mid (abx, Y) \in F, b \notin X\}.$$

Proposition 14. *F' in the previous definition is a failure-set.*

The relations $\xrightarrow{a, X}$ turn Fl into a deterministic transition system with label set $A \times \mathcal{P}(A)$. To obtain a transition system with label set A put $F \xrightarrow{a} F'$ if $F \xrightarrow{a, X} F'$ for some X. This defines a dynamics $\zeta_{Fl} : Fl \to \mathcal{P}(Fl)^A$ and consequently a transition system $\mathbb{F} = \langle Fl, \zeta_{Fl} \rangle$.

Lemma 2. *Let F be a failure-set.*

1. $I(F) = C_F(\varepsilon)$.
2. *If* $F \xrightarrow{a,X} F'$, *then* $I(F') = C_F(a) - X$.

Proof. For the first statement, it is easy to see that $F \xrightarrow{a}$ iff $(a, \emptyset) \in F$ for all $a \in A$, so $I(F) = \{a \in A \mid (a, \emptyset) \in F\} = C_F(\varepsilon)$. For the second statement, if $F \xrightarrow{a,X} F'$, then $I(F') = \{b \in A \mid (b, \emptyset) \in F'\} = \{b \in A \mid (ab, \emptyset) \in F, b \notin X\} = C_F(a) - X$.

Proposition 15. *For all* $F \in Fl$, $F = fl_{\mathbb{F}}(F)$.

This result allows to prove simple properties by going back and forth between failure-sets F and their sets of failures $fl_{\mathbb{F}}(F)$ as convenient. An example with an obvious proof is: If $F \xrightarrow{x} F'$ and $(y, Y) \in F'$, then $(xy, Y) \in F$.

5.2 Failure-Systems

A *failure-system* is a transition system $\mathbb{S} = \langle S, \psi \rangle$ such that:

FS1 if $s \xrightarrow{a} t_1$, $s \xrightarrow{a} t_2$ and $I(t_1) = I(t_2)$, then $t_1 = t_2$;
FS2 if $s \xrightarrow{a} t$ and $I(t) \subseteq J \subseteq C_s(a)$, then $s \xrightarrow{a} t'$ and $I(t') = J$ for some t';
FS3 if $s \xrightarrow{a} t_i \xrightarrow{b} u_i$ $(i = 1, 2)$, then $t_i \xrightarrow{b} u_{3-i}$ $(1 = 1, 2)$.

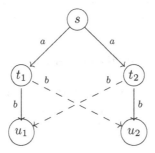

We shall write $s \xrightarrow{a,X} t$ if $s \xrightarrow{a} t$ and $X = C_s(a) - I(t)$, and $s \xrightarrow{a,X}$ if $s \xrightarrow{a,X} t$ for some t. The relations $\xrightarrow{a,X}$ are deterministic (functional), because $s \xrightarrow{a,X} t_1$ and $s \xrightarrow{a,X} t_2$ imply $I(t_1) = C_s(a) - X = I(t_2)$, hence $t_1 = t_2$, by **FS1**. Let $\mathbf{D}_{Fl}$ be the full subcategory of $\mathbf{C}$ of failure-systems.

Proposition 16. *The transition system* $\mathbb{F} = \langle Fl, \zeta_{Fl} \rangle$ *is a failure-system.*

Theorem 6. *The failure-system* $\mathbb{F} = \langle Fl, \zeta_{Fl} \rangle$ *is final in* $\mathbf{D}_{Fl}$. *For any* $\mathbb{S} = \langle S, \psi \rangle$ *in* $\mathbf{D}_{Fl}$, *the unique morphism from* $\mathbb{S}$ *to* $\mathbb{F}$ *is* $fl_{\mathbb{S}} : S \to Fl$.

Proof. Assuming that $fl_{\mathbb{S}}$ is a morphism, to prove uniqueness suppose there was another morphism $f : S \to Fl$. By Proposition 11, $fl_{\mathbb{S}}(s) = fl_{\mathbb{S}}(f(s))$ for every $s \in S$. But by Proposition 15, $fl_{\mathbb{S}}(F) = F$ for every failure-set F, so $fl_{\mathbb{S}}(f(s)) = f(s)$. Therefore, $fl_{\mathbb{S}}(s) = f(s)$ for all $s \in S$, that is, $f = fl_{\mathbb{S}}$.

To show that $fl_\mathbb{S}$ is a morphism we prove two statements: (i) if $s \xrightarrow{a,X} t$, then $fl_\mathbb{S}(s) \xrightarrow{a,X} fl_\mathbb{S}(t)$; (ii) if $fl_\mathbb{S}(s) \xrightarrow{a,X} F$, there is t such that $s \xrightarrow{a,X} t$ and $fl_\mathbb{S}(t) = F$. Assuming the first statement, to prove the second one we only need to show that $fl_\mathbb{S}(s) \xrightarrow{a,X}$ implies $s \xrightarrow{a,X}$, since $\xrightarrow{a,X}$ is deterministic in both systems; to prove this last condition note that $fl_\mathbb{S}(s) \xrightarrow{a,X}$ implies by definition that (a, X) is a primary failure of s, so there is t such that $s \xrightarrow{a} t$ and $I(t) \cap X = \emptyset$; since (a, X) is a primary failure, $X \subseteq C_s(a) - I(t)$, so that $I(t) \subseteq C_s(a) - X \subseteq C_s(a)$; by **FS2**, there is t' such that $s \xrightarrow{a} t'$ and $I(t') = C_s(a) - X$, hence $s \xrightarrow{a,X} t'$.

To prove statement (i) above note that $s \xrightarrow{a,X} t$ implies by definition that (a, X) is a primary failure of s, so $fl_\mathbb{S}(s) \xrightarrow{a,X} F$ for some F; we show that $F = fl_\mathbb{S}(t)$; specifically, we show by induction on the length of y that $(y, Y) \in F$ iff $(y, Y) \in fl_\mathbb{S}(t)$ for all Y. By Proposition 13, we may assume that (y, Y) is a primary failure.

Suppose first that $(y, Y) \in F$. If $y = \varepsilon$, then $Y = \emptyset$, and we have immediately $(\varepsilon, \emptyset) \in fl_\mathbb{S}(t)$. If $y = bx$, then $(abx, Y) \in fl_\mathbb{S}(s)$ and $b \notin X$, by definition of the transition $fl_\mathbb{S}(s) \xrightarrow{a,X} F$. By definition of failure of s, there exist u, v and w such that $s \xrightarrow{a} u \xrightarrow{b} v \xrightarrow{x} w$ and $I(w) \cap Y = \emptyset$. But $s \xrightarrow{a,X} t$ and $b \in C_s(a) - X = I(t)$ imply $t \xrightarrow{b}$. By **FS3**, $t \xrightarrow{b} v$, so $(bx, Y) \in fl_\mathbb{S}(t)$, again by definition of failure.

Conversely, suppose $(y, Y) \in fl_\mathbb{S}(t)$. If $y = \varepsilon$, we conclude as before that $(y, Y) \in F$. If $y = bx$, then $s \xrightarrow{a,X} t \xrightarrow{bx} u$ for some u such that $I(u) \cap Y = \emptyset$, hence $(abx, Y) \in fl_\mathbb{S}(s)$. But $b \notin X$, since $X = C_s(a) - I(t)$, so $(bx, Y) \in F$. This ends the proof.

We shall need later to consider the quotient of a transition system by the bisimilarity relation. For future reference, we prove here that the quotient of a failure-system is again a failure-system. Recall that the quotient of a transition system $\mathbb{S} = \langle S, \psi \rangle$ by a bisimulation equivalence R is the system $\mathbb{S}/R = \langle S/R, \psi/R \rangle$ where S/R is the set of equivalence classes $[s]_R$ for $s \in S$ and ψ/R is given by the transitions $[s]_R \xrightarrow{a} [t]_R$ such that $s \xrightarrow{a} t$ in $\mathbb{S}$. It is easy to see that $I([s]_R) = I(s)$ and $C_{[s]_R}(x) = C_s(x)$ for all s and x.

Proposition 17. *If $\mathbb{S}$ is a failure-system and R is a bisimulation equivalence, $\mathbb{S}/R$ is a failure-system.*

Proof. The axioms of failure-systems are easy to check. Consider **FS2**, for example. Suppose $[s]_R \xrightarrow{a} [t]_R$ and $I([s]_R) \subseteq J \subseteq C_{[s]_R}(a)$. We may assume without loss of generality that $s \xrightarrow{a} t$, and we also have $I(s) \subseteq J \subseteq C_s(a)$. By **FS2** applied to $\mathbb{S}$, there is t' such that $s \xrightarrow{a} t'$ and $I(t') = J$. But then $[s]_R \xrightarrow{a} [t']_R$ and $I([t']_R) = J$.

When R is the bisimilarity relation $\sim$, we write simply $[s]$ instead of $[s]_\sim$; we also abbreviate $S/\sim$ to $\tilde{S}$, and similarly for $\mathbb{S}$ and ψ. The assignment $\mathbb{S} \mapsto \tilde{\mathbb{S}}$ is the object function of a functor that maps a morphism $f : \mathbb{S}_1 \to \mathbb{S}_2$ to the function $\tilde{f} : \tilde{\mathbb{S}}_1 \to \tilde{\mathbb{S}}_2$ such that $\tilde{f}([s]) = [f(s)]$.

5.3   From Transition Systems to Failure-Systems

We now define an endofunctor T_{Fl} on **C**, as we did for traces and ready-traces. The functor T_{Fl} is the composition of a functor F_{Fl} with the functor that quotients with respect to the bisimilarity relation. On objects, F_{Fl} associates a failure-system $\hat{\mathbb{S}} = \langle \hat{S}, \hat{\psi} \rangle$ with a transition system $\mathbb{S} = \langle S, \psi \rangle$ as follows. We first extend the notation $I(-)$ to subsets of S; specifically, put $I(M) = \bigcup_{s \in M} I(s)$ for any $M \subseteq S$. We can now define

$$\hat{S} = \{(M, J) \mid M \subseteq_{\text{ne}} S, I(s) \subseteq J \subseteq I(M) \text{ for some } s \in M\}.$$

Note that $(M, I(M)) \in \hat{S}$ for every $M \subseteq_{\text{ne}} S$, and $(M, K) \in \hat{S}$ whenever $(M, I) \in \hat{S}$ and $I \subseteq K \subseteq I(M)$. Given $(M, J), (N, K) \in \hat{S}$ and $a \in J$, put

$$(M, J) \overset{a}{\to} (N, K)$$

iff $M \overset{a}{\to} N$ in $\bar{\mathbb{S}} = T_{Tr}\mathbb{S}$, that is, $N = \{s' \mid s \overset{a}{\to} s' \text{ for some } s \in M\}$; this defines $\hat{\psi}$. Note that if we also have $(M, J) \overset{a}{\to} (N', K')$, then $N = N'$, since $\bar{\mathbb{S}}$ is deterministic; and if $(M, J) \in \hat{S}$, $a \in J$ and $M \overset{a}{\to} N$, then $(M, J) \overset{a}{\to} (N, I(N))$. This implies that $I(M, J) = J$ and $C_{(M,J)}(a) = I(N)$.

Lemma 3. $\hat{\mathbb{S}} = \langle \hat{S}, \hat{\psi} \rangle$ *is a failure-system.*

To conclude the definition of F_{Fl} we describe its effect on arrows $f : \mathbb{S}_1 \to \mathbb{S}_2$. Specifically, put $F_{Fl}(f) = \hat{f}$, where $\hat{f} : \hat{\mathbb{S}}_1 \to \hat{\mathbb{S}}_2$ is given by $\hat{f}(M, J) = (N, J)$, with $N = \bar{f}(M) = \{f(s) \mid s \in m\}$.

Proposition 18. F_{Fl} *is well-defined, is a functor and its image is contained in* $\mathbf{D}_{Fl}$.

Proof. To show that F_{Fl} is well-defined and a functor it is enough to check that $\hat{f}$ is a morphism of transition systems, since it is then obvious that F_{Fl} preserves identity morphisms and the composition of morphisms. The first step is to check that $\hat{f}(M, J) = (\bar{f}(M), J)$ is indeed in $\hat{S}_2$. We must show that $\bar{f}(M) \neq \emptyset$ and $I(f(s)) \subseteq J \subseteq I(\bar{f}(M))$ for some $s \in M$; this is so because $M \neq \emptyset$, $I(s) \subseteq J \subseteq I(M)$ for some $s \in M$, $I(f(s)) = I(s)$ and $I(\bar{f}(M)) = I(M)$. Next note that if $(M, J) \overset{a}{\to} (N, K)$, then $(\bar{f}(M), J) \overset{a}{\to} (\bar{f}(N), K)$, because $a \in J$ and $\bar{f}(M) \overset{a}{\to} \bar{f}(N)$, by Proposition 5. By the same reason, if $(\bar{f}(M), J) \overset{a}{\to} (N', K)$, then $N' = \bar{f}(N)$ such that $M \overset{a}{\to} N$, and it is easy to see that $(M, J) \overset{a}{\to} (N, K)$. Finally, the image of T_{Fl} is contained in $\mathbf{D}_{Fl}$ by Lemma 3.

We may now define T_{Fl} as the functor that takes quotients with respect to the bisimilarities composed with F_{Fl}. The following result is an immediate consequence of Propositions 17 and 18.

Proposition 19. *The image of* T_{Fl} *is contained in* $\mathbf{D}_{Fl}$.

In the sequel we shall write $T_{Fl}\mathbb{S} = \mathbb{S}^\dagger = \langle S^\dagger, \psi^\dagger \rangle$ and $T_{Fl}f = f^\dagger$ for any morphism f, thus abbreviating $\hat{\mathbb{S}}$ to $\mathbb{S}^\dagger$ and similarly to S, ψ and f. We shall also abbreviate $[(\{s\}, I(s))] \in S^\dagger$ to $\langle s \rangle$ for every $s \in S$. Note that if f is any morphism on $\mathbb{S}$, then $f^\dagger(\langle s \rangle) = [(\{f(s)\}, I(s))] = [(\{f(s)\}, I(f(s)))] = \langle f(s) \rangle$.

The following result is the reason for using T_{Fl} rather than F_{Fl}.

Lemma 4. *If $\mathbb{S}$ is a failure-system, the assignment $s \mapsto \langle s \rangle$ is a morphism from $\mathbb{S}$ to $\mathbb{S}^\dagger$.*

Proof. We first prove a preliminary result: If $M = \{t' \mid s \xrightarrow{a} t'\}$ and $t \in M$, then $\langle t \rangle = [(M, I(t))]$. To prove that $(\{t\}, I(t)) \sim (M, I(t))$ we show more specifically that $(\{t\}, I(t)) \xrightarrow{b} (N, K)$ iff $(M, I(t)) \xrightarrow{b} (N, K)$ for all b and all (N, K). Writing $(M, I(t)) \xrightarrow{b} (N', K')$, it is enough to show that $N = N'$. We have $N = \{u \mid t \xrightarrow{b} u\}$ and $N' = \{u' \mid \exists t' \in M, t' \xrightarrow{b} u'\}$, and it is clear that $N \subseteq N'$ since $t \in M$. In the other direction, if $u' \in N'$, we have $s \xrightarrow{a} t' \xrightarrow{b} u'$ for some $t' \in M$; but $b \in I(t)$, so $s \xrightarrow{a} t \xrightarrow{b} u$ for some u; by **FS3**, $t \xrightarrow{b} u'$, so $u' \in N$.

Now consider $s \xrightarrow{a} t$ in $\mathbb{S}$. Then $\langle s \rangle \xrightarrow{a} [(M, I(t))]$, where $t \in M = \{t' \mid s \xrightarrow{a} t'\}$. By the auxiliary result, $\langle t \rangle = [(M, I(t))]$, hence $\langle s \rangle \xrightarrow{a} \langle t \rangle$. On the other hand, suppose $\langle s \rangle \xrightarrow{a} [(M, J)]$; we must find t such that $s \xrightarrow{a} t$ and $\langle t \rangle = [(M, J)]$. We have $M = \{t' \mid s \xrightarrow{a} t'\}$ and $I(t') \subseteq J \subseteq I(M)$ for some $t' \in M$. Since $s \xrightarrow{a} t'$ and $C_s(a) = I(M)$, there is t such that $s \xrightarrow{a} t$ and $I(t) = J$, by **FS2**. By the auxiliary result, $\langle t \rangle = [(M, I(t))] = [(M, J)]$.

The natural transformation $\eta_{Fl} : U \to UT_{Fl}$ is defined by $\eta_{Fl\mathbb{S}}(s) = \langle s \rangle$ for all transition systems $\mathbb{S}$ and states s of $\mathbb{S}$.

Proposition 20. *$\eta = \eta_{Fl}$ is a natural transformation and $\eta_\mathbb{S} : \mathbb{S} \to \mathbb{S}^\dagger$ is a morphism whenever $\mathbb{S}$ is in $\mathbf{D}_{Tr}$.*

Proof. The structure of the proof is similar to that of the proof of Proposition 6. Let $f : \mathbb{S}_1 \to \mathbb{S}_2$ be a morphism of transition systems. We have $\eta_{\mathbb{S}_2} \circ f = f^\dagger \circ \eta_{\mathbb{S}_1}$, since both sides apply any $s \in S_1$ to $\langle f(s) \rangle$; thus, η is a natural transformation. If $\mathbb{S}$ is a failure-system, that $\eta_\mathbb{S}$ is a morphism is the statement in Lemma 4.

Lemma 5. *Let $\mathbb{S}$ be a transition system. Then $fl_\mathbb{S}(s) = fl_{S^\dagger}(\langle s \rangle)$ for all $s \in S$.*

Theorem 7. *For any transition system $\mathbb{S}$, $fl_\mathbb{S} = beh_\mathbb{S}^\mathbb{F}$, the unique T_{Fl}-morphism from $\mathbb{S}$ to $\mathbb{F}$.*

Proof. This result is similar to Theorems 3 and 5, as is its proof, using Lemma 5.

6   Comparing the Behaviour Equivalences

We now compare the behaviour equivalences associated with traces, ready-traces and failures.

Proposition 21. *The behaviour equivalences for traces, ready-traces and fail-ures satisfy* $\overset{\mathbb{R}}{=}_{\mathbb{S}} \subseteq \overset{\mathbb{F}}{=}_{\mathbb{S}} \subseteq \overset{\mathbb{T}}{=}_{\mathbb{S}}$ *for any transition system* $\mathbb{S}$.

Proof. For every $s \in S$, it is easy to see that s and $\eta_{Fl\mathbb{S}}(s) = \langle s \rangle$ have the same traces, and s and $\eta_{Rt\mathbb{S}}(s) = \{s\}$ have the same failures. The conclusion follows from Proposition 2.

7   Concluding Remarks

It is our belief that the coalgebraic reconstruction of the behaviours in the linear time – branching time spectrum is an important contribution to our under-standing of those behaviours. In the process of doing so we had to enlarge our understanding of the notion of behaviour of a coalgebra (system), and we wish to explore this extended notion of behaviour in other situations. Some ways to continue the work reported herein are to try to extend the outlined approach to the remaining cases in the spectrum, to take into account systems with τ-transitions, to work with categories based on other categories than **Set**, like categories of algebras or presheaf categories, and to relate our approach with the work in [3, 4, 5].

Acknowledgements. I am grateful to the anonymous referees for their sug-gestions, some of which have not been considered in this paper by lack of space.

References

[1] Rutten, J.: Universal coalgebra: a theory of systems. Theoretical Computer Sci-ence 249(1), 3–80 (2000)

[2] van Glabbeek, R.: The linear time–branching time spectrum I: the semantics of con-crete, sequential processes. In: Bergstra, J., Ponse, A., Smolka, S. (eds.) Handbook of process algebra, pp. 3–99. Elsevier, Amsterdam (2001)

[3] Power, J., Turi, D.: A coalgebraic foundation for linear time semantics. In: Hofmann, M., Rosolini, G., Pavlovic, D. (eds.) CTCS 1999, Conference on Cat-egory Theory and Computer Science. Electronic Notes in Theoretical Computer Science, vol. 29, pp. 259–274. Elsevier, Amsterdam (1999)

[4] Jacobs, B.: Trace semantics for coalgebras. In: Adamek, J., Milius, S. (eds.) Coal-gebraic Methods in Computer Science. Electronic Notes in Theoretical Computer Science, vol. 106, pp. 167–184. Elsevier, Amsterdam (2004)

[5] Hasuo, I., Jacobs, B., Sokolova, A.: Generic trace semantics via coinduction. Logical Methods in Computer Science 3(4:11), 1–36 (2007)

[6] Monteiro, L.: A coalgebraic characterization of behaviours in the linear time – branching time spectrum. Technical Report UNL-DI 4-2008, Universidade Nova de Lisboa (2008), http://ctp.di.fct.unl.pt/~lm/publications/tr4-2008.pdf

Heterogeneous Logical Environments for Distributed Specifications*

Till Mossakowski[1,2] and Andrzej Tarlecki[3,4]

[1] Institute for Computer Science, Albert-Ludwigs-Universität Freiburg
[2] Safe and Secure Cognitive Systems, DFKI GmbH, Bremen
[3] Institute of Informatics, University of Warsaw
[4] Institute of Computer Science, Polish Academy of Sciences

Abstract. We use the theory of institutions to capture the concept of a heterogeneous logical environment as a number of institutions linked by institution morphisms and comorphisms. We discuss heterogeneous specifications built in such environments, with inter-institutional specification morphisms based on both institution morphisms and comorphisms. We distinguish three kinds of heterogeneity: (1) specifications in logical environments with universal logic (2) heterogeneous specifications focused at a particular logic, and (3) heterogeneous specifications distributed over a number of logics.

1   Introduction

The theory of *institutions* [GB92] provides an excellent framework where the theory of specification and formal software development may be presented in an adequately general and abstract way [ST88a, ST97, Tar03]. The initial work within this area captured specifications built and developments carried out in an arbitrary but fixed logical system formalised as an institution. However, the practice of software specification and development goes much beyond this. Different logical systems may be appropriate or most convenient for specification of different modules of the same system, of different aspects of system behaviour, or of different stages of system development. This leads to the need for a number of logical systems to be used in the same specification and development project, linked by appropriate notions of morphisms between institutions [GR02]. This observation spurred a substantial amount of research work already, and motivates the research presented here.

In such a framework, one works in a *heterogeneous logical environment* formed by a number of logical systems formalised as institutions and linked with each other in a way captured by various maps between institutions. One such logical environment is the HETS family of institutions [Mos05], supported by a tool to build and work with heterogeneous specifications [MML07].

---

* This work has been partially supported by European projects IST-2005-015905 MO-BIUS and IST-2005-016004 SENSORIA, by a visiting grant to the University of Illinois at Urbana-Champaign (AT) and by the German Federal Ministry of Education and Research (Project 01 IW 07002 FormalSafe) and by the DFG-funded SFB/TR 8 "Spatial cognition" (TM).

A. Corradini and U. Montanari (Eds.): WADT 2008, LNCS 5486, pp. 266–289, 2009.

Given a heterogeneous logical environment, there are several possible ways of using it to build heterogeneous specifications:

1. In some logical environments, we have a single logical system (usually coming with good tool support) that can be used as a *universal logic* into which all other systems are mapped. Then the maps between logics are used for mapping specifications from all logics in the environment into the universal logic, where they can be further combined then as usual. Various logical systems have been proposed and used as such universal logics, including higher-order logic in various versions [NPW02], Edinburgh LF [HHP93], rewriting logic [MOM02], fork algebra [PF06], etc.

2. *Focused heterogeneous specifications* are more liberal: parts of a specification may be written in different logics (also exploiting the availability of specialised tools for these logics). However, these parts ultimately are assembled in one logical system, where the models of interest live. This is made possible by extending the repertoire of *specification-building operations* with ones that move specifications from one logic to another using various maps between logical systems, as perhaps first mentioned in the context similar to what we use here in [ST88b] and further developed in [Tar96, Tar00, Mos03, Dia02].

3. *Distributed heterogeneous specifications* involve a number of specifications in different logical systems, with compatibility links between them given by logic maps, but not necessarily with a single specification in a particular logic providing an overall integration. *Heterogeneous development graphs* [Mos02b] offer a first hint in this direction.

While the first two methodologies have been studied in the literature, the third one seems to have attracted only little attention from the formal specification community so far, although it is clear that in frameworks like UML, distributed heterogeneous specifications arise rather naturally.

In this paper we largely set up a framework for further work, collecting the ideas, concepts and facts put forward earlier at other places (by us and others). No new big results are to be expected at this stage. However, a new overall view of heterogeneous logical environments and distributed specifications in such environment seems to be emerging here.

We introduce a notion of a heterogeneous logical environment, and study to what extent such environments can be made *uniform*, i.e., based on one kind of a mapping between institutions. We discuss various ways of building focused heterogeneous specifications in such environments. Then, given heterogeneous, inter-institutional (co)morphisms between such specifications and specification categories they define, we introduce distributed specifications as specification diagrams. These come with a natural notion of a distributed model, and so also other standard concepts like consistency, consequence, implementations, etc. Finally, we show that these concepts apply in the context of heterogeneous specification categories built over any heterogeneous logical environment.

2 Heterogeneous Logical Environments

Let us begin by recalling the notion of an institution, as a formalisation of an arbitrary logical system [GB92], assuming that the reader is familiar with all the intuitions that this notion brings in.

Definition 2.1. *An* institution $\mathcal{I}$ *consists of:*

- *a category* $\mathbf{Sign}_{\mathcal{I}}$ *of signatures;*
- *a functor* $\mathbf{Sen}_{\mathcal{I}} \colon \mathbf{Sign}_{\mathcal{I}} \to \mathbf{Set}$,[1] *giving a set* $\mathbf{Sen}(\Sigma)$ *of* Σ-*sentences for each signature* $\Sigma \in |\mathbf{Sign}_{\mathcal{I}}|$, *and a function* $\mathbf{Sen}(\sigma) \colon \mathbf{Sen}(\Sigma) \to \mathbf{Sen}(\Sigma')$, *denoted by* σ, *that yields* σ-*translation of* Σ-*sentences to* Σ'-*sentences for each signature morphism* $\sigma \colon \Sigma \to \Sigma'$;
- *a functor* $\mathbf{Mod}_{\mathcal{I}} \colon \mathbf{Sign}_{\mathcal{I}}^{op} \to \mathbf{Set}$,[2] *giving a set* $\mathbf{Mod}(\Sigma)$ *of* Σ-*models for each signature* $\Sigma \in |\mathbf{Sign}_{\mathcal{I}}|$, *and a functor* $\mathbf{Mod}(\sigma) \colon \mathbf{Mod}(\Sigma') \to \mathbf{Mod}(\Sigma)$, *denoted by* $\_|_{\sigma}$, *that yields* σ-*reducts of* Σ'-*models for each signature morphism* $\sigma \colon \Sigma \to \Sigma'$; *and*
- *for each* $\Sigma \in |\mathbf{Sign}_{\mathcal{I}}|$, *a satisfaction relation* $\models_{\mathcal{I},\Sigma} \subseteq \mathbf{Mod}_{\mathcal{I}}(\Sigma) \times \mathbf{Sen}_{\mathcal{I}}(\Sigma)$

such that for any signature morphism $\sigma \colon \Sigma \to \Sigma'$, Σ-*sentence* $\varphi \in \mathbf{Sen}_{\mathcal{I}}(\Sigma)$ *and* Σ'-*model* $M' \in \mathbf{Mod}_{\mathcal{I}}(\Sigma')$:

$$M' \models_{\mathcal{I},\Sigma'} \sigma(\varphi) \iff M'|_{\sigma} \models_{\mathcal{I},\Sigma} \varphi \qquad [\text{Satisfaction condition}]$$

Whenever convenient, we avoid spelling out the standard notations for institution components, and allow primes, subscripts and superscripts to determine which institution is referred to.

The next concept we need is a mapping between institutions. We concentrate here on *institution morphisms* [GB92] and on *institution comorphisms* (named so in [GR02]; see "plain maps of institutions" in [Mes89] and "institution representations" in [Tar87, Tar96]).

Definition 2.2. *Let* $\mathcal{I}$ *and* $\mathcal{I}'$ *be institutions. An* institution morphism $\mu \colon \mathcal{I} \to \mathcal{I}'$ *consists of:*

- *a functor* $\mu^{Sign} \colon \mathbf{Sign} \to \mathbf{Sign}'$;
- *a natural transformation* $\mu^{Sen} \colon \mu^{Sign} ; \mathbf{Sen}' \to \mathbf{Sen}$,[3] *that is, a family of functions* $\mu_{\Sigma}^{Sen} \colon \mathbf{Sen}'(\mu^{Sign}(\Sigma)) \to \mathbf{Sen}(\Sigma)$, *natural in* $\Sigma \in |\mathbf{Sign}|$; *and*
- *a natural transformation* $\mu^{Mod} \colon \mathbf{Mod} \to (\mu^{Sign})^{op} ; \mathbf{Mod}'$, *that is, a family of functions* $\mu_{\Sigma}^{Mod} \colon \mathbf{Mod}(\Sigma) \to \mathbf{Mod}'(\mu^{Sign}(\Sigma))$, *natural in* $\Sigma \in |\mathbf{Sign}|$,

such that for any signature $\Sigma \in |\mathbf{Sign}|$, *the translations* $\mu_{\Sigma}^{Sen} \colon \mathbf{Sen}'(\rho^{Sign}(\Sigma)) \to \mathbf{Sen}(\Sigma)$ *of sentences and* $\mu_{\Sigma}^{Mod} \colon \mathbf{Mod}(\Sigma) \to \mathbf{Mod}'(\rho^{Sign}(\Sigma))$ *of models preserve the satisfaction relation, i.e., for any* $\varphi' \in \mathbf{Sen}'(\mu^{Sign}(\Sigma))$ *and* $M \in \mathbf{Mod}(\Sigma)$:

---

[1] The category **Set** has all sets as objects and all functions as morphisms.

[2] To keep things simple, we work with the version of institutions where morphisms between models, not needed here, are disregarded. To capture standard examples, we should allow here for the use of classes, rather than just sets of models — but again, we will disregard such foundational subtleties here.

[3] We write composition of morphisms in any category in the diagrammatic order and denote it by ";" (semicolon).

$$M \models_\Sigma \mu_\Sigma^{Sen}(\varphi') \iff \mu_\Sigma^{Mod}(M) \models'_{\mu^{Sign}(\Sigma)} \varphi' \quad [Satisfaction\ condition]$$

Institution morphisms compose in the obvious, component-wise manner. The category of institutions with institution morphisms is denoted by $\mathcal{INS}$.

An institution comorphism $\rho \colon \mathcal{I} \to \mathcal{I}'$ consists of:

- *a functor $\rho^{Sign} \colon \mathbf{Sign} \to \mathbf{Sign}'$;*
- *a natural transformation $\rho^{Sen} \colon \mathbf{Sen} \to \rho^{Sign} \, ; \mathbf{Sen}'$, that is, a family of functions $\rho_\Sigma^{Sen} \colon \mathbf{Sen}(\Sigma) \to \mathbf{Sen}'(\rho^{Sign}(\Sigma))$, natural in $\Sigma \in |\mathbf{Sign}|$; and*
- *a natural transformation $\rho^{Mod} \colon (\rho^{Sign})^{op} \, ; \mathbf{Mod}' \to \mathbf{Mod}$, that is, a family of functions $\rho_\Sigma^{Mod} \colon \mathbf{Mod}'(\rho^{Sign}(\Sigma)) \to \mathbf{Mod}(\Sigma)$, natural in $\Sigma \in |\mathbf{Sign}|$,*

such that for any $\Sigma \in |\mathbf{Sign}|$, the translations $\rho_\Sigma^{Sen} \colon \mathbf{Sen}(\Sigma) \to \mathbf{Sen}'(\rho^{Sign}(\Sigma))$ of sentences and $\rho_\Sigma^{Mod} \colon \mathbf{Mod}'(\rho^{Sign}(\Sigma)) \to \mathbf{Mod}(\Sigma)$ of models preserve the satisfaction relation, i.e., for any $\varphi \in \mathbf{Sen}(\Sigma)$ and $M' \in \mathbf{Mod}'(\rho^{Sign}(\Sigma))$:

$$M' \models'_{\rho^{Sign}(\Sigma)} \rho_\Sigma^{Sen}(\varphi) \iff \rho_\Sigma^{Mod}(M') \models_\Sigma \varphi \quad [Satisfaction\ condition]$$

Institution comorphisms compose in the obvious, component-wise manner. The category of institutions with institution comorphisms is denoted by $co\mathcal{INS}$.

Whenever no confusion may arise, the superscripts identifying the components of an institution morphism will be omitted, so that all components of an institution morphism μ will be written as μ, and similarly for institution comorphisms.

Even though the only essential difference between institution morphisms and comorphisms is in the direction of sentence and model translations w.r.t. signature translation, the intuition they capture is quite different. Very informally, an institution morphism $\mu \colon \mathcal{I} \to \mathcal{I}'$ shows how a "richer" institution $\mathcal{I}$ is "projected" onto a "poorer" institution $\mathcal{I}'$ (by removing some parts of signatures and models of $\mathcal{I}$ to obtain the simpler signatures and models of $\mathcal{I}'$, and by embedding simpler $\mathcal{I}'$-sentences into more powerful $\mathcal{I}$-sentences). Then, an institution comorphism $\mathcal{I} \to \mathcal{I}'$ shows how a "simpler" institution $\mathcal{I}$ is represented in a "more complex" institution $\mathcal{I}'$ (by representing the simpler signatures and sentences of $\mathcal{I}$ as signatures and sentences of $\mathcal{I}'$, and extracting simpler $\mathcal{I}$-models from more complex $\mathcal{I}'$-models).[4]

Given the two possible ways to link institutions with each other, a notion of a *heterogeneous logical environment* may be formalised as a collection of institutions linked by institution morphisms and comorphisms.

Definition 2.3. *A heterogeneous logical environment $\mathcal{HLE}$ is a collection of institutions and of institution morphisms and comorphisms between them, that is, a pair of diagrams $\langle \mathcal{HLE}^\mu \colon \mathcal{G}^\mu \to \mathcal{INS}, \mathcal{HLE}^\rho \colon \mathcal{G}^\rho \to co\mathcal{INS} \rangle$[5] in the category*

---

[4] Variants of comorphisms are also used to encode "more complex" institutions into "simpler" ones: e.g. in [GR02], a so-called simple theoroidal comorphism is used to code first-order logic with equality in first-order logic without equality. See also [MDT09] for discussion of relative strength of logical systems in a similar context.

[5] We assume that $\mathcal{G}^\mu$ is a graph that gives the shape of the diagram; its nodes $n \in |\mathcal{G}^\mu|$ carry institutions $\mathcal{HLE}^\mu(n)$ linked by institution morphisms $\mathcal{HLE}^\mu(e) \colon \mathcal{HLE}^\mu(n) \to \mathcal{HLE}^\mu(m)$ for each edge $e \colon n \to m$ in $\mathcal{G}$. Similar notation is used for diagrams in other categories.

$\mathcal{INS}$ of institutions and their morphisms and co$\mathcal{INS}$ of institutions and their comorphisms, respectively, such that the two underlying graphs have no common edges and diagrams coincide on common nodes, i.e., for all nodes $n \in |\mathcal{G}^\mu| \cap |\mathcal{G}^\rho|$, $\mathcal{HLE}^\mu(n) = \mathcal{HLE}^\rho(n)$.

We write $\mathcal{G}$ for the union of $\mathcal{G}^\mu$ and $\mathcal{G}^\rho$, and w.l.o.g. assume that all the nodes of the underlying graphs are common, $|\mathcal{G}| = |\mathcal{G}^\mu| = |\mathcal{G}^\rho|$.

Such a heterogeneous logical environment is morphism-uniform if $\mathcal{G}^\rho$ is discrete (has no edges); we can then identify $\mathcal{HLE}$ with $\mathcal{HLE}^\mu \colon \mathcal{G} \to \mathcal{INS}$. Similarly, $\mathcal{HLE}$ is comorphism-uniform if $\mathcal{G}^\mu$ is discrete; we can identify it then with $\mathcal{HLE}^\rho \colon \mathcal{G} \to \mathcal{INS}$.

The lack of uniformity in linking institutions in heterogeneous environments (we use both institution morphisms and comorphisms here; other kinds of maps between institutions may be considered as well) may be somewhat surprising and certainly is technically cumbersome. Morphism-uniform environments, where only institution morphisms are used, are conveniently captured by a single diagram in the appropriate institution category, and similarly for comorphism-uniform environments. These concepts coincide with what was studied in the literature as *indexed institutions* [Dia02] and *indexed coinstitutions* [Mos02a].

One way to make logical environments uniform is by noticing that in fact a link of each kind may be captured by links of the other kind, albeit in general a span of those may be needed. This has been noticed already in [Mos03] and spelled out in [Mos05]; see also [MW98] for similar ideas with institution forward morphisms (called transformations there) as a primary notion.

Definition 2.4. *Consider an institution morphism* $\mu \colon \mathcal{I} \to \mathcal{I}'$. *We build an "intermediate institution" by re-indexing* $\mathcal{I}'$ *using the signature translation:* $\mathcal{I}'_0 = \langle \mathbf{Sign}, \mu^{Sign} \, ; \mathbf{Sen}', \mu^{Sign} \, ; \mathbf{Mod}', \langle \models'_{\mu^{Sign}(\Sigma)} \rangle_{\Sigma \in |\mathbf{Sign}|} \rangle$. *Two comorphisms emerge then*[6]: $\rho_{\mu,1} = \langle id, \mu^{Sen}, \mu^{Mod} \rangle \colon \mathcal{I}'_0 \to \mathcal{I}$ *and* $\rho_{\mu,2} = \langle \mu^{Sign}, id, id \rangle \colon \mathcal{I}'_0 \to \mathcal{I}'$. *The comorphism span for* μ, *written* $span(\mu)$, *is the following span of institution comorphisms:* $\mathcal{I} \xleftarrow{\rho_{\mu,1}} \mathcal{I}'_0 \xrightarrow{\rho_{\mu,2}} \mathcal{I}'$.

Consider an institution comorphism $\rho \colon \mathcal{I} \to \mathcal{I}'$. *We build an "intermediate institution":* $\mathcal{I}'_0 = \langle \mathbf{Sign}, \rho^{Sign} \, ; \mathbf{Sen}', \rho^{Sign} \, ; \mathbf{Mod}', \langle \models'_{\rho^{Sign}(\Sigma)} \rangle_{\Sigma \in |\mathbf{Sign}|} \rangle$. *Two institution morphisms emerge then:* $\mu_{\rho,1} = \langle id, \rho^{Sen}, \rho^{Mod} \rangle \colon \mathcal{I}'_0 \to \mathcal{I}$ *and* $\mu_{\rho,2} = \langle \mu^{Sign}, id, id \rangle \colon \mathcal{I}'_0 \to \mathcal{I}'$. *The morphism span for* ρ, *written* $span(\rho)$, *is the following span of institution morphisms:* $\mathcal{I} \xleftarrow{\mu_{\rho,1}} \mathcal{I}'_0 \xrightarrow{\mu_{\rho,2}} \mathcal{I}'$.

Informally, the span of comorphisms $span(\mu)$ captures exactly the same relationship between the components of $\mathcal{I}$ and $\mathcal{I}'$ as the original institution morphism $\mu \colon \mathcal{I} \to \mathcal{I}'$; and similarly, the span of morphisms $span(\rho)$ captures exactly the same relationship between the components of $\mathcal{I}$ and $\mathcal{I}'$ as the original institution comorphism $\rho \colon \mathcal{I} \to \mathcal{I}'$.

---

[6] We write id for identities in any category, in particular, here for the identity functor as well as the identity natural transformations.

This essentially allows us to concentrate on heterogeneous logical environments that are uniform in the sense that only institution morphisms (or comorphisms) are used to link with each other the institutions involved.

Definition 2.5. *Let* $\mathcal{HLE} = \langle \mathcal{HLE}^\mu \colon \mathcal{G}^\mu \to \mathcal{INS}, \mathcal{HLE}^\rho \colon \mathcal{G}^\rho \to co\mathcal{INS} \rangle$ *be a heterogeneous logical environment. By* $span^\mu(\mathcal{HLE}) \colon span^\mu(\mathcal{G}) \to \mathcal{INS}$ *we denote the morphism-uniform environment obtained from* $\mathcal{HLE}$ *by replacing each comorphism* ρ *in* $\mathcal{HLE}^\rho$ *by* $span(\rho)$. *Similarly, by* $span^\rho(\mathcal{HLE}) \colon span^\rho(\mathcal{G}) \to co\mathcal{INS}$ *we denote the comorphism-uniform environment obtained from* $\mathcal{HLE}$ *by replacing each morphism* μ *in* $\mathcal{HLE}^\mu$ *by* $span(\mu)$.

Note that the "uniformisation" described above typically will change the shape of the graph underlying the heterogeneous logical environment: while building $span^\mu(\mathcal{HLE})$ we remove each edge in $\mathcal{G}^\rho$ adding a corresponding span of edges in $\mathcal{G}^\mu$, with a new node that carries the new "intermediate" institution, and similarly for $span^\rho(\mathcal{HLE})$.

A very rough intuition about various ways of linking institutions by spans and sinks of (co)morphisms is that in a span of comorphisms $\mathcal{I} \xleftarrow{\rho} \mathcal{I}_0 \xrightarrow{\rho'} \mathcal{I}'$ and in a sink of morphisms $\mathcal{I} \xrightarrow{\mu} \mathcal{I}_0 \xleftarrow{\mu'} \mathcal{I}'$, the intermediate institution $\mathcal{I}_0$ captures the common features of $\mathcal{I}$ and $\mathcal{I}'$, and so this relationship may be used to express some "sharing" requirements between models of $\mathcal{I}$ and $\mathcal{I}'$. Dually, in a sink of comorphisms $\mathcal{I} \xrightarrow{\rho} \mathcal{I}_0 \xleftarrow{\rho'} \mathcal{I}'$ and in a span of morphisms $\mathcal{I} \xleftarrow{\mu} \mathcal{I}_0 \xrightarrow{\mu'} \mathcal{I}'$, the intermediate institution $\mathcal{I}_0$ is richer than both $\mathcal{I}$ and $\mathcal{I}'$ and combines the features present in them, and therefore may be used to express some "consistency" properties between models of $\mathcal{I}$ and $\mathcal{I}'$.

3 Specifications and Their Heterogeneous Categories

The original purpose of introducing the notion of institution (under the name of a *language* in [BG80]) was to free the theory of specifications from dependency on any particular logical system. We follow [ST88a] and for an arbitrary institution $\mathcal{I}$ consider a class $Spec_\mathcal{I}$ of specifications built in $\mathcal{I}$ starting from *basic specifications* (*presentations*, which essentially consist of a signature and a set of sentences over this signature) by means of a number of *specifications-building operations*, including *union* of specifications with common signature (written $SP_1 \cup SP_2$), *translation* along a signature morphism (written $\sigma(SP)$), *hiding* (or "*derive*") w.r.t. a signature morphism (written $SP'|_\sigma$), etc. We will not dwell here on the particular choice of these operations, as usual assuming though that specifications come with their basic semantics given in terms of model classes. That is, for each specification $SP \in Spec_\mathcal{I}$, we have its *signature* $Sig[SP] \in |\mathbf{Sign}|$ and its class of *models* $Mod[SP] \subseteq \mathbf{Mod}(Sig[SP])$. In particular, we have $Mod[SP_1 \cup SP_2] = Mod[SP_1] \cap Mod[SP_2]$, $Mod[\sigma(SP)] = \{M' \mid M'|_\sigma \in Mod[SP]\}$, and $Mod[SP'|_\sigma] = \{M'|_\sigma \mid M' \in Mod[SP']\}$. The semantics also determines the obvious notion of specification equivalence: $SP_1 \equiv SP_2$ iff $Sig[SP_1] = Sig[SP_2]$ and $Mod[SP_1] = Mod[SP_2]$.

Working in a heterogeneous logical environment, where we have a number of institutions linked by institution morphisms and comorphisms (no uniformity assumption necessary at this stage), we can enrich the collection of specification-building operations by translation along institution comorphisms and hiding w.r.t. institution morphisms, see [ST88b, Tar96]. Somewhat less naturally, we can also define translation along institution morphisms and hiding w.r.t. institution comorphisms, but the target signature has to be given explicitly then:

Definition 3.1. *Let* $\mu \colon \mathcal{I} \to \mathcal{I}'$ *be an institution morphism. Given a specification* $SP \in Spec_{\mathcal{I}}$, *we write* $SP|_{\mu}$ *for a new specification in* $Spec_{\mathcal{I}'}$ *with the semantics given by* $Sig[SP|_{\mu}] = \mu^{Sign}(Sig[SP])$ *and* $Mod[SP|_{\mu}] = \mu^{Mod}(Mod[SP])$ *(* $= \{\mu^{Mod}_{Sig[SP]}(M) \mid M \in Mod[SP]\}$*).*

Given a specification $SP' \in Spec_{\mathcal{I}'}$ *and signature* $\Sigma \in |\mathbf{Sign}|$ *such that* $\mu^{Sign}(\Sigma) = Sig[SP']$, *we write* $\mu(SP')^{\Sigma}$ *for a new specification in* $Spec_{\mathcal{I}}$ *with the semantics given by* $Sig[\mu(SP')^{\Sigma}] = \Sigma$ *and* $Mod[\mu(SP')^{\Sigma}] = (\mu^{Mod})^{-1}(Mod[SP'])$ *(* $= \{M \in \mathbf{Mod}(\Sigma) \mid \mu^{Mod}_{\Sigma}(M) \in Mod[SP']\}$*).*

Let $\rho \colon \mathcal{I} \to \mathcal{I}'$ *be an institution comorphism. Given a specification* $SP \in Spec_{\mathcal{I}}$, *we write* $\rho(SP)$ *for a new specification in* $Spec_{\mathcal{I}'}$ *with the semantics given by* $Sig[\rho(SP)] = \rho^{Sign}(Sig[SP])$ *and* $Mod[\rho(SP)] = (\rho^{Mod})^{-1}(Mod[SP])$ *(* $= \{M' \in \mathbf{Mod}'(\rho^{Sign}(Sig[SP])) \mid \rho^{Mod}_{Sig[SP]}(M') \in Mod[SP]\}$*).*

Given a specification $SP' \in Spec_{\mathcal{I}'}$ *and signature* $\Sigma \in |\mathbf{Sign}|$ *such that* $\rho^{Sign}(\Sigma) = Sig[SP']$, *we write* $SP'|_{\rho}^{\Sigma}$ *for a new specification in* $Spec_{\mathcal{I}}$ *with the semantics given by* $Sig[SP'|_{\rho}^{\Sigma}] = \Sigma$ *and* $Mod[SP'|_{\rho}^{\Sigma}] = \rho^{Mod}(Mod[SP'])$ *(* $= \{\rho^{Mod}_{\Sigma}(M') \mid M' \in Mod[SP']\}$*).*

These new, inter-institutional specification-building operations may be arbitrarily mixed with other (intra-institutional) operations, yielding heterogeneous specifications. Parts of such specifications may be given in different institutions of the heterogeneous logical environment we work in. However, each such a specification as a whole eventually focuses on a particular institution in this environment, where its overall semantics (signature and the class of models) is given. In essence, viewed from a certain perspective, such *focused heterogeneous specifications* do not differ much from the structured specifications built within a single institution. For instance, the view of a software specification and development process as presented in [ST88b, ST97] directly adapts to the use of such specifications without much (semantic) change. For each institution $\mathcal{I}$, we still denote the class of such heterogeneous specifications focused on $\mathcal{I}$ by $Spec_{\mathcal{I}}$.

The standard notion of a specification morphism carries over to heterogeneous specifications focused at the same institution without any change: a specification morphism between specifications $SP, SP' \in Spec_{\mathcal{I}}$ is a signature morphism $\sigma \colon Sig[SP] \to Sig[SP']$ such that for all models $M' \in Mod[SP']$, $M'|_{\sigma} \in Mod[SP]$. This yields the category $\mathbf{Spec}_{\mathcal{I}}$ of specifications focused on $\mathcal{I}$, with the model-class semantics that extends to the functor $\mathbf{Mod} \colon \mathbf{Spec}_{\mathcal{I}}^{op} \to \mathbf{Set}$.

To generalise this definition to a truly heterogeneous case, with specifications involved focused on different institutions, we first have to appropriately generalise the notion of a signature morphism. Of course, we need some link (given by an institution morphism or comorphism) between the institutions involved.

Definition 3.2. *Consider institutions $\mathcal{I}$ and $\mathcal{I}'$ and signatures $\Sigma \in |\mathbf{Sign}|$ and $\Sigma' \in |\mathbf{Sign}'|$.*

A heterogeneous signature morphism is a pair $\langle \mu, \sigma \rangle \colon \Sigma \to \Sigma'$ that consists of an institution morphism $\mu \colon \mathcal{I}' \to \mathcal{I}$ and a signature morphism $\sigma \colon \Sigma \to \mu^{Sign}(\Sigma')$ in $\mathbf{Sign}$. It induces the heterogeneous reduct $\_|_{\langle \mu, \sigma \rangle} \colon \mathbf{Mod}'(\Sigma') \to \mathbf{Mod}(\Sigma)$ defined as the composition $\mu_{\Sigma'}^{Mod} ; \mathbf{Mod}(\sigma)$, i.e., $M'|_{\langle \mu, \sigma \rangle} = \mu_{\Sigma'}^{Mod}(M')|_\sigma$, for all $M' \in \mathbf{Mod}'(\Sigma')$.

A heterogeneous signature comorphism is a pair $\langle \rho, \sigma' \rangle \colon \Sigma \to \Sigma'$ that consists of an institution comorphism $\rho \colon \mathcal{I} \to \mathcal{I}'$ and a signature morphism $\sigma' \colon \rho^{Sign}(\Sigma) \to \Sigma'$ in $\mathbf{Sign}'$. It induces the heterogeneous reduct $\_|_{\langle \rho, \sigma' \rangle} \colon \mathbf{Mod}'(\Sigma') \to \mathbf{Mod}(\Sigma)$ defined as the composition $\mathbf{Mod}'(\sigma') ; \rho_\Sigma^{Mod}$, i.e., $M'|_{\langle \rho, \sigma' \rangle} = \rho_\Sigma^{Mod}(M'|_{\sigma'})$, for all $M' \in \mathbf{Mod}'(\Sigma')$.

Heterogeneous signature morphisms compose as expected: $\langle \mu_1, \sigma_1 \rangle ; \langle \mu_2, \sigma_2 \rangle = \langle \mu_2 ; \mu_1, \sigma_1 ; \mu_1^{Sign}(\sigma_2) \rangle$. For any morphism-uniform heterogeneous logical environment $\mathcal{HLE}^\mu \colon \mathcal{G}^\mu \to \mathcal{INS}$ this yields the heterogeneous category $\mathbf{Sign}(\mathcal{HLE}^\mu)$ of signatures in institutions in $\mathcal{HLE}^\mu$ with heterogeneous morphisms that involve institution morphisms in $\mathcal{HLE}^\mu$ (and their compositions, and identities). Then model functors extend to $\mathbf{Mod}(\mathcal{HLE}^\mu) \colon \mathbf{Sign}(\mathcal{HLE}^\mu)^{op} \to \mathbf{Set}$ using the reducts defined above.

Heterogeneous signature comorphisms compose as expected: $\langle \rho_1, \sigma_1 \rangle ; \langle \rho_2, \sigma_2 \rangle = \langle \rho_1 ; \rho_2, \rho_2^{Sign}(\sigma_1) ; \sigma_2 \rangle$. For any comorphism-uniform heterogeneous logical environment $\mathcal{HLE}^\rho \colon \mathcal{G}^\rho \to co\mathcal{INS}$ this yields the heterogeneous category $\mathbf{Sign}(\mathcal{HLE}^\rho)$ of signatures in institutions in $\mathcal{HLE}^\rho$ with heterogeneous comorphisms that involve institution comorphisms in $\mathcal{HLE}^\rho$ (and their compositions, and identities). Then model functors extend to $\mathbf{Mod}(\mathcal{HLE}^\rho) \colon \mathbf{Sign}(\mathcal{HLE}^\rho)^{op} \to \mathbf{Set}$ using the reducts defined above.

We stop short here of defining translation of sentences and proving the satisfaction condition. Otherwise though, the above follows the construction of the category of signatures and model reducts in the Grothendieck institution given in [Dia02] for the morphism-uniform case and in [Mos02a] for the comorphism-uniform case (heterogeneous signature (co)morphisms were called Grothendieck signature morphisms there). For full formality, signatures in the heterogeneous categories of signatures defined above should really be written as pairs $\langle \mathcal{I}, \Sigma \rangle$, marking them explicitly with the institution they come from (or even the nodes in the institution diagram) but we continue relying on the reader's good will to decipher the institution from the context.

Note that we retain the overall informal idea that a signature morphism goes from the simpler to more complex signature — hence the contravariance with the use of institution morphisms in the definition of heterogeneous signature morphism. This intuition also dictated the choice of the placement and direction of signature morphism components in heterogeneous signature (co)morphisms.

Note also that the inter-institutional specification-building operations given in Def. 3.1 arise now as hiding w.r.t. and translation along heterogeneous signature morphisms and comorphisms (with the identities as signature morphisms).

In a (non-uniform) heterogeneous logical environment $\mathcal{HLE}$, Def. 3.2 yields two heterogeneous signature categories, one for the morphism-uniform, the other for the comorphism-uniform part of $\mathcal{HLE}$. The two categories share all objects, but have different morphisms. We can put these categories together by formally adding compositions of morphisms of the two kinds involved, modulo the expected identification of morphisms that arise from intra-institutional signature morphisms by adding identity institution (co)morphisms.

Definition 3.3. *Let $\mathcal{HLE} = \langle \mathcal{HLE}^\mu, \mathcal{HLE}^\rho \rangle$ be a heterogeneous logical environment, and consider the disjoint union of the signature categories of all institutions in $\mathcal{HLE}$, which embeds in the obvious way into both $\mathbf{Sign}(\mathcal{HLE}^\mu)$ and $\mathbf{Sign}(\mathcal{HLE}^\rho)$. We write $\mathbf{Sign}(\mathcal{HLE})$ for the heterogeneous category of signatures in $\mathcal{HLE}$ and their generalised heterogeneous morphisms, defined as the pushout (in $\mathbf{Cat}^7$) of the two embedding functors. The model functors extend to $\mathbf{Mod}(\mathcal{HLE})$: $\mathbf{Sign}(\mathcal{HLE})^{op} \to \mathbf{Set}$ using the compositions of reducts from Def. 3.2.*

We again stop short from extending this definition to a complete construction of a *Bi-Grothendieck institution* for $\mathcal{HLE}$ with $\mathbf{Sign}(\mathcal{HLE})$ as its signature category and $\mathbf{Mod}(\mathcal{HLE})$ as its model functor, as spelled out in [Mos03].

Given the above, the definitions of inter-institutional specification morphisms are now obvious (cf. [Dia98] for a similar notion of an extra theory morphism, and the notions of specification morphisms arising in Grothendieck (co)institutions).

Definition 3.4. *Consider a heterogeneous logical environment $\mathcal{HLE}$, institutions $\mathcal{I}$ and $\mathcal{I}'$ in $\mathcal{HLE}$ and specifications $SP \in Spec_\mathcal{I}$ and $SP' \in Spec_{\mathcal{I}'}$.*

A generalised heterogeneous signature morphism $\zeta: Sig[SP] \to Sig[SP'] \in \mathbf{Sign}(\mathcal{HLE})$ is a heterogeneous specification morphism $\zeta: SP \to SP'$ if for all models $M' \in Mod[SP']$, $M'|_\zeta \in Mod[SP]$. Heterogeneous specification morphisms compose, which yields the heterogeneous category $\mathbf{Spec}(\mathcal{HLE})$ of specifications focused on institutions in $\mathcal{HLE}$. The model functions extend to the functor $\mathbf{Mod}(\mathcal{HLE})$: $\mathbf{Spec}(\mathcal{HLE})^{op} \to \mathbf{Set}$, using heterogeneous reducts.

Given a heterogeneous logical environment $\mathcal{HLE} = \langle \mathcal{HLE}^\mu, \mathcal{HLE}^\rho \rangle$, the heterogeneous category of specifications $\mathbf{Spec}(\mathcal{HLE})$ has subcategories $\mathbf{Spec}(\mathcal{HLE}^\mu)$ and $\mathbf{Spec}(\mathcal{HLE}^\rho)$, given by heterogeneous signature morphisms of the form $\langle \mu, \sigma \rangle \in \mathbf{Sign}(\mathcal{HLE}^\mu)$ and, respectively, by heterogeneous signature comorphisms of the form $\langle \rho, \sigma \rangle \in \mathbf{Sign}(\mathcal{HLE}^\rho)$.

The category $\mathbf{Spec}_\mathcal{I}$ of specifications focused on a particular institution in an environment is a subcategory of the heterogeneous category of specifications built in the environment (via the obvious embedding which adds identity institution (co)morphisms).

Directly from the definitions:

---

[7] **Cat** is the (quasi-)category of all categories, as usual.

Lemma 3.5. *Consider institutions $\mathcal{I}$, $\mathcal{I}'$ and specifications $SP \in Spec_{\mathcal{I}}$, $SP' \in Spec_{\mathcal{I}'}$.*

A heterogeneous signature morphism $\langle \mu, \sigma \rangle \colon Sig[SP] \to Sig[SP']$ (so that $\mu \colon \mathcal{I}' \to \mathcal{I}$, $\sigma \colon Sig[SP] \to \mu^{Sign}(Sig[SP'])$) is a heterogeneous specification morphism $\langle \mu, \sigma \rangle \colon SP \to SP'$ if and only if $\sigma \colon SP \to SP'|_{\mu}$ is a specification morphism (in $\mathbf{Spec}_{\mathcal{I}}$).

A heterogeneous signature comorphism $\langle \rho, \sigma' \rangle \colon Sig[SP] \to Sig[SP']$ (so that $\rho \colon \mathcal{I} \to \mathcal{I}'$, $\sigma' \colon \rho^{Sign}(Sig[SP]) \to Sig[SP'])$ is a heterogeneous specification morphism $\langle \rho, \sigma' \rangle \colon SP \to SP'$ if and only if $\sigma' \colon \rho(SP) \to SP'$ is a specification morphism (in $\mathbf{Spec}_{\mathcal{I}'}$).

In all the specification categories we consider, equivalent specifications are isomorphic (identity signature morphisms being isomorphisms between them).

One observation about the heterogeneous categories of specifications as defined above is that they reveal a potential problem with making heterogeneous logical environments uniform. Replacing institution morphisms by spans of comorphisms, and vice-versa, replacing institution comorphisms by spans of morphisms, changes the inter-institutional specification-building operations that are available in the logical environment. Fortunately, in view of symmetry in Def. 3.1, this is not much of a problem:

Lemma 3.6. *Let $\mu : \mathcal{I} \to \mathcal{I}'$ be an institution morphism, and let $span(\mu)$ be $\mathcal{I} \xleftarrow{\rho_{\mu,1}} \mathcal{I}'_0 \xrightarrow{\rho_{\mu,2}} \mathcal{I}'$.*

- $SP|_{\mu} \equiv \rho_{\mu,2}(SP|_{\rho_{\mu,1}}^{\Sigma})$ *for any $SP \in Spec_{\mathcal{I}}$ with $Sig[SP] = \Sigma$.*
- $\mu(SP')^{\Sigma} \equiv \rho_{\mu,1}(SP'|_{\rho_{\mu,2}}^{\Sigma})$ *for any $SP' \in Spec_{\mathcal{I}'}$ with $\mu^{Sign}(\Sigma) = Sig[SP']$.*

Dually then, let $\rho : \mathcal{I} \to \mathcal{I}'$ be an institution comorphism, and let $span(\rho)$ be $\mathcal{I} \xleftarrow{\mu_{\rho,1}} \mathcal{I}'_0 \xrightarrow{\mu_{\rho,2}} \mathcal{I}'$.

- $\rho(SP) \equiv \mu_{\rho,1}(SP)^{\Sigma}|_{\mu_{\rho,2}}$ *for any $SP \in Spec_{\mathcal{I}}$ with $Sig[SP] = \Sigma$.*
- $SP'|_{\rho}^{\Sigma} \equiv \mu_{\rho,2}(SP')^{\Sigma}|_{\mu_{\rho,1}}$ *for any $SP' \in Spec_{\mathcal{I}'}$ with $\rho^{Sign}(\Sigma) = Sig[SP']$.*

Together with Lemma 3.5, this yields a useful characterisation of heterogeneous specification (co)morphisms, see [Mos03]:

Proposition 3.7. *Let $\mu \colon \mathcal{I} \to \mathcal{I}'$ be an institution morphism, where $span(\mu)$ is $\mathcal{I} \xleftarrow{\rho_{\mu,1}} \mathcal{I}'_0 \xrightarrow{\rho_{\mu,2}} \mathcal{I}'$. Consider specifications $SP \in Spec_{\mathcal{I}}$ and $SP' \in Spec_{\mathcal{I}'}$ and heterogeneous signature morphism $\langle \mu, \sigma \rangle \colon Sig[SP'] \to Sig[SP]$. Then $\langle \mu, \sigma \rangle \colon SP' \to SP$ is a specification morphism if and only if $\sigma \colon SP' \to \rho_{\mu,2}(SP|_{\rho_{\mu,1}}^{Sig[SP]})$ is a specification morphism (in $\mathbf{Spec}_{\mathcal{I}'}$).*

Let $\rho \colon \mathcal{I} \to \mathcal{I}'$ be an institution morphism, where $span(\rho)$ is $\mathcal{I} \xleftarrow{\mu_{\rho,1}} \mathcal{I}'_0 \xrightarrow{\mu_{\rho,2}} \mathcal{I}'$. Consider specifications $SP \in Spec_{\mathcal{I}}$ and $SP' \in Spec_{\mathcal{I}'}$ and heterogeneous signature comorphism $\langle \rho, \sigma \rangle \colon Sig[SP] \to Sig[SP']$. Then $\langle \rho, \sigma \rangle \colon SP \to SP'$ is a specification morphism if and only if $\sigma \colon \mu_{\rho,1}(SP)^{Sig[SP]}|_{\mu_{\rho,2}} \to SP'$ is a specification morphism (in $\mathbf{Spec}_{\mathcal{I}'}$).

This proposition is the central motivation for the use of spans. It means that all proof obligations arising in a mixed heterogeneous logical environment can be properly expressed already in a uniform heterogeneous logical environment, using spans (even if one cannot express them directly as a theory morphism between original specifications).

However, specification morphisms that arise from institution morphisms and spans of comorphisms that replace them are quite different, and similarly for comorphisms and spans of morphisms. In particular, given a heterogeneous specification morphism $\langle \mu, \sigma \rangle \colon SP' \to SP$, with an institution morphism $\mu \colon \mathcal{I} \to \mathcal{I}'$, there seems to be no natural way to link specifications SP and SP' by heterogeneous specification comorphisms built over the comorphism span $span(\mu)$ (but see Sect. 5.5 for more on this). Consequently, for a heterogeneous logical environment $\mathcal{HLE}$, the heterogeneous specification categories $\mathbf{Spec}(\mathcal{HLE})$, $\mathbf{Spec}(span^\mu(\mathcal{HLE}))$ and $\mathbf{Spec}(span^\rho(\mathcal{HLE}))$ are quite different in general.

4   Uniformity via Signature Adjunctions

The discrepancy between the categories of heterogeneous specifications built on institution morphisms and the one built on the spans of comorphisms that could replace them, pointed out in Sect. 3, may seem a bit disturbing and suggests a search for "better" ways of making heterogeneous logical environment uniform. One such possibility arises in most practical examples, when institution morphisms involve "forgetful" signature functors that have left adjoints (which restores the "forgotten" structure of signatures in the source institution in the free way) and/or when institution comorphisms involve signature functors that have right adjoints (that forget the extra structure added to signatures of the source institution to encode them in the target institution). Under such circumstances, institution morphisms can be turned into comorphisms, and vice versa, see [AF96, Dia08]:

Theorem 4.1. *Let $\mu \colon \mathcal{I}' \to \mathcal{I}$ be an institution morphism and let $\mu^{Sign} \colon \mathbf{Sign}' \to \mathbf{Sign}$ have a left adjoint $\rho^{Sign} \colon \mathbf{Sign} \to \mathbf{Sign}'$ with unit $\eta \colon id_{\mathbf{Sign}} \to \rho^{Sign} ; \mu^{Sign}$. Then $L(\mu) = \langle \rho^{Sign}, \rho^{Sen}, \rho^{Mod} \rangle$, where for $\Sigma \in |\mathbf{Sign}|$, $\rho^{Sen}_\Sigma = \mathbf{Sen}(\eta_\Sigma) ; \mu^{Sen}_{\rho^{Sign}(\Sigma)}$ and $\rho^{Mod}_\Sigma = \mu^{Mod}_{\rho^{Sign}(\Sigma)} ; \mathbf{Mod}(\eta_\Sigma)$, is an institution comorphism $L(\mu) \colon \mathcal{I} \to \mathcal{I}'$.*

Let $\rho \colon \mathcal{I} \to \mathcal{I}'$ be an institution comorphism and let $\rho^{Sign} \colon \mathbf{Sign} \to \mathbf{Sign}'$ have a right adjoint $\mu^{Sign} \colon \mathbf{Sign}' \to \mathbf{Sign}$ with counit $\varepsilon \colon \mu^{Sign} ; \rho^{Sign} \to id_{\mathbf{Sign}'}$. Then $R(\rho) = \langle \mu^{Sign}, \mu^{Sen}, \mu^{Mod} \rangle$, where for $\Sigma' \in |\mathbf{Sign}'|$, $\mu^{Sen}_{\Sigma'} = \rho^{Sen}_{\mu^{Sign}(\Sigma')} ; \mathbf{Sen}'(\varepsilon_{\Sigma'})$ and $\mu^{Mod}_\Sigma = \mathbf{Mod}'(\varepsilon_{\Sigma'}) ; \rho^{Mod}_{\mu^{Sign}(\Sigma')}$, is an institution morphism $R(\rho) \colon \mathcal{I}' \to \mathcal{I}$.

Moreover, R and L can be chosen so that $R(L(\mu)) = \mu$ and $L(R(\rho)) = \rho$.

Now, given any heterogeneous logical environment $\mathcal{HLE} = \langle \mathcal{HLE}^\mu, \mathcal{HLE}^\rho \rangle$, if all institution comorphisms in $\mathcal{HLE}^\rho$ have signature functors with right adjoints, we can build a morphism-uniform heterogeneous logical environment $adj^\mu(\mathcal{HLE})$ with each institution comorphism ρ in $\mathcal{HLE}^\rho$ replaced by the institution morphism $R(\rho)$. Then the specification categories $\mathbf{Spec}(\mathcal{HLE})$ and $\mathbf{Spec}(adj^\mu(\mathcal{HLE}))$ are equivalent. Similarly, if all institution morphisms in $\mathcal{HLE}^\mu$ have signature functors with left adjoints, we can build a comorphism-uniform heterogeneous logical environment $adj^\rho(\mathcal{HLE})$ with each institution morphism

μ in $\mathcal{HLE}^\mu$ replaced by the institution comorphism $L(\mu)$. Then the specification categories $\mathbf{Spec}(\mathcal{HLE})$ and $\mathbf{Spec}(adj^\rho(\mathcal{HLE}))$ are equivalent. These equivalences follow essentially from results in [Mos02a, Dia08], which show that switching between institution morphisms and comorphisms as in Thm. 4.1, and so between indexed institutions and indexed coinstitutions, does not affect the resulting Grothendieck institution that can be built.

More explicitly, focused heterogeneous specifications that one can build as in Def. 3.1 using an institution morphism coincide (up to equivalence) with those one can build using the institution comorphism determined by the left adjoint to the signature functor of the institution morphism, and vice versa. Moreover, in each case, heterogeneous specification morphisms determine each other (generalising the original result of [AF96] for theories).

Lemma 4.2. *Let μ and ρ be as in Thm. 4.1, with $L(\mu) = \rho$ and $R(\rho) = \mu$. Consider specifications $SP \in Spec_\mathcal{I}$, $SP' \in Spec_{\mathcal{I}'}$ with $Sig[SP] = \Sigma$ and $Sig[SP'] = \Sigma'$. Then:*

- *$SP'|_\mu \equiv (SP'|_{\varepsilon_{\Sigma'}})|_\rho^{\mu(\Sigma')}$*
- *$\rho(SP) \equiv \mu(\eta_\Sigma(SP))^{\rho(\Sigma)}$*
- *$\mu(SP)^{\Sigma'_0} \equiv \varepsilon_{\Sigma'_0}(\rho(SP))$, for any $\Sigma'_0 \in |\mathbf{Sign}'|$ with $\mu^{Sign}(\Sigma'_0) = \Sigma$,*
- *$SP'|_\rho^{\Sigma_0} \equiv (SP|_\mu)|_{\eta_{\Sigma_0}}$, for any $\Sigma_0 \in |\mathbf{Sign}|$ with $\rho^{Sign}(\Sigma_0) = \Sigma'$.*

Moreover, if $\sigma \colon \Sigma \to \mu^{Sign}(\Sigma')$ and $\sigma' \colon \rho^{Sign}(\Sigma) \to \Sigma'$ are signature morphisms corresponding to each other under bijection given by the adjunction between signature categories (i.e., such that $\eta_\Sigma ; \mu^{Sign}(\sigma') = \sigma$) then $\langle \mu, \sigma \rangle$ is a heterogeneous specification morphism $\langle \mu, \sigma \rangle \colon SP \to SP'$ iff $\langle \rho, \sigma' \rangle$ is a heterogeneous specification comorphism $\langle \rho, \sigma' \rangle \colon SP \to SP'$.

Overall this means that then when in a heterogeneous logical environment the institution morphisms or comorphisms link signature categories by adjunctions, we can gracefully make the environment uniform by replacing institution morphisms by the corresponding comorphisms or, respectively, by replacing institution comorphisms by the corresponding morphisms, with losing neither specifications that can be built nor heterogeneous (co)morphisms between them.

5   Distributed Specifications

Heterogeneity of the logical environment was used in focused heterogeneous specifications to build various parts of specifications in various logical systems (institutions) and then put them together to end up in one logical system, where the models of interest are. However, quite often, for instance in UML [BRJ98], heterogeneous specifications are presented rather differently, by simply giving a number of specifications in various logical systems, and then (implicitly or explicitly) linking them with each other to ensure the expected compatibility properties. This leads to the idea of *distributed specifications* which we will present in this section; see also [CKTW08] for an earlier sketch of this to provide an understanding of distributed heterogeneous UML specifications.

5.1 Distributed Specifications and Their Models

We will work in the context of a *specification frame*: a category **Spec** of (abstract) specifications with semantics given by a model functor **Mod**: **Spec**^{op} → **Set**. The terminology follows [CBEO99], the concept appeared earlier as "specification logic" in [EBCO92, EBO93]). As before, functions **Mod**(σ), for σ: SP → SP' in **Spec**, will be called reducts and denoted by $\_|_\sigma$. Moreover, specification frames can be linked by morphisms and comorphisms much in the same way as institutions, by just leaving out the sentence translation component.

For a while we will not need to discuss how such a specification frame was built: for instance, it may be the category of specifications built in an institution (which is perhaps the prime example) or a heterogeneous category of specifications built over a heterogeneous logical framework (which are examples of interest here).

Definition 5.1. *Let* $\mathcal{F} = \langle$**Spec**, **Mod**: **Spec**^{op} → **Set**$\rangle$ *be a specification frame. A* distributed specification *in* $\mathcal{F}$ *is a collection of specifications linked by specification morphisms, that is, a diagram* DSP: $\mathcal{G}$ → **Spec** *in* **Spec**.

A distributed model *of DSP is a family of models* $\langle M_n \rangle_{n \in |\mathcal{G}|}$ *that is compatible with morphisms in DSP (i.e., for each edge* e: $n \to m$ *in* $\mathcal{G}$, $M_m|_{DSP(e)} = M_n$) *and such that for each node* $n \in |\mathcal{G}|$, $M_n \in$ **Mod**($DSP(n)$). *We write* $Mod[DSP]$ *for the collection of all such distributed models of DSP.*

A specification morphism *between distributed specifications* DSP: $\mathcal{G}$ → **Spec** *and* DSP': $\mathcal{G}'$ → **Spec** *is a pair* (F, τ), *where* F: $\mathcal{G}$ → $\mathcal{G}'$ *is a functor, and* τ: DSP → F; DSP' *a natural transformation. Such morphisms compose as usual, which yields the category* **DSpec**($\mathcal{F}$) *of distributed specifications in* $\mathcal{F}$.

Distributed model reducts w.r.t. such morphisms are defined in the obvious way: for $\mathcal{M}' = \langle M'_n \rangle_{n \in |\mathcal{G}'|} \in Mod[DSP']$, $\mathcal{M}'|_{(F,\tau)} = \mathcal{M}$, *where* $\mathcal{M} = \langle M'_{F(m)}|_{\tau_m} \rangle_{m \in |\mathcal{G}|} \in Mod[DSP]$.

This defines a new specification frame $\mathcal{DSP}(\mathcal{F})$ *of distributed specifications in* $\mathcal{F}$ *and their distributed models.*

Similar concepts were introduced for instance already in [Cla93]. The definition of $\mathcal{DSP}(\mathcal{F})$ resembles the construction of the institution of structured theories in [DM03], but differs from it somewhat by using whole specifications (linked by specification morphisms) as the building blocks for our distributed specifications, whereas the institution of structured theories relies on the use of collections of sentences distributed over signature diagrams.

Given the notion of a distributed specification and its class of models, many concepts and terminology carry over from standard to distributed specifications. For instance, consistency: a distributed specification DSP is *consistent* if $Mod[DSP] \neq \emptyset$.

5.2 Removing Distributivity

The literature so far focused largely on specification frames that are homogeneous in some sense, for instance arise as the category of theories or of specifications built in a single institution, with the usual semantics. In such a case, the following fact will often apply and could be used to diminish the role of distributed specifications by using a corresponding standard (colimit) specification.

Proposition 5.2. *Let $\mathcal{F} = \langle \mathbf{Spec}, \mathbf{Mod} \colon \mathbf{Spec}^{op} \to \mathbf{Set} \rangle$ be a (finitely) exact specification frame — that is, $\mathbf{Spec}$ is (finitely) cocomplete and $\mathbf{Mod}$ preserves (finite) limits.*[8]

Then for any (finite) distributed specification $DSP \colon \mathcal{G} \to \mathbf{Spec}$ in $\mathcal{F}$, there exists a (colimit of DSP) specification $SP \in |\mathbf{Spec}|$ with specification morphisms $\langle \iota_n \colon DSP(n) \to SP \rangle_{n \in |\mathcal{G}|}$ such that each model $M \in Mod[SP]$ determines uniquely a distributed model of DSP, $\langle M|_{\iota_n} \rangle_{n \in |\mathcal{G}|} \in Mod[DSP]$, and vice versa: each distributed model $\langle M_n \rangle_{n \in |\mathcal{G}|} \in Mod[DSP]$ determines a unique model $M \in Mod[SP]$ such that $M_n = M|_{\iota_n}$ for $n \in |\mathcal{G}|$.

This construction can easily be turned into both a morphism and a comorphism of specification frames:

Proposition 5.3. *Let $\mathcal{F} = \langle \mathbf{Spec}, \mathbf{Mod} \colon \mathbf{Spec}^{op} \to \mathbf{Set} \rangle$ be an exact specification frame. Then there is a specification frame (co)morphism $Colim \colon DSP(\mathcal{F}) \to \mathcal{F}$, taking a distributed specification to its colimit, that is an isomorphism on model classes.*

A similar fact, although not stated explicitly there, is already present in the proof of Thm. 20 in [DM03].

The assumption necessary for Props. 5.2 and 5.3 holds for instance for specification frames given by the categories of theories or of specifications (closed under translation and union) built in any (finitely) *exact* institution, where the category of signatures is (finitely) cocomplete and the model functor preserves (finite) limits, see [GB92, ST88a, DGS93]. It is well-known that practically all institutions that capture many-sorted logics are in fact exact. For single-sorted case, the model functors tend not to preserve coproducts, but typically such institutions are *semi-exact* (signature pushouts exists and the model functor maps them to pullbacks in **Set**). Then the resulting specification frames are semi-exact in the analogous sense, which is enough to establish the fact as above for finite connected distributed specifications.

However, this is quite in contrast with specification frames typically arising in heterogeneous logical environments, where (semi-)exactness is very rare. What one essential would need then is the (semi-)exactness of Grothendieck institutions built over (uniform) heterogeneous logical environments. See [Dia02] for results than ensure this for the morphism-uniform case, and [Mos02a] for the comorphism-uniform case. Unfortunately, albeit mathematically interesting and elegant, these results tend to rely on assumptions that are rarely met by the environments arising in practice — for instance, they require the shape of the heterogeneous logical environment to be (co)complete. This essentially would imply that in our logical environment there already is a single "maximal" institution capable of expressing all the specifications built in other institutions in

---

[8] That is, $\mathbf{Mod}$ maps (finite) colimits in $\mathbf{Spec}$ to limits in $\mathbf{Set}$.

the environment via a unique representation. For example, in the Heterogeneous Tool Set HETS [MML07], there is no such "maximal" institution, rather, there are "local maxima", like the logic of Isabelle/HOL, which is used to encode many other logics. But even when restricting to a subgraph of logics represented in Isabelle/HOL, each logic is typically represented in it in more than one way, and so this is not a colimit (indeed, a colimit would have to make identifications that turn it into a rather artificial institution). Moreover, not all of the comorphisms involved in HETS are exact, but this would be needed to make the Grothendieck institution exact.

This is why distributed specifications become of real interest and relevance in the context of heterogeneous logical environments.

As stated above, the exactness assumption of Prop. 5.3 is unrealistically strong. A somewhat more realistic assumption is the following:

Definition 5.4. *A specification frame* $\mathcal{F} = \langle \mathbf{Spec}, \mathbf{Mod} \colon \mathbf{Spec}^{op} \to \mathbf{Set} \rangle$ *is quasi-exact if each diagram* $D \colon \mathcal{G} \to \mathbf{Spec}$ *has a cocone* $\langle \iota_n \colon D(n) \to SP \rangle_{n \in |\mathcal{G}|}$ *that, moreover, is weakly amalgamable. The latter means that any compatible family of models* $\langle M_n \in \mathbf{Mod}(D(n)) \rangle_{n \in |\mathcal{G}|}$ *can be amalgamated to a (not necessarily unique) model* $M \in \mathbf{Mod}(SP)$ *with* $M|_{\iota_n} = M_n$ *for* $n \in |\mathcal{G}|$.

This notion leads to a mathematically less elegant, but practically somewhat more applicable variant of Prop. 5.3:

Proposition 5.5. *Let* $\mathcal{F} = \langle \mathbf{Spec}, \mathbf{Mod} \colon \mathbf{Spec}^{op} \to \mathbf{Set} \rangle$ *be a quasi-exact specification frame. Let* $Discr(\mathcal{DSP}(\mathcal{F}))$ *be the sub-specification frame of* $\mathcal{DSP}(\mathcal{F})$ *where all non-identity specification morphisms are removed. Then there is at least one specification frame comorphism* $WeakAmalg \colon Discr(\mathcal{DSP}(\mathcal{F})) \to \mathcal{F}$ *that is surjective on models, taking a distributed specification to the tip of a weakly amalgamable cocone.*

Note that the need of the move to discrete specification categories (via $Discr(\_)$) is caused by the construction not being functorial.

5.3 Implementing Distributed Specifications

Working in a specification frame $\mathcal{F} = \langle \mathbf{Spec}, \mathbf{Mod} \rangle$, in this section we adapt to distributed specifications the standard view of the process of systematic software development, as presented using implementation steps involving constructors, see [ST88b, ST97]. Recall that for (standard) specifications SP and SP', a *constructor* from SP' to SP is simply a function $\kappa \colon Mod[SP'] \to Mod[SP]$. Given such a constructor, we say that SP' *implements* SP *via* κ, written $SP \underset{\kappa}{\leadsto} SP'$.[9] To generalise this to distributed specifications, we also have to "distribute" the constructor:

---

[9] The definition in the framework of an institution is a bit more delicate: κ is a partial function between model classes over the signatures of SP' and SP, respectively, and then for $SP \underset{\kappa}{\leadsto} SP'$ one requires that on models in $Mod[SP']$, κ is defined and yields models in $Mod[SP]$.

Definition 5.6. *To implement a distributed specification* $DSP: \mathcal{G} \to$ **Spec** *by* $DSP': \mathcal{G}' \to$ **Spec**, *one needs to provide a* covering function $f: |\mathcal{G}| \to |\mathcal{G}'|$ *and a family of* constructors $K = \langle \kappa_n : Mod[DSP'(f(n))] \to Mod[DSP(n)] \rangle_{n \in |\mathcal{G}|}$.

Then DSP' implements DSP via f and K, written $DSP \underset{f,K}{\rightsquigarrow} DSP'$, if for each distributed model $\langle M_{n'} \rangle_{n' \in |\mathcal{G}'|} \in Mod[DSP']$, the family $\langle \kappa_n(M_{f(n)}) \rangle_{n \in |\mathcal{G}|}$ is compatible with morphisms in DSP.

Of course, if $DSP \underset{f,K}{\rightsquigarrow} DSP'$ then for each distributed model $\langle M_{n'} \rangle_{n' \in |\mathcal{G}'|} \in Mod[DSP']$, $\langle \kappa_n(M_{f(n)}) \rangle_{n \in |\mathcal{G}|}$ is a model of DSP.

As can be seen directly from the definition, to establish $DSP \underset{f,K}{\rightsquigarrow} DSP'$ we first have to show that for all $n \in |\mathcal{G}|$, $DSP(n) \underset{\kappa_n}{\rightsquigarrow} DSP'(f(n))$ (which is just as in the implementation steps for standard specifications) and then add that the constructors in K on the respective models from any family satisfying (and hence compatible with) DSP' yield a family of models compatible with DSP. The latter requirement is essentially new and, in general, may require new proof techniques. However, in some simple cases it can be shown using standard categorical reasoning:

Proposition 5.7. *Consider any distributed specifications* $DSP: \mathcal{G} \to$ **Spec** *and* $DSP': \mathcal{G}' \to$ **Spec**, *and let* $(F, \tau): DSP \to DSP'$ *be a specification morphism in* **DSpec**$(\mathcal{F})$. *Then* $DSP \underset{f,K}{\rightsquigarrow} DSP'$, *where* f *is the object (node) part of* F *and* $K = \langle \_|_{\tau_n} \rangle_{n \in |\mathcal{G}|}$ *is the family of reducts w.r.t.* τ_n, $n \in |\mathcal{G}|$.

Note that the above is just an instance (in $\mathcal{DSP}(\mathcal{F})$) of the well-known general fact that for any specification morphism $\sigma: SP \to SP'$, the reduct w.r.t. σ yields a correct implementation of SP by SP', $SP \underset{\_|_\sigma}{\rightsquigarrow} SP'$, cf. [ST88b].

In the case captured by the proposition above, we ensure that the family of constructors given as reducts w.r.t. specification morphisms preserves compatibility of model families in the most simple and expected categorical way. The use of reducts as constructors here may seem very restrictive, but in fact, if one works in a sufficiently rich specification frame, for instance based on institutions with "derived" signature morphisms, then reducts may cover essentially all relevant constructors. Here, a very general concept of a derived signature morphism may be used. Informally, a derived signature morphism $\delta: \Sigma \to \Sigma'$ maps each symbol in Σ to its definition in terms of symbols in Σ'; then for any Σ'-model which interprets the symbols in Σ', its reduct w.r.t. δ is a Σ-model built using the definitions for the symbols in Σ given by δ. In an institution-independent setting, a derived signature morphism could be defined to be an ordinary signature into a definitional extension (see Def. 5.8 below).

Finally, we should stress here that the above notion of implementation covers all possible (and necessary in the development process) changes. First, as usual, individual specifications may be refined, by adding more requirements and "implementation decisions". Second, the structure of the distributed specification may change here: ultimately, we may even arrive at a single standard specification. Finally, in the case when we are working in the heterogeneous category of specifications built in a heterogeneous logical framework, institutions in which individual specifications are built may be changed as well!

5.4 Comparing Distributed Specifications

For usual (homogeneous or focused heterogeneous) specifications, we have introduced the basic notion of equivalence as a way to identify specifications with the same model classes. For distributed specifications, this cannot be so simple. The point is that, very informally, some of the nodes in a distributed specification may play only an auxiliary role, so that in any distributed model of such a distributed specification, the individual models given for such nodes are always uniquely determined by the rest of the family. Effectively, such nodes and their corresponding individual models may be disregarded when comparing distributed models of distributed specifications. This leads to a generalisation of the notion of equivalence of specifications in any specification frame $\mathcal{F} = \langle \mathbf{Spec}, \mathbf{Mod} \rangle$.

Definition 5.8. *A specifications SP' is a* definitional extension *of a specification SP along a specification morphism $\sigma\colon SP \to SP'$ if any SP-model has a unique σ-expansion to an SP'-model, i.e., the reduct $\_|_\sigma\colon Mod[SP'] \to Mod[SP]$ is a bijection.*

Definition 5.9. *Two specifications SP_1 and SP_2 are* pre-equivalent, *written $SP_1 \cong SP_2$, iff there is a common definitional extension SP of SP_1 and SP_2.* Derived equivalence[10] *of specification is defined to be transitive closure of pre-equivalence.*

Proposition 5.10. *Derived equivalence is an equivalence. In exact specification frames, pre-equivalence is transitive, hence derived equivalence and pre-equivalence coincide.*

Now, two distributed specifications $DSP_1\colon \mathcal{G}_1 \to \mathbf{Spec}$ and $DSP_2\colon \mathcal{G}_2 \to \mathbf{Spec}$ are pre-equivalent (in $\mathcal{DSP}(\mathcal{F})$) iff there is a distributed specification $DSP\colon \mathcal{G} \to \mathbf{Spec}$ with distributed specification morphisms $(F_1, \tau_1)\colon DSP_1 \to DSP$ and $(F_2, \tau_2)\colon DSP_2 \to DSP$ such that DSP is a definitional extension of DSP_1 along (F_1, τ_1) and of DSP_2 along (F_2, τ_2). In other words, any distributed model $\mathcal{M}_1 \in Mod[DSP_1]$ extends then uniquely along (F_1, τ_1) to a distributed model of DSP, which in turn reduces (uniquely) w.r.t. (F_2, τ_2) to a distributed model of DSP_2, and vice versa, yielding a "natural" bijection between distributed models of DSP_1 and DSP_2, respectively.

5.5 Distributed Heterogeneous Specifications

The machinery developed above may now be employed to deal with distributed heterogeneous specifications in a heterogeneous logical environment $\mathcal{HLE}$,

---

[10] This terminology is meant to reflect the comments concerning derived specification morphisms in Sect. 5.3 above.

understood as collections of specifications in $\mathcal{HLE}$ linked by (generalised) heterogeneous specification morphisms. Formally, such a distributed heterogeneous specification is just a distributed specification in the sense of Def. 5.1 in the specification frame $\mathcal{HSF}(\mathcal{HLE}) = \langle \mathbf{Spec}(\mathcal{HLE}), \mathbf{Mod}(\mathcal{HLE}) \rangle$, given by Def. 3.4. This yields the specification frame $\mathcal{DSP}(\mathcal{HSF}(\mathcal{HLE}))$ of distributed heterogeneous specifications. We can extend it further to an institution:

Definition 5.11. *Let $\mathcal{HLE}$ be a heterogeneous logical environment. Then the institution $\mathcal{DHSI}(\mathcal{HLE})$ has the category $\mathbf{DSpec}(\mathcal{HSF}(\mathcal{HLE}))$ of distributed heterogeneous specifications as its "signature" category; the model functor is inherited from $\mathcal{DSP}(\mathcal{HSF}(\mathcal{HLE}))$. Given a distributed specification $DSP: \mathcal{G} \to \mathbf{Spec}(\mathcal{HLE})$ in $|\mathbf{DSpec}(\mathcal{HSF}(\mathcal{HLE}))|$, a DSP-sentence is of the form $\langle n, \varphi \rangle$ for $n \in |\mathcal{G}|$ and $\varphi \in \mathbf{Sen}_{\mathcal{I}_n}(\Sigma_n)$, where $Sig[DSP(n)] = \langle \mathcal{I}_n, \Sigma_n \rangle$. A distributed model $\langle M_k \rangle_{k \in |\mathcal{G}|} \in Mod[DSP]$ satisfies such a sentence $\langle n, \varphi \rangle$ if $M_n \models \varphi$ in $\mathcal{I}_n$. For a distributed specification morphism $(F, \tau): (DSP: \mathcal{G} \to \mathbf{Spec}(\mathcal{HLE})) \to (DSP': \mathcal{G}' \to \mathbf{Spec}(\mathcal{HLE}))$ in $\mathbf{DSpec}(\mathcal{HSF}(\mathcal{HLE}))$, translation of such a sentence is given by $(F, \tau)(\langle n, \varphi \rangle) = \langle F(n), \tau_n(\varphi) \rangle$, where for each $n \in |\mathcal{G}|$, the translation $\tau_n(\varphi)$ of φ along the generalised heterogeneous specification morphism τ_n is defined by composing in the natural order the translations along the signature morphism and the institution (co)morphism involved in τ_n. The satisfaction condition follows easily.*

$\mathcal{DHSI}(\mathcal{HLE})$ leads, in the expected way, to a notion of logical *consequences* of a distributed specification (sentences that hold in all models of the distributed specifications). Spelling this out: for $DSP: \mathcal{G} \to \mathbf{Spec}(\mathcal{HLE})$, $n \in |\mathcal{G}|$, $Sig[DSP(n)] = \langle \mathcal{I}_n, \Sigma_n \rangle$, and $\varphi \in \mathbf{Sen}_{\mathcal{I}_n}(\Sigma_n)$), we say that $\langle n, \varphi \rangle$ is a consequence of DSP, written $DSP \models_n \varphi$, iff for all distributed models $\langle M_k \rangle_{k \in |\mathcal{G}|} \in Mod[DSP]$, $M_n \models_{\mathcal{I}_n, \Sigma_n} \varphi$. Note that such consequences include, in general properly, the usual consequences of the individual specifications involved.

$\mathcal{DHSI}(\mathcal{HLE})$ also gives a notion of a *(distributed heterogeneous) theory* of a distributed heterogeneous specification: for $DSP: \mathcal{G} \to \mathbf{Spec}(\mathcal{HLE})$ and $n \in |\mathcal{G}|$ with $Sig[DSP(n)] = \langle \mathcal{I}_n. \Sigma_n \rangle$, we have $Th(DSP)(n) = \{\varphi \in \mathbf{Sen}_{\mathcal{I}_n}(\Sigma_n) \mid DSP \models_n \varphi\}$. Then for each $e: m \to n$ in $\mathcal{G}$, the signature morphism $DSP(e)$ is a theory morphism, $DSP(e): Th(DSP)(m) \to Th(DSP)(n)$, so that we get a diagram in the usual (heterogeneous) category of theories of $\mathcal{HLE}$. Note though that the individual theories $Th(DSP)(n)$ need not be in general finitely presentable in $\mathcal{I}_n$, even if all the individual specifications in DSP are finite presentations.

An interesting alternative way to present distributed heterogeneous specifications would be to first define an institution that differs from $\mathcal{DHSI}(\mathcal{HLE})$ by taking as signatures diagrams in the category $\mathbf{Sign}(\mathcal{HLE})$ of heterogeneous signatures (which would coincide with the institution of structured theories, as defined in [DM03], built for the Bi-Grothendieck institution of $\mathcal{HLE}$). It is easy to see that each distributed heterogeneous specification could be then obtained as a structured specification built in this institution. Moreover, although structured specifications in this institution would also correspond to families of specifications with their signatures linked by signature morphisms that are not necessarily specification morphisms, it can be shown that for such structured specifications, at least when their signatures are finite directed diagrams, we can always give an equivalent distributed heterogeneous specification as defined here.

If the specification frame $\mathcal{HSF}(\mathcal{HLE})$ is quasi-exact, Prop. 5.5 can be used for $\mathcal{DSP}(\mathcal{HSF}(\mathcal{HLE}))$. Moreover, it can be checked that the specification frame morphism $WeakAmalg$ defined on $Discr(DSP(\mathcal{HSF}(\mathcal{HLE})))$ there can be extended to an institution comorphism from $\mathcal{DHSI}(\mathcal{HLE})$ to the Bi-Grothendieck institution build on $\mathcal{HLE}$. The importance of this fact lies in the possibility of transferring logical consequence:

Proposition 5.12. *For any institution comorphism $\rho\colon \mathcal{I} \to \mathcal{I}'$ that is surjective on models, and set of Σ-sentences $\Gamma \cup \{\varphi\}$ in $\mathcal{I}$, we have:*

$$\Gamma \models_{\Sigma}^{\mathcal{I}} \varphi \text{ iff } \rho_{\Sigma}^{Sen}(\Gamma) \models^{\mathcal{I}'} \rho_{\Sigma}^{Sen}(\varphi)$$

That is, given any proof calculus or theorem prover capturing logical consequence in $\mathcal{I}'$, we can re-use it to capture logical consequence in $\mathcal{I}$. When combined with Prop. 5.5, this means that for quasi-exact (Bi-Grothendieck) institutions, logical consequence for distributed heterogeneous specifications can be reduced to logical consequence for focused heterogeneous specifications.

In a heterogeneous setting, the property of quasi-exactness for the specification frame (or institution) of heterogeneous specifications remains quite a strong requirement. However, if one restricts attention to distributed specifications with particular shapes of diagram (namely, so-called connected finitely bounded inf-complete diagrams), then it can be obtained under rather realistic assumptions. For details, see Corollaries 30 and 31 of [CM08].

Finally, we can return in this setting to the issue of making heterogeneous logical environments uniform. It turns out that for any heterogeneous logical environment $\mathcal{HLE}$, even though the heterogeneous specification categories $\mathbf{Spec}(\mathcal{HLE})$, $\mathbf{Spec}(span^{\mu}(\mathcal{HLE}))$ and $\mathbf{Spec}(span^{\rho}(\mathcal{HLE}))$ are quite different, the distributed specifications we can build in each of these categories are essentially the same.

Proposition 5.13. *Given a heterogeneous logical environment $\mathcal{HLE}$, consider any distributed heterogeneous specification $DSP \in |\mathbf{DSpec}(\mathcal{HLE})|$. There exists then a comorphism-uniform distributed heterogeneous specification $DSP^{\rho} \in |\mathbf{DSpec}(span^{\rho}(\mathcal{HLE}))|$ such that $DSP^{\rho} \cong DSP$. Similarly, there is a morphism-uniform distributed heterogeneous specification $DSP^{\mu} \in |\mathbf{DSpec}(span^{\mu}(\mathcal{HLE}))|$ such that $DSP^{\mu} \cong DSP$.*

The proof is related to that of Thm. 11 of [Mos03], relying on Prop. 3.7. For instance, consider an institution morphism $\mu\colon \mathcal{I} \to \mathcal{I}'$, where $span(\mu)$ is $\mathcal{I} \xleftarrow{\rho_{\mu,1}} \mathcal{I}_0' \xrightarrow{\rho_{\mu,2}} \mathcal{I}'$, specifications $SP \in Spec_{\mathcal{I}}$, with $Sig[SP] = \Sigma$, and $SP' \in Spec_{\mathcal{I}'}$, and a signature morphism $\sigma\colon Sig[SP'] \to \mu^{Sign}(\Sigma)$. Then a heterogeneous specification morphism $SP' \xrightarrow{\langle \mu, \sigma \rangle} SP$ in a distributed heterogeneous specification may be replaced by a sequence of heterogeneous specifications comorphisms $SP' \xrightarrow{\langle id, \sigma \rangle} \rho_{\mu,2}(SP|_{\rho_{\mu,1}}^{\Sigma}) \xleftarrow{\langle \rho_{\mu,2}, id \rangle} SP|_{\rho_{\mu,1}}^{\Sigma} \xrightarrow{\langle \rho_{\mu,1}, id \rangle} SP$.

6   Final Remarks

The sentence part of the institution morphisms and comorphisms has rarely played any role in the considerations in this paper (that is, after sentences have been used to build basic specifications). Consequently, we could replace the use of institution morphisms and comorphisms by institution semi-morphisms and semi-comorphisms, respectively (semi-(co)morphisms are just like (co)morphisms but without the translation of sentences, and hence without caring about the satisfaction at all, see [ST88b, Tar96]). With the obvious projection from the category of institutions and their (co)morphisms to the category of institutions and their semi-(co)morphisms, essentially all we presented here would be a special case of a formally more general (but in the presentation basically identical) development using semi-morphisms and semi-comorphisms. Of course, the sentences and satisfaction start matter when it comes to consideration of consequence and proofs in the framework presented here. Remarks on theories for distributed specifications and discussion of Prop. 5.5 in Sect. 5.5 give but the first hints in this direction. Then full institution morphisms and comorphisms provide considerably more possibilities then their "semi-" versions. A proof calculus for focused heterogeneous specifications has been developed in [Mos02a, Mos05]. Using Prop. 5.5, it can be extended to distributed heterogeneous specifications under suitable conditions using weakly amalgamable cocones, which are not unrealistic to be met in practice. The exact tuning of these conditions remains a topic for further research. In cases without weak amalgamation, probably there is no better way than to resolve the proof problems on a case-by-case basis, for each specific link between institutions.

A simple analysis of possible mutual directions of translations involved in maps between institutions leads to further notions of maps between institutions, as suggested in [Tar96] and then studied in [GR02] (see also [MW98]). In particular, when all translations go in the same direction, we obtain institution *forward morphisms*, and when both sentences and models are translated contravariantly w.r.t. signatures, we obtain *forward comorphisms*. It turns out that the span construction helps here again: with spans of morphisms, we can simulate forward (co)morphisms (as well as semi-(co)morphisms) much in the same way as we have been able to simulate comorphisms, see [Mos05] for details. (A similar remark holds for spans of comorphisms.) It may be a bit more difficult to bring into the picture institution (co)morphisms in their *theoroidal* versions, where signatures of one institution are mapped to theories, rather than just signatures, of the other institution [Mes89, GR02]. A technically easy way to achieve this is to add to the heterogeneous logical environment enough infrastructure to allow for expressing theoroidal institution (co)morphisms as plain institution (co)morphisms: for each institution $\mathcal{I}$, its institution of theories $\mathcal{I}^{th}$ needs to be added, along with the obvious morphism $\mathcal{I}^{th} \to \mathcal{I}$ and comorphism $\mathcal{I} \to \mathcal{I}^{th}$.[11]

---

[11] Even generalised theoroidal comorphisms in the sense of [Cod] can then be expressed as semi-comorphisms between institutions of theories.

While the general theory works also for this extended heterogeneous logical environment, it remains to be checked which properties of the heterogeneous logical environment are preserved under this extension, and whether the duplication of $\mathcal{I}$ into $\mathcal{I}$ and $\mathcal{I}^{th}$ can be eliminated, possibly using techniques of [Mos96]. At least it is clear that a theorem prover for $\mathcal{I}$ can easily be lifted to $\mathcal{I}^{th}$.

While non-uniform heterogeneous logical environments naturally arise and can be used in practice, we also offer two ways to make them uniform. The first way, via adjunctions between signature categories, leaves the resulting category of heterogeneous specifications essentially untouched. However, adjunctions are not always available. The second way, via the construction involving spans, is completely general, but leads to a certain modification of the category of heterogeneous specifications. While the same focused heterogeneous specifications can be expressed, we do not directly obtain the same heterogeneous specification (co)morphisms. Nevertheless, we can capture the proof obligations that the morphisms in the non-uniform environment carry by considering logically equivalent specification diagrams. The same method shows that making a heterogeneous logical environment uniform preserves (up to equivalence) the set of distributed heterogeneous specifications.

Another possibility would be to consider an even more general category of institutions, where both morphisms and comorphisms (as well as their semi- and forward versions) can be placed together. One obvious candidate could be based on a notion of institution *relational links*, where the categories of signatures are linked e.g. by distributors (also called profunctors) [Bor94], which are a relational version of functors. Then for any two related signatures, a relation between the sentences over them and a relation between models over them would be given, natural in the related signature morphisms. Generalising the satisfaction condition for institution (co)morphisms, we would of course require these relations to preserve satisfaction. Such relational links clearly compose and cover all kinds of maps between institutions we considered. Hence, in this way we would obtain a category of institutions with relational links between them, into which each of the categories of institutions considered so far could be faithfully embedded. However, as far as we can see, such a category brings little benefit: the notion so obtained seems a bit artificial, and does not ensure any of the expected properties (e.g., entailment is in general neither preserved nor reflected by relational links, the category is neither finitely complete nor finitely cocomplete, etc).

One consequence may be that we have to live with non-uniform environments, where the maps considered do not compose in general, and so we cannot view them simply as diagrams in a category of institutions. In fact, this is what is really happening in HETS [MML07, Mos05], where both institution morphisms and comorphisms are used, while the projection (via spans) to a comorphism-uniform environment is applied for theorem proving. Future work will apply this approach to the heterogeneous logical environment arising from UML (see [CKTW08] for initial promising steps in this direction).

Acknowledgements. Many thanks to the anonymous referees for detailed comments.

References

[AF96]   Arrais, M., Fiadeiro, J.-L.: Unifying theories in different institutions. In: Haveraaen, M., Dahl, O.-J., Owe, O. (eds.) Abstract Data Types 1995 and COMPASS 1995. LNCS, vol. 1130, pp. 81–101. Springer, Heidelberg (1996)

[BG80]   Burstall, R.M., Goguen, J.A.: The semantics of CLEAR, a specification language. In: Bjorner, D. (ed.) Abstract Software Specifications. LNCS, vol. 86, pp. 292–332. Springer, Heidelberg (1980)

[Bor94]   Borceux, F.: Handbook of Categorical Algebra I. Cambridge University Press, Cambridge (1994)

[BRJ98]   Booch, G., Rumbaugh, J., Jacobson, I.: The Unified Modeling Language User Guide. Addison-Wesley, Reading (1998)

[CBEO99]   Cornelius, F., Baldamus, M., Ehrig, H., Orejas, F.: Abstract and behaviour module specifications. Mathematical Structures in Computer Science 9(1), 21–62 (1999)

[CKTW08]   Cengarle, M.V., Knapp, A., Tarlecki, A., Wirsing, M.: A heterogeneous approach to UML semantics. In: Degano, P., De Nicola, R., Meseguer, J. (eds.) Concurrency, Graphs and Models. LNCS, vol. 5065, pp. 383–402. Springer, Heidelberg (2008)

[Cla93]   Classen, I.: Compositionality of application oriented structuring mechanisms for algebraic specification languages with initial algebra semantics. Phd thesis, Technische Universität Berlin (1993)

[CM08]   Codescu, M., Mossakowski, T.: Heterogeneous colimits. In: Boulanger, F., Gaston, C., Schobbens, P.-Y. (eds.) MoVaH 2008 Workshop on Modeling, Validation and Heterogeneity. IEEE press, Los Alamitos (2008)

[Cod]   Codescu, M.: Generalized theoroidal institution comorphisms. This volume

[DGS93]   Diaconescu, R., Goguen, J., Stefaneas, P.: Logical support for modularisation. In: Huet, G., Plotkin, G. (eds.) Logical Environments, pp. 83–130. Cambridge Univ. Press, Cambridge (1993)

[Dia98]   Diaconescu, R.: Extra theory morphisms for institutions: Logical semantics for multi-paradigm languages. J. Applied Categorical Structures 6, 427–453 (1998)

[Dia02]   Diaconescu, R.: Grothendieck institutions. J. Applied Categorical Structures 10, 383–402 (2002)

[Dia08]   Diaconescu, R.: Institution-independent model theory. Birkhäuser, Basel (2008)

[DM03]   Durán, F., Meseguer, J.: Structured theories and institutions. Theor. Comput. Sci. 309(1-3), 357–380 (2003)

[EBCO92]   Ehrig, H., Baldamus, M., Cornelius, F., Orejas, F.: Theory of algebraic module specification including behavioral semantics and constraints. In: Nivat, M., Rattray, C., Rus, T., Scollo, G. (eds.) Algebraic Methodology and Software Technology AMAST 1991, Proc. 2nd Intl. Conf., Iowa City, 1991, Workshops in Computing, pp. 145–172. Springer, Heidelberg (1992)

[EBO93]   Ehrig, H., Baldamus, M., Orejas, F.: New concepts of amalgamation and extension for a general theory of specifications. In: Bidoit, M., Choppy, C. (eds.) Abstract Data Types 1991 and COMPASS 1991. LNCS, vol. 655, pp. 199–221. Springer, Heidelberg (1993)

[GB92]   Goguen, J.A., Burstall, R.M.: Institutions: Abstract model theory for specification and programming. Journal of the ACM 39(1), 95–146 (1992)

[GR02]   Goguen, J.A., Rosu, G.: Institution morphisms. Formal Aspects of Compututing 13(3-5), 274–307 (2002)

[HHP93]   Harper, R., Honsell, F., Plotkin, G.D.: A framework for defining logics. Journal of the ACM 40(1), 143–184 (1993)

[MDT09]   Mossakowski, T., Diaconescu, R., Tarlecki, A.: What is a logic translation? In: Beziau, J.-Y. (ed.) Logica Universalis, Birkhäuser, Basel (to appear, 2009)

[Mes89]   Meseguer, J.: General logics. In: Logic Colloquium 1987, pp. 275–329. North Holland, Amsterdam (1989)

[MML07]   Mossakowski, T., Maeder, C., Lüttich, K.: The Heterogeneous Tool Set. In: Grumberg, O., Huth, M. (eds.) TACAS 2007. LNCS, vol. 4424, pp. 519–522. Springer, Heidelberg (2007)

[MOM02]   Martí-Oliet, N., Meseguer, J.: Rewriting logic: roadmap and bibliography. Theor. Comput. Sci. 285(2), 121–154 (2002)

[Mos96]   Mossakowski, T.: Different types of arrow between logical frameworks. In: Meyer auf der Heide, F., Monien, B. (eds.) ICALP 1996. LNCS, vol. 1099, pp. 158–169. Springer, Heidelberg (1996)

[Mos02a]   Mossakowski, T.: Comorphism-based Grothendieck logics. In: Diks, K., Rytter, W. (eds.) MFCS 2002. LNCS, vol. 2420, pp. 593–604. Springer, Heidelberg (2002)

[Mos02b]   Mossakowski, T.: Heterogeneous development graphs and heterogeneous borrowing. In: Nielsen, M., Engberg, U. (eds.) FOSSACS 2002. LNCS, vol. 2303, pp. 326–341. Springer, Heidelberg (2002)

[Mos03]   Mossakowski, T.: Foundations of heterogeneous specification. In: Wirsing, M., Pattinson, D., Hennicker, R. (eds.) WADT 2003. LNCS, vol. 2755, pp. 359–375. Springer, Heidelberg (2003)

[Mos05]   Mossakowski, T.: Heterogeneous Specification and the Heterogeneous Tool Set. Habilitation thesis, Universität Bremen (2005)

[MW98]   Martini, A., Wolter, U.: A single perspective on arrows between institutions. In: Haeberer, A.M. (ed.) AMAST 1998. LNCS, vol. 1548, pp. 486–501. Springer, Heidelberg (1998)

[NPW02]   Nipkow, T., Paulson, L.C., Wenzel, M.: Isabelle/HOL — A Proof Assistant for Higher-Order Logic. LNCS, vol. 2283. Springer, Heidelberg (2002)

[PF06]   López Pombo, C., Frias, M.F.: Fork algebras as a sufficiently rich universal institution. In: Johnson, M., Vene, V. (eds.) AMAST 2006. LNCS, vol. 4019, pp. 235–247. Springer, Heidelberg (2006)

[ST88a]   Sannella, D., Tarlecki, A.: Specifications in an arbitrary institution. Information and Computation 76, 165–210 (1988)

[ST88b]   Sannella, D., Tarlecki, A.: Toward formal development of programs from algebraic specifications: Implementations revisited. Acta Informatica 25, 233–281 (1988)

[ST97]   Sannella, D., Tarlecki, A.: Essential concepts of algebraic specification and program development. Formal Aspects of Computing 9, 229–269 (1997)

[Tar87]   Tarlecki, A.: Institution representation. Unpublished note, Dept. of Computer Science, University of Edinburgh (1987)

[Tar96]   Tarlecki, A.: Moving between logical systems. In: Haveraaen, M., Dahl, O.-J., Owe, O. (eds.) Abstract Data Types 1995 and COMPASS 1995. LNCS, vol. 1130, pp. 478–502. Springer, Heidelberg (1996)

[Tar00]   Tarlecki, A.: Towards heterogeneous specifications. In: Gabbay, D., de Rijke, M. (eds.) Frontiers of Combining Systems 2. Studies in Logic and Computation, pp. 337–360. Research Studies Press (2000)

[Tar03]   Tarlecki, A.: Abstract specification theory: An overview. In: Broy, M., Pizka, M. (eds.) Models, Algebras, and Logics of Engineering Software. NATO Science Series — Computer and System Sciences, vol. 191, pp. 43–79. IOS Press, Amsterdam (2003)

Term-Generic Logic*

Andrei Popescu** and Grigore Roşu

Department of Computer Science,
University of Illinois at Urbana-Champaign
{popescu2,grosu}@cs.uiuc.edu

Abstract. *Term-generic logic (TGL)* is a first-order logic parameterized with terms defined axiomatically (rather than constructively), by requiring them to only provide generic notions of *free variable* and *substitution* satisfying reasonable properties. TGL has a complete Gentzen system generalizing that of first-order logic. A certain fragment of TGL, called HORN², possesses a much simpler Gentzen system, similar to traditional typing derivation systems of λ-calculi. HORN² appears to be sufficient for defining a whole plethora of λ-calculi as *theories* inside the logic. Within intuitionistic TGL, a HORN² specification of a calculus is likely to be *adequate by default*. A bit of extra effort shows adequacy w.r.t. classic TGL as well, endowing the calculus with a complete loose semantics.

1 Introduction

First-order logic (FOL) does not allow variables to be bound in terms (but only in formulae, via quantifiers), thus providing a straightforward notion of substitution in terms. On the other hand, most calculi that are used in the domain of programming languages, and not only, are crucially based on the notion of *binding* of variables *in terms*: terms "export" only a subset of their variables, the *free* ones, that can be substituted. Because of their complex formulation for terms, these calculi cannot be naturally defined as FOL theories. Consequently, they need to define their own models and deduction rules, and to state their own theorems of completeness, not always easy to prove. In other words, they are presented as entirely *new logics*, as opposed to *theories in an existing logic*, thus incurring all the drawbacks (and boredom) of repeating definitions and proofs following generic, well-understood patterns, but facing new "details".

In this paper we define *term-generic first-order logic*, or simply *term-generic logic (TGL)*, as a first-order logic parameterized by any terms that come with abstract notions of *free variable* and *substitution*. More precisely, in TGL terms are elements in a generic set *Term* (including a subset *Var* whose elements are called variables) that comes with functions $FV : Term \rightarrow \mathcal{P}_f(Var)$ and $Subst : Term \times Term^{Var} \rightarrow Term$ for *free variables* and *substitution*, respectively, satisfying some expected properties. TGL models provide interpretations of terms that

---

* Supported in part by NSF grants CCF-0448501, CNS-0509321 and CNS-0720512, by NASA contract NNL08AA23C, by the Microsoft/Intel funded Universal Parallel Computing Research Center at UIUC, and by several Microsoft gifts.
** Also: Institute of Mathematics "Simion Stoilow" of the Romanian Academy.

A. Corradini and U. Montanari (Eds.): WADT 2008, LNCS 5486, pp. 290–307, 2009.
© Springer-Verlag Berlin Heidelberg 2009

satisfy, again, some reasonable properties. We show that TGL admits a *complete* Gentzen-like deduction system, which is syntactically very similar to that of FOL; its proof of completeness modifies the classic proof of completeness for FOL to use the generic notions of term, free variables and substitution.

Because of not committing to any particular definition of term, TGL can be instantiated to different types of terms, such as standard FOL terms or different categories of (typed or untyped) λ-terms. When instantiated to standard FOL terms, TGL becomes, as expected, precisely FOL. However, when instantiated to more complex terms, e.g., the terms of λ-calculus, TGL becomes a logic where a particular calculus is a particular theory. For example, the TGL axiom for typing abstractions in simply-typed λ-calculus can be

$$(\forall x.\, x : t \Rightarrow X : t') \Rightarrow (\lambda x{:}t.X) : t \to t'$$

where x and t, t' denote data and type variables, respectively, X denotes an arbitrary data term, $\Rightarrow$ is the logical implication, and $\to$ is the arrow type construct (binary operator on types). The above is an axiom-scheme, parameterized by any choice of variables x, t, t' and term X (and, as customary, each of its instances is implicitly assumed universally quantified over all its free variables). The colons in $x : t$ and $X : t'$ and the outermost colon in $(\lambda x : t.\, X) : t \to t'$ refer to a binary relation symbol in TGL, while the colon in $\lambda x{:}t.X$ is part of the term syntax. The term X may contain the free variable x, which is bound by $\forall$ in the lefthand side of the outermost implication, and by λ in the righthand side. Both these capturings of x from X are *intended* – in fact, the migration of x between the two scopes is at the heart of the intended typing mechanism: x is an *actual*, but *arbitrary* input to the function described by X in the former case, and a *formal* parameter in the latter; the type $t \to t'$ is assigned to the abstraction $\lambda x{:}t.X$ by "experimenting" with arbitrary x's of type t and "observing" if the result has type t'. (Using the same notation for actual as for formal parameters of functional expressions is well-established mathematical practice.)

A possible instance of the above axiom-scheme, taking, e.g., $\lambda y{:}t''.\, y\,x$ as X and spelling out all the universal quantifications, is

$$\forall t, t', t''.\, (\forall x.\, x : t \Rightarrow (\lambda y{:}t''.\, y\,x) : t') \Rightarrow (\lambda x{:}t.\, \lambda y{:}t''.\, y\,x) : t \to t',$$

which implies in TGL, instantiating t'' with $t \to t'''$ and t' with $(t \to t''') \to t'''$,

$$\forall t, t'''.\, (\forall x.\, x : t \Rightarrow (\lambda y{:}t \to t'''.\, y\,x) : (t \to t''') \to t''') \Rightarrow$$
$$(\lambda x{:}t.\, \lambda y{:}t \to t'''.\, y\,x) : t \to (t \to t''') \to t'''.$$

Moreover, we can prove in TGL, using again the above axiom-scheme and another axiom for application, that the hypothesis (i.e., the lefthand side of the outermost $\Rightarrow$) in the latter sentence is true for all t', t''', hence we obtain a TGL derivation of $\forall t, t'''.\, (\lambda x{:}t.\, \lambda y{:}t \to t'''.\, y\,x) : t \to (t \to t''') \to t'''$.

A specification of a calculus in TGL brings a *meaningful complete semantics* for that calculus, because the axioms are stated *about some models*, the content of the axioms making the models "desirable". Indeed, TGL models are initially "blank", in that they are only required to interpret the terms consistently with substitution – it is the axioms that customize the models. For instance, the previously discussed description of x as an "actual, but arbitrary parameter" is not merely an informal idea to help the intuition, but a mathematical fact within the TGL semantics: when "escaped" from the scope of λ into the scope of $\forall$, x indeed denotes an actual, but arbitrary inhabitant of a desirable model.

Even though the completeness (being equivalent to semi-decidability) of a fragment of a logic (whose syntax is decidable) follows from the completeness of the richer logic, there are good reasons to develop complete proof systems for certain particular sublogics as well. Besides a better understanding and self-containment of the sublogic, one important reason is the *granularity of proofs*. Indeed, proofs of goals in the sublogic that use the proof-system of the larger logic may be rather long and "junkish" and may look artificial in the context of the sublogic. For example, equational logic admits a very intuitive complete proof system [5], that simply "replaces equals by equals", thus avoiding the more intricate first-order proofs. An important goal of this paper is to also investigate conditions under which sublogics of TGL admit specialized coarse-granularity proof systems.

It appears that a certain fragment of TGL, that we call HORN^2, is sufficient for calculi-specification purposes. HORN^2 consists of TGL sentences of the form

$$\forall \overline{y}.(\forall \overline{x}.\bigwedge_{i=1}^{n} a_i(\overline{x}, \overline{y}) \Rightarrow b_i(\overline{x}, \overline{y})) \Rightarrow c(\overline{y})$$

with a_i, b_i, c atomic formulae ($\overline{x}$ and $\overline{y}$ denote tuples of variables), i.e., generalized Horn implications whose conditions are themselves (single-hypothesis)[1] Horn implications. We show that, under a reasonable restriction that we call *amenability*, a HORN^2 theory admits a *complete* Gentzen system that "implements" each HORN^2 formula as above into a deduction rule of the form

$$\frac{\Gamma, a_i(\overline{z}, \overline{T}) \triangleright b_i(\overline{z}, \overline{T}) \text{ for all } i \in \{1, \ldots, n\}}{\Gamma \triangleright c(\overline{T})}$$

where $\overline{T}$ is a tuple of terms substituting $\overline{y}$ and $\overline{z}$ is a fresh tuple of variables substituting $\overline{x}$. The (multiple-formulae antecedent, single-formula succedent) structure of this system follows the style of intuitionistic logic, and indeed we show that it specializes the Gentzen system of the intuitionistic version of TGL. Thus we obtain for the HORN^2 fragment an intuitionistic proof system which is complete w.r.t. classical models! Moreover, this "lower-level" Gentzen system, extracted from the higher-level notation used in the HORN^2 theory, recovers the original calculus itself (bringing what in syntactic encoding frameworks is usually referred to as *adequacy* of the representation). For instance, the HORN^2 deduction rule corresponding to the aforementioned typing axiom for typed λ-calculus is *precisely* the familiar context-based typing rule for abstractions:

$$\frac{\Gamma, z:T \triangleright Z : T'}{\Gamma \triangleright (\lambda z:T. Z) : T \to T'}[z \text{ fresh for } \Gamma]$$

By substitution, x from the typing axiom became a fresh z in the deductive system, the variables t, t' became arbitrary terms T, T', and X became a term Z

---

[1] Single hypothesis, in the sense that each $a_i(\overline{x}, \overline{y})$ has to be an atomic formula, as opposed to being a conjunction of atomic formulae.

such that the positions in which x occurred in X are the same as the positions in which z now occurs in Z (because the term X and the positioning of x in X were arbitrary in the typing axiom, it follows that the term Z and the positioning of z in Z are also arbitrary in the resulted deduction rule). This transformation is prescribed uniformly, i.e., calculus-independently, for any HORN^2 theory.

The remainder of this paper is structured as follows. Section 2 introduces classic TGL (syntax, models, institutional structure, Gentzen system and completeness theorem) and intuitionistic TGL. Section 3 discusses the HORN^2 fragment and its specialized Gentzen systems, whose completeness "prepares" the logic for future adequacy results. Section 4 illustrates the TGL *adequate by default* specification style for λ-calculi, taking System F as a working example. Section 5 discusses related work and draws conclusions. More details regarding the topics addressed in this paper, including proofs of the stated facts, can be found in the technical report [23] – the main part of the report has the same content as this paper, while the appendix of the report contains further details and proofs.

2   Term-Generic First-Order Logic

We introduce a generic notion of first-order term, axiomatized by means of free variables and substitution, purposely not committing to any concrete syntax for terms. Then we show our first novel result in this paper, namely a development of first-order logic that does not depend on the syntax of terms, but *only on the properties of substitution*. We first develop the logic in an unsorted form and without equality, and later sketch equality and order-sorted extensions, as well as an intuitionistic variant.

Definition 1. *Let Var be a countably infinite set of variables. A* term syntax *over Var consists of the following data:*

(a) A (countably infinite) set Term such that $Var \subseteq Term$, whose elements are called terms;

(b) A mapping $FV: Term \to \mathcal{P}_f(Var)$ (where $\mathcal{P}_f$ means "the set of finite sets of"); the elements of $FV(T)$ are called free *variables, or simply* variables, *of T;*

(c) A mapping $Subst: Term \times Term^{Var} \to Term$, called substitution.

These are subject to the following requirements (where x ranges over variables, T, T' over terms, and θ, θ' over maps in $Term^{Var}$):

(1) $Subst(x, \theta) = \theta(x)$;

(2) $Subst(T, Var \hookrightarrow Term) = T$;

(3) If $\theta\restriction_{FV(T)} = \theta'\restriction_{FV(T)}$, then $Subst(T, \theta) = Subst(T, \theta')$;[2]

(4) $Subst(Subst(T, \theta), \theta') = Subst(T, \theta; \theta')$, where $\theta; \theta' : Var \to Term$ is defined by $(\theta; \theta')(y) = Subst(\theta(y), \theta')$;

(5) $FV(x) = \{x\}$;

(6) $FV(Subst(T, \theta)) = \bigcup\{FV(\theta(x)) : x \in FV(T)\}$.

---

[2] Here and later, if $f : U \to V$ and $U' \subseteq U$, $f\restriction_{U'}$ denotes the restriction of f to U'.

Note that we assume the notion of term coming together with a notion of substitution which is *composable* (condition (4) above). In general, for a syntax with bindings, composability of substitution does not hold for raw terms, but only for α-equivalence classes – therefore, in the concrete instances of our logic to calculi with bindings, TGL terms will be α-equivalence classes of what are usually called (raw) "terms" in these calculi. Conditions (1)-(6) from Definition 1 are natural (and well-known) properties of substitution holding for virtually all notions of terms with static binding (modulo α-equivalence).

For *distinct* variables $x_1, \dots, x_n$, we write $[T_1/x_1, \dots, T_n/x_n]$ for the function $Var \to Term$ that maps x_i to T_i for $i = \overline{1, n}$ and all the other variables to themselves, and $T[T_1/x_1, \dots, T_n/x_n]$ for $Subst(T, [T_1/x_1, \dots, T_n/x_n])$.

Definition 2. *A term-generic language consists of a term syntax $(Term, FV, Subst)$ over a set Var and an at most countable ranked set $\Pi = (\Pi_n)_{n \in \mathbb{N}}$, of relation symbols. A TGL model for a language as above is a triple of the form $(A, (A_T)_{T \in Term}, (A_{(n,\pi)})_{n \in \mathbb{N}, \pi \in \Pi_n})$, where:*

(a) A is a set, called the carrier set.
(b) For each $T \in Term$, A_T is a mapping $A^{Var} \to A$ such that the following hold for all $x \in Var$, $T \in Term$, $\rho, \rho' \in A^{Var}$, and $\theta \in Term^{Var}$:
  — (b.i) $A_x(\rho) = \rho(x)$;
  — (b.ii) If $\rho\!\restriction_{FV(T)} = \rho'\!\restriction_{FV(T)}$, then $A_T(\rho) = A_T(\rho')$;
  — (b.iii) $A_{Subst(T, \theta)}(\rho) = A_T(A_\theta(\rho))$, where $A_\theta : A^{Var} \to A^{Var}$ is defined by $A_\theta(\rho)(y) = A_{\theta(y)}(\rho)$.
(c) For each $n \in \mathbb{N}$ and $\pi \in \Pi_n$, $A_{(n,\pi)}$ is an n-ary relation on A.

Thus, unlike in classic FOL models where the interpretation of terms is *built* from operations, in TGL models the interpretation of terms is *assumed* (in the style of Henkin models). It turns out that condition (b.ii) is redundant (follows from the other conditions and Definition 1 – see Section F.1 in [23] for a proof)[3] – we keep it though as part of the definition of a model for the sake of symmetry with Definition 1.

In what follows, we let x, x_i, y, u, v, etc., range over variables, T, T_i, T', etc., over terms, θ, θ', etc., over maps in $Term^{Var}$, ρ, ρ', etc., over valuations in A^{Var}, and π, π', etc., over relation symbols. Sometimes we simply write $Term$ for term syntaxes $(Term, FV, Subst)$ and $(Term, \Pi)$ for term-generic languages.

Formulae are defined as usual, starting from *atomic formulae* $\pi(T_1, \dots, T_n)$ and applying connectives $\wedge, \Rightarrow$ and quantifier $\forall$. (Other connectives and quantifiers may of course be also considered, but we omit them since they shall not be needed for our specifications in this paper.) For each formula φ, the set $A_\varphi \subseteq A^{Var}$, of *valuations that make φ true in A*, is defined recursively on the structure of formulae as follows: $\rho \in A_{\pi(T_1, \dots, T_n)}$ iff $(A_{T_1}(\rho), \dots, A_{T_n}(\rho)) \in A_\pi$; $\rho \in A_{\varphi \Rightarrow \psi}$ iff $\rho \in A_\varphi$ implies $\rho \in A_\psi$; $\rho \in A_{\varphi \wedge \psi}$ iff $\rho \in A_\varphi$ and $\rho \in A_\psi$; $\rho \in A_{\forall x. \varphi}$ iff $\rho[x \leftarrow a] \in A_\varphi$ for all $a \in A$. If $\rho \in A_\varphi$ we say that A *satisfies φ under valuation ρ* and write $A \models_\rho \varphi$. If $A_\varphi = A^{Var}$ we say that A *satisfies φ* and write $A \models \varphi$. Given a set of formulae Γ, $A \models \Gamma$ means $A \models \varphi$ for all $\varphi \in \Gamma$. Above,

---

[3] We are indebted to one of the referees for bringing this to our attention.

and from now on, we let φ, ψ, χ range over formulae and A, B over models (sometimes, when we want to distinguish models from their carrier set A, B, we write $\mathcal{A}, \mathcal{B}$ for the models). For formulae, the notions of free variables, α-equivalence, and substitution are the natural ones, defined similarly to the case of FOL, but on top of our generic terms rather than FOL terms. For substitution in formulae we adopt notational conventions similar to the ones about substitution in terms, e.g., $\varphi[T/x]$. Note that TGL is a logic generic only w.r.t. terms - formulae are "concrete" first-order formulae built over generic terms, with a "concrete" (and not generic) notion of α-equivalence, standardly defined using the bindings from quantifiers, which preserves satisfaction and the free variables and is compatible with substitution and the language constructs. Hereafter we identify formulae modulo α-equivalence. Let $\overline{x} = (x_1, \dots, x_n)$ be a tuple of variables and $J = \{y_1, \dots, y_m\}$ a set of variables. Then $Vars(\overline{x})$ denotes the set $\{x_1, \dots, x_n\}$, $\forall \overline{x}. \varphi$ denotes $\forall x_1 \dots \forall x_n.\varphi$, and $\forall J. \varphi$ denotes $\forall y_1 \dots \forall y_m.\varphi$ (the latter notation making an (immaterial for our purposes) choice of a total ordering on J). A *sentence* is a formula with no free variables. The *universal closure* of a formula φ is the sentence $\forall FV(\varphi). \varphi$. (See Section A.1 in [23] for more details.)

The inclusion of an emphasized equality symbol in our logic, interpreted in all models as equality, yields *TGL with equality*. *Many-sorted* and *order-sorted* variants of TGL (in the style of [13]) can also be straightforwardly obtained, by extending Definition 1 to term syntaxes involving multiple sorts (syntactic categories) and Definition 2 to models having as carriers sort-indexed sets. For example, in the case of order-sorted TGL, a poset $(S, <)$ of sorts is fixed and carriers of models are families of sets $(A_s)_{s \in S}$ such that $s < s'$ implies $A_s \subseteq A_{s'}$ for all $s, s' \in S$. (See Sections A.2 and A.3 in [23] for details.) All the concepts and results about TGL in this paper, including completeness of various proof systems for various fragments of the logic, can be easily (but admittedly tediously) extended to the many-sorted and order-sorted cases.

FOL. As expected, classic FOL is an instance of TGL. Indeed, let (Var, Σ, Π) be a classic first-order language, where $\Sigma = (\Sigma_n)_{n \in \mathbb{N}}$ and $\Pi = (\Pi_n)_{n \in \mathbb{N}}$ are ranked sets of operation and relation symbols. Let *Term* be the term syntax consisting of ordinary first-order terms over Σ and Var with $FV : Term \to \mathcal{P}_f(Var)$ giving *all* the variables in each term and $Subst : Term \times Term^{Var} \to Term$ the usual substitution on FOL terms. Define a term-generic language as $(Term, \Pi)$. A classic FOL model $(A, (A_\sigma)_{\sigma \in \Sigma}, (A_{(n,\pi)})_{n \in \mathbb{N}, \pi \in \Pi_n})$ yields a TGL model $(A, (A_T)_{T \in Term}, (A_{(n,\pi)})_{n \in \mathbb{N}, \pi \in \Pi_n})$ by defining the meaning of terms as derived operations. Conversely, a TGL model $(A, (A_T)_{T \in Term}, (A_{(n,\pi)})_{n \in \mathbb{N}, \pi \in \Pi_n})$ yields an FOL model by defining $A_\sigma : A^n \to A$ as $A_\sigma(a_1, \dots, a_n) = A_{\sigma(x_1, \dots, x_n)}(\rho)$, where $x_1, \dots, x_n$ are distinct variables and ρ is a valuation that maps each x_i to a_i. The two model mappings are mutually inverse and preserve satisfaction. Thus, for this particular choice of terms, TGL yields FOL.

A Formula-Typed Logic (an "Extremely-Typed" λ-Calculus). FOL is a simple instance of TGL. However, TGL terms may be arbitrarily exotic. Besides terms of λ-calculi (discussed in Section 4), one may also have terms that interfere with formulae in non-trivial ways, where, e.g., terms may abstract variables having formulae as types (thinking of types as sets defined by comprehension). Let (Var, Σ, Π) be a classic first-order language and:

$Term ::= Var \mid \Sigma(Term, \dots, Term) \mid Term\ Term \mid \lambda\,Var : Fmla.\ Term$

$Fmla ::= \Pi(Term, \dots, Term) \mid Fmla \Rightarrow Fmla \mid Fmla \wedge Fmla \mid \forall\,Var.\ Fmla,$

with the natural restrictions w.r.t. the rank of operations/relations. Free variables and substitution are as expected, making terms up to α-equivalence a TGL term syntax. Moreover, although formulae and terms were defined mutually recursively, the former are still nothing but first-order formulae over terms, hence fall under TGL. Does the interpretation of formulae inside terms match their first-order interpretation at the top level? The answer is "no, unless axioms require it" (remember that TGL models are blank, but customizable). Here, postulating $(\forall x.\,(\varphi \Leftrightarrow \psi) \wedge (\varphi \Rightarrow T = T')) \Rightarrow \lambda x : \varphi.T = \lambda x : \psi.T'$ does the job.

The TGL institution. Next we submit TGL to a standard well-behaving test for a logical system, organizing it as an institution [12]. By doing so, we present TGL terms and models in a more structural light, and, more importantly, create a framework for λ-calculi with different flavors to cohabitate with each other and with classic FOL *under different signatures and axioms, but within the same logic*, connected through the railway system of signature morphisms.

Let Var be a fixed set of variables. The *signatures* are term-generic languages $(Term, \Pi)$ over Var. The $(Term, \Pi)$- *sentences*, *models* and *satisfaction relation* were already defined for TGL. Given $\theta : Var \to Term$, we write $\bar{\theta}$ for the map $T \mapsto Subst(T, \theta)$. Moreover, for any model $(A, (A_T)_{T \in Term}, (A_{(n,\pi)})_{n \in \mathbb{N}, \pi \in \Pi_n})$ and map $\rho : Var \to A$, we write $\bar{\rho}^A$ for the map $T \mapsto A_T(\rho)$. A model can be alternatively presented as a tuple $\mathcal{A} = (A, \dot{}^A, (A_{(n,\pi)})_{n \in \mathbb{N}, \pi \in \Pi_n})$ where the $A_{(n,\pi)}$'s are relations as before and $\dot{}^A : A^{Var} \to A^{Term}$ is such that $\bar{\rho}^A \circ (Var \hookrightarrow Term) = \rho$, $[\bar{\rho}^A(T) = \bar{\rho'}^A(T)$ whenever $\rho \restriction_{FV(T)} = \rho' \restriction_{FV(T)}]$, and $\bar{\rho}^A \circ \bar{\theta} = \overline{\bar{\rho}^A \circ \theta}^A$. A *model homomorphism* between $\mathcal{A} = (A, \dot{}^A, (A_{(n,\pi)})_{n \in \mathbb{N}, \pi \in \Pi_n})$ and $\mathcal{B} = (B, \dot{}^B, (B_{(n,\pi)})_{n \in \mathbb{N}, \pi \in \Pi_n})$ is a map $h : A \to B$ that commutes with the relations in the usual way and has the property that $h \circ \bar{\rho}^A = \overline{h \circ \rho}^B$ for all $\rho \in A^{Var}$. (Note the structural similarity between the conditions defining the three concepts of term syntax, model and model homomorphism, which allows one to easily see that $(Term, \dot{}, (Term_{(n,\pi)})_{n \in \mathbb{N}, \pi \in \Pi_n})$ is a model for any choice of relations $Term_{(n,\pi)}$ and that $(Term, \dot{}, (\emptyset)_{\pi \in \Pi})$ is freely generated by Var.)

A *signature morphism* between $(Term, \Pi)$ and $(Term', \Pi')$ is a pair (u, v) with $v = (v_n : \Pi_n \to \Pi'_n)_{n \in \mathbb{N}}$ and $u : Term \to Term'$ such that $u \circ (Var \hookrightarrow Term) = (Var \hookrightarrow Term')$, $FV(u(T)) = FV(T)$ for all $T \in Term$ and $u \circ \bar{\theta} = \overline{u \circ \theta}' \circ u$ (where $\dot{}'$ is the map $Term' \to Term'$ associated to the term syntax $Term'$) for all $\theta : Var \to Term$. (Intuition for the last condition on signature morphism: say we have concrete terms, like the ones of FOL or λ-calculus, θ maps x to S and all other variables to themselves (thus $\bar{\theta}(T) = T[S/x]$ for all terms T), and u maps each T to the term $T[g/f]$ obtained by replacing an operation symbol f with an operation symbol g of the same arity; then one has

$$\frac{\cdot}{\Gamma \rhd \Delta} \text{(Ax)} \quad [\Gamma \cap \Delta \neq \emptyset]$$

$$\frac{\Gamma \rhd \Delta, \varphi \quad \Gamma, \psi \rhd \Delta}{\Gamma, \varphi \Rightarrow \psi \rhd \Delta} \text{(Left}\Rightarrow) \qquad \frac{\Gamma, \varphi \rhd \Delta, \psi}{\Gamma \rhd \Delta, \varphi \Rightarrow \psi} \text{(Right}\Rightarrow)$$

$$\frac{\Gamma, \varphi, \psi \rhd \Delta}{\Gamma, \varphi \wedge \psi \rhd \Delta} \text{(Left}\wedge) \qquad \frac{\Gamma \rhd \Delta, \varphi \quad \Gamma \rhd \Delta, \psi}{\Gamma \rhd \Delta, \varphi \wedge \psi} \text{(Right}\wedge)$$

$$\frac{\Gamma, \forall x.\varphi, \varphi[T/x] \rhd \Delta}{\Gamma, \forall x.\varphi \rhd \Delta} \text{(Left}\forall) \qquad \frac{\Gamma \rhd \Delta, \varphi[y/x]}{\Gamma \rhd \Delta, \forall x.\varphi} \begin{array}{l} \text{(Right}\forall) \\ [y \text{ fresh}] \end{array}$$

Fig. 1. Gentzen System $\mathcal{G}$

$T[S/x][g/f] = T[g/f][S[g/f]/x]$, i.e., $(u \circ \bar{\theta})(T) = (\overline{u \circ \theta}' \circ u)(T)$, for all terms T.) To any signature morphism (u, v), we associate:

- A *translation* map between the sentences of $(Term, \Pi)$ and $(Term', \Pi')$, that replaces the terms and relation symbols with their images through u and v.

- The following *reduct functor* between the categories of models of $(Term', \Pi')$ and $(Term, \Pi)$: On objects, it maps any $(Term', \Pi')$-model $\mathcal{A}' = (A', \bar{}^{A'}, (A'_{(n,\pi')})_{n \in \mathbb{N}, \pi' \in \Pi'_n})$ to the model $\mathcal{A} = (A, \bar{}^A, (A_{(n,\pi)})_{n \in \mathbb{N}, \pi \in \Pi_n})$ where $A = A'$, $A_{(n,\pi)} = A'_{(n,v_n(\pi))}$ and $\bar{}^A = \rho \mapsto \bar{\rho}^{A'} \circ u$. On morphisms, it maps a function representing a model homomorphism to the same function regarded as an homomorphism between the reduct models. (Details and pictures in [23], Sec. A.4.)

Theorem 1. *TGL as organized above forms an institution that extends conservatively the institution of FOL.*

TGL Gentzen System and Completeness. The axiomatic properties of the generic notions of free variable and substitution in TGL provide enough infrastructure to obtain generic versions of classic FOL results. We are interested in a completeness theorem here (but other model-theoretic results could be also generalized). We shall use a generalization of the cut-free system in [10].

We fix a term-generic language $(Term, \Pi)$. A *sequent* is a pair written $\Gamma \rhd \Delta$, with *antecedent* Γ and *succedent* Δ (at most) countable sets of formulae, assumed to have *finite support*, in that $FV(\Gamma)$ and $FV(\Delta)$ are finite, where $FV(\Gamma) = \bigcup_{\varphi \in \Gamma} FV(\varphi)$ (and likewise for Δ). (In standard Gentzen systems for FOL, Γ and Δ are typically assumed finite, which of course implies finite support.) The sequent $\Gamma \rhd \Delta$ is called *tautological*, written $\models \Gamma \rhd \Delta$, if $\bigcap_{\varphi \in \Gamma} A_\varphi \subseteq \bigcup_{\psi \in \Delta} A_\psi$ for all models A; it is called *E-tautological* (where E is a set of sentences), written $E \models (\Gamma \rhd \Delta)$, if $A \models E$ implies $\bigcap_{\varphi \in \Gamma} A_\varphi \subseteq \bigcup_{\psi \in \Delta} A_\psi$ for all models A. If $\Gamma = \emptyset$, we write $E \models \Delta$ instead of $E \models (\Gamma \rhd \Delta)$.

We consider the Gentzen system, say $\mathcal{G}$, given by the rule schemes in Figure 1, meant to deduce TGL tautological sequents (we write Γ, φ instead of $\Gamma \cup \{\varphi\}$). Note that these rules make sense in our generic framework, since concrete syntax of terms is *not* required; all that is needed here are abstract notions of term and substitution. We write $\vdash_{\mathcal{G}} \Gamma \rhd \Delta$ to mean that $\Gamma \rhd \Delta$ is deducible in $\mathcal{G}$. Similar notation will be used for the other proof systems hereafter.

Theorem 2. $\mathcal{G}$ *is sound and complete for TGL.*

Intuitionistic Term-Generic Logic (ITGL). It has the same syntax as TGL, and its Gentzen system $\mathcal{GI}$ is obtained by modifying $\mathcal{G}$ so that the succedents in sequents are no longer sets of formulae, but single formulae, as follows: **(1)** Δ is deleted from all the Right rules and is replaced by a single formula χ in (Ax) and in all the Left rules except for (Left⇒). **(2)** The rule (Left⇒) is replaced by

$$\frac{\Gamma \rhd \varphi \quad \Gamma, \psi \rhd \chi}{\Gamma, \varphi \Rightarrow \psi \rhd \chi} \qquad \text{(Section A.5 in [23] gives more details.)}$$

3 The Horn² Fragment of Term-Generic Logic

We next consider a fragment of TGL, called HORN² because it only allows formulae which are universally quantified implications whose conditions are themselves universally quantified implications of atomic formulae. A whole plethora of λ-calculi can be specified by HORN² formulae (see Section 4 and [23], Section C). As shown in the sequel, we can associate uniformly to these specifications complete intuitionistic proof systems that turn out to *coincide* with the originals.

In what follows, $\overline{x}$ denotes variable tuples $(x_1, \ldots, x_n)$, $\overline{T}$ term tuples $(T_1, \ldots, T_n)$, and, for a formula φ, $\varphi(\overline{x})$ indicates that φ has all its free variables *among* $\{x_1, \ldots, x_n\}$, with $\varphi(\overline{T})$ then denoting $\varphi[T_1/x_1, \ldots, T_n/x_n]$. Since variables are particular terms, given $\overline{y} = (y_1, \ldots, y_n)$, we may use the notation $\varphi(\overline{y})$ with two different meanings, with disambiguation coming from the context: either to indicate that φ has its variables among $\{y_1, \ldots, y_n\}$, case in which $\varphi(\overline{y})$ is the same as φ, or to denote the formula obtained from $\varphi(\overline{x})$ by substituting the variables $\overline{x}$ with $\overline{y}$. (Thus, e.g., in the property $(*)$ below, $a_i(\overline{x}, \overline{y})$ is the same as a_i, where in addition we have indicated that the free variables of a_i are among $Vars(\overline{x}, \overline{y})$ (where $(\overline{x}, \overline{y})$ is the concatenation of the tuples $\overline{x}$ and $\overline{y}$). Later, given the tuples $\overline{z}$ and $\overline{T}$ of appropriate lengths, $a_i(\overline{z}, \overline{T})$ denotes the formula obtained from a_i by substituting the variables of $\overline{x}$ correspondingly with those of $\overline{z}$ and the variables of $\overline{y}$ correspondingly with the terms of $\overline{T}$.) For convenience, we assume the logic also contains $\top$ (meaning "true") as an atomic formula.

Let HORN² be the TGL fragment given by the sentences:

$$\forall \overline{y}. \left(\forall \overline{x}. \bigwedge_{i=1}^{n} (a_i(\overline{x}, \overline{y}) \Rightarrow b_i(\overline{x}, \overline{y})) \right) \Rightarrow c(\overline{y}) \qquad (*)$$

where a_i, b_i, c are atomic formulae and we assume $Vars(\overline{x}) \cap Vars(\overline{y}) = \emptyset$. We call these HORN² *sentences* (sometimes we shall refer to them as HORN² *formulae*, not forgetting though that they have no free variables). When one of the above a_i's is $\top$ we write only $b_i(\overline{x}, \overline{y})$ instead of $a_i(\overline{x}, \overline{y}) \Rightarrow b_i(\overline{x}, \overline{y})$, and when all the a_i's are $\top$ we call the sentence $(*)$ *extensional* [25]; if, in addition, $\overline{x}$ has length 0, we obtain a Horn sentence (that is, a universal closure of a Horn formula). When all b_i's are $\top$ or $n = 0$, the whole sentence $(*)$ becomes $\forall \overline{y}. c(\overline{y})$. A theory E is called HORN², extensional, or Horn if it consists of such sentences.

Fix a term-generic language and a HORN[2] theory (i.e., specification) E over this language. In what follows, we focus on *Horn sequents*, i.e., sequents $\Gamma \rhd d$ with Γ finite set of atomic formulae and d atomic formula, which can be deduced from E. Only Horn consequences are usually relevant for λ-calculi, and, moreover, the other more syntactically complicated consequences can be deduced from these. We let $\mathcal{K}_E$ denote the following Gentzen system for Horn sequents:

$\dfrac{\cdot}{\Gamma \rhd d}$ (Axiom) $[d \in \Gamma]$	$\dfrac{\Gamma, a_i(\overline{z},\overline{T}) \rhd b_i(\overline{z},\overline{T}) \text{ for } i = \overline{1,n}}{\Gamma \rhd c(\overline{T})}$ (Inst-e)

In the rule (Inst-e) above (the "instance of e" rule), e is a sentence in E of the form $(*)$ (thus a_i, b_i, c are the atomic formulae that build e), $\overline{T}$ is a tuple of terms with the same length as $\overline{y}$, and $\overline{z}$ is a *fresh* tuple of variables with the same length as $\overline{x}$ (where "fresh" (without further attributes) means, as usual, "fresh for everything in that context", namely: for Γ, the a_i's, the b_i's, c and $\overline{T}$). Thus (Inst-e) is a rule (more precisely, a rule-scheme) parameterized not only by e, but also by $\overline{z}$ and $\overline{T}$ as above (and by Γ, too). (More details in [23], Sec. B.)

A first result is that $\mathcal{K}_E$ deduces all intuitionistic Horn consequences of E:

Theorem 3. $\vdash_{\mathcal{GI}} (E \cup \Gamma) \rhd d$ *iff* $\vdash_{\mathcal{K}_E} \Gamma \rhd d$ *for all Horn sequents* $\Gamma \rhd d$.

Now consider the following family of rules (Drop) $=$ (Drop-(e,a))$_{e,a}$, parameterized by formulae $e \in E$ of the form $(*)$ and by atomic formulae a such that a is one of the a_i's (for some $i \in \{1, \ldots, n\}$):

$$\dfrac{\Gamma, a(\overline{z},\overline{T}) \rhd d}{\Gamma \rhd d} \text{ (Drop-}(e,a))$$

($\overline{T}$ is a tuple of terms of the same length as $\overline{y}$ and $\overline{z}$ a tuple of variables of the same length as $\overline{x}$ *fresh for d* (where $\overline{y}$ and $\overline{x}$ are the ones referred in $(*)$).)

From the point of view of forward proofs, (Drop) effectively drops $a(\overline{z},\overline{T})$. More interesting than the actual usage of (Drop) is its admissibility in a system. In a specification of a type system for a λ-calculus, $a(\overline{z},\overline{T})$ will typically have the form $z\!:\!T$, and closure of the system under $a(\overline{z},\overline{T})$ will be a *condensing lemma* [3]: the assumption $z\!:\!T$ is useless provided z is not in the succedent.

Next are our main results of this section, exploring closure under (Drop). The first gives a sufficient criterion ensuring completeness of $\mathcal{K}_E$ w.r.t. TGL models. The second gives a stronger fully syntactic criterion.

Theorem 4. *Assume that:*
-(i) If a_i is not $\top$, then $Vars(\overline{x}) \cap FV(a_i) \neq \emptyset$, for all formulae $e \in E$ of the form $(*)$ and all $i \in \overline{1,n}$.
-(ii) (Drop) is admissible in $\mathcal{K}_E$.
   Then $\vdash_{\mathcal{K}_E} \Gamma \rhd d$ iff $E \models (\Gamma \rhd d)$ for all Horn sequents $\Gamma \rhd d$.

Theorem 5. *Assume that all e in E of the form $(*)$ satisfy the following for all $i \in \overline{1,n}$:*
-(i) If a_i is not $\top$, then $Vars(\overline{x}) \cap FV(a_i) \neq \emptyset$.
-(ii) $Vars(\overline{y}) \cap FV(b_i) \subseteq FV(c)$.
   Then (Drop) is admissible in $\mathcal{K}_E$, hence the conclusion of Theorem 4 holds.

Definition 3. *We call a* HORN^2 *theory E:*
- amenable, if it satisfies the hypotheses of Theorem 4;
- syntax-amenable, if it satisfies hypothesis (i) of Theorem 4 (same as hypothesis (i) of Theorem 5);
- strongly syntax-amenable, if it satisfies the hypotheses of Theorem 5.
(Thus strong syntax-amenability implies amenability.)

If E is a Horn theory, Theorems 4, 5 yield the completeness result for a well-known Hilbert system of Horn logic. More generally, amenability, hence completeness, holds trivially for extensional theories, since they have no (Drop) rules.

Thus classic TGL has, with respect to amenable theories and Horn consequences, *the same deductive power as intuitionistic TGL*. This fact will prove useful for adequacy results and completeness of the TGL models for various calculi. Because these calculi are traditionally specified following an intuitionistic pattern, an amenable HORN^2 specification E of a calculus will recover, in the system $\mathcal{K}_E$, *the represented calculus itself* – we discuss this phenomenon next.

4   Specifying Calculi in Term-Generic Logic

This section illustrates the TGL λ-calculi specification style. Our running example is the typing system and reduction of *System F*, an impredicative polymorphic typed λ-calculus introduced independently in [11] and [24]. (Many other examples can be found in [23], Section C.) Its syntax modulo α-equivalence clearly forms a two-sorted TGL term syntax. The sorts are *type* and *data*, and we write *TVar* for *Var*$_{type}$ (ranged over by t, t') and *DVar* for *Var*$_{data}$ (ranged over by x, y), as well as *TTerm* for *Term*$_{type}$ (ranged over by T, T') and *DTerm* for *Term*$_{data}$ (ranged over by X, Y). Here is the grammar for (the raw terms out of which, by factoring to standard α-equivalence, one obtains) the terms:

$$T ::= t \mid T \to T' \mid \Pi t.T$$
$$X ::= x \mid \lambda x : T.X \mid XY \mid \lambda t.X \mid XT$$

A *typing context* Γ is a finite set $\{x_1 : T_1, \ldots, x_n : T_n\}$ (written $x_1 : T_1, \ldots, x_n : T_n$ for brevity), where the x_i's are data variables, the T_i's are type terms, and no data variable appears twice. The typing system for System F, denoted TSF, deriving sequents $\Gamma \triangleright X : T$, is the following:

$$\frac{\cdot}{\Gamma \triangleright x : T} \text{ (SF-InVar)} \quad [(x : T) \in \Gamma]$$

$$\frac{\Gamma, x : T \triangleright X : T'}{\Gamma \triangleright (\lambda x : T.X) : T \to T'} \text{ (SF-Abs)} \quad [x \text{ fresh for } \Gamma]$$

$$\frac{\Gamma \triangleright X : T \to T' \quad \Gamma \triangleright Y : T}{\Gamma \triangleright XY : T'} \text{(SF-App)}$$

$$\frac{\Gamma \triangleright X : T}{\Gamma \triangleright (\lambda t.X) : \Pi t.T} \text{ (SF-T-Abs)} \quad [t \text{ fresh for } \Gamma]$$

$$\frac{\Gamma \triangleright X : \Pi t.T}{\Gamma \triangleright XT' : T[T'/t]} \text{(SF-T-App)}$$

We specify TSF as a HORN^2 theory by identifying the implicit universal quantifications and implications involved in the original system. For example, we read

(SF-Abs) as: if one can type X to T' *uniformly on* x assuming x has type T, i.e., *for all* x of type T, then $\lambda x{:}T.\,X$ receives type $T \to T'$. But this is HORN^2! (T and T' above are not involved in any bindings relevant here, hence we can use TGL variables instead.) Below is the whole theory, $\mathcal{TSF}$, in a term-generic language over the indicated term syntax and having the infixed relation symbol ":" with arity *data* × *type*. (The colons denoting this relation, although related with, should not be confounded with the colons used as part of the term syntax – our poly-semantic usage of ":" mirrors the usage in the original system TSF.)

$(\forall x.\, x : t \Rightarrow X : t')$ $\Rightarrow (\lambda x{:}t.X) : t \to t'$	(Abs)	$(\forall t.\, X : T)$ $\Rightarrow (\lambda t.X) : (\Pi t.\,T)$	(T-Abs)
$x : t \to t' \wedge y : t \Rightarrow (x\,y) : t'$	(App)	$x : (\Pi t.\,T) \Rightarrow (x\,t) : T$	(T-App)

(Abs), (T-Abs) and (T-App) are axiom-schemes, parameterized by arbitrary terms X, T. In (Abs), a presumptive occurrence of x in the leftmost X is in the scope of the universal quantifier, and in the rightmost X in the scope of the λ-abstraction; similarly for t versus X and t versus T in (T-Abs). This migration of the variables x and t between scopes may look surprising at first – note however that the same situation appears in the corresponding rules ((SF-Abs) and (SF-T-Abs)) from the familiar system TSF. Thus, in (SF-Abs), any occurrence of x in the term X from the succedent of the conclusion sequent $\Gamma \rhd (\lambda x{:}T.\,X) : T \to T'$ is in the scope of the λ-abstraction, while *the same* occurrence of x in X when part of the antecedent of the hypothesis sequent $\Gamma, x : T \rhd X : T'$ is not in the scope of any binder (more precisely, is in the scope of the implicit outer binder of the sequent).

Both in the original system and in our HORN^2 specification, the assumption that T, X, etc. are terms *modulo α-equivalence* is consistent with their usage in combination with binding constructs, since, for example, the syntactic operator $(\lambda_{-}{:}\_.\_) : DVar \times TTerm \times DTerm \to DTerm$ is well defined on α-equivalence classes. Note that a concrete HORN^2 specification cannot be stated solely in terms of the logic's constructs (as is the case of representations in a *fixed* logic, like HOL) simply because TGL does not specify the term syntax, but *assumes* it. Consequently, our examples of specifications employ, at the meta-level, constructs like the above $(\lambda_{-}{:}\_.\_)$, not "purely TGL". (This paper does *not* discuss how to define and represent term syntaxes conveniently, but how to represent the structure of a calculus *on top* of a given term syntax – see also Section 5.)

One should think of the above HORN^2 axioms *semantically*, as referring to items called *data* and *types* that inhabit TGL models – hence our terminology, which distinguishes between *data* terms and variables on the one hand and *type* terms and variables on the other (compare this with the more standard terminology distinguishing between *terms* and *types* from purely syntactic presentations of λ-calculi). As usual, focussing on the semantics allows one to state the desired properties without worrying about syntactic details such as typing contexts and side-conditions; all such lower-level details can nevertheless become available when one "descends" into the deductive system of TGL.

What is the formal relationship between the original typing system TSF and the HORN^2 theory $\mathcal{TSF}$? TSF is precisely $\mathcal{K}_{\mathcal{TSF}}$ from Section 3, the Gentzen system associated to a HORN^2 theory in a uniform way. (Namely, referring to the notations of Section 3: (SF-InVar) is (Axiom), (SF-Abs) is (Inst-Abs), (SF-T-Abs) is (Inst-T-Abs), (SF-App) is (Inst-App), and (SF-T-App) is (Inst-T-App)) Therefore, not only that $\mathcal{TSF}$ specifies TSF, but also TSF implements $\mathcal{TSF}$ as its specialized deductive system. Consequently, the following adequacy result w.r.t. intuitionistic TGL is *built in* the representation (via Theorem 3):

Proposition 1. *Let* $x_1, \ldots, x_n$ *be distinct data variables,* X *data term and* $T, T_1, \ldots, T_n$ *type terms. Then the following are equivalent:*
(a) $\vdash_{TSF} x_1 : T_1, \ldots, x_n : T_n \triangleright X : T$.
(b) $\vdash_{\mathcal{K}_{\mathcal{TSF}}} x_1 : T_1, \ldots, x_n : T_n \triangleright X : T$.
(c) $\vdash_{\mathcal{GI}} \mathcal{TSF}, x_1 : T_1, \ldots, x_n : T_n \triangleright X : T$
(where $\mathcal{TSF}, x_1 : T_1, \ldots, x_n : T_n$ *is a notation for* $\mathcal{TSF} \cup \{x_1 : T_1, \ldots, x_n : T_n\}$*).*

In order to obtain adequacy w.r.t. classic TGL as well, we further need to notice:

Lemma 1. $\mathcal{TSF}$ *satisfies (a many-sorted version of) strong syntax-amenability.*

and then invoke Theorem 5, obtaining:

Proposition 2. *Let* $x_1, \ldots, x_n$ *be distinct data variables,* X *data term and* $T, T_1, \ldots, T_n$ *type terms. Then the following are equivalent:*
(a) $\vdash_{TSF} x_1 : T_1, \ldots, x_n : T_n \triangleright X : T$.
(b) $\mathcal{TSF} \models (x_1 : T_1, \ldots, x_n : T_n \triangleright X : T)$.

Next, we consider the following standard Hilbert system for reduction in System F [11,24] (obtained from the one for the untyped λ-calculus [4] by ignoring the type annotations), denoted RSF:

$$\frac{\cdot}{(\lambda x : T.\, Y) X \rightsquigarrow Y[X/x]}(\text{SF-}\beta) \qquad \frac{X \rightsquigarrow X'}{\lambda x : T.\, X \rightsquigarrow \lambda x : T.\, X'}(\text{SF-}\xi) \qquad \frac{X \rightsquigarrow X'}{X\,Y \rightsquigarrow X'\,Y}(\text{SF-AppL})$$

$$\frac{Y \rightsquigarrow Y'}{X\,Y \rightsquigarrow X\,Y'}(\text{SF-AppR})$$

$$\frac{\cdot}{(\lambda t.\, Y) T \rightsquigarrow Y[T/t]}(\text{SF-T-}\beta) \qquad \frac{X \rightsquigarrow X'}{\lambda t.\, X \rightsquigarrow \lambda t.\, X'}(\text{SF-T-}\xi) \qquad \frac{X \rightsquigarrow X'}{X\,T \rightsquigarrow X'\,T}(\text{SF-T-App})$$

Our HORN^2 specification, denoted $\mathcal{RSF}$, uses relation $\rightsquigarrow$ of arity *data* $\times$ *data*.

$(\lambda x : t.\, Y) x \rightsquigarrow Y$	(β)
$(\lambda t.\, Y) t \rightsquigarrow Y$	$(\text{T-}\beta)$
$(\forall x.\, X \rightsquigarrow X') \Rightarrow \lambda x : t.\, X \rightsquigarrow \lambda x : t.\, X'$	(ξ)
$(\forall t.\, X \rightsquigarrow X') \Rightarrow \lambda t.\, X \rightsquigarrow \lambda t.\, X'$	$(\text{T-}\xi)$

$x \rightsquigarrow x' \Rightarrow x\,y \rightsquigarrow x'\,y$	(AppL)
$y \rightsquigarrow y' \Rightarrow x\,y \rightsquigarrow x\,y'$	(AppR)
$x \rightsquigarrow x' \Rightarrow x\,t \rightsquigarrow x'\,t$	(T-App)

Particularly interesting are our axioms for β-reduction. In (β), we employ the same variable x to indicate both the *formal parameter* of the functional expression $\lambda x : t.\, Y$ and its *actual parameter* (the occurrence of x on the right of the application from the left side of $\rightsquigarrow$). Indeed, in the latter case, as well as in any presumptive occurrences in the rightmost Y, x is exposed to the environment, hence denotes an (arbitrary) actual value in a model.

Again, $\mathcal{K}_{\mathcal{RSF}}$ is the same as RSF (modulo a standard identification of Hilbert systems with simple Gentzen systems where antecedents remain unchanged). Moreover, $\mathcal{RSF}$ is an extensional theory, hence trivially amenable, hence both intuitionistically and classically adequate:

Proposition 3. *Let X and Y be data terms. Then the following are equivalent:*
(a) $\vdash_{RSF} X \rightsquigarrow Y$.
(b) $\vdash_{\mathcal{GI}} \mathcal{RSF} \triangleright X \rightsquigarrow Y$.
(c) $\mathcal{RSF} \models X \rightsquigarrow Y$.

One can readily see that, since the relation symbols of $\mathcal{STF}$ and $\mathcal{RSF}$ are distinct, putting these theories together preserves adequacy – in other words, Propositions 1 and 2 remain true after replacing $\mathcal{SFT}$ with $\mathcal{SFT} \cup \mathcal{RSF}$ and Proposition 3 remains true after replacing $\mathcal{RSF}$ with $\mathcal{SFT} \cup \mathcal{RSF}$. In the union language, we can express relevant properties such as *type preservation*: $\forall x, y, t.\, x : t \land x \rightsquigarrow y \Rightarrow y : t$. The proof of such properties requires reasoning *about the calculus*, hence transcends the realm of adequate representations. To handle them, TGL needs to be extended with inductive proof schemes, such as:

$$\frac{\varphi(x) \land \varphi(y) \Rightarrow \varphi(x\,y) \quad ((\forall x.\, \varphi(X)) \Rightarrow \varphi(\lambda x\!:\!t.\, X))_{X \in DTerm}}{\forall x.\, \varphi(x)}\, (\mathsf{Induct}_{data})$$

The problem of meta-reasoning in a framework where object-level calculi are represented without explicitly encoding free variables and substitution (currently still open in frameworks such as HOAS) is not addressed in this paper, but is left as important future work.

Intuitionistic TGL adequacy (Proposition 1) holds immediately (for the same reason as for System F) for all calculi specified in [23], Section C. Classic TGL adequacy, on the other hand, while trivial for System F (in the context of our a priori proof theory), is not so in other calculi, where strong syntax-amenability does not hold, but only syntax amenability does, and closure under (Drop), while intuitive, is not obvious to prove. Fortunately however, for most of these calculi this property coincides with a known result called the *condensing lemma* (see [3]): in a typing context $\Gamma \triangleright U : V$, an assumption $x : T$ from Γ with x fresh for U and V may be dropped without losing provability. Note that, via the propositions-as-types correspondence, representing adequately type systems in TGL also implies representing adequately proof systems for structural logics.

Sometimes a calculus does not come with a reduction relation, but with an equational theory. (Notably, a standard formulation of untyped λ-calculus [4] is equational.) For these situations, a version of TGL with equality seems a more elegant choice, but adequacy proofs along our lines seem to require more effort, since the TGL equality axioms raise problems regarding amenability (not to mention that type preservation needs to be proved beforehand for the calculus). Alternatively, one may provide semantic proofs for adequacy, taking advantage of the equivalence between the TGL models and some ad hoc models for which the calculus is known to be complete (see Section E in [23] for this approach).

5   Concluding Remarks

Summing up the contribution of this paper:

(1) We showed that the development of first-order logic is largely orthogonal to the particular syntax of terms by defining a logic, TGL, that considers terms as "black-boxes" exporting substitution and free variables and requires models to represent terms consistently. TGL forms an institution, hence allows in principle for well-structured logical specifications.

(2) TGL provides a convenient notation and intuition for defining λ-calculi, that encourages a semantic specification style. We developed some proof theory to support this specification style. Intuitionistic TGL allows immediately adequate specifications, while for classic TGL adequacy, if provable, endows the specified calculus with a *default complete semantics*.

The idea of developing first-order logic on top of an abstract term syntax, as well as our proof-theoretic results that prepare the logic in advance for adequate representations of λ-calculi, seem new.[4] We separate the discussion of related work into two (non-disjoint) topics.

One concerns semantics. The semantics that TGL offers to the specified calculi for free falls into the category of *loose*, or *logical* semantics. Examples of loose semantics for λ-calculi include: (so called) "syntactic" models for untyped λ-calculus, Henkin models for simply-typed λ-calculus, Kripke-style models for recursive types, and Girard's qualitative domains and Bruce-Meyer-Mitchell models for System F, not to mention all their categorical variants. The monographs [4,14,20] contain extensive presentations of these and many other loose semantics for various calculi. For a particular calculus defined as a TGL theory, the attached TGL semantics has all the advantages, but, naturally, also all the drawbacks, of loose semantics. It was not the concern of this paper to advocate for a loose or for a fixed-model semantics, especially because we believe that there is no absolute answer. What we consider to be a particularly appealing aspect of TGL semantics though is its uniform, *calculus-independent* nature. (We argue in [23] (Section E), with untyped λ-calculus and System F as witnesses, that the "general-purpose" TGL semantics of a calculus tends to be equivalent to the set-theoretic "domain-specific" one whose completeness theorem is typically worked out separately with substantial mathematical effort in the literature.)

The other topic concerns existing specification frameworks in the literature:

- Purely first-order encodings, such as combinatory logic [4], de Bruijn-style representations [6], and the calculus with explicit substitution [1]. Part of the motivation of TGL was to avoid the degree of awkwardness and auxiliary proof or execution overhead of such encodings.

- Higher-order abstract syntax (HOAS) [15,17,21]. This approach encodes (in a binding-preserving fashion) object-level terms into terms of a *fixed meta logic* (usually HOL or an other type theory) – consequently, the interpretation of the object syntax into presumptive models of the meta logic would be indirect, filtered through the encoding. To the contrary, TGL is a *parameterized logic*, and

---

[4] But see below the related work on HOAS and categorical models of syntax.

gets *instantiated* to various calculi by importing the original term syntax *as is* and relating models to this syntax *directly* through valuations. Moreover, usually model-theoretic considerations are not the concern of HOAS, which aims at proof-theoretic adequacy alone, a property that so far seemed to require an intuitionistic meta logic; here we also developed for TGL a technique for establishing adequacy within a classic logic.

Yet, TGL representations have important similarities with HOAS encodings in variants of HOL (in the style of, e.g., [17]). For instance, our axiom-scheme (Abs) from the HORN[2] theory $\mathcal{TSF}$ may be in such an encoding $\forall X. (\forall x. x : t \Rightarrow X(x) : t') \Rightarrow Lam(X) : (t \rightarrow t')$, where $X : data \longrightarrow data$ is a second-order variable and $Lam : (data \longrightarrow data) \longrightarrow data$ is a third-order constant. A HOAS encoding has typically two parts, each requiring its own adequacy result: one deals with representing the syntax of terms, and one with representing the deductive mechanism. Because TGL does not provide a representation of syntax (but assumes one already), some of our axioms, namely those changing variable scopes, such as (Abs), are (still) axiom-schemes, just like the rules of the original calculus are rule-schemes; to the contrary, the above HOAS axiom would be a single statement. On the other hand, for the same reason (of not dealing with term syntax representation), we were able to discuss the second part, of representing the deductive mechanism, *generically, for any term syntax,* and have created a theoretical framework where adequacy for the deductive mechanisms requires minimal proof effort. "Pasting" various solutions offered by HOAS to representing terms into the TGL framework for representing deduction could allow a HOAS setting to benefit from our theorems in Section 3, as well as allow a HOAS representation of an effective-syntax fragment of TGL to bypass the need of axiom-schemes in specifications.

Categorical models of syntax in the style of [9,16] also fall within HOAS. Typing contexts are explicitly modeled as possible worlds, types becoming presheaves. The presheaf structure of λ-terms from [16] and the substitution algebras from [9] are roughly equivalent to our term syntaxes (whose presheaf structure would come, just like in the concrete cases, from classifying terms by their sets of free variables). The model theory of the these categorical settings follows a different approach than ours though – they require the models to support *substitution within themselves and between each other* (hence to be inhabited by syntactic items such as (abstract) terms and variables), while we require the models to allow *valuations from a fixed term model.*

- Nominal logic (NL) [22]. It stands somewhere in between purely first-order encodings and HOAS, as it captures object-level bindings, but not substitution, by corresponding meta-level mechanisms. The NL terms with bindings form term syntaxes in our sense. Like in the categorical approaches mentioned above and unlike TGL models, NL models are inhabited by abstract syntactic objects (having, e.g., free names that can be swapped/permuted) rather than constituting "pure" FOL-like semantics.

- Explicitly closed families of functionals (ECFFs) [2] (a.k.a. binding algebras [27]). In the tradition of HOL a la Church, all bindings are reduced there to functional abstraction. Their terms form term syntaxes in our sense, and ECFFs are particular cases of TGL models.

- Binding logic (BL) [8]. It is a first-order logic defined on top of a general notion of syntax with binding, allowing bindings in both operations and atomic predicates. BL models reflect the bindings *functionally* (similarly to [2], [27]). While BL terms form TGL term syntaxes, it appears that the class of BL models is strictly embedded in that of TGL models for TGL terms syntax instantiated to a BL language of terms.

- Hereditary Harrop Formulae (HHF). For the FOL and HOL instances of TGL, HORN[2] formulae are particular cases of such formulae, advocated in [19] for logic programming. Our proof-theoretic results from Section 3 seem HORN[2]-specific, a generalization to HHF not being apparent.

- In the general realm of logical and algebraic specifications, a salient framework is that of institutions [12,7]. Like TGL, the notion of institution does not represent logical systems by *encoding* them, but by becoming *instantiated* to them. Since we showed that TGL is itself an institution,[5] our work in this paper offers to the λ-calculi adequately specifiable in TGL institutional citizenship, hence the algebraic arsenal of tools and techniques from institution theory [26].

References

1. Abadi, M., Cardelli, L., Curien, P.-L., Lévy, J.-J.: Explicit substitutions. J. Funct. Program. 1(4), 375–416 (1991)
2. Aczel, P.: Frege structures and the notions of proposition, truth and set. In: The Kleene Symposium, pp. 31–59. North Holland, Amsterdam (1980)
3. Barendregt, H.: Introduction to generalized type systems. J. Funct. Program. 1(2), 125–154 (1991)
4. Barendregt, H.P.: The Lambda Calculus. North-Holland, Amsterdam (1984)
5. Birkhoff, G.: On the structure of abstract algebras. Proceedings of the Cambridge Philosophical Society 31, 433–454 (1935)
6. Bruijn, N.: λ-calculus notation with nameless dummies, a tool for automatic formula manipulation, with application to the Church-Rosser theorem. Indag. Math. 34(5), 381–392 (1972)
7. Diaconescu, R.: Institution-independent Model Theory. Birkhauser, Basel (2008)
8. Dowek, G., Hardin, T., Kirchner, C.: Binding logic: Proofs and models. In: Baaz, M., Voronkov, A. (eds.) LPAR 2002. LNCS, vol. 2514, pp. 130–144. Springer, Heidelberg (2002)
9. Fiore, M., Plotkin, G., Turi, D.: Abstract syntax and variable binding. In: Proc. 14th LICS Conf., pp. 193–202. IEEE, Los Alamitos (1999)
10. Gallier, J.H.: Logic for computer science. Foundations of automatic theorem proving. Harper & Row (1986)
11. Girard, J.-Y.: Une extension de l'interpretation de Gödel a l'analyse, et son application a l'elimination des coupure dans l'analyse et la theorie des types. In: Fenstad, J. (ed.) 2nd Scandinavian Logic Symposium, pp. 63–92. North Holland, Amsterdam (1971)

---

[5] In fact, we showed that TGL is an institution endowed with a (sound and complete) proof system, making it a *logical system* in the sense of [18].

12. Goguen, J., Burstall, R.: Institutions: Abstract model theory for specification and programming. Journal of the ACM 39(1), 95–146 (1992)
13. Goguen, J., Meseguer, J.: Order-sorted algebra I. Theoretical Computer Science 105(2), 217–273 (1992)
14. Gunter, C.A.: Semantics of Programming Languages. MIT Press, Cambridge (1992)
15. Harper, R., Honsell, F., Plotkin, G.: A framework for defining logics. In: Proc. 2nd LICS Conf., pp. 194–204. IEEE, Los Alamitos (1987)
16. Hofmann, M.: Semantical analysis of higher-order abstract syntax. In: Proc. 14th LICS Conf., pp. 204–213. IEEE, Los Alamitos (1999)
17. McDowell, R.C., Miller, D.A.: Reasoning with higher-order abstract syntax in a logical framework. ACM Trans. Comput. Logic 3(1), 80–136 (2002)
18. Meseguer, J.: General logics. In: Ebbinghaus, H.-D., et al. (eds.) Proceedings, Logic Colloquium 1987, pp. 275–329. North-Holland, Amsterdam (1989)
19. Miller, D., Nadathur, G., Pfenning, F., Scedrov, A.: Uniform proofs as a foundation for logic programming. Ann. Pure Appl. Logic 51(1-2), 125–157 (1991)
20. Mitchell, J.C.: Foundations for Programming Languages. MIT Press, Cambridge (1996)
21. Pfenning, F., Elliot, C.: Higher-order abstract syntax. In: PLDI 1988, pp. 199–208. ACM Press, New York (1988)
22. Pitts, A.M.: Nominal logic: A first order theory of names and binding. In: Kobayashi, N., Pierce, B.C. (eds.) TACS 2001. LNCS, vol. 2215, pp. 219–242. Springer, Heidelberg (2001)
23. Popescu, A., Roşu, G.: Term-generic logic. Tech. Rep. Univ. of Illinois at Urbana-Champaign UIUCDCS-R-2009-3027 (2009)
24. Reynolds, J.C.: Towards a theory of type structure. In: Robinet, B. (ed.) Programming Symposium. LNCS, vol. 19, pp. 408–423. Springer, Heidelberg (1974)
25. Roşu, G.: Extensional theories and rewriting. In: Díaz, J., Karhumäki, J., Lepistö, A., Sannella, D. (eds.) ICALP 2004. LNCS, vol. 3142, pp. 1066–1079. Springer, Heidelberg (2004)
26. Sannella, D., Tarlecki, A.: Foundations of Algebraic Specifications and Formal Program Development. To appear in Cambridge University Press (Ask authors for current version at tarlecki@mimuw.edu.pl)
27. Sun, Y.: An algebraic generalization of Frege structures - binding algebras. Theoretical Computer Science 211(1-2), 189–232 (1999)

Declarative Debugging of Rewriting Logic Specifications*

Adrian Riesco, Alberto Verdejo, Rafael Caballero, and Narciso Martí-Oliet

Facultad de Informática, Universidad Complutense de Madrid, Spain
ariesco@fdi.ucm.es, {alberto,rafa,narciso}@sip.ucm.es

Abstract. Declarative debugging is a semi-automatic technique that starts from an incorrect computation and locates a program fragment responsible for the error by building a tree representing this computation and guiding the user through it to find the wrong statement. This paper presents the fundamentals for the declarative debugging of rewriting logic specifications, realized in the Maude language, where a wrong computation can be a reduction, a type inference, or a rewrite. We define appropriate debugging trees obtained as the result of collapsing in proof trees all those nodes whose correctness does not need any justification. Since these trees are obtained from a suitable semantic calculus, the correctness and completeness of the debugging technique can be formally proved. We illustrate how to use the debugger by means of an example and succinctly describe its implementation in Maude itself thanks to its reflective and metalanguage features.

1 Introduction

In this paper we present a declarative debugger for *Maude specifications*, including equational functional specifications and concurrent systems specifications. Maude [10] is a high-level language and high-performance system supporting both equational and rewriting logic computation for a wide range of applications. Maude modules correspond to specifications in *rewriting logic* [14], a simple and expressive logic which allows the representation of many models of concurrent and distributed systems. This logic is an extension of equational logic; in particular, Maude *functional modules* correspond to specifications in *membership equational logic* [1, 15], which, in addition to equations, allows the statement of *membership axioms* characterizing the elements of a sort. In this way, Maude makes possible the faithful specification of data types (like sorted lists or search trees) whose data are not only defined by means of constructors, but also by the satisfaction of additional properties. Rewriting logic extends membership equational logic by adding rewrite rules, that represent transitions in a concurrent system. Maude *system modules* are used to define specifications in this logic.

---

* Research supported by MEC Spanish projects *DESAFIOS* (TIN2006-15660-C02-01) and *MERIT-FORMS* (TIN2005-09027-C03-03), and Comunidad de Madrid program *PROMESAS* (S-0505/TIC/0407).

A. Corradini and U. Montanari (Eds.): WADT 2008, LNCS 5486, pp. 308–325, 2009.

The Maude system supports several approaches for debugging Maude programs: tracing, term coloring, and using an internal debugger [10, Chap. 22]. The tracing facilities allow us to follow the execution of a specification, that is, the sequence of applications of statements that take place. The same ideas have been applied to the functional paradigm by the tracer *Hat* [9], where a graph constructed by graph rewriting is proposed as suitable trace structure. Term coloring consists in printing with different colors the operators used to build a term that does not fully reduce. The Maude debugger allows to define break points in the execution by selecting some operators or statements. When a break point is found the debugger is entered. There, we can see the current term and execute the next rewrite with tracing turned on. The Maude debugger has as a disadvantage that, since it is based on the trace, it shows to the user every small step obtained by using a single statement. Thus the user can lose the general view of the *proof* of the incorrect inference that produced the wrong result. That is, when the user detects an unexpected statement application it is difficult to know where the incorrect inference started. Here we present a different approach based on declarative debugging that solves this problem for Maude specifications.

Declarative debugging, also known as algorithmic debugging, was first introduced by E. Y. Shapiro [23]. It has been widely employed in the logic [12, 16, 25], functional [18, 19, 20], multi-paradigm [3, 7, 13], and object-oriented [4] programming languages. Declarative debugging starts from a computation considered incorrect by the user (error symptom) and locates a program fragment responsible for the error. The declarative debugging scheme [17] uses a *debugging tree* as logical representation of the computation. Each node in the tree represents the result of a computation step, which must follow from the results of its child nodes by some logical inference. Diagnosis proceeds by traversing the debugging tree, asking questions to an external oracle (generally the user) until a so-called *buggy node* is found. A buggy node is a node containing an erroneous result, but whose children have all correct results. Hence, a buggy node has produced an erroneous output from correct inputs and corresponds to an erroneous fragment of code, which is pointed out as an error. From an explanatory point of view, declarative debugging can be described as consisting of two stages, namely the debugging tree *generation* and its *navigation* following some suitable strategy [24].

The application of declarative debugging to Maude functional modules was already studied in our previous papers [5, 6]. The executability requirements of Maude functional modules mean that they are assumed to be confluent, terminating, and sort-decreasing[1] [10]. These requirements are assumed in the form of the questions appearing in the debugging tree. In this paper, we considerably extend that work by also considering system modules. Now, since the specifications described in this kind of modules can be non-terminating and non-confluent, their handling must be quite different.

The debugging process starts with an incorrect computation from the initial term to an unexpected one. The debugger then builds an appropriate debugging

---

[1] All these requirements must be understood *modulo* some axioms such as associativity and commutativity that are associated to some binary operations.

tree which is an abbreviation of the corresponding proof tree obtained by applying the inference rules of membership equational logic and rewriting logic. The abbreviation consists in collapsing those nodes whose correctness does not need any justification, such as those related with transitivity or congruence. Since the questions are located in the debugging tree, the answers allow the debugger to discard a subset of the questions, leading and shortening the debugging process. In the case of functional modules, the questions have the form "Is it correct that T fully reduces to T'?", which in general are easier to answer. However, in the absence of confluence and termination, these questions do not make sense; thus, in the case of system modules, we have decided to develop two different trees whose nodes produce questions of the form "Is it correct that T is rewritten to T'?" where the difference consists in the number of steps involved in the rewrite. While one of the trees refers only to one-step rewrites, which are often easier to answer, the other one can also refer to many-steps rewrites that, although may be harder to answer, in general discard a bigger subset of nodes. The user, depending on the debugged specification or his "ability" to answer questions involving several rewrite steps, can choose between these two kinds of trees.

Moreover, exploiting the fact that rewriting logic is *reflective* [11], a key distinguishing feature of Maude is its systematic and efficient use of reflection through its predefined META-LEVEL module [10, Chap. 14], a feature that makes Maude remarkably extensible and powerful, and that allows many advanced metaprogramming and metalanguage applications. This powerful feature allows access to metalevel entities such as specifications or computations as usual data. Therefore, we are able to generate and navigate the debugging tree of a Maude computation using operations in Maude itself. In addition, the Maude system provides another module, LOOP-MODE [10, Chap. 17], which can be used to specify input/output interactions with the user. However, instead of using this module directly, we extend Full Maude [10, Chap. 18], that includes features for parsing, evaluating, and pretty-printing terms, improving the input/output interaction. Moreover, Full Maude allows the specification of concurrent object-oriented systems, that can also be debugged. Thus, our declarative debugger, including its user interactions, is implemented in Maude itself.

The rest of the paper is structured as follows. Sect. 2 provides a summary of the main concepts of both membership equational logic and rewriting logic, and how their specifications are realized in Maude functional and system modules, respectively. Sect. 3 describes the theoretical foundations of the debugging trees for inferences in both logics. Sect. 4 shows how to use the debugger by means of an example, while Sect. 5 comments some aspects of the Maude implementation. Finally, Sect. 6 concludes and mentions some future work.

Detailed proofs of the results, additional examples, and much more information about the implementation can be found in the technical report [21], which, together with the Maude source files for the debugger, is available from the webpage http://maude.sip.ucm.es/debugging

2 Rewriting Logic and Maude

As mentioned in the introduction, Maude modules are executable rewriting logic specifications. Rewriting logic [14] is a logic of change very suitable for the specification of concurrent systems that is parameterized by an underlying equational logic, for which Maude uses membership equational logic (*MEL*) [1, 15], which, in addition to equations, allows the statement of membership axioms characterizing the elements of a sort.

2.1 Membership Equational Logic

A *signature* in *MEL* is a triple (K, Σ, S) (just Σ in the following), with K a set of *kinds*, $\Sigma = \{\Sigma_{k_1\ldots k_n,k}\}_{(k_1\ldots k_n,k)\in K^*\times K}$ a many-kinded signature, and $S = \{S_k\}_{k\in K}$ a pairwise disjoint K-kinded family of sets of *sorts*. The kind of a sort s is denoted by $[s]$. We write $T_{\Sigma,k}$ and $T_{\Sigma,k}(X)$ to denote respectively the set of ground Σ-terms with kind k and of Σ-terms with kind k over variables in X, where $X = \{x_1 : k_1, \ldots, x_n : k_n\}$ is a set of K-kinded variables. Intuitively, terms with a kind but without a sort represent undefined or error elements.

The atomic formulas of *MEL* are either *equations* $t = t'$, where t and t' are Σ-terms of the same kind, or *membership axioms* of the form $t : s$, where the term t has kind k and $s \in S_k$. *Sentences* are universally-quantified Horn clauses of the form $(\forall X)\, A_0 \Leftarrow A_1 \wedge \ldots \wedge A_n$, where each A_i is either an equation or a membership axiom, and X is a set of K-kinded variables containing all the variables in the A_i. A *specification* is a pair (Σ, E), where E is a set of sentences in *MEL* over the signature Σ.

Models of *MEL* specifications are Σ-*algebras* $\mathcal{A}$ consisting of a set A_k for each kind $k \in K$, a function $A_f : A_{k_1} \times \cdots \times A_{k_n} \longrightarrow A_k$ for each operator $f \in \Sigma_{k_1\ldots k_n,k}$, and a subset $A_s \subseteq A_k$ for each sort $s \in S_k$. The meaning $[\![t]\!]_{\mathcal{A}}$ of a term t in an algebra $\mathcal{A}$ is inductively defined as usual. Then, an algebra $\mathcal{A}$ satisfies an equation $t = t'$ (or the equation holds in the algebra), denoted $\mathcal{A} \models t = t'$, when both terms have the same meaning: $[\![t]\!]_{\mathcal{A}} = [\![t']\!]_{\mathcal{A}}$. In the same way, satisfaction of a membership is defined as: $\mathcal{A} \models t : s$ when $[\![t]\!]_{\mathcal{A}} \in A_s$.

A *MEL* specification (Σ, E) has an initial model $\mathcal{T}_{\Sigma/E}$ whose elements are E-equivalence classes of terms $[t]$. We refer to [1, 15] for a detailed presentation of (Σ, E)-algebras, sound and complete deduction rules (that we adapt to our purposes in Fig. 1 in Sect. 3.1), as well as the construction of initial and free algebras. Since the *MEL* specifications that we consider are assumed to satisfy the executability requirements of confluence, termination, and sort-decreasingness, their equations $t = t'$ can be oriented from left to right, $t \to t'$. Such a statement holds in an algebra, denoted $\mathcal{A} \models t \to t'$, exactly when $\mathcal{A} \models t = t'$, i.e., when $[\![t]\!]_{\mathcal{A}} = [\![t']\!]_{\mathcal{A}}$. Moreover, under those assumptions an equational condition $u = v$ in a conditional equation can be checked by finding a common term t such that $u \to t$ and $v \to t$. The notation we will use in the inference rules studied in Sect. 3 for this situation is $u \downarrow v$.

2.2   Rewriting Logic

Rewriting logic extends equational logic by introducing the notion of *rewrites* corresponding to transitions between states; that is, while equations are interpreted as equalities and therefore they are symmetric, rewrites denote changes which can be irreversible. A rewriting logic specification, or *rewrite theory*, has the form $\mathcal{R} = (\Sigma, E, R)$, where (Σ, E) is an equational specification and R is a set of *rules* as described below. From this definition, one can see that rewriting logic is built on top of equational logic, so that rewriting logic is parameterized with respect to the version of the underlying equational logic; in our case, Maude uses *MEL*, as described in the previous section. A rule in R has the general conditional form[2]

$$(\forall X)\, t \Rightarrow t' \Leftarrow \bigwedge_{i=1}^{n} u_i = u_i' \wedge \bigwedge_{j=1}^{m} v_j : s_j \wedge \bigwedge_{k=1}^{l} w_k \Rightarrow w_k'$$

where the head is a rewrite and the conditions can be equations, memberships, and rewrites; both sides of a rewrite must have the same kind. From these rewrite rules, one can deduce rewrites of the form $t \Rightarrow t'$ by means of general deduction rules introduced in [2, 14], that we have adapted to our purposes.

Models of rewrite theories are called $\mathcal{R}$-*systems* in [14]. Such systems are defined as categories that possess a (Σ, E)-algebra structure, together with a natural transformation for each rule in the set R. More intuitively, the idea is that we have a (Σ, E)-algebra, as described in Sect. 2.1, with transitions between the elements in each set A_k; moreover, these transitions must satisfy several additional requirements, including that there are identity transitions for each element, that transitions can be sequentially composed, that the operations in the signature Σ are also appropriately defined for the transitions, and that we have enough transitions corresponding to the rules in R. Then, if we keep in this context the notation $\mathcal{A}$ to denote an $\mathcal{R}$-system, a rewrite $t \Rightarrow t'$ is satisfied by $\mathcal{A}$, denoted $\mathcal{A} \models t \Rightarrow t'$, when there is a transition $[\![t]\!]_{\mathcal{A}} \Rightarrow_{\mathcal{A}} [\![t']\!]_{\mathcal{A}}$ in the system between the corresponding meanings of both sides of the rewrite, where $\Rightarrow_{\mathcal{A}}$ will be our notation for such transitions. The rewriting logic deduction rules introduced in [14] are sound and complete with respect to this notion of model. Moreover, they can be used to build initial and free models; see [14] for details.

2.3   Maude Modules

Maude functional modules [10, Chap. 4], introduced with syntax `fmod` ... `endfm`, are executable membership equational specifications and their semantics is given by the corresponding initial membership algebra in the class of algebras satisfying the specification. In a functional module we can declare sorts

---

[2] Note that we use the notation $\Rightarrow$ for rewrites (as in Maude) and $\rightarrow$ for *oriented* equations and reductions using such equations. Other papers on rewriting logic use instead the notation $\rightarrow$ for rewrites.

(by means of keyword sort(s)); subsort relations between sorts (subsort); operators (op) for building values of these sorts, giving the sorts of their arguments and result, and which may have attributes such as being associative (assoc) or commutative (comm), for example; memberships (mb) asserting that a term has a sort; and equations (eq) identifying terms. Both memberships and equations can be conditional (cmb and ceq).

Maude system modules [10, Chap. 6], introduced with syntax mod ... endm, are executable rewrite theories and their semantics is given by the initial system in the class of systems corresponding to the rewrite theory. A system module can contain all the declarations of a functional module and, in addition, declarations for rules (rl) and conditional rules (crl).

The executability requirements for equations and memberships are confluence, termination, and sort-decreasingness. With respect to rules, the satisfaction of all the conditions in a conditional rewrite rule is attempted sequentially from left to right, solving rewrite conditions by means of search; for this reason, we can have new variables in such conditions but they must become instantiated along this process of solving from left to right (see [10] for details). Furthermore, the strategy followed by Maude in rewriting with rules is to compute the normal form of a term with respect to the equations before applying a rule. This strategy is guaranteed not to miss any rewrites when the rules are *coherent* with respect to the equations [10, 26].

The following section describes an example of a Maude system module with both equations and rules.

2.4   An Example: Knight's Tour Problem

A knight's tour is a journey around the chessboard in such a way that the knight lands on each square exactly once. The legal move for a knight is two spaces in one direction, then one in a perpendicular direction. We want to solve the problem for a 3×4 chessboard with the knight initially located in one corner.

We represent positions in the chessboard as pairs of integers and journeys as lists of positions.

```
(mod KNIGHT is
  protecting INT .
  sorts Position Movement Journey Problem .
  subsort Position < Movement .
  subsorts Position < Journey < Problem .
  op [_,_] : Int Int -> Position .
  op nil : -> Journey .
  op __ : Journey Journey -> Journey [assoc id: nil] .
  vars N X Y : Int .  vars P Q : Position .  var J : Journey .
```

The term move P represents a position reachable from position P. Since the reachable positions are not unique, this operation is defined by means of rewrite rules, instead of equations. The reachable positions can be outside the chessboard, so we define the operation legal, that checks if a position is inside the 3×4 chessboard.

```
op move_ : Position -> Movement .
rl [mv1] : move [X, Y] => [X + 2, Y + 1] .
...
rl [mv8] : move [X, Y] => [X - 1, Y - 2] .
op legal : Position -> Bool .
eq [leg] : legal([X, Y]) = X >= 1 and Y >= 1 and X <= 3 and Y <= 4 .
```

The function contains(J, P) checks if position P occurs in the journey J.

```
op contains : Journey Position -> Bool .
eq [con1] : contains(P J, P) = true .
eq [con2] : contains(J, P) = false [otherwise] .
```

knight(N) represents a journey where the knight has performed N hops. When no hops are taken, the knight remains at the first position [1, 1]. When N > 0 the problem is recursively solved (using backtracking in an implicit way) as follows: first a legal journey of N - 1 steps is found, then a new hop from the last position of that journey is performed, and finally it is checked that this last hop is legal and compatible with the other ones.

```
op knight : Nat -> Problem .
rl [k1] : knight(0) => [1, 1] .
crl [k2] : knight(N) => J P Q
  if N > 0
  /\ knight(N - 1) => J P
  /\ move P => Q
  /\ legal(Q)
  /\ not(contains(J P, Q)) .
endm)
```

The solution to the 3 × 4 chessboard can be found by looking for a journey with 11 hops, but we obtain the following unexpected, wrong result, where the journey contains repeated positions. We will show how to debug it in Sect. 4.

```
Maude> (rew knight(11) .)
result Journey :
    [1,1] [2,3] [3,1] [2,3] [3,1] [2,3] [3,1] [2,3] [3,1] [2,3] [3,1] [2,3]
```

3   Debugging Trees for Maude Specifications

Now we will describe debugging trees for both *MEL* specifications and rewriting logic specifications. Since a *MEL* specification coincides with a rewrite theory with an empty set of rules, our treatment will simply be at the level of rewrite theories. Our proof and debugging trees will include statements for reductions $t \to t'$, memberships $t : s$, and rewrites $t \Rightarrow t'$, and in the following sections we will describe how to build the debugging trees from the proof trees taking into account each kind of statement.

3.1   Proof Trees

Before defining the debugging trees employed in our declarative debugging framework we introduce the semantic rules defining the semantics of a rewrite theory $\mathcal{R}$. The inference rules of the calculus can be found in Fig. 1, where θ denotes a substitution. The rules allow to deduce statements of the three kinds and are an adaptation of the rules presented in [1, 15] for *MEL* and in [2, 14] for rewriting logic. With respect to *MEL*, because of the executability assumptions, we have a more operational interpretation of the equations, which are oriented from left to right. With respect to rewriting logic, we work with terms (as in [2]) instead of equivalence classes of terms (as in [14]); moreover, unlike [2], replacement is not nested. Both changes make the logical representation closer to the way the Maude system operates. As usual, we represent deductions in the calculus as *proof trees*, where the premises are the child nodes of the conclusion at each inference step. We assume that the inference labels $(Rep_\Rightarrow)$, $(Rep_\rightarrow)$, and (Mb) decorating the inference steps contain information about the particular rewrite rule, equation, and membership axiom, respectively, applied during the inference. This information will be used by the debugger in order to present to the user the incorrect fragment of code causing the error.

In our debugging framework we assume the existence of an *intended interpretation* $\mathcal{I}$ of the given rewrite theory $\mathcal{R} = (\Sigma, E, R)$. The intended interpretation must be an $\mathcal{R}$-system corresponding to the model that the user had in mind while writing the specification $\mathcal{R}$. Therefore the user expects that $\mathcal{I} \models t \Rightarrow t'$, $\mathcal{I} \models t \rightarrow t'$, and $\mathcal{I} \models t : s$ for each rewrite $t \Rightarrow t'$, reduction $t \rightarrow t'$, and membership $t : s$ computed w.r.t. the specification $\mathcal{R}$. We will say that a statement $t \Rightarrow t'$ (respectively $t \rightarrow t'$, $t : s$) is *valid* when it holds in $\mathcal{I}$, and *invalid* otherwise. Declarative debuggers rely on some external oracle, normally the user, in order to obtain information about the validity of some nodes in the debugging tree. The concept of validity can be extended to distinguish *wrong rules*, *wrong equations*, and *wrong membership axioms*, which are those specification pieces that can deduce something invalid from valid information.

Definition 1. *Let* $r \equiv (af \Leftarrow \bigwedge_{i=1}^{n} u_i = u_i' \wedge \bigwedge_{j=1}^{m} v_j : s_j \wedge \bigwedge_{k=1}^{l} w_k \Rightarrow w_k')$ *where af denotes an atomic formula, that is, r is either a rewrite rule, an oriented equation, or a membership axiom (in the last two cases $l = 0$) in some rewrite theory $\mathcal{R}$. Then:*

- *$\theta(r)$ is a* wrong rewrite rule instance *(respectively* wrong equation instance *and* wrong membership axiom instance*) w.r.t. an intended interpretation $\mathcal{I}$ when*
  1. *There exist terms $t_1, \ldots, t_n$ such that $\mathcal{I} \models \theta(u_i) \rightarrow t_i$, $\mathcal{I} \models \theta(u_i') \rightarrow t_i$ for $i = 1 \ldots n$.*
  2. *$\mathcal{I} \models \theta(v_j) : s_j$ for $j = 1 \ldots m$.*
  3. *$\mathcal{I} \models \theta(w_k) \Rightarrow \theta(w_k')$ for $k = 1 \ldots l$.*
  4. *$\theta(af)$ does not hold in $\mathcal{I}$.*
- *r is a* wrong rewrite rule *(respectively,* wrong equation *and* wrong membership axiom*) if it admits some wrong instance.*

(**Reflexivity**)

$$\overline{t \Rightarrow t} \;^{(Rf_\Rightarrow)} \qquad\qquad \overline{t \rightarrow t} \;^{(Rf_\rightarrow)}$$

(**Transitivity**)

$$\frac{t_1 \Rightarrow t' \quad t' \Rightarrow t_2}{t_1 \Rightarrow t_2} \;(Tr_\Rightarrow) \qquad\qquad \frac{t_1 \rightarrow t' \quad t' \rightarrow t_2}{t_1 \rightarrow t_2} \;(Tr_\rightarrow)$$

(**Congruence**)

$$\frac{t_1 \Rightarrow t'_1 \quad \dots \quad t_n \Rightarrow t'_n}{f(t_1, \dots, t_n) \Rightarrow f(t'_1, \dots, t'_n)} \;(Cong_\Rightarrow) \qquad \frac{t_1 \rightarrow t'_1 \quad \dots \quad t_n \rightarrow t'_n}{f(t_1, \dots, t_n) \rightarrow f(t'_1, \dots, t'_n)} \;(Cong_\rightarrow)$$

(**Replacement**)

$$\frac{\{\theta(u_i) \downarrow \theta(u'_i)\}_{i=1}^n \; \{\theta(v_j) : s_j\}_{j=1}^m \; \{\theta(w_k) \Rightarrow \theta(w'_k)\}_{k=1}^l}{\theta(t) \Rightarrow \theta(t')} \;(Rep_\Rightarrow)$$

if $t \Rightarrow t' \Leftarrow \bigwedge_{i=1}^n u_i = u'_i \wedge \bigwedge_{j=1}^m v_j : s_j \wedge \bigwedge_{k=1}^l w_k \Rightarrow w'_k$

$$\frac{\{\theta(u_i) \downarrow \theta(u'_i)\}_{i=1}^n \; \{\theta(v_j) : s_j\}_{j=1}^m}{\theta(t) \rightarrow \theta(t')} \;(Rep_\rightarrow) \text{ if } t \rightarrow t' \Leftarrow \bigwedge_{i=1}^n u_i = u'_i \wedge \bigwedge_{j=1}^m v_j : s_j$$

(**Equivalence Class**) (**Subject Reduction**)

$$\frac{t \rightarrow t' \quad t' \Rightarrow t'' \quad t'' \rightarrow t'''}{t \Rightarrow t'''} (EC) \qquad \frac{t \rightarrow t' \quad t' : s}{t : s} (SRed)$$

(**Membership**)

$$\frac{\{\theta(u_i) \downarrow \theta(u'_i)\}_{i=1}^n \; \{\theta(v_j) : s_j\}_{j=1}^m}{\theta(t) : s} \;(Mb) \qquad \text{if } t : s \Leftarrow \bigwedge_{i=1}^n u_i = u'_i \wedge \bigwedge_{j=1}^m v_j : s_j$$

Fig. 1. Semantic calculus for Maude modules

The general schema of [17] presents declarative debugging as the search of *buggy nodes* (invalid nodes with all children valid) in a debugging tree representing an erroneous computation. In our scheme instance, the proof trees constructed by the inferences of Fig. 1 seem natural candidates for debugging trees. Although this is a possible option, we will use instead a suitable abbreviation of these trees. This is motivated by the following result:

Proposition 1. *Let N be a buggy node in some proof tree in the calculus of Fig. 1 w.r.t. an intended interpretation $\mathcal{I}$. Then:*

1. *N is the result of either a* membership *or a* replacement *inference step.*
2. *The statement associated to N is either a wrong rewrite rule, a wrong equation, or a wrong membership axiom.*

Both points are a consequence of the definition of the semantic calculus. The first result states that all the inference steps different from *membership* and *replacement* are logically sound w.r.t. the definition of $\mathcal{R}$-system, i.e., they always produce valid results from valid premises. The second result can be checked by observing that any *membership* or *replacement* buggy node satisfies the requirements of Def. 1: the valid premises correspond to the points 1-3 of the definition, while the invalid conclusion fulfills the last point.

3.2   Abbreviated Proof Trees

Our goal is to find a buggy node in any proof tree T rooted by the initial error symptom detected by the user. This could be done simply by asking questions to the user about the validity of the nodes in the tree according to the following *top-down* strategy:

Input: A tree T with an invalid root.
Output: A buggy node in T.
Description: Consider the root N of T. There are two possibilities: if all the children of N are valid, then finish pointing out at N as buggy; otherwise, select the subtree rooted by any invalid child and use recursively the same strategy to find the buggy node.

Proving that this strategy is complete is straightforward by using induction on the height of T.

However, we will not use the proof tree T as debugging tree, but a suitable abbreviation which we denote by $APT(T)$ (from *Abbreviated Proof Tree*), or simply APT if the proof tree T is clear from the context. The reason for preferring the APT to the original proof tree is that it reduces and simplifies the questions that will be asked to the user while keeping the soundness and completeness of the technique. In particular the APT essentially contains only nodes related to the *replacement* and *membership* inferences using statements included in the specification, which are the only possible buggy nodes as Prop. 1 indicates. Thus, in order to minimize the number of questions asked to the user the debugger should consider the validity of $(Rep_{\Rightarrow})$, $(Rep_{\rightarrow})$, or (Mb). The APT rules can be seen in Fig. 2.

The rules are assumed to be applied top-down: if several APT rules can be applied at the root of a proof tree, we must choose the first one, that is, the rule of least number. As a matter of fact, the figure includes rules for two different possible $APTs$, which we call *one-step* abbreviated proof tree (in short $APT^o(T)$), defined by all the rules in the figure excluding $(\mathbf{APT}_4^m)$, and *many-steps* abbreviated proof tree (in short $APT^m(T)$), defined by all the rules in the figure excluding $(\mathbf{APT}_4^o)$. Analogously, we will use the notation $APT'^o(T)$ (resp. $APT'^m(T)$) for the subset of rules of APT' excluding $(\mathbf{APT}_4^m)$ (resp. $(\mathbf{APT}_4^o)$).

The one-step debugging tree follows strictly the idea of keeping only nodes corresponding to the *replacement* and *membership* inference rules. However, the many-steps debugging tree also keeps nodes corresponding to the *transitivity* inference rule for rewrites. The user will choose which debugging tree (one-step or many-steps) will be used for the declarative debugging session, taking into account that the many-steps debugging tree usually leads to shorter debugging sessions (in terms of the number of questions) but with likely more complicated questions. The number of questions is usually reduced because keeping

$$(\mathbf{APT}_1)\ APT\left(\frac{T_1\dots T_n}{af}{}_{(R)}\right) = \frac{APT'\left(\frac{T_1\dots T_n}{af}{}_{(R)}\right)}{af}$$

$$(\mathbf{APT}_2)\ APT'\left(\frac{}{t \Rightarrow t}{}^{(Rf_\Rightarrow)}\right) = \emptyset$$

$$(\mathbf{APT}_3)\ APT'\left(\frac{\dfrac{T_1\dots T_n}{t_1 \to t'}{}^{(Rep_\to)}\quad T'}{t_1 \to t_2}{}_{(Tr_\to)}\right) =$$

$$\left\{\frac{APT'(T_1)\dots APT'(T_n)\ APT'(T')}{t_1 \to t_2}{}_{(Rep_\to)}\right\}$$

$$(\mathbf{APT}_4^o)\ APT'\left(\frac{T_1\quad T_2}{t_1 \Rightarrow t_2}{}_{(Tr_\Rightarrow)}\right) = APT'(T_1) \bigcup APT'(T_2)$$

$$(\mathbf{APT}_4^m)\ APT'\left(\frac{T_1\quad T_2}{t_1 \Rightarrow t_2}{}_{(Tr_\Rightarrow)}\right) = \left\{\frac{APT'(T_1)\ APT'(T_2)}{t_1 \Rightarrow t_2}{}_{(Tr_\Rightarrow)}\right\}$$

$$(\mathbf{APT}_5)\ APT'\left(\frac{T_1\dots T_n}{t_1 \Rightarrow t_2}{}_{(Cong_\Rightarrow)}\right) = APT'(T_1) \bigcup\dots \bigcup APT'(T_n)$$

$$(\mathbf{APT}_6)\ APT'\left(\frac{T_1\quad T_2}{t : s}{}_{(SRed)}\right) = APT'(T_1) \bigcup APT'(T_2)$$

$$(\mathbf{APT}_7)\ APT'\left(\frac{T_1\dots T_n}{t : s}{}_{(Mb)}\right) = \left\{\frac{APT'(T_1)\dots APT'(T_n)}{t : s}{}_{(Mb)}\right\}$$

$$(\mathbf{APT}_8)\ APT'\left(\frac{T_1\dots T_n}{t_1 \Rightarrow t_2}{}_{(Rep_\Rightarrow)}\right) = \left\{\frac{APT'(T_1)\dots APT'(T_n)}{t_1 \Rightarrow t_2}{}_{(Rep_\Rightarrow)}\right\}$$

$$(\mathbf{APT}_9)\ APT'\left(\frac{T'\quad\dfrac{T_1\dots T_n}{t \Rightarrow t'}{}^{(Rep_\Rightarrow)}\quad T''}{t_1 \Rightarrow t_2}{}_{(EC)}\right) =$$

$$\left\{\frac{APT'(T')\ APT'(T_1)\dots APT'(T_n)\ APT'(T'')}{t_1 \Rightarrow t_2}{}_{(Rep_\Rightarrow)}\right\}$$

$$(\mathbf{APT}_{10})\ APT'\left(\frac{T_1\dots T_n}{t_1 \Rightarrow t_2}{}_{(EC)}\right) = APT'(T_1) \bigcup \dots \bigcup APT'(T_n)$$

(R) any inference rule $\Rightarrow$ either $\to$ or $\Rightarrow$
af either $t_1 \to t_2$, $t : s$ or $t_1 \Rightarrow t_2$

Fig. 2. Transforming rules for obtaining abbreviated proof trees

the transitivity nodes for rewrites shapes some parts of the debugging tree as a balanced binary tree (each transitivity inference has two premises, i.e., two child subtrees), and this allows the debugger to use very efficient navigation strategies [23, 24]. On the contrary, removing the transitivity inferences for rewrites (as rule $(\mathbf{APT}_4^o)$ does) produces flattened trees where this strategy is no longer efficient. On the other hand, in rewrites $t \Rightarrow t'$ appearing as conclusion of the transitivity inference rule, the term t' can contain the result of rewriting several subterms of t, and determining the validity of such nodes can be complicated, while in the one-step debugging tree each rewrite node $t \Rightarrow t'$ corresponds to a single rewrite applied at t and checking its validity is usually easier. The user must balance the pros and cons of each option, and choose the best one for each debugging session.

The rules $(\mathbf{APT}_3)$ and $(\mathbf{APT}_9)$ deserve a more detailed explanation. They keep the corresponding label $(Rep_\Rightarrow)$ but changing the conclusion of the replacement inference in the lefthand side. For instance, $(\mathbf{APT}_3)$ replaces $t_1 \to t'$ by the conclusion of the next transitivity inference $t_1 \to t_2$. We do this as a pragmatic way of simplifying the structure of the $APTs$, since t_2 is obtained from t' and hence likely simpler (the root of the tree T' in $(\mathbf{APT}_3)$ must be necessarily of the form $t' \to t_2$ by the structure of the inference rule for transitivity in Fig. 1). A similar reasoning explains the form of $(\mathbf{APT}_9)$. We will formally state now that these changes are safe from the point of view of the debugger.

Theorem 1. *Let T be a finite proof tree representing an inference in the calculus of Fig. 1 w.r.t. some rewrite theory $\mathcal{R}$. Let $\mathcal{I}$ be an intended interpretation of $\mathcal{R}$ such that the root of T is invalid in $\mathcal{I}$. Then:*

- *Both $APT^o(T)$ and $APT^m(T)$ contain at least one buggy node (completeness).*
- *Any buggy node in $APT^o(T)$, $APT^m(T)$ has an associated wrong statement in $\mathcal{R}$ (correctness).*

The theorem states that we can safely employ the abbreviated proof tree as a basis for the declarative debugging of Maude system and functional modules: the technique will find a buggy node starting from any initial symptom detected by the user. Of course, these results assume that the user answers correctly all the questions about the validity of the APT nodes asked by the debugger.

4   A Debugging Session

The debugger is initiated in Maude by loading the file **dd.maude** (available from **http://maude.sip.ucm.es/debugging**). This starts an input/output loop that allows the user to interact with the tool. Then, the user can enter Full Maude modules and commands, as well as commands for the debugger. The current version supports all kinds of modules. When debugging a rewrite computation, two different debugging trees can be built: one whose questions are related to one-step rewrites and another whose questions are related to several steps. The latter tree is partially built so that any node corresponding to a one-step rewrite is expanded only when the navigation process reaches it.

The debugger provides two strategies to traverse the debugging tree: *top-down*, that traverses the tree from the root asking each time for the correctness of all the children of the current node, and then continues with one of the incorrect children; and *divide and query*, that each time selects the node whose subtree's size is the closest one to half the size of the whole tree, keeping only this subtree if its root is incorrect, and deleting the whole subtree otherwise. Note that, although the navigation strategy can be changed during the debugging session, the construction strategy is selected before the tree is built and cannot be changed.

The user can select a module containing only correct statements. By checking the correctness of the inferences with respect to this module (i.e., using this module as oracle) the debugger can reduce the number of questions. The debugger allows us to debug specifications where some statements are suspicious and have been labeled. Only these labeled statements generate nodes in the proof tree, being the user in charge of this labeling. The user can decide to use all the labeled statements as suspicious or can use only a subset by trusting labels and modules. Moreover, the user can answer that he trusts the statement associated with the currently questioned inference; that is, statements can be trusted "on the fly." The user can also give the answer "don't know," that postpones the answer to that question by asking alternative questions. An **undo** command, allowing the user to return to the previous state, is also provided. We refer the reader to [21, 22] for further information.

In Sect. 2.4 we described a system module that simulates a knight's tour. However, this system module contains a bug and the knight repeats some positions in its tour. This error is also obtained when looking for a 3 steps journey:

```
Maude> (rew knight(3) .)
result List :  [1,1][2,3][3,1][2,3]
```

Thus, we debug this smaller computation. Moreover, after inspecting the rewrite rules describing the eight possible moves, we are sure that they are not responsible for the error; therefore, we trust them by using commands that allow us to select the suspicious statements.

```
Maude> (set debug select on .)
Maude> (debug select con1 con2 leg k1 k2 .)
Maude> (debug knight(3) =>* [1,1][2,3][3,1][2,3]  .)
```

The default one-step tree construction strategy is used and the tree shown below is built, where every operation has been abbreviated with its first letter.

$$\frac{\overline{k(0) \Rightarrow_1 [1,1]}^{k1} \quad \overline{1([2,3]) \to t}^{leg} \quad \overline{c([1,1],[2,3]) \to f}^{con2}}{k(1) \Rightarrow_1 J2}_{k2}$$

$$\frac{\overline{1([3,1]) \to t}^{leg} \quad \overline{c(J2,[3,1]) \to f}^{con2}}{k(2) \Rightarrow_1 J1}_{k2}$$

$$\frac{\overline{1([2,3]) \to t}^{leg} \quad \overline{c(J1,[2,3]) \to f}^{con2}}{k(3) \Rightarrow_1 [1,1][2,3][3,1][2,3]}_{k2}$$

where J1 denotes the journey [1,1][2,3][3,1] and J2 denotes [1,1][2,3].

Since the tree is navigated by using the default divide and query strategy, the first two questions asked by the debugger are

```
Is this rewrite (associated with the rule k2) correct?
knight(1) =>1 [1,1][2,3]
Maude> (yes .)
Is this rewrite (associated with the rule k2) correct?
knight(2) =>1 [1,1][2,3][3,1]
Maude> (yes .)
```

Notice the form =>1 of the arrow in the rewrites appearing in the questions, to emphasize that they are one-step rewrites.

In both cases the answer is **yes** because these paths are possible, legal behaviors of the knight when it can do one or two hops. These two subtrees are removed and the current tree looks as follows:

$$\frac{\overline{1([2,3]) \to t}^{leg} \quad \overline{c(J1,[2,3]) \to f}^{con2}}{k(3) \Rightarrow_1 [1,1][2,3][3,1][2,3]}_{k2}$$

The next question is

```
Is this reduction (associated with the equation con2) correct?
contains([1,1][2,3][3,1],[2,3]) -> false
Maude> (no .)
```

Clearly, this is not a correct reduction, since position [2,3] is already in the path [1,1][2,3][3,1]. With this answer this subtree is selected and, since it is a single node, the bug is located:

```
The buggy node is:
contains([1,1][2,3][3,1],[2,3]) -> false
with the associated equation: con2
```

Looking at the definition of the **contains** operation, we realize that it defines the membership operation for *sets*, not for lists. A correct definition of the **contains** operation is as follows:

```
  eq [con1] : contains(nil, P) = false .
  eq [con2] : contains(Q J, P) = P == Q or contains(J, P) .
```

5   The Implementation

As mentioned in the introduction, a key distinguishing feature of Maude is its systematic and efficient use of reflection through its predefined META-LEVEL module [10, Chap. 14]. This powerful feature allows access to metalevel entities such as specifications or computations as usual data. Therefore, we are able to generate and navigate the debugging tree of a Maude computation using operations in Maude itself. In addition, the Maude system provides another module, LOOP-MODE [10, Chap. 17], which can be used to specify input/output interactions with the user. Thus, our declarative debugger, including its user interface, is implemented in Maude itself, as an extension of Full Maude [10, Chap. 18]. Instead of creating the complete proof tree and then abbreviating it, we build the abbreviated proof tree directly. Since navigation is done by asking questions to the user, this stage has to handle the navigation strategy together with the input/output interaction with the user. The technical report [21] provides a full explanation of this implementation, including the user interaction.

The way in which the debugging trees for reductions and memberships or rewrites are built is completely different. In the first case, we use the facts that equations and membership axioms are both terminating and confluent, which allow us to build the debugging tree in a "greedy" way, selecting at each moment the first equation applicable to the current term. However, we have to use a different methodology in the construction of the debugging tree for incorrect rewrites. We use breadth-first search to find from the initial term the wrong term introduced by the user, and then we use the found path to build the debugging tree in the two possible ways described in previous sections.

The functions in charge of building the debugging trees, that correspond to the *APT* function from Fig. 2, have a common initial behavior. They receive the module where the wrong inference took place, a correct module (or a special constant when no such module is provided) to prune the tree, the initial term, the (erroneous) result obtained, and the set of suspicious statements labels. They keep the initial inference as the root of the tree and generate the forest of abbreviated trees corresponding to the inference with functions that, in addition to the arguments above, receive the initial module "cleaned" of suspicious statements and correspond to the *APT'* function from Fig. 2. This transformed module is used to improve the efficiency of the tree construction, because we can use it to check if an inference can be obtained by using only trusted statements, thus avoiding to build a tree that will be finally empty.

The function that builds debugging trees for wrong reductions works with the same innermost strategy as the Maude interpreter: it first fully reduces the subterms recursively building their debugging trees (it mimics a specific behavior of the *congruence* rule in Fig. 1), and once all the subterms have been reduced, if the result is not the final one, it tries to reduce at the top to reach the final result by *transitivity*. Reduction at the top tries to apply one equation,[3] by

---

[3] Since the module is assumed to be confluent, we can choose any equation and the final result should be the same.

using the *replacement* rule from Fig. 1. Debugging trees for the conditions of the equation are also built and placed as children of the replacement rule. The construction of debugging trees for wrong memberships mimics the *subject reduction* rule from Fig. 1 by computing the tree for the full reduction of the term and then computing the tree for the membership inference of its least sort by using the operator declarations and the membership axioms, which corresponds to a concrete application of the *membership* inference rule.

The one-step tree for wrong rewrites computes the tree for the reduction from the initial term to normal form and then computes the rest of the tree, that corresponds to a rewrite from a fully reduced term (this corresponds to a concrete application of the *equivalence class* inference rule from Fig. 1). The debugging tree for this rewrite is computed from the trace, that is obtained with the predefined function `metaSearchPath`. Each step of the trace corresponds to the application of one rule, that generates a tree, with the trees corresponding to the conditions of the rule as its children (reproducing the *replacement* rule). Note that although the information in the trace is related to the whole rewritten term, the application of a rule can be in a subterm, which corresponds with the *congruence* inference rule, so only the rewritten subterms appear in the debugging tree. Other children are generated for the reduction to normal form due to the *equivalence class* inference rule. Finally, all the steps are put together as children of the same root by using the *transitivity* inference rule. The many-steps debugging tree is built *by demand*, so that the debugging subtrees corresponding to one-step rewrites are only generated when they are pointed out as wrong. These one-step nodes are used to create a balanced binary tree, by dividing them into two forests of approximately the same size, recursively creating their trees, and then using them as children of a new binary tree that has as root the combination by *transitivity* of the rewrites in their roots.

6   Concluding Remarks

In this paper we have developed the foundations of declarative debugging of executable rewriting logic specifications, and we have applied them to implement a debugger for Maude modules. The work encompasses and extends previous presentations [5, 6] on the declarative debugging of Maude functional modules, which constitute now a particular case of a more general setting.

We have formally described how debugging trees can be obtained from Maude proof trees, proving the correctness and completeness of the debugging technique. The tool based on these ideas allows the user to concentrate on the logic of the program disregarding the operational details. In order to deal with the possibly complex questions associated to rewrite statements, the tool offers the possibility of choosing between two different debugging trees: the one-step trees, with simpler questions and likely longer debugging sessions, and the many-steps trees, which in general require fewer but more complex questions before finding the bug. The experience will show the user which one must be chosen in each case depending on the complexity of the specification.

In our opinion, this debugger provides a complement to existing debugging techniques for Maude, such as tracing and term coloring. An important advantage of our debugger is the help provided by the tool in locating the buggy statements, assuming the user answers correctly the corresponding questions.

As future work we want to provide a graphical interface, that allows the user to navigate the tree with more freedom. We are also investigating how to improve the questions done in the presence of the `strat` operator attribute, that allows the specifier to define an evaluation strategy. This can be used to represent some kind of laziness. Finally, we plan to study how to debug *missing answers* [8, 16] in addition to the wrong answers we have treated thus far.

References

1. Bouhoula, A., Jouannaud, J.-P., Meseguer, J.: Specification and proof in membership equational logic. Theoretical Computer Science 236, 35–132 (2000)
2. Bruni, R., Meseguer, J.: Semantic foundations for generalized rewrite theories. Theoretical Computer Science 360(1), 386–414 (2006)
3. Caballero, R.: A declarative debugger of incorrect answers for constraint functional-logic programs. In: Proceedings of the 2005 ACM SIGPLAN Workshop on Curry and Functional Logic Programming (WCFLP 2005), pp. 8–13. ACM Press, Tallinn (2005)
4. Caballero, R., Hermanns, C., Kuchen, H.: Algorithmic debugging of Java programs. In: López-Fraguas, F.J. (ed.) 15th Workshop on Functional and (Constraint) Logic Programming, WFLP 2006, Madrid. Electronic Notes in Theoretical Computer Science, vol. 177, pp. 75–89. Elsevier, Amsterdam (2007)
5. Caballero, R., Martí-Oliet, N., Riesco, A., Verdejo, A.: Declarative debugging of membership equational logic specifications. In: Degano, P., De Nicola, R., Meseguer, J. (eds.) Concurrency, Graphs and Models. LNCS, vol. 5065, pp. 174–193. Springer, Heidelberg (2008)
6. Caballero, R., Martí-Oliet, N., Riesco, A., Verdejo, A.: A declarative debugger for Maude functional modules. In: Roşu, G. (ed.) Proceedings Seventh International Workshop on Rewriting Logic and its Applications, WRLA 2008, Budapest. Electronic Notes in Computer Science, vol. 238, pp. 63–81. Elsevier, Amsterdam (2009)
7. Caballero, R., Rodríguez-Artalejo, M.: DDT: A declarative debugging tool for functional-logic languages. In: Kameyama, Y., Stuckey, P.J. (eds.) FLOPS 2004. LNCS, vol. 2998, pp. 70–84. Springer, Heidelberg (2004)
8. Caballero, R., Rodríguez-Artalejo, M., del Vado Vírseda, R.: Declarative diagnosis of missing answers in constraint functional-logic programming. In: Garrigue, J., Hermenegildo, M.V. (eds.) FLOPS 2008. LNCS, vol. 4989, pp. 305–321. Springer, Heidelberg (2008)
9. Chitil, O., Luo, Y.: Structure and properties of traces for functional programs. In: Mackie, I. (ed.) Proceedings of the Third International Workshop on Term Graph Rewriting (TERMGRAPH 2006). Electronic Notes in Theoretical Computer Science, vol. 176, pp. 39–63. Elsevier, Amsterdam (2007)
10. Clavel, M., Durán, F., Eker, S., Lincoln, P., Martí-Oliet, N., Meseguer, J., Talcott, C.: All About Maude: A High-Performance Logical Framework. LNCS, vol. 4350. Springer, Heidelberg (2007)

11. Clavel, M., Meseguer, J., Palomino, M.: Reflection in membership equational logic, many-sorted equational logic, Horn logic with equality, and rewriting logic. Theoretical Computer Science 373(1-2), 70–91 (2007)
12. Lloyd, J.W.: Declarative error diagnosis. New Generation Computing 5(2), 133–154 (1987)
13. MacLarty, I.: Practical declarative debugging of Mercury programs. Master's thesis, University of Melbourne (2005)
14. Meseguer, J.: Conditional rewriting logic as a unified model of concurrency. Theoretical Computer Science 96(1), 73–155 (1992)
15. Meseguer, J.: Membership algebra as a logical framework for equational specification. In: Parisi-Presicce, F. (ed.) WADT 1997. LNCS, vol. 1376, pp. 18–61. Springer, Heidelberg (1998)
16. Naish, L.: Declarative diagnosis of missing answers. New Generation Computing 10(3), 255–286 (1992)
17. Naish, L.: A declarative debugging scheme. Journal of Functional and Logic Programming 1997(3) (1997)
18. Nilsson, H.: How to look busy while being as lazy as ever: the implementation of a lazy functional debugger. Journal of Functional Programming 11(6), 629–671 (2001)
19. Nilsson, H., Fritzson, P.: Algorithmic debugging of lazy functional languages. Journal of Functional Programming 4(3), 337–370 (1994)
20. Pope, B.: A Declarative Debugger for Haskell. PhD thesis, The University of Melbourne, Australia (2006)
21. Riesco, A., Verdejo, A., Caballero, R., Martí-Oliet, N.: Declarative debugging of Maude modules. Technical Report SIC-6/2008, Dpto. Sistemas Informáticos y Computación, Universidad Complutense de Madrid (2008), http://maude.sip.ucm.es/debugging
22. Riesco, A., Verdejo, A., Martí-Oliet, N., Caballero, R.: A declarative debugger for Maude. In: Meseguer, J., Roşu, G. (eds.) AMAST 2008. LNCS, vol. 5140, pp. 116–121. Springer, Heidelberg (2008)
23. Shapiro, E.Y.: Algorithmic Program Debugging. In: ACM Distinguished Dissertation. MIT Press, Cambridge (1983)
24. Silva, J.: A comparative study of algorithmic debugging strategies. In: Puebla, G. (ed.) LOPSTR 2006. LNCS, vol. 4407, pp. 143–159. Springer, Heidelberg (2007)
25. Tessier, A., Ferrand, G.: Declarative diagnosis in the CLP scheme. In: Deransart, P., Hermenegildo, M.V., Maluszynski, J. (eds.) DiSCiPl 1999. LNCS, vol. 1870, pp. 151–174. Springer, Heidelberg (2000)
26. Viry, P.: Equational rules for rewriting logic. Theoretical Computer Science 285(2), 487–517 (2002)

Translating a Dependently-Typed Logic to First-Order Logic

Kristina Sojakova and Florian Rabe

Jacobs University, Bremen

Abstract. DFOL is a logic that extends first-order logic with dependent types. We give a translation from DFOL to FOL formalized as an institution comorphism and show that it admits the model expansion property. This property together with the borrowing theorem implies the soundness of borrowing — a result that enables us to reason about entailment in DFOL by using automated tools for FOL. In addition, the translation permits us to deduce properties of DFOL such as completeness, compactness, and existence of free models from the corresponding properties of FOL, and to regard DFOL as a fragment of FOL. We give an example that shows how problems about DFOL can be solved by using the automated FOL prover Vampire. Future work will focus on the integration of the translation into the specification and translation tool HeTS.

1 Introduction and Related Work

Dependent type theory, DTT, ([ML75]) provides a very elegant language for many applications ([HHP93, NPS90]). However, its definition is much more involved than that of simple type theory because all well-formed terms, types, and their equalities must be defined in a single joint induction. Several quite complex model classes, mainly related to locally cartesian closed categories, have been studied to provide a model theory for DTT (see [Pit00] for an overview).

Many of the complications disappear if dependently-typed extensions of first-order logic are considered, i.e., systems that have dependent types, but no (simple or dependent) function types. Such systems were investigated in [Mak97], [Rab06], and [Bel08]. They provide very elegant axiomatizations of many important mathematical theories such as those of categories or linear algebra while retaining completeness with respect to straightforward set-theoretic models.

However, these systems are of relatively little practical use because no automated reasoning tools, let alone efficient ones, are available. Therefore, our motivation is to translate one of these systems into first-order logic, FOL. Such a translation would translate a proof obligation to FOL and discharge it by calling existing FOL provers. This is called borrowing ([CM93]).

In principle, there are two ways how to establish the soundness of borrowing: proof-theoretically by translating the obtained proof back to the original logic,

A. Corradini and U. Montanari (Eds.): WADT 2008, LNCS 5486, pp. 326–341, 2009.

or model-theoretically by exhibiting a model-translation between the two logics. Proof-theoretical translations of languages with dependent types have been used in [JM93] to translate parts of DTT to simple type theory, in [Urb03] to translate Mizar ([TB85]) into FOL, and in Scunak [Bro06] to translate parts of DTT into FOL. The Scunak translation is only partial as for example the translation of lambda expressions is omitted. Similar partial translations, but in the simply-typed case, are used in Omega ([BCF+97]), Leo-II ([BPTF07]) and in the sledgehammer tactic of Isabelle ([Pau94]). If the FOL prover succeeds, the reconstruction of the FOL proof term is possible in practice but somewhat tricky: For example, sledgehammer uses the output of a strong prover to guide a second, weaker prover, from whose output the proof term is reconstructed. In the cases of Mizar and Scunak, it is not done at all. Furthermore, the more complex the translation of proof goals is, the more difficult it becomes to translate the FOL proof term back into the original logic.

Here we take the model-theoretic approach and formulate a translation from the system introduced in [Rab06] to FOL within the framework of institutions ([GB92]). Mathematically, our main results can be summarized as follows. We use the institution DFOL as given in [Rab06] and give an institution comorphism from DFOL into FOL. Every DFOL-signature is translated to a FOL-theory whose axioms are used to express the typing properties of the translated symbols. The signature translation uses an $n+1$-ary FOL-predicate P_s for every dependent type constructor s with n arguments. Then the formulas quantifying over x of type $s(t_1, \ldots, t_n)$ can be translated by relativizing (see [Obe62]) using the predicate $P_s(t_1, \ldots, t_n, x)$. Finally, we show that this comorphism admits model expansion. Using the borrowing theorem ([CM93]), this yields the soundness of the translation.

Thus, we provide a simple way to write problems in the conveniently expressive DFOL syntax and solve them by calling FOL theorem provers. It is also possible to extend FOL theorem provers with dependently typed input languages, or to integrate DFOL seamlessly into existing implementations of institution-based algebraic specification languages such as OBJ ([GWM+93]) and CASL ([ABK+02]). Finally, our result provides easier proofs of the free model and completeness theorems given in [Rab06].

2   Definitions

We now present some definitions necessary for our work. We assume that the reader is familiar with the basic concepts of category theory and logic. For introduction to category theory, see [Lan98].

Using categories and functors we can define an *institution*, which is a formalization of a logical system abstracting from notions such as formulas, models, and satisfaction. Institutions structure the variety of different logics and allow us to formulate institution-independent theorems for the general theory of logic. For more on institutions, see [GB92].

Definition 1 (Institution). *An institution is a 4-tuple* $(Sig, Sen, Mod, \models)$ *where*

- *Sig is a category,*
- *Sen : Sig → Set is a functor,*
- *Mod : Sig → Cat^{op} is a functor,*
- $\models$ *is a family of relations* $\models_{\Sigma}$ *for* $\Sigma \in |Sig|$, $\models_{\Sigma} \subseteq Sen(\Sigma) \times |Mod(\Sigma)|$

such that for each morphism $\sigma : \Sigma \to \Sigma'$, *sentence* $F \in Sen(\Sigma)$, *and model* $M' \in |Mod(\Sigma')|$ *we have*

$$Mod(\sigma)(M') \models_{\Sigma} F \quad \text{iff} \quad M' \models_{\Sigma'} Sen(\sigma)(F)$$

The category Sig is called the *category of signatures*. The morphisms in Sig are called *signature morphisms* and represent notation changes. The functor Sen assigns to each signature Σ a set of *sentences* over Σ and to each morphism $\sigma : \Sigma \to \Sigma'$ the induced *sentence translation* along σ. Similarly, the functor Mod assigns to each signature Σ a category of *models* for Σ and to each morphism $\sigma : \Sigma \to \Sigma'$ the induced *model reduction* along σ. For a signature Σ, the relation $\models_{\Sigma}$ is called a *satisfaction relation*.

We now define what *entailment* and *theory* are in the context of institutions.

Definition 2 (Entailment). *Let* $(Sig, Sen, Mod, \models)$ *be an institution. For a fixed* Σ, *let* $T \subseteq Sen(\Sigma)$ *and* $F \in Sen(\Sigma)$. *Then we say that* T *entails* F, *denoted* $T \models_{\Sigma} F$, *if for any model* $M \in |Mod(\Sigma)|$ *we have that*

$$\text{if } M \models_{\Sigma} G \text{ for all } G \in T \text{ then } M \models_{\Sigma} F$$

Definition 3 (Category of theories). *Let* $I = (Sig, Sen, Mod, \models)$ *be an institution. We define the category of theories of* I *to be the category* Th^I *where*

- *The objects are pairs* (Σ, T), *with* $\Sigma \in |Sig|$, $T \subseteq Sen(\Sigma)$
- σ *is a morphism from* (Σ, T) *to* (Σ', T') *iff* σ *is a signature morphism from* Σ *to* Σ' *in* I *and for each* $F \in T$ *we have that* $T' \models_{\Sigma} Sen(\sigma)(F)$

The objects in Th^I are called *theories* of I, and for each theory $Th = (\Sigma, T)$, the set T is called the set of *axioms* of Th. The morphisms in Th^I are called *theory morphisms*. For a theory (Σ, T) and a sentence F over Σ, we say $(\Sigma, T) \models F$ in place of $T \models_{\Sigma} F$.

For a given institution I, we sometimes need to construct another institution I^{Th}, whose signatures are the theories of I. We have the following lemma.

Lemma 1 (Institution of theories). *Let* $I = (Sig, Sen, Mod, \models)$ *be an institution. Denote by* I^{Th} *the tuple* $(Th^I, Sen^{Th}, Mod^{Th}, \models^{Th})$ *where*

- $Sen^{Th}(\Sigma, T) = Sen(\Sigma)$ *and* $Sen^{Th}(\sigma) = Sen(\sigma)$ *for* $\sigma : (\Sigma, T) \to (\Sigma', T')$.
- $Mod^{Th}(\Sigma, T)$ *is the full subcategory of* $Mod(\Sigma)$ *whose objects are those models* M *in* $|Mod(\Sigma)|$ *for which we have* $M \models_{\Sigma} G$ *whenever* $G \in T$. *For a theory morphism* $\sigma : (\Sigma, T) \to (\Sigma', T')$, $Mod^{Th}(\sigma)$ *is the restriction of* $Mod(\sigma)$ *to* $Mod^{Th}(\Sigma', T')$.

– $\models^{Th}_{(\Sigma,T)}$ is the restriction of $\models_\Sigma$ to $|Mod^{Th}(\Sigma,T)| \times Sen^{Th}(\Sigma,T)$.

Then I^{Th} is an institution, called the institution of theories of I.

We are now ready to define a certain kind of translation between two institutions.

Definition 4 (Institution comorphism). *Let* $I = (Sig^I, Sen^I, Mod^I, \models^I)$, $J = (Sig^J, Sen^J, Mod^J, \models^J)$ *be two institutions. An institution comorphism from* I *to* J *is a triple* (Φ, α, β) *where*

– $\Phi : Sig^I \to Sig^J$ *is a functor,*
– $\alpha : Sen^I \to \Phi; Sen^J$ *is a natural transformation,*
– $\beta : Mod^I \leftarrow \Phi; Mod^J$ *is a natural transformation*

such that for each $\Sigma \in |Sig^I|$, $F \in Sen^I(\Sigma)$, *and* $M' \in |Mod^J(\Phi(\Sigma))|$ *we have*

$$\beta_\Sigma(M') \models^I_\Sigma F \quad iff \quad M' \models^J_{\Phi(\Sigma)} \alpha_\Sigma(F)$$

where β_Σ *is regarded as a morphism from* $Mod^J(\Phi(\Sigma))$ *to* $Mod^I(\Sigma)$ *in the category* Cat.

Institution comorphisms are particularly useful if they have the following property.

Definition 5 (Model expansion property). *Let* (Φ, α, β) *be an institution comorphism from* I *to* J. *We say that the comorphism has the model expansion property if each functor* β_Σ *for* $\Sigma \in Sig^I$ *is surjective on objects.*

The following lemma is then applicable.

Lemma 2 (Borrowing). *Let* (Φ, α, β) *be an institution comorphism from* I^{Th} *to* J^{Th} *having the model expansion property. Then for any theory* (Σ,T) *in* I *and a sentence* F *over* Σ, *we have that*

$$(\Sigma,T) \models^I F \quad iff \quad \Phi(\Sigma,T) \models^J \alpha_\Sigma(F)$$

In other words, we can use the institution J to reason about theories in I. For more on borrowing, see [CM93].

3   DFOL and FOL as Institutions

The formal definition of a dependent type theory is typically very complex and long because both for the syntax and for the semantics a joint induction over signatures, contexts, terms, and types must be used. Therefore, in [Rab06], the syntax of DFOL is defined within the Edinburgh logical framework (LF, [HHP93]), thus saving one induction. In [Rab08], a model theory for LF is given so that both inductions can be done once and for all in the logical framework, thus permitting a very elegant and compact definition of DFOL.

Here, to be self-contained, we give the syntax directly, but omit the precise definition of well-formed expressions. Then the semantics is given by a partial interpretation function defined only for well-formed expressions. This has the advantage of making the main concepts intuitively clear while being short and precise.

3.1   Signatures

In DFOL, we have three base types, defined as follows:

$$\mathbf{S} : type \qquad Univ : \mathbf{S} \rightarrow type \qquad o : type$$

Here $\mathbf{S}$ is the type of sorts (semantically: names of universes). The type $Univ$ is an operator assigning to each sort the type of its terms (semantically: its universe of individuals). The type o is the type of formulas (semantically: the values $true$ and $false$).

A DFOL signature consists of a finite sequence of declarations of the form

$$c : \Pi x_1 : Univ(S_1), \ldots, \Pi x_n : Univ(S_n). T$$

meaning that c is a function taking n arguments of types $S_1, \ldots, S_n$ respectively, and returning an argument of type T, where T is one of the three base types. Here $\Pi x_i : Univ(S_i)$ denotes the domain of a dependent function type, i.e., x_i may occur in $S_{i+1}, \ldots, S_n, T$.

When the return type of c is o, we say that c is a *predicate symbol*. Likewise, if the return type is $\mathbf{S}$ or $Univ(S)$, we say that c is a *sort symbol* or a *function symbol* respectively. We abbreviate $\Pi x : Univ(S)$ as $\Pi x : S$ and $\Pi x : A. B$ as $A \rightarrow B$ if the variable x does not occur in B.

We define DFOL signatures Σ inductively on the number of declarations. Let Σ_k be a DFOL signature consisting of k declarations, $k \geq 0$. We define a *function* over Σ_k as follows:

- Any variable symbol is a function over Σ_k
- If f in Σ_k is a function symbol of arity n and $\mu_1, \ldots, \mu_n$ are functions over Σ_k, then $f(\mu_1, \ldots, \mu_n)$ is a function over Σ_k

If s in Σ_k is a sort symbol of arity n and $\mu_1, \ldots, \mu_n$ are functions over Σ_k, then $s(\mu_1, \ldots, \mu_n)$ is a *sort* over Σ_k. Similarly, if p in Σ_k is a predicate symbol of arity n and $\mu_1, \ldots, \mu_n$ are functions over Σ_k, then $p(\mu_1, \ldots, \mu_n)$ is a *predicate* over Σ_k. The word *term* refers to either a function, a sort, or a predicate.

Clearly, not all terms are well-formed in DFOL. A context Γ for a signature Σ in DFOL has the form $\Gamma = x_1 : S_1, \ldots, x_n : S_n$, where $S_1, \ldots, S_n$ are sorts and S_i contains no variables except possibly $x_1, \ldots, x_{i-1}$. Given a valid context Γ, a DFOL term is well-formed with respect to Γ only if it is well-typed in the LF type theory. For details we refer the reader to [Rab06].

Now the $k + 1$-th declaration has one of the following forms:

- $s : \Pi x_1 : S_1, \ldots, \Pi x_n : S_n. \mathbf{S}$
  where $S_1, \ldots, S_n$ are sorts over Σ_k and S_i contains no variables except possibly $x_1, \ldots, x_{i-1}$. We say that s is a sort symbol.
- $f : \Pi x_1 : S_1, \ldots, \Pi x_n : S_n. S_{n+1}$
  where $S_1, \ldots, S_{n+1}$ are sorts over Σ_k and S_i contains no variables except possibly $x_1, \ldots, x_{i-1}$.
- $p : \Pi x_1 : S_1, \ldots, \Pi x_n : S_n. o$
  where $S_1, \ldots, S_n$ are sorts over Σ_k and S_i contains no variables except possibly $x_1, \ldots, x_{i-1}$. We say that p is a predicate symbol.

As with terms, the declaration must be a well-typed according to the rules of LF.

Running example. The theory of categories has the following DFOL signature

$Ob : \mathbf{S}$

$Mor : Ob \rightarrow Ob \rightarrow \mathbf{S}$

$id : \Pi A : Ob.\ Mor(A, A)$

$\circ : \Pi A, B, C : Ob.\ Mor(A, B) \rightarrow Mor(B, C) \rightarrow Mor(A, C)$

$term : Ob \rightarrow o$

$isom : Ob \rightarrow Ob \rightarrow o$ For simplicity, we declare the signature model morphisms in DFOL to just be the identity morphisms.

3.2 Sentences

The set of DFOL formulas over a signature Σ can be described as follows:

- If P is a predicate over Σ, then P is a Σ-formula
- If μ_1, μ_2 are functions over Σ, then $\mu_1 \doteq \mu_2$ is a Σ-formula
- If F is a Σ-formula, then $\neg F$ is a Σ-formula
- If F, G are Σ-formulas, then $F \wedge G$, $F \vee G$, and $F \Rightarrow G$ are Σ-formulas
- If F is a Σ-formula and S is a sort term over Σ, then $\forall x : S.\ F$ is a Σ-formula
- If F is a Σ-formula and S is a sort term over Σ, then $\exists x : S.\ F$ is a Σ-formula

Closed and atomic formulas are defined in the obvious way analogous to first-order logic. As with terms, DFOL formulas are well-formed only if they are well-typed in the LF type theory. For a precise definition, see [Rab06].

Running example. We have the following axioms for the theory of categories, with equivalence defined as usual

$I1 : \forall A, B : Ob.\ \forall f : Mor(A, B).\ id(A) \circ f \doteq f$

$I2 : \forall A, B : Ob.\ \forall f : Mor(B, A).\ f \circ id(A) \doteq f$

$A1 : \forall A, B, C, D : Ob.\ \forall f : Mor(A, B).\ \forall g : Mor(B, C).\ \forall h : Mor(C, D).$
$f \circ (g \circ h) \doteq (f \circ g) \circ h$

$D1 : \forall A : Ob.\ (term(A) \iff \forall B : Ob.\ \exists f : Mor(B, A).\ \forall g : Mor(B, A).\ f \doteq g)$

$D2 : \forall A, B : Ob.\ (isom(A, B) \iff \exists f : Mor(A, B).\ \exists g : Mor(B, A).$
$(f \circ g \doteq id(A) \ \wedge \ g \circ f \doteq id(B)))$

3.3 Models

A model of a DFOL signature Σ is an interpretation function I. Since the declaration of a symbol may depend on symbols declared before, we define I inductively on the number of declarations.

Suppose I is defined for the first k declarations, $k \geq 0$. An *assignment function* φ for I is a function mapping each variable to an element of any set defined by I as an interpretation of a sort symbol.

Let μ be a term over Σ. We define the *interpretation* of μ induced by φ to be $I_\varphi(\mu)$, where I_φ is given by:

- $I_\varphi(x) = \varphi(x)$ for any variable x
- $I_\varphi(d(\mu_1, \ldots, \mu_k)) = d^I(I_\varphi(\mu_1), \ldots, I_\varphi(\mu_k))$ for a sort, predicate, or function symbol d if

- each of the interpretations $I_\varphi(\mu_1), \ldots, I_\varphi(\mu_n)$ exists and
- d^I is defined for the tuple $(I_\varphi(\mu_1), \ldots, I_\varphi(\mu_n))$

Otherwise we say $I_\varphi(d(\mu_1, \ldots, \mu_k))$ does not exist.

Given a valid context $\Gamma = x_1 : S_1, \ldots, x_n : S_n$ and an assignment function φ, we say that φ is an *assignment function for* Γ if for each i we have that $I_\varphi(S_i)$ exists and $\varphi(x_i) \in I_\varphi(S_i)$. From now on, we will only talk about assignment functions for a context; the general definition was introduced only to avoid some technical difficulties.

Now the $k + 1$-st declaration has one of the following forms:

- $s : \Pi x_1 : S_1, \ldots, \Pi x_n : S_n.\ \mathbf{S}$
  Let φ be any assignment function for the context $\Gamma = x_1 : S_1, \ldots, x_n : S_n$. Then

  $$s^I(\varphi(x_1), \ldots, \varphi(x_n)) \text{ is a (possibly empty) set}$$

  disjoint from any other set defined by I as an interpretation of a sort symbol.
- $f : \Pi x_1 : S_1, \ldots, \Pi x_n : S_n.\ S$
  Let φ be any assignment function for the context $\Gamma = x_1 : S_1, \ldots, x_n : S_n$. Then

  $$f^I(\varphi(x_1), \ldots, \varphi(x_n)) \in I_\varphi(S)$$

- p is a predicate symbol, $p : \Pi x_1 : S_1, \ldots, \Pi x_n : S_n.\ o$
  Let φ be any assignment function for the context $\Gamma = x_1 : S_1, \ldots, x_n : S_n$. Then

  $$p^I(\varphi(x_1), \ldots, \varphi(x_n)) \in \{true, false\}$$

Running example. An example model I for the signature of categories is given by any small category C. Then we have $Ob^I = |C|$, $Mor^I(A, B) = C(A, B)$, and the obvious interpretations for composition and identity. Furthermore, we can put $term^I(A) = true$ iff A is a terminal element and $isom^I(A, B) = true$ iff A and B are isomorphic.

3.4   Satisfaction Relation

To define the satisfaction relation, we first define the interpretation of formulas. Let Σ be a DFOL signature, I be a DFOL model for Σ, and Γ be a valid context over Σ. Furthermore, let φ be an assignment function for Γ and F be a well-formed DFOL formula for Γ. Then we define $I_\varphi(F)$ recursively on the structure of F:

- F is a predicate. Then $I_\varphi(F)$ is true if and only if $p^I(I_\varphi(\mu_1), \ldots, I_\varphi(\mu_n)) = true$.
- F is of the form $\mu_1 \doteq \mu_2$. Then $I_\varphi(F)$ is true if and only if $I_\varphi(\mu_1) = I_\varphi(\mu_2)$.
- F is of the form $\neg G$. Then $I_\varphi(F)$ is true if and only if $I_\varphi(G)$ is false.

- F is of the form $F_1 \wedge F_2$. Then $I_\varphi(F)$ is true if and only if both $I_\varphi(F_1)$ and $I_\varphi(F_2)$ are true.
- F is of the form $F_1 \vee F_2$. Then $I_\varphi(F)$ is true if and only if $I_\varphi(F_1)$ is true or $I_\varphi(F_2)$ is true.
- F is of the form $F_1 \implies F_2$. Then $I_\varphi(F)$ is true if and only if $I_\varphi(F_1)$ is false or $I_\varphi(F_2)$ is true.
- F is of the form $\exists x : S.\ G$. Then $I_\varphi(F)$ is true if and only if $I_{\varphi[x/a]}(G)$ is true for some $a \in I_\varphi(S)$.
- F is of the form $\forall x : S.\ G$. Then $I_\varphi(F)$ is true if and only if $I_{\varphi[x/a]}(G)$ is true for any $a \in I_\varphi(S)$.

Now if F is in fact a closed formula, its interpretation is independent of φ. Hence, we define that I satisfies F if and only if $I_\varphi(F)$ is true for some φ.

Running example. It is easy to see that the example model for the signature of categories satisfies the axioms given in section 3.2.

Putting our previous definitions together, we have the following lemma.

Lemma 3. $DFOL = (Sig, Sen, Mod, \models)$ *is an institution.*

The FOL institution is then obtained from DFOL by restricting the signatures to contain a unique sort symbol, having arity 0. Any other symbols are either function or predicate symbols. (Technically, this does not yield FOL because DFOL permits empty universes. But our FOL signatures will always have a nullary function symbol so that this does not constitute a problem). A FOL model is then denoted as (U, I), where I is the interpretation function and U is the universe corresponding to the unique sort symbol.

4   Translation of DFOL to FOL

The main idea of the translation is to associate with each n-ary sort symbol in DFOL an $n+1$-ary predicate in FOL and relativize the universal and existential quantifiers (the technique of relativization was first introduced by Oberschelp in [Obe62]).

Formally, the translation will be given as an institution comorphism from $DFOL$ to FOL^{Th}. We specify a functor Φ, mapping DFOL signatures to FOL theories and DFOL signature morphisms to FOL theory morphisms. For each DFOL signature Σ, we give a function α_Σ mapping DFOL sentences over Σ to FOL sentences over the translated signature $\Phi(\Sigma)$, and show that the family of functions α_Σ defines a natural transformation. Similarly, for each DFOL signature Σ we give a functor β_Σ mapping FOL models for the translated signature $\Phi(\Sigma)$ to DFOL models for Σ, and show that the family of functors β_Σ defines a natural transformation. Finally, we prove the satisfaction condition for (Φ, α, β) and show that the comorphism has the model expansion property.

Definition 6 (Signature translation). *Let Σ be a DFOL signature. We define $\Phi(\Sigma)$ to be the FOL theory (Σ', T'), where Σ' and T' are specified as follows. Σ' contains:*

- an n-ary function symbol f for each n-ary function symbol f in Σ,
- an n-ary predicate symbol p for each n-ary predicate symbol p in Σ,
- an $n + 1$-ary predicate symbol s for each n-ary sort symbol s in Σ,
- a special constant symbol $\bot$, different from any of the above symbols,
- no other symbols besides the above

T' contains:

S1. *Axioms ensuring that no element can belong to the universe of more than one sort. For any two sort symbols* s_1, s_2 *with* s_1 *different from* s_2, *we have the axiom*

$$\forall x_1, \ldots, x_n, y_1, \ldots, y_m, z. \left(s_1(x_1, \ldots, x_n, z) \implies \neg s_2(y_1, \ldots, y_m, z) \right)$$

and for each sort symbol s_1 *we have the axiom*

$$\forall x_1, \ldots, x_n, y_1, \ldots, y_n, z. \left(s_1(x_1, \ldots, x_n, z) \wedge s_1(y_1, \ldots, y_n, z) \right.$$
$$\left. \implies x_1 \doteq y_1 \wedge \ldots \wedge x_n \doteq y_n \right)$$

S2. *An axiom ensuring that each element different from* $\bot$ *belongs to the universe of at least one sort. If* $s_1, \ldots, s_k$ *are the sort symbols, then we have the axiom*

$$\forall y. \left(\neg y \doteq \bot \implies \exists x_1, \ldots, x_{n_1}. s_1(x_1, \ldots, x_{n_1}, y) \vee \ldots \vee \right.$$
$$\left. \exists x_1, \ldots, x_{n_k}. s_k(x_1, \ldots, x_{n_k}, y) \right)$$

S3. *Axioms ensuring that the special symbol* $\bot$ *is not contained in the universe of any sort. For each sort symbol* s, *we have the axiom*

$$\forall x_1, \ldots, x_n. \neg s(x_1, \ldots, x_n, \bot)$$

S4. *Axioms ensuring that if the arguments to a sort constructor are not of the correct types, the resulting sort has an empty universe. For each sort symbol* $s : \Pi x_1 : s_1(\mu_1^1, \ldots, \mu_{k_1}^1), \ldots, \Pi x_n : s_n(\mu_1^n, \ldots, \mu_{k_n}^n). \mathbf{S}$, *we have the axiom*

$$\forall x_1, \ldots, x_n. \left(\neg s_1(\mu_1^1, \ldots, \mu_{k_1}^1, x_1) \vee \ldots \vee \neg s_n(\mu_1^n, \ldots, \mu_{k_n}^n, x_n) \right.$$
$$\left. \implies \forall y. \neg s(x_1, \ldots, x_n, y) \right)$$

F1. *Axioms ensuring that if the arguments to a function are of the correct types, the function returns a value of the correct type. For each function symbol* $f : \Pi x_1 : s_1(\mu_1^1, \ldots, \mu_{k_1}^1), \ldots, \Pi x_n : s_n(\mu_1^n, \ldots, \mu_{k_n}^n). s(\mu_1, \ldots, \mu_k)$, *we have the axiom*

$$\forall x_1, \ldots, x_n. \left(s_1(\mu_1^1, \ldots, \mu_{k_1}^1, x_1) \wedge \ldots \wedge s_n(\mu_1^n, \ldots, \mu_{k_n}^n, x_n) \implies \right.$$
$$\left. s(\mu_1, \ldots, \mu_k, f(x_1, \ldots, x_n)) \right)$$

F2. *Axioms ensuring that if the arguments to a function are not of the correct types, the function returns the special symbol* $\bot$. *For each function symbol* $f : \Pi x_1 : s_1(\mu_1^1, \ldots, \mu_{k_1}^1), \ldots, \Pi x_n : s_n(\mu_1^n, \ldots, \mu_{k_n}^n). s(\mu_1, \ldots, \mu_k)$, *we have the axiom*

$$\forall x_1, \ldots, x_n. \left(\neg s_1(\mu_1^1, \ldots, \mu_{k_1}^1, x_1) \vee \ldots \vee \neg s_n(\mu_1^n, \ldots, \mu_{k_n}^n, x_n) \right.$$
$$\left. \implies f(x_1, \ldots, x_n) \doteq \bot \right)$$

P1. *Axioms ensuring that if the arguments to a predicate are not of the correct types, the predicate is false. For each predicate symbol*
$$p : \Pi x_1 : s_1(\mu_1^1, \ldots, \mu_{k_1}^1), \ \ldots, \ \Pi x_n : s_n(\mu_1^n, \ldots, \mu_{k_n}^n).\ o,\ \textit{we have the axiom}$$

$$\forall x_1, \ldots, x_n. \left(\neg s_1(\mu_1^1, \ldots, \mu_{k_1}^1, x_1) \ \lor \ \ldots \ \lor \ \neg s_n(\mu_1^n, \ldots, \mu_{k_n}^n, x_n) \right.$$
$$\left. \implies \ \neg p(x_1, \ldots, x_n) \right)$$

N. No other axioms besides the above

Defining Φ on signature morphisms is trivial since by our definition the only signature morphisms in DFOL are the identity morphisms. From this it follows immediately that Φ is a functor.

Running example. Denote the translated signature of categories by the theory (Σ', T'). Then Σ' contains the following symbols:

- Function symbols:
  - id of arity 1
  - ∘ of arity 5
  - ⊥ of arity 0
- Predicate symbols:
  - ob of arity 1
  - mor of arity 3
  - term of arity 1
  - isom of arity 2

The theory T' consists of the axioms $S1.1$ up to $P1.2$ in Fig.1.

Definition 7 (Sentence translation). *Let Σ be a DFOL signature. We define the function α_Σ on the set of all DFOL formulas over Σ. We do this recursively on the structure of the formula F:*

- *If F is of the form $p(\mu_1, \ldots, \mu_n)$, we set $\alpha_\Sigma(F) = F$*
- *If F is of the form $\mu_1 \doteq \mu_2$, we set $\alpha_\Sigma(F) = F$*
- *If F is of the form $\neg G$, we set $\alpha_\Sigma(F) = \neg \alpha_\Sigma(G)$*
- *If F is of the form $F_1 \land F_2$, we set $\alpha_\Sigma(F) = \alpha_\Sigma(F_1) \land \alpha_\Sigma(F_2)$*
- *If F is of the form $F_1 \lor F_2$, we set $\alpha_\Sigma(F) = \alpha_\Sigma(F_1) \lor \alpha_\Sigma(F_2)$*
- *If F is of the form $F_1 \implies F_2$, we set $\alpha_\Sigma(F)$ to be the formula*

$$\alpha_\Sigma(F_1) \implies \alpha_\Sigma(F_2)$$

- *If F is of the form $\forall x : s(\mu_1, \ldots, \mu_n). G$, we set $\alpha_\Sigma(F)$ to be the formula*

$$\forall x.\ s(\mu_1, \ldots, \mu_n, x) \implies \alpha_\Sigma(G)$$

- *If F is of the form $\exists x : s(\mu_1, \ldots, \mu_n). G$, we set $\alpha_\Sigma(F)$ to be the formula*

$$\exists x.\ s(\mu_1, \ldots, \mu_n, x) \land \alpha_\Sigma(G)$$

It is easy to see that α_Σ maps closed formulas to closed formulas. Hence, we can restrict α_Σ to the set of DFOL sentences over Σ to obtain our desired translation map. The naturality of α_Σ follows immediately since the only signature morphisms in DFOL are the identity morphisms.

$S1.1 : \forall A_1, B_1, f.\ (mor(A_1, B_1, f) \implies \forall A_2, B_2.\ (mor(A_2, B_2, f)$
$\implies A_2 \doteq A_1 \wedge B_2 \doteq B_1))$
$S1.2 : \forall y, B, C.\ (ob(y) \implies \neg mor(B, C, y))$
$S2 : \forall y.\ (\neg y \doteq \bot \implies ob(y) \vee \exists A, B.\ mor(A, B, y))$
$S3.1 : \neg ob(\bot)$
$S3.2 : \forall A, B.\ \neg mor(A, B, \bot)$
$S4 : \forall A, B.\ ((\neg ob(A) \vee \neg ob(B)) \implies \forall f.\ \neg mor(A, B, f))$
$F1.1 : \forall A.\ (ob(A) \implies mor(A, A, id(A)))$
$F1.2 : \forall A, B, C, f, g.\ (ob(A) \wedge ob(B) \wedge ob(C) \wedge mor(A, B, f) \wedge mor(B, C, g) \implies$
$mor(A, C, f; g))$
$F2.1 : \forall A.\ (\neg ob(A) \implies id(A) \doteq \bot)$
$F2.2 : \forall A, B, C, f, g.\ (\neg ob(A) \vee \neg ob(B) \vee \neg ob(C) \vee \neg mor(A, B, f) \vee$
$\neg mor(B, C, g) \implies f; g \doteq \bot)$
$P1.1 : \forall A.\ (\neg ob(A) \implies \neg term(A))$
$P1.2 : \forall A, B.\ (\neg ob(A) \vee \neg ob(B) \implies \neg isom(A, B))$

$I1 : \forall A, B, f.\ (ob(A) \wedge ob(B) \wedge mor(A, B, f) \implies id(A); f \doteq f)$
$I2 : \forall A, B, f.\ (ob(A) \wedge ob(B) \wedge mor(B, A, f) \implies f; id(A) \doteq f)$
$A1 : \forall A, B, C, D, f, g, h.\ (ob(A) \wedge ob(B) \wedge ob(C) \wedge ob(D) \wedge mor(A, B, f) \wedge$
$mor(B, C, g) \wedge mor(C, D, h) \implies f; (g; h) \doteq (f; g); h)$
$D1 : \forall A.\ (ob(A) \implies (term(A) \iff \forall B.\ (ob(B) \implies \exists f.\ (mor(B, A, f) \wedge$
$\forall g.\ (mor(B, A, g) \implies f \doteq g)))))$
$D2 : \forall A, B.\ (ob(A) \wedge obj(B) \implies (isom(A, B) \iff \exists f, g.\ (mor(A, B, f) \wedge$
$mor(B, A, g) \wedge f; g \doteq id(A) \wedge g; f \doteq id(B))))$

$Conjecture : \forall A, B.\ (ob(A) \wedge obj(B) \wedge term(A) \wedge term(B) \implies isom(A, B))$

Fig. 1. Translation of the running example

Running example. The translated axioms of the theory of categories are the axioms $I1$ up to $D2$ in Fig.1.

Definition 8 (Model reduction). *Let Σ be a DFOL signature and $M = (U, I)$ be a FOL model for $\Phi(\Sigma)$. We define the translated DFOL model $\beta_\Sigma(M)$ for Σ to be the interpretation function J, defined inductively on the number of declarations in Σ.*
Suppose J is defined for the first k symbols in Σ, $k \geq 0$. Then the $(k + 1)$-st declaration has one of the following forms:

- $s : \Pi x_1 : S_1, \ldots, \Pi x_n : S_n.\ \mathbf{S}$
  Let φ be any assignment function for the context $\Gamma = x_1 : S_1, \ldots, x_n : S_n$. We set

$$s^J(\varphi(x_1), \ldots, \varphi(x_n)) = \{u \in U \mid s^I(\varphi(x_1), \ldots, \varphi(x_n), u)\}$$

- $f : \Pi x_1 : S_1, \ldots, \Pi x_n : S_n.\ S$
  Let φ be any assignment function for the context $\Gamma = x_1 : S_1, \ldots, x_n : S_n$. We set
$$f^J(\varphi(x_1), \ldots, \varphi(x_n)) = f^I(\varphi(x_1), \ldots, \varphi(x_n))$$

- p is a predicate symbol, $c : \Pi x_1 : S_1, \ldots, \Pi x_n : S_n. o$

  Let φ be any assignment function for the context $\Gamma = x_1 : S_1, \ldots, x_n : S_n$. We set

$$p^J(\varphi(x_1), \ldots, \varphi(x_n)) \quad \text{iff} \quad p^I(\varphi(x_1), \ldots, \varphi(x_n))$$

We note here how the axioms introduced earlier are needed to ensure that J is indeed a DFOL model for Σ. We now turn to the proof of the satisfaction condition.

Theorem 1 (Satisfaction condition). (Φ, α, β) is an institution comorphism.

Proof. We have already shown that Φ is a functor and α, β are natural transformations. It remains to show that the satisfaction condition holds.

Let Σ be a DFOL signature, Γ be a valid context for Σ, φ be an assignment function for Γ, and F be a well-formed DFOL sentence for Γ. Furthermore, let $M = (U, I)$ be a FOL model for the translated signature $\Phi(\Sigma)$, and J be the translated model $\beta_\Sigma(M)$. We first observe the following two facts:

- φ is also an assignment function for M
- if μ is a well-formed function term for Γ, then $J_\varphi(\mu) = I_\varphi(\mu)$

Both of these facts follow directly from the construction of J. We now show that we have

$$J_\varphi(F) \quad \text{iff} \quad I_\varphi(\alpha_\Sigma(F))$$

To prove the claim, we proceed recursively on the structure of F:

- F is of the form $p(\mu_1, \ldots, \mu_n)$. Then $J_\varphi(F)$ is true if and only if $p^J(J_\varphi(\mu_1), \ldots, J_\varphi(\mu_n))$. By the construction of J, we have

$$p^J(J_\varphi(\mu_1), \ldots, J_\varphi(\mu_n)) \quad \text{iff} \quad p^I(J_\varphi(\mu_1), \ldots, J_\varphi(\mu_n))$$

  As noted above, $J_\varphi(\mu_i) = I_\varphi(\mu_i)$ for each i, hence

$$p^J(J_\varphi(\mu_1), \ldots, J_\varphi(\mu_n)) \quad \text{iff} \quad p^I(I_\varphi(\mu_1), \ldots, I_\varphi(\mu_n))$$

  Thus we have $J_\varphi(F)$ if and only if $I_\varphi(F)$. Since $F = \alpha_\Sigma(F)$, this proves the claim.
- F is of the form $\mu_1 \doteq \mu_2$. Then $J_\varphi(F)$ is true if and only if $J_\varphi(\mu_1) = J_\varphi(\mu_2)$. As noted above, $J_\varphi(\mu_1) = I_\varphi(\mu_1)$ and $J_\varphi(\mu_2) = I_\varphi(\mu_2)$, hence

$$J_\varphi(\mu_1) = J_\varphi(\mu_2) \quad \text{iff} \quad I_\varphi(\mu_1) = I_\varphi(\mu_2)$$

  Thus we have $J_\varphi(F)$ if and only if $I_\varphi(F)$. Since $F = \alpha_\Sigma(F)$, this proves the claim.
- F is of the form $\neg G$. Then $J_\varphi(F)$ is true if and only if $J_\varphi(G)$ is false. By the induction hypothesis, we have $J_\varphi(G)$ iff $I_\varphi(\alpha_\Sigma(G))$. Thus $J_\varphi(F)$ is true if and only if $I_\varphi(\alpha_\Sigma(G))$ is false, or equivalently

$$J_\varphi(F) \quad \text{iff} \quad I_\varphi(\neg\alpha_\Sigma(G))$$

Since $\neg \alpha_\Sigma(G) = \alpha_\Sigma(F)$, this proves the claim.

- F is of the form $F_1 \wedge F_2$. Then $J_\varphi(F)$ is true if and only if both $J_\varphi(F_1)$ and $J_\varphi(F_2)$ are true. By the induction hypothesis, we have $J_\varphi(F_1)$ iff $I_\varphi(\alpha_\Sigma(F_1))$ and $J_\varphi(F_2)$ iff $I_\varphi(\alpha_\Sigma(F_2))$. Hence, $J_\varphi(F)$ is true if and only if both $I_\varphi(\alpha_\Sigma(F_1))$ and $I_\varphi(\alpha_\Sigma(F_2))$ are true. Equivalently,

$$J_\varphi(F) \quad \text{iff} \quad I_\varphi(\alpha_\Sigma(F_1) \wedge \alpha_\Sigma(F_2))$$

Since $\alpha_\Sigma(F_1) \wedge \alpha_\Sigma(F_2) = \alpha_\Sigma(F)$, this proves the claim.

- F is of the form $F_1 \vee F_2$. Since F is equivalent to the formula $\neg(\neg F_1 \wedge \neg F_2)$, the claim follows from the previous steps.

- F is of the form $F_1 \implies F_2$. Since F is equivalent to the formula $\neg F_1 \vee F_2$, the claim follows from the previous steps.

- F is of the form $\exists x : s(\mu_1, \ldots, \mu_n). \ G$. By definition, $J_\varphi(F)$ is true if and only if there exists an $a \in J_\varphi(s(\mu_1, \ldots, \mu_n))$ such that $J_{\varphi[x/a]}(G)$ is true. Again by definition,

$$J_\varphi(s(\mu_1, \ldots, \mu_n)) = s^J(J_\varphi(\mu_1), \ldots, J_\varphi(\mu_n))$$

Since $J_\varphi(\mu_i) = I_\varphi(\mu_i)$ for each i, we have

$$s^J(J_\varphi(\mu_1), \ldots, J_\varphi(\mu_n)) = s^J(I_\varphi(\mu_1), \ldots, I_\varphi(\mu_n))$$

By the construction of J, we have that a belongs to $s^J(I_\varphi(\mu_1), \ldots, I_\varphi(\mu_n))$ if and only if a belongs to U and $s^I(I_\varphi(\mu_1), \ldots, I_\varphi(\mu_n), a) = true$. Now since μ_i does not contain x for any i, we have that

$$s^I(I_\varphi(\mu_1), \ldots, I_\varphi(\mu_n), a) = I_{\varphi[x/a]}(s(\mu_1, \ldots, \mu_n, x))$$

Also, by the induction hypothesis we have that

$$J_{\varphi[x/a]}(G) \quad \text{iff} \quad I_{\varphi[x/a]}(\alpha_\Sigma(G))$$

Combining this, we get precisely that

$$J_\varphi(F) \quad \text{iff} \quad I_\varphi(\exists x. \ s(\mu_1, \ldots, \mu_n, x) \ \wedge \ \alpha_\Sigma(G))$$

Since $\exists x. \ s(\mu_1, \ldots, \mu_n, x) \wedge \alpha_\Sigma(G) = \alpha_\Sigma(F)$, this proves the claim.

- F is of the form $\forall x : s(\mu_1, \ldots, \mu_n). \ G$. Since F is equivalent to the formula $\neg \exists x : s(\mu_1, \ldots, \mu_n). \ \neg G$, the claim follows from the previous steps.

At last, we prove the model expansion property.

Theorem 2 (Model expansion property). *The institution comorphism (Φ, α, β) has the model expansion property.*

Proof. Let Σ be a DFOL signature and J be a DFOL model for Σ. We construct a FOL model $M = (U, I)$ for the translated signature $\Phi(\Sigma)$ such that $J = \beta_\Sigma(M)$.

To define U, let $s_1, \ldots, s_k$ be the sort symbols of Σ. For s_i of arity n_i, set

$$U_i = \bigcup_{(x_1, \ldots, x_{n_i})} s_i(x_1, \ldots, x_{n_i})$$

where $(x_1, \ldots, x_{n_i})$ ranges through all n_i-tuples for which s_i is defined. Set

$$U = \{\bot\} \cup U_1 \cup \ldots \cup U_n$$

We now define I as follows.

- Let p be a predicate symbol in Σ, $p : \Pi x_1 : S_1, \ldots, \Pi x_n : S_n. o$. Let φ be an assignment function for M. If φ is also an assignment function for the context $\Gamma = x_1 : S_1, \ldots, x_n : S_n$, we set

$$p^I(\varphi(x_1), \ldots, \varphi(x_n)) \quad \text{iff} \quad p^J(\varphi(x_1), \ldots, \varphi(x_n))$$

  otherwise we set $p^I(\varphi(x_1), \ldots, \varphi(x_n))$ to be false.
- Let f be a function symbol in Σ, $f : \Pi x_1 : S_1, \ldots, \Pi x_n : S_n. S$. Let φ be an assignment function for M. If φ is also an assignment function for the context $\Gamma = x_1 : S_1, \ldots, x_n : S_n$, we set

$$f^I(\varphi(x_1), \ldots, \varphi(x_n)) = f^J(\varphi(x_1), \ldots, \varphi(x_n))$$

  otherwise we set $f^I(\varphi(x_1), \ldots, \varphi(x_n)) = \bot$.
- Let s be a sort symbol in Σ, $s : \Pi x_1 : S_1, \ldots, \Pi x_n : S_n. S$. Let φ be an assignment function for M. If φ is also an assignment function for the context $\Gamma = x_1 : S_1, \ldots, x_n : S_n$, we set

$$s^I(\varphi(x_1), \ldots, \varphi(x_n), \varphi(y)) \quad \text{iff} \quad \varphi(y) \in s^J(\varphi(x_1), \ldots, \varphi(x_n))$$

  otherwise we set $s^I(\varphi(x_1), \ldots, \varphi(x_n), \varphi(y)) = false$.

It is easy to see that $M = (U, I)$ satisfies all the axioms in the translated signature $\Phi(\Sigma)$ and that we have $J = \beta_\Sigma(M)$.

Hence, the institution comorphism (Φ, α, β) permits borrowing and we have that a DFOL theory entails a sentence if and only if the translated FOL theory entails the translated sentence.

5   Conclusion and Future Work

We have given an institution comorphism from a dependently-typed logic to FOL and have shown that it admits model expansion. Together with the borrowing theorem [CM93] this implies the soundness of borrowing.

This result is important for several reasons. The need for dependent types arises in several areas of mathematics such as linear algebra and category theory. DFOL provides a more natural way of formulating mathematical problems while staying close to FOL formally and intuitively. On the other hand, for FOL we have machine support in the form of automated theorem-provers and model-finders. The translation enables us to formulate a DFOL problem, translate it to FOL, and then use the known automated methods for FOL (e.g., theorem-provers such as Vampire [RV02] or SPASS [WAB+99], and model finders such as Paradox [CS03]) to find a solution.

First experiments with the translation have proved successful: For example, Vampire was able to prove instantaneously that the translation of our running example is a FOL theorem. It remains to be seen how much the encoding of type information in predicates and the addition of axioms in the translation affects the performance of FOL provers on larger theories. In the future we will integrate our translation into HeTS ([MML07]), a CASL-based application that provides a framework for the implementation of institutions and institution translations. That will provide the infrastructure to create and translate big, structured DFOL theories, and thus to apply our translation on a larger scale.

Since DFOL is defined within LF, we will also treat it as a running example for an implementation of the framework introduced in [Rab08]. That will permit to define arbitrary institutions and institution translations in LF and then incorporate these definitions into HeTS.

On the theoretical side, the translation shows that DFOL can be regarded as a fragment of FOL, which generalizes the well-known results for many-sorted first-order logic. In particular, we are able to derive properties of DFOL such as completeness, compactness, and the existence of free models immediately from the corresponding properties of FOL.

References

[ABK+02]   Astesiano, E., Bidoit, M., Kirchner, H., Krieg-Brückner, B., Mosses, P., Sannella, D., Tarlecki, A.: CASL: The Common Algebraic Specification Language. Theoretical Computer Science (2002)

[BCF+97]   Benzmüller, C., Cheikhrouhou, L., Fehrer, D., Fiedler, A., Huang, X., Kerber, M., Kohlhase, M., Konrad, K., Melis, E., Meier, A., Schaarschmidt, W., Siekmann, J., Sorge, V.: ΩMEGA: Towards a mathematical assistant. In: McCune, W. (ed.) Proceedings of the 14th Conference on Automated Deduction, pp. 252–255. Springer, Heidelberg (1997)

[Bel08]   Belo, J.: Dependently Sorted Logic. In: Miculan, M., Scagnetto, I., Honsell, F. (eds.) TYPES 2008, pp. 33–50. Springer, Heidelberg (2008)

[BPTF07]   Benzmller, C., Paulson, L., Theiss, F., Fietzke, A.: The LEO-II Project. In: Automated Reasoning Workshop (2007)

[Bro06]   Brown, C.: Combining Type Theory and Untyped Set Theory. In: Furbach, U., Shankar, N. (eds.) IJCAR 2006. LNCS, vol. 4130, pp. 205–219. Springer, Heidelberg (2006)

[CM93]   Cerioli, M., Meseguer, J.: May I Borrow Your Logic? In: Borzyszkowski, A., Sokolowski, S. (eds.) Mathematical Foundations of Computer Science, pp. 342–351. Springer, Heidelberg (1993)

[CS03]   Claessen, K., Sorensson, N.: New techniques that improve MACE-style finite model finding. In: 19th International Conference on Automated Deduction (CADE-19) Workshop on Model Computation - Principles, Algorithms, Applications (2003)

[GB92]   Goguen, J., Burstall, R.: Institutions: Abstract model theory for specification and programming. Journal of the Association for Computing Machinery 39(1), 95–146 (1992)

[GWM+93]   Goguen, J., Winkler, T., Meseguer, J., Futatsugi, K., Jouannaud, J.: Introducing OBJ. In: Goguen, J. (ed.) Applications of Algebraic Specification using OBJ, Cambridge (1993)

[HHP93]   Harper, R., Honsell, F., Plotkin, G.: A framework for defining logics. Journal of the Association for Computing Machinery 40(1), 143–184 (1993)

[JM93]   Jacobs, B., Melham, T.: Translating dependent type theory into higher order logic. In: Bezem, M., Groote, J. (eds.) Typed Lambda Calculi and Applications, pp. 209–229 (1993)

[Lan98]   Mac Lane, S.: Categories for the Working Mathematician. Springer, Heidelberg (1998)

[Mak97]   Makkai, M.: First order logic with dependent sorts, FOLDS (1997) (Unpublished)

[ML75]   Martin-Löf, P.: An Intuitionistic Theory of Types: Predicative Part. In: Proceedings of the Logic Colloquium 1973, pp. 73–118 (1975)

[MML07]   Mossakowski, T., Maeder, C., Lüttich, K.: The Heterogeneous Tool Set. In: Grumberg, O., Huth, M. (eds.) TACAS 2007. LNCS, vol. 4424, pp. 519–522. Springer, Heidelberg (2007)

[NPS90]   Nordström, B., Petersson, K., Smith, J.: Programming in Martin-Löf's Type Theory: An Introduction. Oxford University Press, Oxford (1990)

[Obe62]   Oberschelp, A.: Untersuchungen zur mehrsortigen Quantorenlogik. Mathematische Annalen 145, 297–333 (1962)

[Pau94]   Paulson, L.: Isabelle: A Generic Theorem Prover. In: Paulson, L.C. (ed.) Isabelle. LNCS, vol. 828. Springer, Heidelberg (1994)

[Pit00]   Pitts, A.: Categorical Logic. In: Abramsky, S., Gabbay, D., Maibaum, T. (eds.) Handbook of Logic in Computer Science ch. 2. Algebraic and Logical Structures, vol. 5, pp. 39–128. Oxford University Press, Oxford (2000)

[Rab06]   Rabe, F.: First-Order Logic with Dependent Types. In: Furbach, U., Shankar, N. (eds.) IJCAR 2006. LNCS, vol. 4130, pp. 377–391. Springer, Heidelberg (2006)

[Rab08]   Rabe, F.: Representing Logics and Logic Translations. PhD thesis, Jacobs University Bremen (2008)

[RV02]   Riazanov, A., Voronkov, A.: The design and implementation of Vampire. AI Communications 15, 91–110 (2002)

[TB85]   Trybulec, A., Blair, H.: Computer assisted reasoning with Mizar. In: Proceedings of the 9th International Joint Conference on Artificial Intelligence, Los Angeles, CA, pp. 26–28 (1985)

[Urb03]   Urban, J.: Translating Mizar for first-order theorem provers. In: Asperti, A., Buchberger, B., Davenport, J.H. (eds.) MKM 2003. LNCS, vol. 2594, pp. 203–215. Springer, Heidelberg (2003)

[WAB+99] Weidenbach, C., Afshordel, B., Brahm, U., Cohrs, C., Engel, T., Keen, E., Theobalt, C., Topić, D.: System description: SPASS version 1.0.0. In: Ganzinger, H. (ed.) Proceedings of the 16th International Conference on Automated Deduction (CADE-16). LNCS (LNAI), vol. 1632, pp. 314–318. Springer, Heidelberg (1999)

Author Index